SURFACE AND COLLOID SCIENCE IN COMPUTER TECHNOLOGY

SURFACE AND COLLOID SCIENCE IN COMPUTER TECHNOLOGY

Edited by
K. L. Mittal
IBM Corporate Technical Institutes
Thornwood, New York

PLENUM PRESS • NEW YORK AND LONDON

Library of Congress Cataloging in Publication Data

Surface and colloid science in computer technology.

"Proceedings of a symposium on surface and colloid science in computer technology, held as part of the Fifth International Conference on Surface and Colloid Science, June 24–28, 1985, in Potsdam, New York"—T.p. verso.
Includes bibliographies and index.
1. Surface chemistry—Congresses. 2. Microelectronics—Congresses. I. Mittal, K. L., 1945-   . II. International Conference on Surface and Colloid Science (5th: 1985: Potsdam, N.Y.)
QD506.A1S86   1987   541.3′453   87-12273
ISBN-13: 978-1-4612-9060-5   e-ISBN-13: 978-1-4613-1905-4
DOI: 10.1007/ 978-1-4613-1905-4

Proceedings of a symposium on Surface and Colloid Science in Computer Technology, held as part of the Fifth International Conference on Surface and Colloid Science, June 24–28,1985, in Potsdam, New York

PREFACE

This volume chronicles the proceedings of the Symposium on
Surface and Colloid Science in Computer Technology held as a part of the
5th International Conference on Surface and Colloid Science, and 59th
Colloid and Surface Science Symposium sponsored by the Division of
Colloid and Surface Chemistry of the American Chemical Society, and the
International Association of Colloid and Interface Scientists at
Clarkson University, Potsdam, N.Y., June 24-28, 1985.

Computer technology has many aspects, e.g., hardware
development, software development, and information processing. However,
in this Symposium we were exclusively concerned with the materials
considerations pertaining to microelectronics; surface and colloid
science plays a very vital role in the materials aspects of
microelectronics/computer technology.

A complete catalog of instances where surface and colloid
science is important would be prohibitively long, so here a few eclectic
examples should suffice to underscore the importance of interfacial
aspects of materials in the wonderful world of computers. As for
colloidal phenomena, the dispersion behavior of ceramic powders (in
making the substrate on which semiconductor chip is placed) and magnetic
particles (for making magnetic tapes and disks) is of critical
importance, This calls for fundamental understanding of stability
behavior in both aqueous and nonaqueous media, and the ways to control
dispersion characteristics. On the other hand, there are myriad
applications of thin films in microelectronics and irrespective of the
function of a thin film, its adhesion to the underlying substrate is of
cardinal importance. If the adhesion is inadequate, it leads to
delamination and associated problems. So the importance of understanding
interfacial interactions between different materials and of devising
means to tailor these interactions is quite patent. As a matter of fact,
many problems/failures can be traced to lack of compatibility or
suboptimum materials interfaces.

In the case of printing, the proper ink-paper interactions are
a desideratum for print quality, and surface and colloidal phenomena
play an important role. Recently, there has been considerable interest
in exploring new monolayer and Langmuir-Blodgett film materials for use
as electron-beam resists for microlithography. Here, fundamental
understanding of the behavior of L-B films, such as stability, radiation
characteristics, etc. are important. This provides a challenge to the
ingenuity of the synthetic chemist to come up with amphiphilic molecules
with desired functional moieties. New vistas in microlithography should
emerge as desired amphiphilic molecules make their presence.

So when this symposium was conceived and initial contacts made
with potential contributors, there was a very gratifying response and

the consensus was that such a symposium was both timely and needed. The final symposium program contained 48 papers by authors who hailed from many and varied institutions and had different backgrounds and interests. But they all were interested in the relevance and importance of interfacial phenomena in microelectronics/computer technology. In essence, this event was truly inter-,multi- and trans-disciplinary in nature, scope, and content.

This symposium was designed to bring together scientists and technologists involved in interfacial aspects of materials germane to microelectronics/computer technology, to discover the latest developments, to provide an opportunity for cross-pollination of ideas, and to identify areas which needed intensified research efforts. If the comments from the authors and attendees are a measure of the success of an event, then this symposium was a great success. Also may I add that many attendees commented that this symposium was the first comprehensive event highlighting the importance and pervasiveness of surface and colloid science in the arena of microelectronics.

Now let me turn to this proceedings volume which contains 27 papers arranged in five parts as follows: Acid-Base Concepts and Colloidal Dispersions; Adhesion of Films and Coatings Including Resists; Adhesion Aspects of Thin Films, and Metal-Polymer Interfaces; Monolayers and Langmuir-Blodgett Films: Relevance to Microelectronics; and Interfacial Aspects in Printing. The topics covered include: importance of acid-base interactions in inorganic powder dispersions, stability of colloids in aqueous and nonaqueous media, colloidal behavior of ceramic and magnetic particles; adhesion of polymeric films, role of silanes in improving polymer-polymer adhesion, resist adhesion and the ways to improve it, stresses in thin films and relevance to adhesion, plasma polymerized films, radiation enhanced adhesion of thin films, metallized polyimides; monolayers and L-B films as electron-beam resists, microlithography, various applications of organized assemblies, magnetic monolayers; paper-ink interactions, surface energetics of papers used in printing, and paper-polymer adhesion. So the topics covered in this volume represent interfacial aspects of many and varied materials germane to microelectronics/computer technology.

It should be recorded for posterity that each paper was peer reviewed by at least two reviewers and the manuscripts were returned to the authors for suitable revisions. So the review process was an integral part of the overall editing process. As for discussion, although no formal discussion is included in this volume, there were brisk and enlightening (not exothermic) discussions, both formal and informal, throughout the symposium.

I certainly hope this proceedings volume, which I believe is the first one on this topic, will be useful to a variety of researchers. Let me mention here that this book contains both overviews and original research contributions. It should provide a source of new ideas and directions for research in surface and colloid science to make this discipline more relevant to microelectronics. Apropos, some of the current ideas, approaches and techniques (e.g., acid-base concepts in dispersions and adhesion, radiation enhancement of adhesion of thin films, fundamental understanding of interfacial interactions, etc.) should provide a quantum jump in our knowledge to enable us to select (or modify) materials so as to make them interfacially compatible.

Acknowledgements: First, I am thankful to the officials of the sponsoring organizations for their interest in this symposium. My special thanks are due to S.B. Korin (IBM Corp.) for allowing me to organize this symposium and to edit these proceedings. I must take this opportunity to express my sincere thanks to the unsung heroes (reviewrs) for their time and very valuable comments, as comments from peers are important in maintaining the quality of publications. In the end, I would like to acknowledge the patience, enthusiasm and contributions of the authors without which this book would not have existed.

K.L. Mittal
IBM Corporate Technical Institutes
500 Columbus Ave.
Thornwood, NY 10594

...material that everyone has to bring when publishing. I wish this little
opportunity to express available appreciation to the young German friends of
the humanities and very valuable members as someday that have very
important in attracting the quality of explanations. In this regard I
would like to especially the assistance organizations contributions of
the enormous book which I in them and of inclusive related.

K. Mikes,
ICM Chemical Technical Engineer
Bratislava
November 27, 1936.

CONTENTS

x

PART I. ACID-BASE CONCEPTS, AND COLLOIDAL DISPERSIONS

ROLE OF ACID-BASE INTERACTIONS IN INORGANIC POWDER DISPERSIONS AND COMPOSITES

Frederick M. Fowkes

Department of Chemistry
Lehigh University, Bethlehem, PA   18015

The dispersion of inorganic powders into polymer solutions or polymer melts requires strong exothermic interfacial interactions between the polymer and surface sites of the powder.   These interactions tend to insure attachment of polymers to the entire surface of each particle, thereby minimizing any tendency to form the clumps which are so deleterious to the performance of magnetic, dielectric, or ceramic dispersions.

The exothermic interactions of polymers with the surface sites of inorganic powders are predominantly of the Lewis acid-base type, and are predictable from measurements of the surface acidity or basicity of the powders and from measurements of the acidity or basicity of the functional sites of polymers.

Acid-base interactions of the surface sites of powders with components of the surrounding organic medium (as in a polymer melt or solution) provides strong attachments and in addition provides electrostatic potentials to the particles and consequent inter-particle repulsive interaction which promotes dispersion stability, especially when adsorbed layers also provide some degree of steric stabilization as well.   The component responsible for electrostatic charging and stabilizing can be the matrix polymer or an added dispersant, but if a dispersant is used it should be compatible with the cast film.

In many practical systems the inorganic powder and the matrix polymer are chosen for their electrical or mechanical properties and they often do not have sufficient mutual acid-base interaction to provide good dispersions.   A practical solution to this problem is to modify the surface of the powder so as to insure strong acid-base interaction with the matrix polymer; we have had the most success in modifying the surface acidity or basicity of inorganic powders by using carefully chosen silane coupling agents.

INTRODUCTION

In the electronics industry one of the general materials problems is how to make concentrated dispersions of inorganic particles in polymer matrices which are uniform, flexible and tough. Such dispersions are used to make polymer films with specific electrical properties provided by the inorganic powder. For instance, magnetic particles are used in magnetic memory storage media such as film or discs, conductive particles such as silver or carbon black are used to provide conductive polymers, and powders of very high dielectric constant are used to provide polymer films of high dielectric constant. In modern ceramic technology, polymer melts or solutions are heavily loaded with ceramic powders and cast or injection-molded into specific shapes for firing at elevated temperatures to drive out the polymer and sinter the powder into the same shape as the cast or extruded form. This technology has become important in manufacture of ceramic substrates for integrated circuit chips, with interconnection patterns in the ceramic which are introduced in the cast films of polymer filled with ceramic powder. The above examples are all considered polymer composites, and they all share the usual problems of composites, such as the quality of dispersion of the powder, and the strength of adhesion of the matrix polymer to the surface sites of the powder.

The bonding of polymers to inorganic surfaces involves three kinds of attractive forces: 1), the general dispersion (or van der Waals) forces which operate in all materials and between all materials; 2), the specific Lewis acid-base interactions between electron-accepting (acid) sites of one material with the electron-donating (basic) sites of another; and 3), electrostatic attractions between positively and negatively charged materials. Hydrogen bonds are also exothermic and specific, but these are only a subset of the Lewis acid-base interactions. Although the interaction of permanent dipoles with one another or with polarizable molecules is important in dense gases and was once thought to be important in condensed media, such interactions are now known to be too weak to be measurable in liquids or solids, or at interfaces[1]. The exothermic nature of acid-base interactions in organic media has been illustrated in much detail by the research papers of Drago and coworkers[2,3], and a correlation of the negative heats of acid-base interaction ($-\Delta H^{ab}$) was developed:

$$-\Delta H^{ab} = C_A C_B + E_A E_B \qquad [1]$$

allowing predictions of heats of acid-base interaction to within about 0.2 kcal/mole. In Equation [1] the acid (A) and base (B) each have two constants; the C constant relates to the covalent character of the bond and the E constant relates to the electrostatic character of the bond. Drago and co-workers used calorimetric and infrared spectral data to evaluate the E and C constants of about forty organic acids and bases, and these constants can be used to estimate the E and C constants of polymers. For instance, polymers with ester groups have about the same basicity as methyl acetate or ethyl acetate, polymers with benzene rings have about the same basicity as benzene, and polymers such as polyvinylchloride have about the same acidity as chloroform, as will be illustrated later in this paper.

The adsorption of polymers from solution onto inorganic powders has been shown to be dominated by interfacial acid-base interactions between the polymer and the surface sites of the powder[4]. Figure 1 illustrates the adsorption of a model basic powder (calcium carbonate, with basic surface sites of carbonate groups) or of a model acidic powder (silica, with acidic SiOH surface sites) in neutral organic solutions of either a model basic polymer (PMMA, polymethylmethacrylate) or a model acidic polymer (CPVC, a post-chlorinated polyvinylchloride).

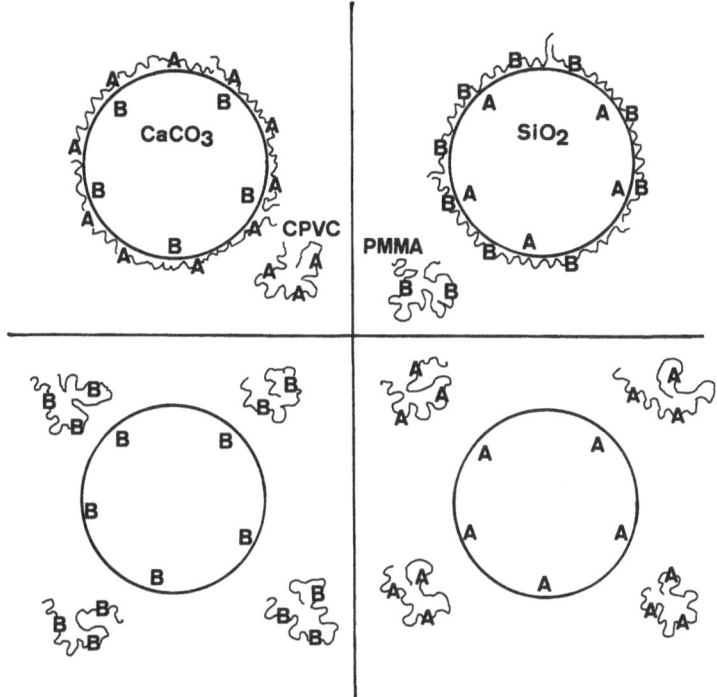

Figure 1.   In neutral solvents acidic polymers such as CPVC adsorb onto
the basic surface of calcium carbonate particles, but not onto the acidic
surface of silica particles.   Basic polymers such as PMMA adsorb only on
the silica particles.

     No adsorption occurred when both powder and polymer were basic, nor
when they were both acidic, but strong adsorption occurred with the
acidic polymer on basic powders or with the basic polymer on acidic
powders.   This study illustrated that dipole forces play no role in such
adsorption, for both polymers have polar groups and both powders have
appreciable surface polarity.   We see that polar groups interact with
each other only when one is acidic and the other basic.   This study then
concentrated on the adsorption of the basic PMMA polymer onto the acidic
silica powders from a wider range of solvents, including acidic solvents
such as methylene chloride and chloroform, and basic solvents such as p-
dioxane and tetrahydrofuran (THF).   In Figure 2 it is seen that the more
basic solvents compete rather successfully against the basic PMMA for
adsorption onto the acidic sites of silica, and that the more acidic
solvents such as chloroform compete rather effectively against the acidic
SiOH sites of silica for the basic sites of PMMA, thus diminishing the
amount of PMMA adsorbed from either the more basic or the more acidic
solvents.
     The specific acid-base interactions which promote adsorption also
provide strong adhesion between matrix polymers and dispersed powders.
Figure 3 illustrates the stiffness (relative modulus) of films of silica-
filled PMMA cast from THF, a strongly basic and competing solvent, and
compares this with the stiffness of films of the same composition cast
from methylene chloride, a more neutral solvent from which PMMA adsorbs
strongly onto silica.   It can be seen that the films cast from methylene
chloride, the non-competing solvent, are much stiffer than those cast
from THF, illustrating that the strong acid-base interactions which
promote the microscopic phenomenon of adsorption also promote the
macroscopic phenomenon of adhesion and stiffening.

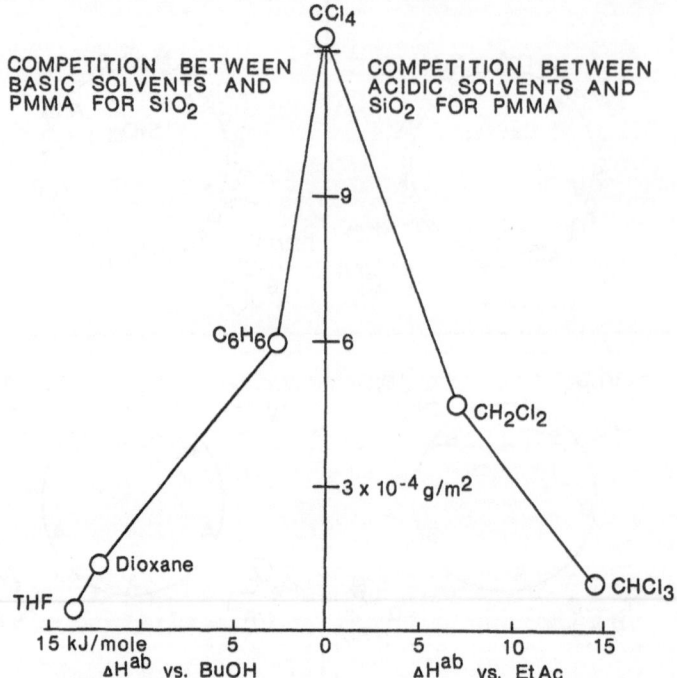

Figure 2. The acidity or basicity of solvents can appreciably reduce the adsorption of the basic polymer PMMA onto the acidic surface sites of silica[4]. Reprinted from IEC Product R&D by permission of the copyright owners, the American Chemical Society.

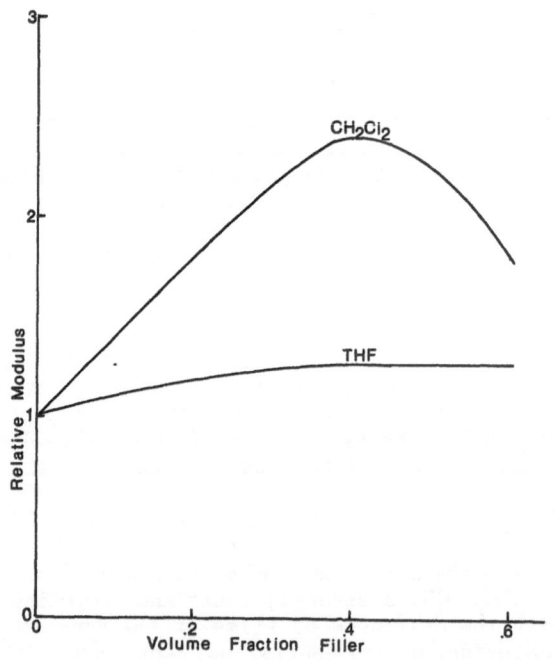

Figure 3. The stiffness of composite films of silica-filled PMMA cast from various solvents depends on the degree of adsorption of the basic PMMA groups onto the acidic surface sites of silica[5]. Films cast from the weakly competing methylene chloride are much stiffer than those cast from the strongly competing THF.

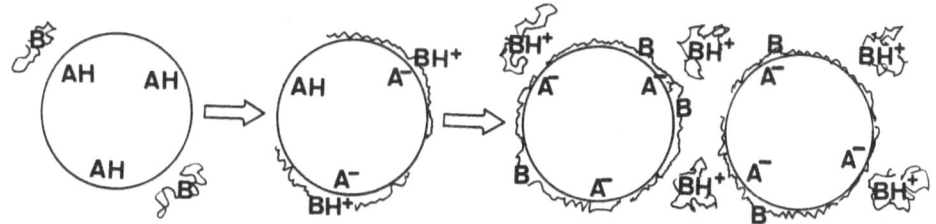

Figure 4. Proton-transfer mechanism of electrostatic charging of
acidic inorganic particles in non-aqueous solutions of basic dispersants.
The adsorbed basic, but uncharged, dispersant becomes charged by accepting
a proton from an acidic site. However the particle becomes charged only
when the dispersant concentration is high enough for the dynamic process
of adsorption-desorption to desorb the charged dispersant molecules.[6]

Interfacial acid-base interactions also promote electrostatic
charging of inorganic particles in organic media. This charge
transfer occurs at the surface of particles, where either proton or
electron transfer provides large enough surface potentials to promote
dispersion-stabilizing electrostatic repulsion between particles (Fig. 4).
In organic media stabilization of dispersions against flocculation
generally requires both electrostatic and steric repulsion, as illustrated
in Figure 5, after a text-book illustration from Shaw.[7] Steric
stabilizers provide an adsorbed film thickness of 5-10 nm, but require
strong anchoring to the surface, usually by acid-base bonds. The most
widely used steric stabilizers have basic nitrogen groups for anchoring,
but some with carboxylic acid anchors are also readily available. The
steric barriers are made up of large organic groups which should be well
solvated for maximum steric repulsion. We find then that acid-base
interactions at the surface of dispersed powders are a requirement for
getting good stable dispersions as well as getting good electrical and
mechanical properties of the resulting composite films.

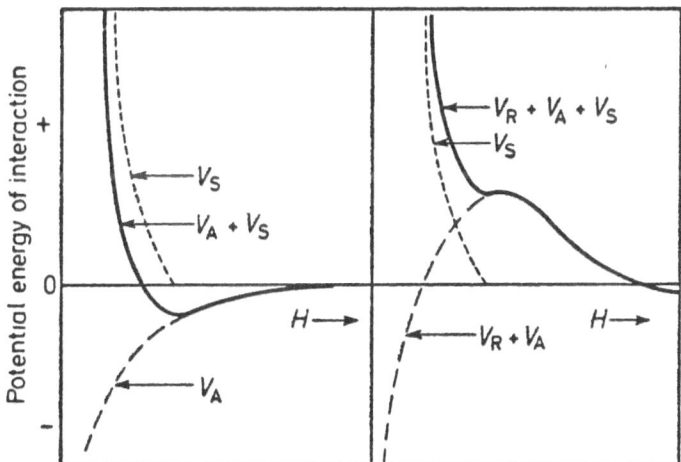

Figure 5. Left: diagram of inter-particle potential energy versus interparticle distance for steric (V_S) repulsion and dispersion force
attraction (V_A). Right: with electrostatic repulsion (V_R) added.

Matrix polymers interact with solvents by two kinds of interactions, the endothermic dispersion-force interactions and the exothermic acid-base interactions. The attractive forces between molecules always include the London dispersion forces which are caused by the mutual perturbation of the electron orbitals of adjacent molecules, and are related to the density of the material. In polymers and solvents the intermolecular forces can be entirely due to dispersion force attractions (as in polyethylene), or shared between dispersion forces and acid-base interactions (as in the hydrogen-bonding of polyamides). The solubility of polymers depends on the extent of interaction with the solvent, with strong contributions from both dispersion forces and acid-base interactions.

London Dispersion Forces of Polymers and Solubility Parameters

The heats of mixing of polymers and solvents which have only London dispersion forces are always positive (endothermic) and can be quantitatively predicted by Hildebrand's solubility parameter equation[8] for endothermic mixing:

$$\Delta H_2^{mix} = V_2 \phi_1^2 (\delta_1 - \delta_2)^2 \qquad [2]$$

where H_2^{mix} is the partial molar heat of mixing per monomer residue of the polymer, V_2 is the molar volume per monomer residue of the polymer, ϕ_1 is the volume fraction of solvent, and where δ_1 and δ_2 are the solubility parameters of the solvent and solute, respectively. Solubility parameters (the square root of the energy of vaporization per unit volume) are easily determined for solvents from their density and temperature-dependence of vapor pressure. However, since the vapor pressures for polymers are negligibly small, there has been no direct method for measuring the solubility parameters for polymers. The usual method for estimating solubility parameters for polymers is to measure solubility, swelling, or intrinsic viscosity of the polymers in a series of solvents of known solubility parameter, as illustrated in Figure 6. The intrinsic viscosity of solutions of a polyisobutene (molecular weight 2400) is plotted as a function of the solubility parameters of the solvents, and a maximum is observed at a solubility parameter of 9.3 (in the usual units of the square root of calories per milliliter). The intrinsic viscosity is a measure of the strength of polymer-solvent interaction, and its maximum at 9.3 indicates the solubility parameter for polyisobutene. The maximum in intermolecular forces indicated by the maximum in intrinsic viscosity in Figure 6 indicates a minimum in the heat of mixing of Equation [2], which is attained when the solubility parameters for polymer and solvent are equal. Polyisobutene was chosen for this example because it has no acidic or basic groups, and fits the criteria postulated by Hildebrand for the use of Equation [2]. Hildebrand warned users that this equation applies to systems in which only dispersion forces are present. He specifically warned against using solubility parameters for systems involving hydrogen-bonds, for these provide exothermic interactions, and Equation [2] is strictly for endothermic interactions. Hildebrand's warning was disregarded by Hansen[9], who tried to fit hydrogen bonds into Equation [2] by adding other endothermic terms which invariably increased the positive magnitude of the heats of mixing, and which never could predict the enhanced solubility which comes from the exothermic interactions of polymer and solvent.

There have been relatively few calorimetric studies of the endothermic polymer-solvent interactions predicted by Equation (2); Prof. Howard Swain has done a few such measurements in our laboratory

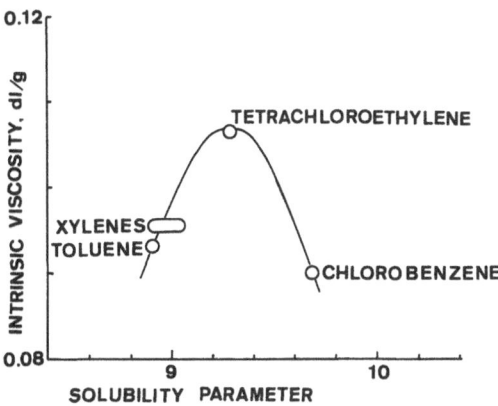

Figure 6. Intrinsic viscosity of solutions of polyisobutene at 25 C in several solvents as a function of their solubility parameter.[10]

with the Tronac Isothermal Microcalorimeter, and these measurements agree well with Equation [2]. However, most calorimetric studies of heats of mixing relate to the much stronger exothermic acid-base interactions.[2,3] The advantage of calorimetric findings is the precision and predictibility of such thermodynamic measurements of intermolecular interactions. Modern spectrometric methods can often give quantitaive spectral shifts which probe intermolecular interactions, and these shifts can be calibrated with calorimetry to provide thermodynamic values such as heats of mixing and solubility parameters. Drago calibrated the infrared shifts of phenol and other hydrogen-donors upon complexing with a wide variety of bases.[2,3] We have recently studied the infrared "solvent shifts" of carbonyl and similar doubly bonded oxygen compounds as they interact with solvents of known dispersion-force character and basicity, and it was found that the infrared shifts are the sum of a dispersion-force term and an acid-base term:[11]

$$\Delta\nu_{C=O} = \Delta\nu_{C=O}^{d} + \Delta\nu_{C=O}^{ab} \qquad [3]$$

which is similar to the heats of mixing:

$$\Delta H_{mix} = \Delta H_{mix}^{d} + \Delta H_{mix}^{ab} \qquad [4]$$

The infrared shift for carbonyl groups interacting by dispersion forces with various solvents is illustrated in Figure 7, in which the "solvent shift" of the carbonyl stretching frequency is plotted for several carbonyl-containing compounds against the surface tension of the solvents, with the vapor frequency plotted at zero surface tension. Surface tension, like solubility parameter, is a measure of intermolecular interaction energy. Surface tension was used in this plot because the correlation coefficient is always better than with solubility parameter.[11] Figure 7 shows that as the surface tension of the solvent increases from 0 to 50 mJ/sq.m. the carbonyl stretch frequency decreases about 30 wave numbers for all C=O bonds and similar shifts were also observed for S=O, N=O and P=O bonds[11]. This shift may allow us to measure solubility parametes of polymers directly, as is illustrated in Figure 8, where the carbonyl stretching frequency of ethyl

9

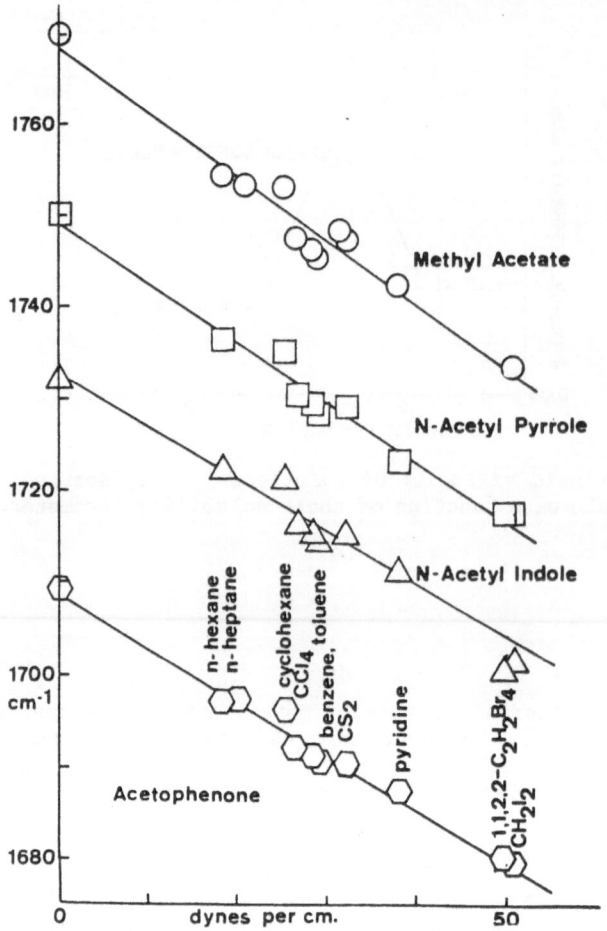

Figure 7. Carbonyl stretching frequency for various carbonyl-containing compounds as a function of the surface tension of the solvent (vapor = 0). Reprinted from the Journal of Polymer Science[11] by permission of the copyright owner.

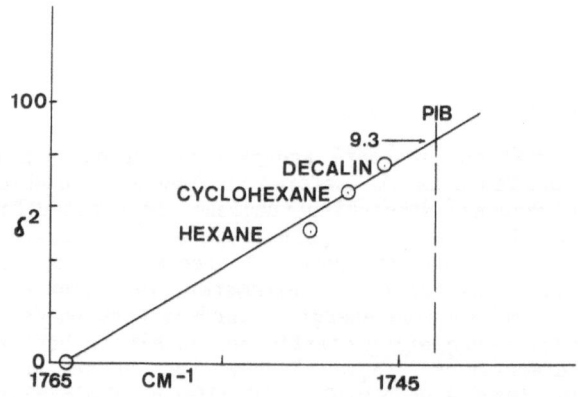

Figure 8. Spectral shift of the carbonyl stretching frequency for an ethyl acetate probe dissolved at 1-5% in various solvents and in poly-isobutene, plotted versus the square of the solubility parameter.[10]

acetate was used as a probe for the solubility parameter of solvents and also for the polyisobutene of Figure 6. The spectral shift of the ethyl acetate dssolved in the polyisobutene is seen to predict a solubility parameter of 9.3, in agreement with the intrinsic viscosity measurements of Figure 6. This finding indicates that infrared spectral shifts can be used to determine directly the solubility parameters of other polymers. In such studies one must use a probe that does not interact with the polymer by acid-base interactions. In the measurements illustrated in Figure 8 we chose ethyl acetate, an oxygen base, as a probe. The polybutene is neither acidic nor basic, and the solvents used for Figure 8 were chosen to exclude acidic media. Ethyl acetate might well be a satisfactory probe for basic polymers such as polyesters, polydienes, polyethers and polyaromatics.

The molar energies of vaporization (U_{vap}) used for determination of the solubility parameters of self-associated solvents (such as water, alcohols, amides and nitriles) include two contributions, the dispersion-force term and the specific interaction term, which we now explain as an acid-base contribution, as previously indicated in Equations [3] and [4]:

$$\Delta U_{vap} = \Delta U_{vap}^{d} + \Delta U_{vap}^{ab} \qquad [5]$$

For self-associated solvents it is desirable that Equation [2] can be used to correctly predict just the dispersion-force contribution to the heats of mixing, and in order to accomplish this goal, a special value for the solubility parameter to account for just the dispersion-force contribution to cohesion is needed. Perhaps the above-described infrared technique will be able to determine this value of the solubility parameter by using a probe which has neither acidity nor basicity. However, at present the best method for determining the dispersion-force contribution to the cohesive energy density of a self-associated solvent is based on measurements of surface tension and work of adhesion of the solvent on an immiscible hydrocarbon, from which the dispersion-force contribution to the surface tension of the solvent (γ_L^d) can be most accurately determined.[12] For instance, for water at 20 C the dispersion force contribution to surface tension is 30.2% of the total surface tension; if the square of the solubility parameter is proportional to surface tension, as proposed by Hildebrand[8] and by Beerbower and co-workers,[13] then the dispersion-force contribution to the solubility parameter for water is about 13 (in the usual units of square root of calories per milliliter). Certain basic solvents, such as benzene, ethyl acetate and pyridine, are found to have no or very little self-association; this is evidence that these basic compounds have negligible acidity. Other basic solvents such as nitriles, sulfoxides, sulfones and dimethyl formamide are found with work of adhesion measurements to be rather strongly self-associated; this is evidence that the electrophilic carbons and sulfurs are fairly acidic, and form acid-base complexes with the basic sites. Acidic compounds with OH or SH groups are self-associated by hydrogen bonds, as are amides or amines with active hydrogens. For all of these self-associated solvents it is most important to measure quantitatively their dispersion-force contributions to the solubility parameter, and to determine their Drago E and C constants of <u>both</u> <u>acidity</u> <u>and</u> <u>basicity</u>.

Acid-Base Interactions of Polymers

Calorimetric determinations of the heats of mixing of polymers with solvents are seldom published, even though thousands of investigators pay lip service to Equation [2] by the use of solubility parameters. Prof. Howard Swain has made some of these measurements with our Tronac Isothermal Microcalorimeter, and though he has invariably found that the better solvents have strong exothermic interactions, it appears from the magnitude of his results that only about half of the acidic or basic sites of the polymers have been neutralized in these experiments, even when a polypropyleneoxide liquid was titrated with stirring directly into chloroform. Perhaps there are steric limitations to the titration of the backbone ethers with chloroform, but the uncertainty of how many sites have been titrated in a calorimetric experiment with polymers makes for equal uncertainty in the calorimetric results and in the conclusions as to the strength of the basic or acidic sites of polymers. For this reason spectroscopic methods of determining heats of acid-base interactions of polymers are strongly favored.

Drago and co-workers used the OH stretch frequencies of phenol and butanol[2,3] and the NH stretch frequency of pyrrole[14] to measure the heats of mixing of acidic and basic solvents in cyclohexane or in carbon tetrachloride. The concentration of phenol was kept below 20 mM to avoid self-association, and the OH stretching peak was consequently very sharp. The OH peak shifts to lower wave numbers upon complexation with a base, and the complex has a very broad peak, as shown in Figure 9. Figure 9 is a difference spectrum made with our Mattson FTIR spectro-photometer; it shows a sharp downward spike where the original OH peak was for uncomplexed phenol, and the broad upwards peak for the complex. The shift in this case is -462 wave-numbers, which indicates a heat of complexation of -7.84 kilocalories/mole, using the Drago calibration for phenol[2]:

$$-\Delta H^{ab} = 3.08 - 0.0103 \text{ kcal/mole} \times \Delta \nu^{ab} \qquad [6]$$

Although the complex peak is broad, the shift can easily be determined to the same wave number in each of several successive measurements, and in a series of measurements with base concentrations varying by a factor of as

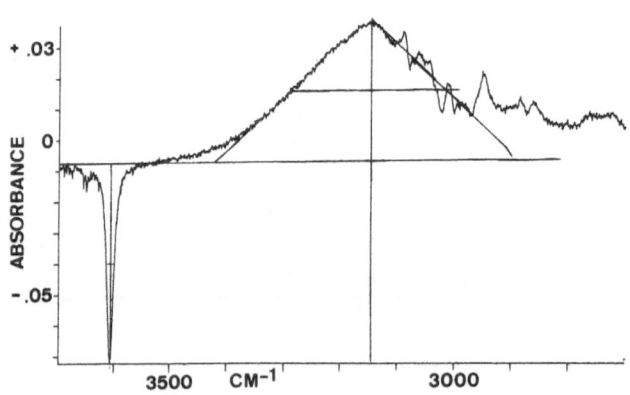

Figure 9. Spectrum of complex of 4-t-butylphenol (14.3 mM) with pyridine (56.06 mM) in carbon tetrachloride minus the spectra of uncomplexed components. Sharp negative peak at 3612 wave-numbers is the original phenolic OH stretch peak and the broad positive peak at 3150 wave-numbers is for the complexed OH peak. Measured with Mattson FTIR spectrometer by Li Guozhen, Visiting Professor from Zhejiang University, China.

much as a hundred-fold, the shift was independent of the concentration of base. If only part of the phenol is complexed, the shift is just the same as when all is complexed; this provides an important advantage over calorimetry, especially with polymers such as polyethers where steric factors can prevent titration of all the backbone ether groups.

In the above studies the interacting acid and base were dilute components in neutral solvents (cyclohexane or carbon tetrachloride); in these systems and in others to be discussed the interactions between two components are taking place in a solvent medium where the dispersion-force component cancels out.[2] This is in sharp contrast with the above-discussed two-component systems of polymer and solvent, or with adsorption studies of a single component adsorbing from the gas phase onto a solid, for in these two-component systems the dispersion forces are major components of the interaction.

We have recently calibrated the infrared shifts of the carbonyl stretching frequencies of esters, ketones and amides[11] versus the known calorimetric heats of acid-base complexation, with the following result:

$$\Delta H^{ab} = 0.236 \text{ kcalories/mole/wave number} \times \Delta \nu^{ab} \qquad [7]$$

Unlike equation [6] and the equations for the OH stretch shift of compounds other than phenol, there is no intercept in this equation, for the entire shift is directly proportional to the heat of acid-base complexation. Figure 10 shows the known calorimetric heats of acid-base complexation for ethyl acetate as a function of the observed spectral shifts of the C=O stretching frequency. The test acids range from the weakly acidic methylene chloride to the very strongly acidic trimethyl aluminum, but the linearity of the relation is quite clear. Figure 10 also shows the spectral shifts of the C=O stretch for PMMA with the same acids, and though these shifts are 30% less than with ethyl acetate, they are directly proportional to the ethyl acetate shifts. It was proposed that the heat of acid-base complexation of any carbonyl compound can be determined with Equation [7], and the line for PMMA in Figure 10 shows the heats of acid-base complexation determined by Equation [7] from the spectral shifts of PMMA observed with same acids (and illustrated with the same symbols) as used with ethyl acetate. Recently the above contention that the heats of acid-base complexation of PMMA are predictable with Equation [7] was tested and confirmed by others.[14]

The heats of acid-base complexation of a candidate base or acid with test acids or bases of known Drago E and C constants may be used to determine the E and C constants of the candidate, as illustrated in Figure 11. This "E and C plot" is a graphical representation of the heats of acid-base interaction of an unknown base with a number of test acids carefully chosen to provide a wide range of C/E values. Equation [1] has been re-arranged to provide a straight-line plot for the heat of complexation measured with each test acid:

$$E_A = -\Delta H^{ab}/E_B - C_A \times (C_B/E_B) \qquad [8]$$

In Figure 11 the slopes C_B/E_B are much greater with the "soft" acids (iodine or $Sb(Cl)_5$) than with the hard hydrogen acids such as chloroform or alcohols, providing statistically significant intersections which are the E_B and C_B values for the unknown base. In Figure 11 we see the intersections for ethyl acetate and for PMMA, together with the their standard deviations. This precise technique could be used to determine the E_B and C_B constants of other carbonyl bases, including polyesters, polyamides and polycarbonates.

The carbonyl shifts of ethyl acetate in complexes with acidic polymers can also be used to determine the heats of acid-base complexation of acidic polymers,[11] such as chlorinated polyvinylchloride (-3.3 kcal/mole), and polyvinylbutyral(-4.8 kcal/mole). In similar

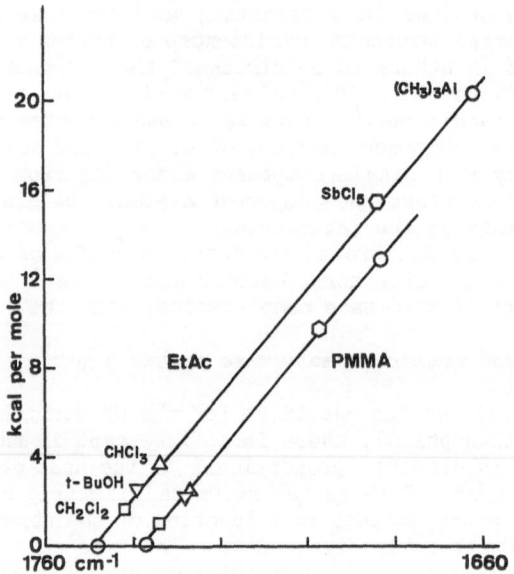

Figure 10. Calibration of the heats of acid-base complexation of ethyl acetate versus the observed infrared shift of complexation in solutions of srface tension 26-29 mN/m. The line for PMMA is drawn parallel,and the measured infrared shifts are shown with the same symbols used for ethyl acetate.[11]

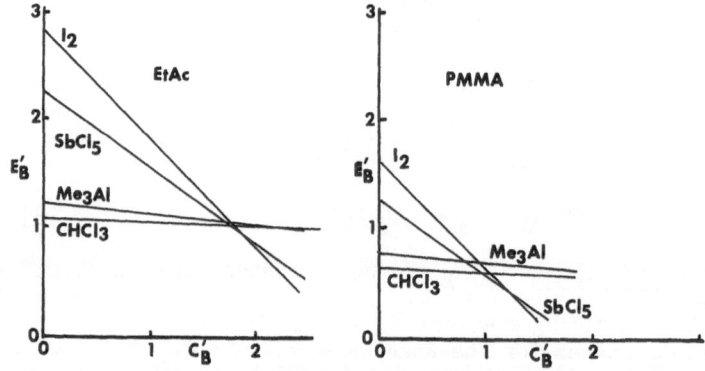

Figure 11. E and C plot for ethyl acetate and for PMMA, using the heats of complexation determined from the spectral shifts for PMMA in Figure 10.

studies with PMMA, the infrared shifts for PMMA showed the heat of complexation of this basic polymer with acidic polymers was -1.9 kcal/mole for polyvinylidenefluoride and -5.2 kcal/mole with polyvinylfluoride.

Now that the acid and base strengths of the interactive sites on polymers can be determined, and the acid-base contribution to the heats of mixing can be measured directly or predicted with Equation [1], it is of interest to see how acid-base interactions affect the solubility of solvents in polymers and of polymers in solvents. We have measured the extent of swelling of polymer powders in solvent vapors,[11] finding that the amount of swelling of amorphous regions is dominated by acid-base interactions, and appears to be independent of solubility parameter.

However, the solubility of polymers in solvents is strongly dependent on both solubility parameter and acid-base interactions. For weakly-complexed polymer-solvent systems the acid-base interaction merely expands the range of solubility parameters. Polystyrene is a good example of a weakly basic polymer which forms weak acid-base complexes with acidic solvents such as methylene chloride or chloroform. Hildebrand[8] has discussed the solubility of polystyrene in a wide variety of solvents, concluding that polystyrene of molecular weight 90,000 is soluble in solvents with solubility parameters between 8.4 and 10.7, a range of 2.3 solubility parameter units. However, in Hildebrand's list are many acidic solvents, and when these are removed, the range of solubility parameters for non-basic polystyrene solvents shrinks to 8.9-10.0, a range of 1.1 solubility parameter units.

Polymers which are strongly complexed with solvents appear to take on new solubility parameters;[11] for instance chlorinated polyvinylchloride has a solubility parameter of about 10, but its solubility parameter in ester solvents is about 8.6. Presumably the polymer chain becomes adorned with strongly bound solvent molecules, thus changing the solubility properties of the polymer from those of the bare polymer to those of the adorned polymer. This may be a general principle which applies even to aqueous systems such as polyethyleneoxide, where the hydrated ether groups provide a water-solubility which disappears with rise of temperature when the hydrate dissociates, giving a negative temperature-coefficient of solubility. Such a negative temperature-coefficient of solubility is also observed with the ester complexes with chlorinated polyvinylchloride.[11]

The same principles may apply to polymer blends, where weakly-interacting acidic and basic polymers have enhanced mutual solubility, but where strongly interacting acidic and basic polymers form brittle complexes.

SURFACE ACIDITY AND BASICITY OF SOLIDS

Inorganic powders interact with solvents, dispersants, plasticizers and polymers by acid-base interactions which are determined by the acidity or basicity of the interacting sites. Magnetic iron oxides usually have both acidic and basic surface sites which can interact with either acidic or basic polymers or dispersants, carbon blacks are predominantly acidic and interact best with weakly basic polymers such as polydienes, and alumina usually has only basic sites and interacts strongly with acidic polymers such as polyvinylbutyrals. Some ceramic powders can have acidic sites and some basic, depending on their composition and on the chemical character of the surface sites.

At the surface of inorganic oxides the nature of the surface sites is often determined by interaction with water or water vapor, for water is easily chemisorbed as surface OH groups. The SiOH sites of silica are widely recognized as having an acidic character,[16] the AlOH sites on alumina have been shown by several investigators to be basic, while the FeOH sites of iron oxides have both acidic and basic character.[17]

15

There are various ways to characterize the acidity or basicity of the surface sites of inorganic powders. Bolger and Michaels[18] used isoelectric points (determined by electrophoresis in aqueous solutions of various pH) as a measure of acidity and basicity. This procedure indeed showed silica to be acidic and alumina to be basic, but it could not determine that iron oxides and similar oxides have both acidic and basic sites. Meguro and Esumi used ESR spectroscopy with strongly electron-accepting tetracyano compounds to measure the basic (electron-donor) properties of alumina surfaces.[19] However, we have proposed to characterize the acidity and basicity of inorganic powders with the heats of adsorption of test acids and bases adsorbing from from neutral organic solvents. With these heats of adsorption we propose to determine the Drago E and C constants of acidity and basicity of surface sites on inorganic powders, for with these constants we can accurately predict the heats of adsorption of a wide range of acidic and basic solvents, dispersants, plasticizers and polymers (Table I).

Powders with surface OH groups tend to adsorb water, and since water is both an acid and a base, the adsorbed moisture tends to reduce the surface acidity or basicity of such oxides. Consequently when these oxides are in dry hydrocarbon solvents they are more acidic or basic than in water, and their degree of acidity depends very much on their degree of dryness. The water content of solvents and solutions must be continually monitored with Karl Fischer analyses. The amount of adsorbed water on the hydrophilic powders can be finely controlled by the moisture level of the solvent and solutions; we try to keep the moisture level at 5–10% of saturation.

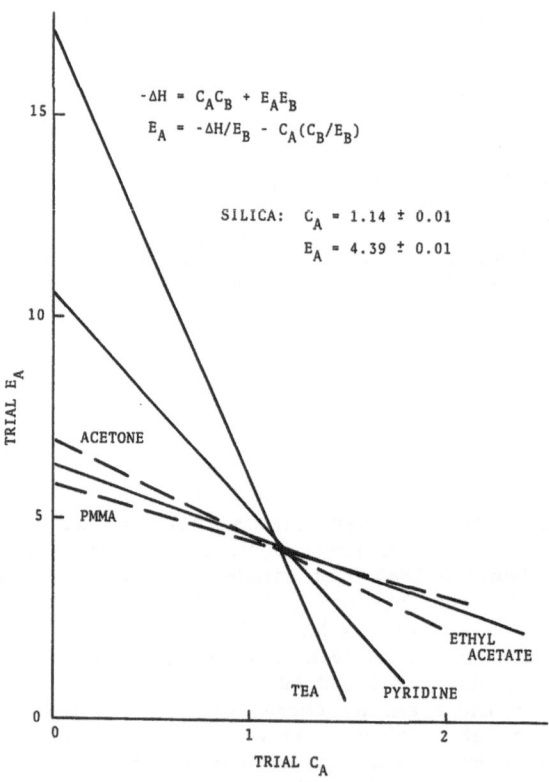

$$-\Delta H = C_A C_B + E_A E_B$$
$$E_A = -\Delta H/E_B - C_A(C_B/E_B)$$

SILICA: $C_A = 1.14 \pm 0.01$
$E_A = 4.39 \pm 0.01$

Figure 12. E and C plot for the determination of the E_A and C_A constants for silica, using heats of adsorption determined by microcalorimetry (solid lines) and by infrared shifts (dashed lines).

Heats of Adsorption by Isothermal Microcalorimetry

Our initial studies have been to characterize the C_A and E_A values for the SiOH surface sites of silica and the TiOH surface sites of rutile.[20]   Figure 12 illustrates some heats of adsorption of test bases on silica in dry cyclohexane solutions, using an "E and C plot" like that of Equation [8], as shown in Figure 11.   The C_B/E_B ratio, which determines the slopes of the lines in Figure 12, is 11.14 for triethylamine, 5.47 for pyridine, and 1.78 for ethylacetate, so the angles of intersection of the straight lines are sufficiently obtuse to provide good statistical significance.   These heats of adsorption were determined calorimetrically, using a Tronac Isothermal Microcalorimeter with dry cyclohexane as solvent.   After thermal equilibration of a well-stirred suspension of 0.3-1 gram of silica (Aerosil 380 or HiSil 233) in 25 ml of cyclohexane, a cyclohexane solution of enough base to neutralize about a third of the surface sites was titrated into the suspension over a fifteen minute period.   The heat of adsorption evolved for more than an hour, after which the equilibrium concentration of base was determined by UV spectrometry in a region of the absorption spectrum which followed Beer's Law.   The resulting heats of adsorption had a standard deviation of about 0.4%, and at the approximately 30% coverage used in these studies we did not find any difference between HiSil 233 and Aerosil 380, even though the HiSil is a precipitated silica and the Aerosil is a pyrolyzed silica.

Each solid line in Figure 12 is the average of two or three measured heats of adsorption, and from the intersection we calculate the Drago constants for the SiOH surface sites of silica to be E_A = 4.39 +/- 0.01 and C_A = 1.14 +/- 0.01.   In parallel studies with rutile (a special preparation without inorganic coating, from duPont) we determined for the TiOH surface sites of rutile that its Drago constants are E_A = 5.66 +/- 0.05 and C_A = 1.02 +/- 0.03.

Heats of Adsorption by Infrared Spectral Shifts

Figure 12 also shows by dashed lines the results of measuring the heats of adsorption by infrared spectral shifts of the carbonyl stretching frequencies of acetone and PMMA, using Equation [7] to calculate the heat of adsorption from the spectral shift.   The spectra were obtained by transmission through a mineral oil mull of silica; the shift was determined from a dilute solution of acetone in the mineral oil before and after the silica was added.   The PMMA was adsorbed onto silica from a dilute solution in methylene chloride, the solvent evaporated, and then a mull in mineral oil was made with the powder. The shift was determined from the C=O stretching frequency of PMMA in toluene, which has the same surface tension as the mineral oil.   It is seen in Figure 12 that the heats determined by infrared spectral shifts agreed very well with those determined calorimetrically.

Infrared spectra of adsorbed species on iron oxide powders are easily determined by transmission spectra, for iron oxides are quite transparent in the infrared region.   Figure 13 shows the infrared spectrum of dimethylformamide in a mineral oil solution and in a mineral oil mull when adsorbed onto a gamma iron oxide having a surface area of 20 square meters per gram.   The shift of the carbonyl stretch frequency from 1693.4 to 1634 wave-numbers shows that the heat of adsorption is -9.4 kilocalories/gram, indicating that the acid sites of this oxide are appreciably stronger than those of hematite (see Table I).   In our most recent infrared studies of adsorbed species on iron oxides we have obtained excellent infrared spectra with photo-acoustic spectroscopy.[21]

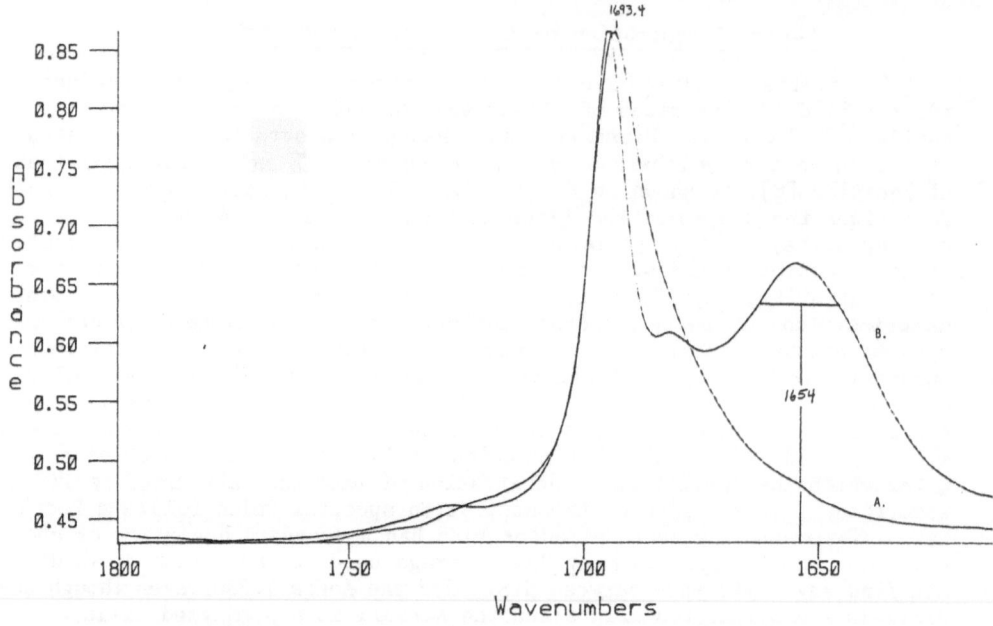

Figure 13. Transmission infrared spectrum of dimethylformamide in mineral oil (C=O stretch at 1693.4 cm^{-1}) and when adsorbed from mineral oil onto gamma iron oxide (C=O stretch at 1654 cm^{-1}). Determined by T. B. Lloyd with Mattson FTIR spectrometer.

Heats of Adsorption by Flow Microcalorimetry

Flow microcalorimetry is liquid chromatography with a sensitive temperature-sensing element in a small adsorption bed. The rates of flow are usually at the low end of the flow rates used in liquid chromatography, and syringe pumps are required, but in other respects the technique and equipment is much the same. Earlier studies used only the thermal detector, and heats of adsorption were determined in joules per gram (or per square meter) of powder, but in our studies we have introduced a standard UV absorption concentration detector, so that the amount of solute adsorbed is determined and the molar heats of adsorption of test acids or bases determined.[22]

Figure 14 is a flow diagram of our flow microcalorimetry system, and Figure 15 shows a typical computer print-out of the observed rates of adsorption and of the distribution of heats of adsorption observed during the adsorption of pyridine from cyclohexane onto alpha hematite, an iron oxide with FeOH surface sites which can adsorb either acids or bases (phenol or pyridine). It is seen that initial exothermic heats of adsorption are quite high, about -20 kcal/mole, but as more pyridine adsorbs, the heats of adsorption decline to less than -10 kcal/mole. ESCA analysis of this oxide (by Prof. Dwight of VPI) showed it to have about 5% of surface sulfate, and the weak basicity of sulfate sites as compared to OH sites is believed to allow the strong acidity of the ferric sulfate sites to be measured by the first pyridine adsorbed. Support for such an explanation was obtained recently by sulfuric acid treatment of a gamma iron oxide which had uniformly acidic sites before treatment and after treatment it had much higher initial heats of pyridine adsorption,[23] similar to the results in Figure 14. In other studies with gamma iron oxides it was found that some gamma iron oxide powders were contaminated with acetone-leachable organics with carboxylate groups; these oxides were found to have very low initial molar heats of

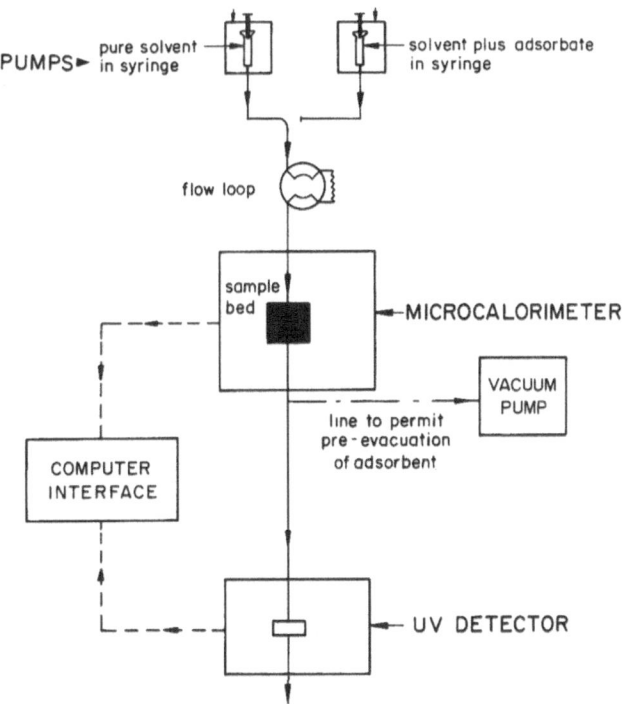

Figure 14. Flow diagram of flow microcalorimetry system, using a 0.15 ml bed of powder in a Microscal calorimeter metal block with thermistor sensors and a Perkin Elmer L-75 UV detector.

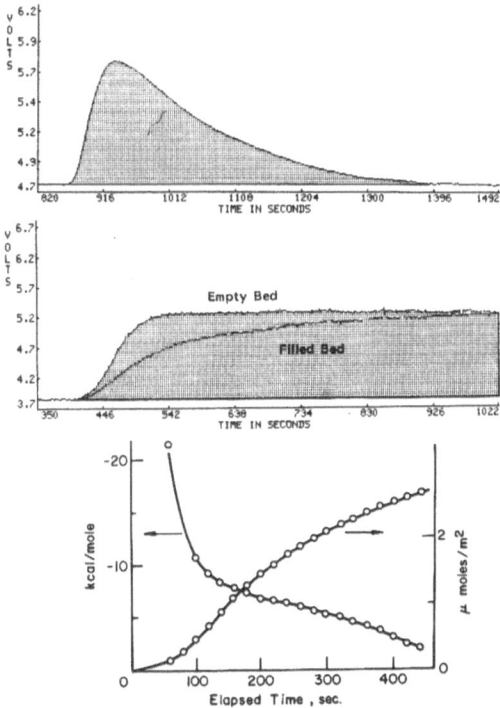

Figure 15.- Computer print-out of the adsorption exotherm (top) for the adsorption of pyridine from cyclohexane at 40 C onto alpha hematite. The middle graph shows the UV detector trace for an empty bed versus a powder-filled bed; the difference is the measure of the amount of pyridine adsorbed. The bottom graph shows the time-dependence of the amount of adsorption and of the heats of adsorption.

pyridine adsorption, perhaps because the endothermic desorption of the carboxylates reduced the exothermic heat of pyridine adsorption.[24] The cumulative heats of adsorption on alpha hematite of triethylamine, pyridine and ethyl acetate were determined by flow microcalorimetry at a surface coverage of one micromole per square meter to be -12.2, -9.5 and -5.2 kilocalories per mole, respectively. From the E and C plot the E_A and C_A were determined to be 3.92 +/- 0.08 and 0.77 +/- 0.01, respectively.

The rates and heats of adsorption of polymers onto oxides are easily observed by flow microcalorimetry.[22] Polyvinylpyridine adsorbed from benzene onto hematite in two steps; in six minutes the pyridine solution emerging from the bed was no longer depleted of pyridine, but it took twenty-six minutes for the full heat of adsorption to be expended. The adsorbed polymer film covered 67% of the surface sites, and about half of all pyridine residues were bonded to the surface FeOH sites of the hematite particles. The heats of adsorption measured by flow microcalorimetry were -6 kcals/mole for the pyridine groups of polyvinyl-pyridine adsorbing onto hematite from benzene, corresponding to -10 kT per bond for pyridine groups replacing the adsorbed benzene molecules on FeOH sites.

Predicting Heats of Adsorption of Bases on Various Oxides

The E_A and C_A constants for silica, rutile and hematite can be used together with the many E_B and C_B constants determined by Drago for nitrogen bases, oxygen bases, sulfur bases and pi-electron bases to predict with accuracy their heats of adsorption on these three oxides, as shown in Table I.

Table I. Heats of Adsorption of Bases on Silica, Rutile and Hematite
(kilocalories per mole)

Bases	Silica	Rutile	Hematite
Nitrogen Bases			
Pyridine	-12.4	-13.1	-9.5
Ethylamine	-12.8	-13.9	-10.0
Diethylamine	-13.9	-13.9	-10.2
Triethylamine	-17.1	-16.9	-12.4
Oxygen Bases			
Ethyl Acetate	-6.2	-7.3	-5.2
Acetone	-6.9	-8.0	-5.7
Ethyl ether	-7.9	-8.8	-6.1
Tetrahydrofuran	-9.1	-9.9	-7.1
Dimethylformamide	-8.1	-9.5	-6.7
Dimethylsulfoxide	-9.0	-10.5	-7.4
Sulfur Bases			
Ethyl sulfide	-10.0	-9.5	-7.0
Pi-Electron Bases			
Benzene	-3.7	-3.9	-3.0
Mesitylene	-5.3	-5.5	-3.9
Polymer Bases			
Polymethylmethacrylate	-4.0	-4.8	-3.4
Polyvinylpyridine	-12.0	-12.7	-9.2

ACID-BASE INTERACTIONS IN DISPERSIBILITY OF POWDERS IN ORGANIC MEDIA

The ability of dispersants and polymers to interact with the surface of powders is important to the degree of dispersion attained during milling and to the stability of the resulting dispersion. In Figure 5 the interparticle potential energy is shown to result from the interparticle attractions due to the London-van der Waals forces, and the interparticle repulsions which result from electrostatic repulsion and steric barriers. The acid-base origin of the electrostatic repulsion is illustrated in Figure 4, where electron-accepting (acidic) particles in an electron-donating (basic) medium are invariably negatively charged, and the magnitude of the resulting zeta-potentials is a measure of the degree of acid-base interaction.

In some systems it is conventional to use dispersants, while in others the matrix polymer is used to provide dispersion stability. A dispersant film on the particles can intervene between the powder and the matrix when the solvent is removed from the dispersion, and such intervention might decrease particle-matrix adhesion. For instance, in magnetic oxide dispersions for magnetic recording media it is normal to use dispersants (such as lecithin or other dispersants with phosphoric acid sites) to minimize the particle clump size, and it is hoped that in the cast film the dispersant does no harm. On the other hand, in ceramic dispersions such as alumina in polyvinylbutyral, no dispersant is used, and the toughness of such films results because of the strong particle-matrix acid-base interaction. The dispersant always requires acid-base interactions with the surface sites of the powder; this requirement is necessary for dispersants which provide either electrostatic or steric repulsive contributions for dispersibility or for dispersion stability.

Magnetic iron oxide particles in a ketone, ether or ester are always negatively charged, for these solvents are oxygen bases which can donate electrons to the acidic FeOH surface sites, or the acidic FeOH surface sites can donate protons to the basic oxygens of these solvents; either mechanism produces negative zeta-potentials. However, a popular dispersant for magnetic iron oxides is an organic partial ester of phosphoric acid, soluble in the above basic solvents, but which makes solutions which are predominantly acidic, and these interact with the basic sites of iron oxide to provide positive zeta-potentials. Dispersions of alumina in ketones or other basic solvents have negligible zeta-potentials, for both medium and the particle surface are basic; however, if methanol (a predominantly acidic solvent) is added, appreciably positive zeta-potentials develop. A favorite binder for alumina is the strongly acidic polyvinylbutyral, and in methanol-ketone solvents this dispersant provides high positive zeta-potentials and good dispersion stability to basic alumina particles.[25]

Steric stabilizers require strong anchoring to the surface of particles, and acid-base interactions are the predominant mechanism for such anchoring. Steric stabilizers must also project into the medium far enough to keep particles apart. Clearly polyvinylpyridine is strongly anchored to acidic particles, but it lies down so flat as to offer no barrier, as evidenced in our flow microcalorimetric measurements of adsorption.[22] Good steric stabilizers must have strong acidity or basicity for anchoring, but the rest of the molecule must interact strongly with the medium and not with the surface. Examples are the polyisobutene (PIB) derivatives with strong basic anchoring groups;[26,27] these provide excellent steric barriers for acidic particles dispersed in solvents with solubility parameters close to that of PIB (9.3, as illustrated in Figure 6).

A recently-recognized steric stabilization mechanism relates to liquid crystal phenomena, and is probably important for lecithin, for this surfactant forms liquid crystalline aggregates (mesophases) in organic solvents in the concentration ranges used in magnetic media formulations. Friberg and co-workers showed that liquid crystals can spread around particles and stabilize them with a steric barrier of several nanometers in thickness.[28] Figure 16 shows the results of particle size measurements of a 1% solution of soya lecithin in THF using a Coulter N4 Sub-Micron Particle Sizer. There are two peaks, one for the inverse spherical micelles (at 2-3 nm diameter), and one for the liquid crystalline particles or droplets (at about 46 nm in diameter). Similar results were obtained with GAF's RE-610, a partial ester of phosphoric acid with a nonionic surfactant like Triton X-100. In 3% solutions of RE-610 in an aromatic ketone the liquid crystal droplets were 460 nm in diameter. These phosphoric acid derivatives probably react with the iron oxide and provide a thick organic coating.

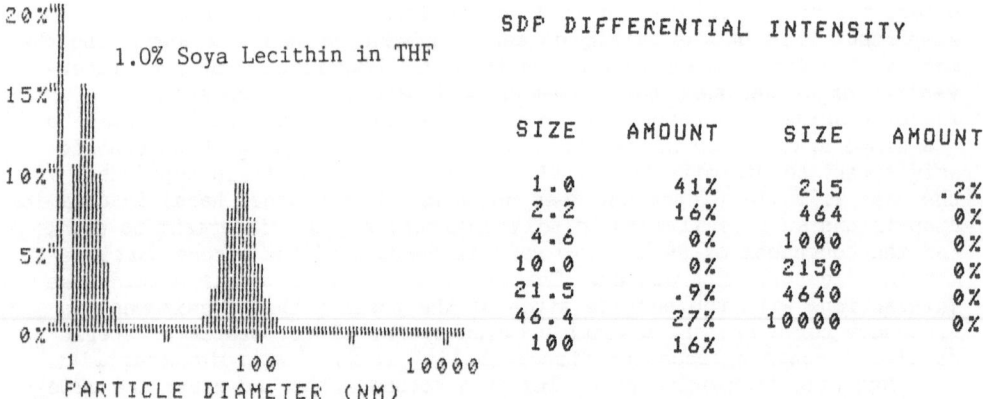

Figure 16. Coulter N4 Sub-Micron Particle Sizer results for 1% solutions of lecithin in THF.[29] The peak on the left is for spherical micelles, and the peak on the right (46 nm) is for the liquid crystal droplets.

SURFACE MODIFICATION OF OXIDE POWDERS TO MAXIMIZE ACID-BASE INTERACTION

In many applications the kind of inorganic powder and the kind of polymer binder for a given application are dictated by electrical, thermal, or mechanical restraints; this makes it difficult to choose materials which will have optimum acid-base interaction, yet such acid-base interaction is also required for the attainment of good dispersions with optimum properties. One way out of this dilemma is to modify the surface of the inorganic powder so that it interacts strongly enough with the binder.

One example of such surface modification is the use of barium titanate powders to raise the dielectric constant of polycarbonate films for energy-storage capacitors.[30] A Lexan polycarbonate was chosen for its excellent high-temperature and mechanical behavior, and its high dielectric strength, while barium titanate powders were chosen for their very high dieletric constant. Both materials are strongly basic, and when films were cast of such dispersions, they were full of clumps, were very brittle, and films exceeding 20 v% could not be cast successfully. It was decided to make a trial with an acidic polymer to see if acid-base interaction between binder and filler would improve the dispersion. The dispersions of barium titanate in dioxane solutions of chlorinated polyvinylchloride (CPVC) were found to contain no clumps and the cast films could be loaded to 67 v% and still be tough and flexible. This polymer had poor electrical and thermal properties, so its use was no solution to the problem, but the advantage of acid-base interaction had been proven. The polycarbonate was required, but it was basic and so was the surface of the barium titanate powder. Neither material could be replaced, so surface modification of the barium titanate was necessary.

The surface modification was accomplished by ion-exchanging the surface barium ions with aluminum ions, by soaking in an aqueous aluminum sulfate solution, filtering and drying. The resulting barium titanate powder was acidic, as evidenced by tests with indicator dyes,[31] and it dispersed well in dioxane solutions of polycarbonate. The resulting cast films were tough and flexible and could be loaded to 67 v% of filler.

Not all powders are as easily modified as barium titanate. A glass powder with negligible surface acidity or basicity was studied for surface modification so that it could be used at high concentrations as a filler for two different acidic polymers (CPVC and polyvinylbutyral, both plasticized with di-esters) and for a basic polymer (unplasticized poly-2-ethylhexylmethacrylate). Little basicity was conferred by ion exchange with inorganic bases or by adsorption of polyvinylpyridine. Much better results were obtained by surface treatments with silane coupling agents, well-known for their enhancement of adhesion, but not recognized as agents for acid-base surface modification.

A series of triethoxypropylsilanes with various functional groups on the terminal propyl carbon were tested for surface modification of the glass powder by measuring zeta-potentials in methanol-MIBK mixed solvents. In this solvent mixture the basic powders became positively charged by acid-base interaction with the methanol, and the acidic powders became negatively charged by acid-base interaction with the ketone. These zeta-potential measurements showed that the silane coupling agents were very effective surface modifiers for control of surface acidity or basicity. Surprisingly, the silane with a methacrylate functional group caused negative zeta-potentials of nearly -100 mV; this indicates that these treated surfaces had become quite acidic, which means that the methacrylate ester groups were bonded directly to the glass and that the new surface has many acidic SiOH groups. This orientation was confirmed by angle-resolved ESCA measurements by Prof. Dwight at VPI which indicated that carbonyl groups were buried,[32] a conclusion which had already been reached by Ishida and co-workers by FTIR measurements with the same coupling agents on another glass surface.[33] The amino-silanes gave positive zeta-potentials, especially the diamino-silanes, which provided about +100 mV, showing that these coupling agents had made the surface quite basic. The angle-resolved ESCA spectra showed that with the diamino-silane one nitrogen was buried and had the higher electron binding energy of an ammonium group, while the second nitrogen was at the surface and had the electron-binding energy typical of a free amine group.

The above glass powders (unmodified, acidified with the methacrylate silane, or made basic with the diamino-silane) were dispersed at high concentrations into solutions of each of the three different polymers (the acidic CPVC and polyvinylbutyral, and the basic poly-2-ethylhexyl-methacrylate). Films were cast and studied for mechanical properties with an Instron tensile tester.[32] The area under the stress-strain curves was largest for the acidic CPVC and acidic polyvinylbutyral when the powders were made basic with the diamino-silane, and highest for the basic poly(2-ethylhexylmethacrylate) when the powders were made acidic with the methacrylate-silane. Figure 17 shows the results with polyvinylbutyral filled with 88% of the unmodified powder in comparison with 88 v% of the diamino-silane modified powder. It is seen that the toughness (the area under the stress-strain curve) increased about six-fold when the surface was made basic. This enhanced toughness was also quite apparent when cutting "dog-bone" specimens for Instron testing; films with unmodified powder were so brittle that 80-90% broke during cutting, but films made with the basic diamino-silane treated powder were easy to handle and cut into shape; none broke.

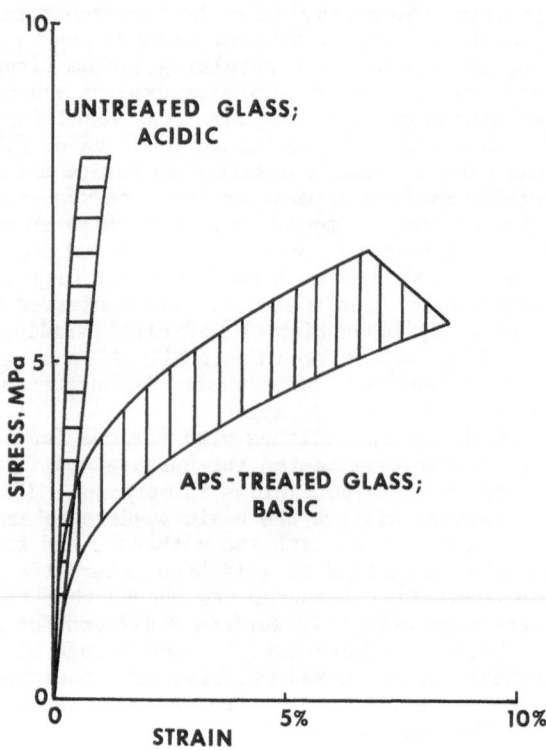

Figure 17. Stress-strain diagram of Instron tensile tests of the acidic polymer polyvinylbutyral filled with 88 w% of untreated glass powder (upper left) or with 88 w% of glass powder treated with amino-silane (lower right). The acid-base interaction increased toughness five-fold.

CONCLUSIONS

Composite films of inorganic powders in polymer matrices are used in the electronics industry for magnetic memory media, for conductive coatings, for high dielectric films and for ceramic processing. The mechanical and electrical quality of these dispersions depends on the magnitude of the exothermic particle-polymer interaction energy, which has been found to be of an acid-base nature, and predictable from the Drago E and C constants for the polymer matrix and for the surface sites of the powder. These E and C constants have been determined from heats of acid-base interaction with test acids and bases, using calorimetry or infrared spectral shifts.

The quality of the initial dispersions used for casting films of filled polymer also depends on acid-base interactions of the powders with the medium or dispersants. Effective dispersion or stabilization by either a steric or electrostatic mechanism requires acid-base interaction with the powder. Dispersions often can be stabilized without dispersants if the matrix polymer has strong enough acid-base interactions with the powder.

The magnitude of acid-base interaction between powders and matrix polymers can sometimes be appreciably enhanced by surface modification of the powder. Silane coupling agents are found to be very effective for this purpose, some providing increased acidity and others increased basicity to glass powders, and thereby appreciably enhancing the toughness of the composite.

REFERENCES

1.- F.M.Fowkes, in "Physicochemical Aspects of Polymer Surfaces", K.L.Mittal, Editor, Vol.2, p.583, Plenum Press, New York, 1983.

2.- R.S.Drago, G.C.Vogel and T.E.Needham, J.Am.Chem.Soc.93, 6014 (1971)

3.- R.S.Drago, L.B.Parr and C.S.Chamberlain, J.Am.Chem.Soc.99,3203 (1977)

4.- F.M.Fowkes and M.A.Mostafa, IEC Product R&D 17, 3 (1978)

5. M.J.Marmo, M.A.Mostafa, H.Jinnai, F.M.Fowkes and J.A. Manson, IEC Product R&D, 15, 206 (1976)

6.- F.M.Fowkes, H.Jinnai, M.A.Mostafa, F.W.Anderson and R.J.Moore, in "Colloids and Surfaces in Reprographic Technology", M.Hair and M.D. Croucher, Editors. ACS Symposium Series 200, p.307, American Chemical Society, Washington, D.C. (1982).

7.- D.J.Shaw,"Introduction to Colloid and Surface Chemistry", p.211 Butterworths, London (1966)

8.- J.H.Hildebrand and R.L. Scott, "Solubility of Non-Electrolytes", 3rd Edition, Reinhold, New York, 1950

9.- C.M.Hansen and K.Skaarup, J.Paint Technol. 39, 505 (1967)

10.- F.M.Fowkes, D.B.Seifert, and M.Palsson, to be published

11.- F.M.Fowkes, D.O.Tischler, J.Wolfe, L.A.Lannigan, C.M.Ademu-John and M.J.Halliwell, J.Polym.Sci., Polym.Chem.Ed. 22, 547 (1984)

12.- F.M.Fowkes, Ind.Eng.Chem. No.12, p.40 (1964)

13.- A.Beerbower, J.Colloid Interface Sci. 35, 126 (1971)

14.- S.Nozari and R.S.Drago, J.Am.Chem.Soc. 92, 7086 (1970)

15.- T.K.Kwei, E.M.Pearce, F.Ren and J.P.Chen, J.Polym.Sci., Polym. Physics Ed. 24, 1597 (1986)

16.- M.L.Hair, "Infrared Spectroscopy in Surface Chemistry", M.Dekker, Inc., New York (1967)

17.- F.M.Fowkes, J.Phys.Chem.64, 726 (1960)

18.- J.C.Bolger and A.S.Michaels, in "Interface Conversion for Polymer Coatings", P.Weiss, Editor, Elsevier, New York, 1968

19.- K.Meguro and K.Esumi, J.Colloid Interface Sci.59, 93 (1977)

20.- F.M.Fowkes, D.C.McCarthy and D.O.Tischler (1986) submitted to J.Polym.Sci.

21.- B.Shah and F.M.Fowkes, unpublished data (1985-6)

22.- S.T.Joslin and F.M.Fowkes, IEC Product R&D 24, 369 (1985)

23.- M.J.Kulp and F.M.Fowkes, unpublished data (1985-6)

24.- C.-Y.Huang and F.M.Fowkes, unpublished data (1985-6)

25.- M.D.Sacks and C.Khadilkar, J.Am.Ceram.Soc.66, 488 (1983)

26.- F.M.Fowkes and R.J.Pugh, in "Polymer Adsorption and Dispersion Stability", ACS Symposium Series 240, 331 (1984)

27.- D.B.Seifert and F.M.Fowkes, manuscript in preparation (1986)

28.- S.E.Friberg and K.Larsson, Adv.Liq.Crystals 2, 173 (1976)

29.- W.-J.Chen and F.M.Fowkes, unpublished data (1985)

30.- F.M.Fowkes, D.C.Kelley and D.Klempner, unpublished data (1968)

31.- F.M.Fowkes, in "Industrial Applications of Surface Analysis", L.A.Casper and C.J.Powell, Editors, ACS Symposium Series 199, p.69, American Chemical Society, Washington, D.C. (1982)

32.- F.M.Fowkes, D.W.Dwight, D.O.Tischler, T.B.Lloyd and B.Shah, manuscript in preparation (1986)

33.- H.Ishida, S.Naviroj and J.L.Koenig, in "Physicochemical Aspects of Polymer Surfaces", K.L.Mittal,Editor, Vol.1, p.91, Plenum Press, N.Y. (1983)

THE RELEVANCE OF LEWIS ACID-BASE CHEMISTRY TO SURFACE INTERACTIONS

William B. Jensen

Department of Chemistry
University of Cincinnati
Cincinnati, OH, 45221

This overview paper summarizes the current status of the Lewis acid-base or generalized donor-acceptor concepts and their application to solubility theory and related problems in the fields of surface and colloid chemistry. Simple second-order perturbation theory is then used to evaluate the pros and cons of various empirical treatments of Lewis acid-base strengths currently found in the literature and to clarify both their limitations and their relation to currently popular empirical approaches to surface and solubility phenomena.

THE LEWIS CONCEPTS

The Lewis acid-base concepts were first formulated by the American physical chemist G. N. Lewis in 1923 [1,2] and define an *acid* as any species (molecule, ion, or nonmolecular solid) capable of accepting a share in a pair of electrons during the course of a chemical reaction and a *base* as any species (molecule, ion, or nonmolecular solid) capable of donating that share. *Neutralization* becomes, in turn, simple coordinate or heterogenic bond formation between the acid and base:

$$A + :B \rightarrow A:B \qquad (1)$$

Figure I summarizes some typical Lewis acid-base interactions, giving examples of acid and base species corresponding not only to neutral molecules but to ions and nonmolecular solids as well. Of particular interest is the last example, which involves a nonmolecular solid (SiO_2) acting as the Lewis acid or electron-pair acceptor (EPA) species. Reaction with a basic oxide anion or electron-pair donor (EPD) species, supplied by an alkali metal or alkaline earth oxide, leads to a progressive depolymerization of the infinite three-dimensional silica framework of SiO_2 and is of importance in the manufacture of glass [3,4] Indeed, glass – or network forming oxides, fluorides, and heavy chalcogens in general are usually good Lewis acids,

$$SnCl_4 + :N\bigcirc \rightleftharpoons Cl_4Sn-N\bigcirc$$

$$|\overset{..}{I}{}^+ + \overset{..}{:}\overset{..}{I}|^- \rightleftharpoons |\overset{..}{I}-\overset{..}{I}|$$

$$Ag^+ + 2:NH_3 \rightleftharpoons Ag(NH_3)_2{}^+$$

$$BF_3 + :\overset{..}{F}|^- \rightleftharpoons BF_4{}^-$$

$$\tfrac{2}{\infty}[SiO_2] + :\overset{..}{O}|^{2-} \rightleftharpoons \tfrac{1}{\infty}[SiO_3]^{2-}$$

Figure 1. Some example Lewis acid-base or EPA-EPD interactions.

whereas, network-modifying species, like the alkali metal oxides, are usually good anion donors or Lewis bases.

Lewis' original definitions were based on the octet rule (so acids were generally identified as octet deficient) and the use of localized two-center, two-electron (2c-2e) bonds and one-center, two-electron (1c-2e) lone pairs typical of simple Lewis dot structures. Though these original definitions are still quite useful and indeed are the only version of the Lewis concepts most chemists encounter in the course of their training, there is a real need for a more sophisticated update that will connect them with currently popular bonding models such as molecular orbital (MO) theory. Such an update has in fact been provided by none other than Mulliken himself in a series of papers beginning in 1951 [5] and forms the basis of our current modernized concepts. This modernization has also led to a substantial broadening of the traditional definitions.

In the idiom of MO theory a Lewis acid is defined as any species employing an empty orbital (be it an atomic orbital, AO; a molecular orbital, MO; or a unfilled band) to initiate an interaction and a Lewis base as any species employing a doubly-occupied orbital. Neutralization, as before, is still heterogenic bond formation between the acid and base, and the term species still includes neutral molecules like BF_3 and NH_3, simple or complex ions like H^+, $Ag(NH_3)_2{}^+$, Cl^- or $NO_3{}^-$, and solids exhibiting nonmolecularity in one or more dimensions, such as the framework structure of SiO_2, mentioned earlier, the layer structures typical of TaS_2 and graphite, or the chain structures found in $ZrCl_3$ and NbI_4. However, despite these similarities, the MO versions of the Lewis definitions are not just a simple rewording of the originals, but have important consequences that are absent or, at best, only implicit in the more traditional versions:

*First, the degree of electron donation or interaction between the acid and base may range over an entire continuum - from nearly zero in the case of weak (but specific) intermolecular interactions and idealized ion associations, to complete transfer of one or more electrons (redox). Regrettably the existence of this continuum of interactions is frequently disguised by our habitual use of approximate limiting-case bonding models to describe differing degrees of electron donation (e.g., the ionic model, the covalent model, various dipole and polarization approximations, etc.). That it actually exists is supported by both surveys of adduct bond strengths and bond lengths [6] and by the direct mapping of donor-acceptor interactions in crystals [7,8]. All of these techniques fail to reveal any discontinuities and so support the supposition that any apparent breaks are artifacts of our approximate bonding models rather than actual phenomena.

*Second, the donor and acceptor orbitals, as well as orbital corresponding to the new bond formed via their interaction, may correspond not only to the two- and one-centered localized bonding components used in traditional Lewis structures, but to delocalized orbitals or to some kind of localized multicentered orbital as well. This allows one to incorporate much of the chemistry of so-called "nonclassical" systems, such as the metallocenes and boranes, within the Lewis acid-base paradigm.

*Third, the donor orbital (which usually corresponds to the highest-occupied MO or HOMO of a species - see Figure 2) need not necessarily be a nonbonding lone pair as in traditional Lewis bases or n-EPD species, but may be bonding in nature, corresponding to either a π or σ bond within the base itself. Likewise, the empty acceptor orbital (which usually corresponds to the lowest-unoccupied MO or LUMO of a species - see Figure

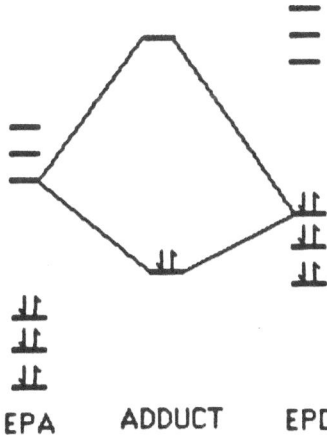

EPA   ADDUCT   EPD

Figure 2. Frontier orbital interactions for a typical EPA -EPD adduct.

2) need not necessarily be nonbonding as in traditional octet-deficient Lewis acids or n-EPA species, but may be antibonding relative to either a π or σ bond within the acid. Thus, in addition to traditional n-EPD and n-EPA species, we have the possiblity of σ-EPD, π-EPD, σ^*-EPA and π^*-EPA species and the nine distinct interactions or adduct types summarized in Table I.

This third point and Table I both underscore the extent to which the MO definitions have broadened Lewis' original concepts, since most traditional Lewis acid-base interactions and adducts belong only to the n-n category. Interactions belonging to the other eight categories can, of course, also give rise to simple acid-base addition and adduct formation provided that the degree of electron donation is not sufficient to completely depopulate a bonding donor orbital or to completely populate an antibonding acceptor orbital. If such extensive donation does occurs, then one will obtain instead either an acid-base displacement reaction, in the case of σ-EPD and σ^*-EPA species, or an acid-base addition across a multiple bond, in the case of π-EPD and π^*-EPA species. However, even in those cases where the degree of donation is insufficient to completely cleave a bond, some degree of bond weakening proportional to the degree of donation will occur within the corresponding b-EPD or a-EPA species. This can generally be followed by monitoring IR stretching frequencies and can serve as a possible measure of the strength of a donor-acceptor interaction.

Example Lewis acids, bases, and adducts corresponding to the categories in Table I are shown in Tables II and III Note that aromatic compounds, such as hexamethylbenzene, act as bases (π-EPD) when they carry electron-donating substituents and as acids (π^*-EPA), such as picric acid, when they carry electron-withdrawing substituents. Also note that in the case of the σ - π^* adducts in Table III, the σ-EPD species correspond to permethyl-polysilanes, which may be viewed as silicon analogs of the alkanes and cycloalkanes in which methyl groups, rather than hydrogen atoms, are used to terminate the ends of the Si-Si frameworks. Since neither lone pairs nor

Table I. Classification of EPA-EPD Interactions in Terms of the Interacting Orbitals

		ACCEPTOR ORBITALS [a]		
		n	σ^*	π^*
	n	n·n	n·σ^*	n·π^*
DONOR ORBITALS [b]	σ	σ·n	σ·σ^*	σ·π^*
	π	π·n	π·σ^*	π·π^*

[a] σ^* and π^* together form the class of antibonding or a–EPA species. [b] σ and π together form the class of bonding or b–EPD species.

Table II. Some Example EPA-EPD Agents (Generalized Lewis Acids and Bases)

CLASS	EXAMPLES
n-EPD	Most virtual Lewis bases, monoatomic and complex anions, neutral molecules with nonbonding lone pairs, such as carbanions, aliphatic amines, amine oxides, sulfides, phosphines, sulfoxides, ketones, ethers, alcohols, halides, etc.
π-EPD	Alkynes, alkenes, and aromatics, particularly with electron-donating substituents.
σ-EPD	Alkanes, borane anions, single bonds of all types, especially those containing hydrogen, highly polar bonds, as in BaO or NaCl, single bonds between heavy atoms, such as the Si-Si bonds in the peralkylsilanes, bent single bonds, as in strained cyclopropane.
n-EPA	Most virtual Lewis acids, monoatomic and complex cations, metastable fragments of large molecules or nonmolecular solids, such as BH_3 (fragment of B_2H_6) and $AlCl_3$ (fragment of the layer structure of solid $AlCl_3$), reaction intermediates with incomplete octets, such as classical carbenium ions, BR_3, etc.
π^*-EPA	N_2, SO_2, CO_2, NO_2^+, BF_3, dienophiles or alkynes, alkenes and aromatics in general with strongly electron-withdrawing substituents.
σ^*-EPA	The halogens (X_2), pseudo and interhalogens, such as ICN and ICl, hydrogen halides and Brønsted acids in general, boranes, alkanes with strongly electron-withdrawing groups, such as $CHCl_3$.

multiple bonds are present, the electron density donated to the π^*-EPA species must originate from the Si-Si σ bonds [9].

Tables II and III also illustrate how hydrogen bonding and the Brønsted or proton definitions of acidity are subsumed by the generalized Lewis concepts. Brønsted acids in general are σ^*-EPA species. Weak donor-acceptor interactions will lead to addition (i.e., hydrogen bonding) with some weakening of the original H-B bond. Stronger interactions will lead to cleavage of the original H-B bond and to proton transfer from HB to the attacking EPD species:

$$EPD: + HB \rightarrow EPD\text{--}H^{\delta+}\text{--}B^{\delta-} \rightarrow EPDH^+ + :B^- \qquad (2)$$

Thus the strength of the H-bonding interaction can be correlated with the shifts in the IR stretches of the original HB bond on complexing and Brønsted proton transfer becomes the limiting case of a strong H-bonding interaction. What the Lewis concepts are claiming, however, is that this scenario is not unique to protonic species. Rather it is characteristic of σ^*-EPA species in

Table III. Some Example Electron-Pair Donor-Acceptor (EPDA) Adducts

CLASSIFICATION	EXAMPLES
n·n	$(CH_3)_3N - BH_3$ $2NH_3 - PtCl_4$
n·σ*	Diethylether $- I_2$ Pyridine $-$ ICl Most hydrogen bonding
n·π*	Ethylamine $-$ picric acid Diethylether $-$ tetracyanoethylene
π·n	Benzene $-$ $AlCl_3$ 1,4 Cyclooctadiene $-$ $PtCl_2$
π·σ*	Hexamethylbenzene $-$ HF Cyclohexene $-$ ICl
π·π*	Anthracene $-$ trinitrobenzene 1,3,5 $-$ Trimethylbenzene $-$ SO_2
σ·n	$(CH_3)_2AlH - BH_3$ KCl $-$ $AlCl_3$
σ·σ*	Cyclohexane $- I_2$ n-Hexane $-$ WF_6
σ·π*	$Si_2(CH_3)_6$ $-$ tetracyanoethylene $[Si(CH_3)_2]_6$ $-$ chloranil

general and should, for example, also be observed in the case of the diatomic halogens. That this is indeed the case is shown in Figure 3, which compares some typical donor-acceptor interactions for HCl and I_2[10].

These conclusions can also be reached by comparing more sophisicated theoretical quantum mechanical treatments of H-bonding, on the one hand, with those used for nonprotonic Lewis acid-base adducts or so-called charge-transfer adducts, on the other - a task performed by Ratajczak and Orville-Thomas [11] in a 1980 review, in which they wrote:

> Thus finally one comes to the important conclusion that fundamentally there is no difference between "charge-transfer" and "hydrogen bond" interactions. Further, the hydrogen bond may be considered as a specific type of electron donor-acceptor interaction which is within the medium to strong range.

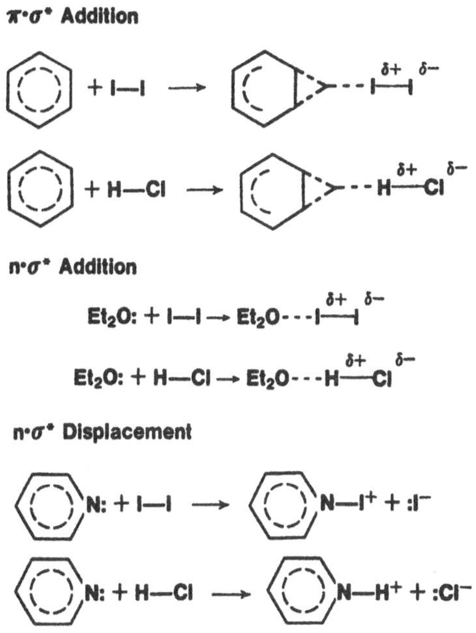

π·σ* Addition

n·σ* Addition

$Et_2O: + I—I \rightarrow Et_2O\text{-}\text{-}\text{-}I\overset{\delta+}{—}\overset{\delta-}{I}$

$Et_2O: + H—Cl \rightarrow Et_2O\text{-}\text{-}\text{-}H\overset{\delta+}{—}\overset{\delta-}{Cl}$

n·σ* Displacement

Figure 3. Parallels between the EPA properties of I_2 and HCl.

Needless to say, consistent use of the generalized Lewis concepts not only provides a common vocabulary and organizational framework for systematizing an enormous amount of previously unrelated chemistry (Table IV), but eliminates a great deal of redundant terminology in the process (Table V). Many of the applications summarized in Table IV are discussed in greater detail in reference 6.

IMPLICATIONS FOR SOLUBILITY THEORY

Perhaps the simplest way to illustrate the implications of the Lewis concepts for surface chemistry is to first illustrate in some detail their impact on the theory of solubility, since an understanding of bulk phase compatibility is in many ways a necessary prerequisite to an understanding of interfacial compatibility.

From a thermodynamic point of view, predicting the solubility of a given species in a given solvent is quite straightforward – at least in principle. One simply writes down an expression for the partial free energy of solution $(\Delta G_2[sol])$ for the solute (species 2),

$$\Delta G_2[sol] = \Delta H_2^\circ - T\Delta S_2^\circ + RT \ln a_2 \qquad (3)$$

Table IV. Chemical Phenomena Subsumed by the Generalized Lewis Concepts

Table IV. Chemical Phenomena Subsumed by the Generalized Lewis Concepts

A) Systems covered by the Arrhenius, Solvent-System, Lux-Flood, and proton acid-base definitions.

B) All of classical and nonclassical coordination chemistry.

C) Solvation, solvolysis, and ionic dissociation phenomena in aqueous and nonaqueous solvents.

D) Electrophilic and nucleophilic reactions in organic and organometallic chemistry.

E) Charge-transfer complexes, molecular addition compounds, weak but specific intermolecular attractions such as H-bonding.

F) Molten salt and glass chemistry.

G) Miscellaneous – such as catalysis, chemisorption, and intercalation of closed-shell species on solids, ionic metathesis reactions, salt formation, etc.

makes use of the standard separation of the activity into a mole fraction contribution and an activity coefficient contribution,

$$RT \ln a_2 = RT \ln x_2 + RT \ln \gamma_2 \qquad (4)$$

and uses standard thermodynamic relations to access the relative contributions of the activity coefficient to the partial enthalpy (ΔH_2[sol])

Table V. Terminology Subsumed by the Generalized Lewis Concepts

Acid or EPA Agent	Acceptor Electophile Cationoid Agent
Base or EPD Agent	Donor Nucleophile Electrodote Anionoid Agent Ligand
Acid-Base or EPDA Complex	Coordination Complex Charge-transfer Complex Donor – Acceptor Adduct Salt Molecular Adduct

and entropy (ΔS_2[sol]) of solution:

$$\Delta H_2[\text{sol}] = T^2(\partial(\Delta G_2[\text{sol}]/T)\partial T) = \Delta H_2° - RT^2(\partial ln\gamma_2/\partial T) \qquad (5)$$

$$\Delta S_2[\text{sol}] = -(\partial G_2[\text{sol}]/\partial T) = \Delta S_2° - Rlnx_2 - Rln\gamma_2 - RT(\partial ln\gamma_2/\partial T) \qquad (6)$$

The maximum solubility is then obtained by substituting either equation (4) or equations (5) and (6) back into equation (3) and applying the condition that ΔG_2[sol] = 0 at equilibrium, which, upon solving for lnx_2[eq] in terms of T, $\Delta H_2°$, $\Delta S_2°$, and $ln\gamma_2$[eq], gives the final result:

$$lnx_2[\text{eq}] = -\Delta H_2°/RT + \Delta S_2°/R - ln\gamma_2[\text{eq}] \qquad (7)$$

The only problem with this delightful scenario is what might be appropriately called "Dirac's Catch 22". This refers, of course, to Dirac's famous – or rather infamous – claim that, with the advent of quantum mechanics, chemistry had been reduced to a branch of applied physics, since it is possible in principle to write down the correct form of the Schrodinger equation for virtually any chemical system of interest [12]. The only difficulty, and this is the Catch 22, is in solving the resulting differential equations. Similarly, in order to bridge the gap between principle and practice with equation (7), one must have a method of explicitly calculating $\Delta H_2°$, $\Delta S_2°$ and $ln\gamma_2$ or of relating them to other easily measured properties of the system.

The first significant step in translating solubility principle into solubility practice was taken nearly 70 years ago by Hildebrand [13] and eventually evolved into what is now called regular solution theory [14,15]. Success was obtained in large part by restricting consideration to that class of solutions (called regular solutions) for which the activity coefficient contributions to the partial entropy of solution cancel:

$$R\, ln\gamma_2 = - RT(\partial\, ln\gamma_2/\partial T) \qquad (8)$$

giving:

$$\Delta S_2[\text{sol}] = \Delta S_2° + \Delta S_2[\text{mix}] = \Delta S_2[\text{tr}] - R\, lnx_2 \qquad (9)$$

$$\Delta H_2[\text{sol}] = \Delta H_2° + \Delta H_2[\text{mix}] = \Delta H_2[\text{tr}] + RT\, ln\gamma_2 \qquad (10)$$

Assuming the absence of specific compound formation between the solute and solvent, $\Delta S_2°$ and $\Delta H_2°$ become identical to the standard entropy and enthalpy of transition, ΔS_2[tr] and ΔH_2[tr], for the phase change required to bring the solute into the same phase as the solvent (equal to ΔS° and ΔH° of fusion or condensation for liquid solutions of solids and gases respectively). Likewise, $\Delta S_2[mix]$ becomes identical to that of an ideal solution and all that is lacking is some way of evaluating $\Delta H_2[mix]$ in terms of RT $ln\gamma_2$.

This remaining problem was eventually solved by recognizing that for the vast majority of liquids forming regular solutions, the cohesive energy, as measured by the molar energy of vaporization, $\Delta U_i[vap]$, is due almost totally to the operation of nondirectional, nonspecific dispersion forces:

$$\Delta U_i[vap] = \Delta E_i[dispersion] \qquad (11)$$

By defining a parameter δ_i (later called the solubility parameter of liquid i) equal to the square root of the liquid's cohesive energy density:

$$\delta_i = (\Delta U_i[vap]/V_i)^{1/2} \qquad (12)$$

it is possible to approximate the rigorous theoretical expression for the change in the dispersion energy on mixing two liquids with the equation:

$$\Delta H_2[mix] = RT \; ln\gamma_2 = V_2\phi_1{}^2(\delta_1 - \delta_2)^2 \qquad (13)$$

where V_2 is the molar volume of the solute, ϕ_1 is the volume fraction of the solvent (species 1) and δ_1 and δ_2 are the solubility parameters of the solvent and solute respectively. Thus equation (7) becomes:

$$ln x_2[eq] = - \Delta H_2[tr]/RT + \Delta S_2[tr] - V_2\phi_1{}^2(\delta_1 - \delta_2)^2/RT \qquad (14)$$

or, in the special case of liquid – liquid solutions, to which, for reasons of simplicity, we will restrict ourselves in what follows:

$$ln x_2[eq] = - V_2\phi_1{}^2(\delta_1 - \delta_2)^2/RT \qquad (15)$$

Both of these equations correlate the degree of solubility with the square of the difference in the solubility parameters of the solute and solvent – a large difference leading to small solubility and a small difference to high solubility. This result is, in effect, a quantification of the old adage that 'like dissolves like", where similarity of the cohesive energy densities as reflected in the solubility parameters is now our quantitative measure of "likeness".

Thus by using rigorous thermodynamics and a reasonable approximation for the partial enthalpy of mixing, Hildebrand and coworkers were able to successfully solve the problem of solubility for the special class of regular solutions. The only blemish to this approach is the fact that regular solutions and dispersion-only liquids form only a small subset of the solutions and solvents commonly employed by chemists. Consequently it is not surprising to discover that a good deal of effort has been expended in the last 30 years or so in attempts to extend Hildebrand's work to other classes of liquids and solutions.

Of these attempted extensions, many of which are summarized in Barton's recent book [16], the most popular and ambitious is probable the extended solubility approach of Hansen [17]. Hansen's model for the cohesive energy of a pure liquid allows for the operation of not only nonspecific dispersion interactions but for specific dipole (or polar) and hydrogen bonding interactions as well:

$$\Delta U_i[\text{vap}] = \Delta E_i[\text{dispersion}] + \Delta E_i[\text{dipole}] + \Delta E_i[\text{H-bond}] \qquad (16)$$

Each of these, in turn, is assigned a partial solubility parameter equal to the square root of the corresponding partial cohesive energy density:

$$V\delta_t^2 = V(\delta_d^2 + \delta_p^2 + \delta_h^2) \qquad (17)$$

and Hansen has also suggested ways of independently approximating each contribution [16]. Finally, to complete the parallel with the more rigorous theory of regular solutions, but with little or no theoretical justification, an analogous three-term equation for $\Delta H_2[\text{mix}]$ is used:

$$\Delta H_2[\text{mix}] = V_2\phi_1^2\{(\delta_1 - \delta_2)_d^2 + (\delta_1 - \delta_2)_p^2 + (\delta_1 - \delta_2)_h^2\} \qquad (18)$$

containing the squared differences of each of the three types of partial solubility parameters. Thus the criteria for high solubility is still one of "likeness", though this measure is now a three-dimensional vector quantity rather than a one-dimensional scalar [19]. Though the resulting model has not been very successful at a rigorous quantitative level, it has been quite useful in qualitatively extending the concepts of regular solution theory to a much boader range of solvents and solutions.

But even at a qualitative level there are some serious problems connected with the use of equation (18), the most important of these being its inability to deal with exothermic heats of mixing and its incorrect modeling of H-bonding interactions The importance of the first of these is illustrated in Figure 4, which shows some typical experimental solubility curves for binary liquid systems. The H_2O - phenol system on the left is the most common type, displaying increasing miscibility as the temperature

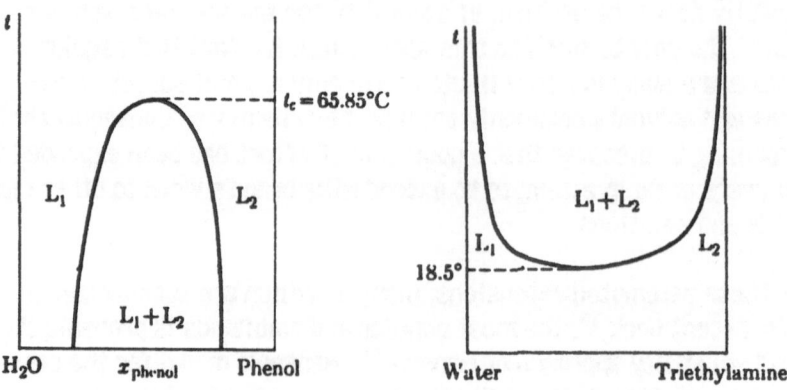

Figure 4. Some typical binary liquid-liquid phase diagrams.

increases, until it reaches a temperature of 65.85°C, called its critical solution temperature (CST), above which complete miscibility occurs [20]. In contrast, the less common H_2O - triethylamine system on the right shows exactly the opposite behavior, leading to increasing miscibility as the temperature is lowered and to a lower critical solution temperature of 18.5°C, below which complete miscibility occurs. Recognizing that the mole fraction of dissolved solute is essentially an equilibrium constant for the solution process and applying LeChatelier's principle, leads at once to the conclusion that ΔH_2[sol] for systems of the H_2O - phenol type is endothermic in nature, whereas for systems of the H_2O - triethylamine type it is exothermic. However, since equation (18) contains only the *squares* of the differences in δ_i, it can only lead to positive or endothermic values of ΔH_2[mix] and is consequently incapable of accounting for systems of the H_2O - Et_3N variety.

This defect is closely related to the second problem, namely the incorrect modeling of the H-bonding interactions. Use of the difference term $(\delta_1 - \delta_2)_h^2$ for H-bonding is equivalent to the claim that, like dispersion interactions, effective solute - solvent H-bonding depends on some *likeness* or similarity of the two liquids, when in reality it depends on the *complementary* matching of the two liquids as illustrated in Table VI. As a consequence the term leads to the incorrect inference that the strongest H-bonding interaction between a solute and solvent will occur when the pure liquids themselves are strongly self-associated by H-bonding, though in fact the opposite is frequently true. Indeed, this is well illustrated by the binary systems in Figure 4, where phenol and water, both of which are strongly self-associated by H-bonding, give rise to an unfavorable enthalpy of solution, whereas triethylamine, which exhibits little self-association due to H-bonding but is a good H-bonding acceptor species, gives a favorable

Table VI. The Failure of Equation (18) to Indicate the Complementary Nature of H-bonding Interactions.

INTERACTION	H-BONDING CHARACTERISTICS	δ_h TERMS
1-1	weak donor - strong acceptor	δ_1^2 small
2-2	strong donor - weak acceptor	δ_2^2 small
1-2	strong donor -strong acceptor	$-2\delta_1\delta_2$ small but really large

enthalpy of solution with water. In short, the ability to act as a strong H-bonding acceptor species in no way implies the necessary existence of the complementary ability to act as a strong H-bonding donor and *vice versa*, though most species are amphoteric to some degree and should be characterized with respect to both properties.

The simplest way of achieving this and of incorporating the inherently complementary nature of H-bonding into the expression for $\Delta H_2[mix]$ is to make $-\Delta E_i[H-bond]$ for a pure liquid the product of two inherently positive numbers, one of which characterizes the liquid's proton donor ability (PD_i) in a H-bonding interaction, and the other its proton acceptor ability (PA_i):

$$-\Delta E_i[H-bond] = PD_i \cdot PA_i \qquad (19)$$

Using a similar expression for H-bonding between different species as well, the change in the H-bonding energy on mixing two liquids can then be approximated using the simplified one-dimensional model in Table VII. Summing the interactions in the table gives:

$$\Delta E_{mix}[H-bond] = PD_1 \cdot PA_1 + PD_2 \cdot PA_2 - PD_1 \cdot PA_2 - PD_2 \cdot PA_1 \quad (20)$$

and application of some high school algebra leads to the final result:

$$\Delta E_{mix}[H-bond] = (PD_1 - PD_2)(PA_1 - PA_2) \qquad (21)$$

first suggested by Small over 30 years ago [21]. As can be seen, equation (21) not only incorporates complementarity by predicting that the most favorable interaction will involve a high PA - low PD liquid and a low PA - high PD liquid, but also resolves the first problem, since the product of the two

39

Table VII. One Dimensional Model for the Mixing of Two Liquids

PROCESS	MODEL	H-BONDING ENERGY
Remove a molecule of liquid 2	$\rightarrow 2 \rightarrow 2 \rightarrow 2 \rightarrow$ $\rightarrow$ $\rightarrow 2$ $2 \rightarrow$ + 2	$2\,PD_2 \cdot PA_2$
Close up the cavity	$\rightarrow 2$ $2 \rightarrow$ $\rightarrow$ $\rightarrow 2 \rightarrow 2 \rightarrow$	$-\,PD_2 \cdot PA_2$
Create a cavity in liquid 1	$\rightarrow 1 \rightarrow 1 \rightarrow 1 \rightarrow$ $\rightarrow$ $\rightarrow 1$ $1 \rightarrow 1 \rightarrow$	$PD_1 \cdot PA_1$
Insert molecule 2 into liquid 1	$\rightarrow 1$ $1 \rightarrow 1 \rightarrow$ + 2 $\rightarrow$ $\rightarrow 1 \rightarrow 2 \rightarrow 1 \rightarrow 1 \rightarrow$	$-PD_1 \cdot PA_2$ $-PD_2 \cdot PA_1$

differences may either be positive (endothermic) or negative (exothermic) in sign.

The first important attempt to use a Small-like equation appears to have been that of Keller *et al.* in 1970 [22], who used a set of corresponding Brønsted base, δ_b (equivalent to PA) and Brønsted acid, δ_a (equivalent to PD) solubility paramenters defined by the equation:

$$-\Delta E_i[\text{H-bond}] = 2V_i \delta_a \cdot \delta_b \qquad (22)$$

and a corresponding expression for $\Delta H_2[\text{mix}]$:

$$\Delta H_2[\text{mix}] = V_2 \phi_1^2 \{ (\delta_1 - \delta_2)_d^2 + (\delta_1 - \delta_2)_p^2 + 2(\delta_1 - \delta_2)_a \cdot (\delta_1 - \delta_2)_b \} \qquad (23)$$

Although this approach incorporates both complementarity and the possibility of exothermic interactions, its use of a composition dependent H-bonding term (via the $V_2 \phi_1^2$ multiplier) and a set of corresponding Brønsted solubility parameters is more open to question. At a naïve level, the use of a composition dependent enthalpy of mixing term for the dispersion-only interactions in the original form of regular solution theory stems from the fact that dispersion interactions are nonspecific. Since the average composition of the fluctuating coordination sphere about a given molecule will vary as the composition of the bulk solution varies, so presumably will the average value of the enthalpy. In contrast, in the case of H-bonding we are dealing with a specific interaction and with adducts of fixed composition. Consequently the enthalpy of this interaction should ideally be invariant to the composition of the bulk solution and, to a first approximation, be determined by a fixed number depending only on the chemical natures of the solute and solvent, as in Small's original suggestion.

40

At a more practical level, difficulties in obtaining a sufficently broad range of δ_b and δ_a parameters for common solvents, as well as problems in obtaining chemically reasonable self-consistency in the values of δ_a, appear to have prevented extensive use of equation (23), though Martin *et. al.* have recently applied it to problems of drug solubility [23].

Having eliminated two of the more serious problems in the Hansen model, what remains to be done? The answer, of course, lies in the further realization that H-bonding is not really a particular kind of intermolecular "force", like a dipole or dispersion force, but rather, as emphasized earlier, an example (albeit, a very important one) of a generalized electron-pair donor – acceptor or Lewis acid – base interaction. Consequently the H-bonding term in equation (23) should be replaced with a generalized EPD – EPA term. Since these interactions are specific and yield adducts (however fleeting) of fixed composition, they should also be modeled along the lines of Small's complementarity equation, with the PA paramenter now being subsumed within a more generalized electron-pair donor number (DN) and the PD parameter within a more generalized electron-pair acceptor number (AN), giving:

$$\Delta H_{mix}[DA] = k(AN_1 - AN_2)(DN_1 - DN_2) \qquad (24)$$

where k is a scaling constant of some sort.

As will be seen later, an examination of both theoretical and empirical measures of Lewis acid – base strengths will suggest yet a third modification of equation (23), as they both indicate that most specific electrostatic or polar interactions are already included within conventional measures of electron-pair donor and acceptor strengths, making the separate polar term in the equation potentially redundant. In short, a consideration of solubility phenomena from the standpoint of the Lewis concepts suggests that the simplest possible chemical models for the cohesive energy of a pure liquid and for the enthalpy of mixing for a binary liquid-liquid solution are two-term equations of the forms:

$$\Delta U_i[vap] = \Delta E_i[dispersion] + \Delta E_i[DA] = V_1 \delta_d^2 + kAN_i \cdot DN_i \qquad (25)$$

$$\Delta H_2[mix] = V_2 \phi_1^2 (\delta_1 - \delta_2)_d^2 + k(AN_1 - AN_2)(DN_1 - DN_2) \qquad (26)$$

IMPLICATIONS FOR SURFACE CHEMISTRY

One reason for reviewing the development of solubility theory in such detail is that the empirical treatment of surface phenomena, and especially that dealing with the work of adhesion, has followed a virtually identical course of development. On the basis of the approximations used for dispersion-only liquids in the theory of regular solutions, it is reasonable to expect a relationship between the work of adhesion of two phases and their surfaces tensions of the form:

$$W_{12}[adh] = 2\phi(\gamma_1\gamma_2)^{1/2} \tag{27}$$

where ϕ is a constant originally thought to depend on the molecular volumes and, indeed, such a relationship was experimentally confirmed by Girifalco and Good in 1957 [24].

In an effort to extend equation (27) to other classes of liquids, Fowkes suggested in 1962 [25,26] that both the work of adhesion and surface tensions could be decomposed into dispersion, dipole and hydrogen bonding components in a manner similar to Hansen's later decomposition of the the solubility parameter:

$$W_{12}[adh] = W_{12}[disp] + W_{12}[polar] + W_{12}[H\text{-bond}] \tag{28}$$

$$\gamma_t = \gamma_d + \gamma_p + \gamma_h \tag{29}$$

and that a relationship similar to equation (27) could be used to reasonably approximate the dispersion component and, to a less rigorous degree, the polar component as well:

$$W_{12}[adh] = 2(\gamma_1\gamma_2)_d^{1/2} + 2(\gamma_1\gamma_2)_p^{1/2} + W_{12}[H\text{-bond}] \tag{30}$$

Later writers attempted to also apply the geometric mean approxiation to the hydrogen bonding component, though, as with the enthalpy of solution, it is incorrect to use an approximation suitable only for nonspecific interactions to model a specific complementary interaction.

In 1978 Fowkes and Mostafa [27] proposed a modified expression for $W_{12}[adh]$ in which the hydrogen bonding component was incorporated within a generalized donor-acceptor term:

$$W_{12}[adh] = 2(\gamma_1\gamma_2)_d^{1/2} + 2(\gamma_1\gamma_2)_p^{1/2} + t\cdot k\Delta H_{12}[DA] \tag{31}$$

where k is a scaling constant that converts enthalpy per mole into free energy per unit area and t is the number of moles of donor-acceptor interactions per unit area. However, use of this model quickly showed that the polar term was redundant and that a two-term equation, paralleling equations (25) and (26) could be used instead [28].

$$W_{12}[adh] = W_{12}[disp] + W_{12}[DA] \qquad (32)$$

Use of Fowkes' expression for the dispersion component along with a generalized Small equation for the donor-acceptor component then gives:

$$W_{12}[adh] = 2(\gamma_1\gamma_2)_d^{1/2} + k[AN_1(xDN_2 - mDN_1) + AN_2(yDN_1 - nDN_2)] \quad (33)$$

where k is the scaling constant, x is the moles of donor sites per unit area at the surface of phase 2 interacting with phase 1, y is the moles of donor sites per unit area at the surface of phase 1 interacting with phase 2, and m and n represent the moles of additional sites for self-association created in phases 1 and 2 upon their separation and are related to the changes in their surface areas upon separation. In the case of gases or solids m and n can be assumed to be vanishingly small, but they may be significant in the case of liquids. Using established relations between $W_{12}[adh]$ and surface tensions, it is also possible to write down similar expressions for the dependency of spreading coefficients and contact angles on the donor - acceptor properties of the interacting phases.

A more complex situation is the adsorption on a solid of a component from a solution, since it involves desolvation of the component as well as possible competition between the absorbate and the solvent for the surface sites of the solid. A generalized Small equation for approximating the donor - acceptor contribution to the enthalpy of adsorption for this situation would be of the form:

$$\Delta H_{ads}[DA] = k\{-xAN_s{\cdot}DN_a - yAN_a{\cdot}DN_s - (t_A - x)AN_s{\cdot}DN_1 - (t_D - y)AN_1{\cdot}DN_s + \ldots$$

$$\ldots (x + y)[AN_a{\cdot}DN_1 + AN_1{\cdot}DN_a] + (t_A + t_D - 2x - 2y)AN_1{\cdot}DN_1\} \qquad (34)$$

where the subscripts s, a, and l stand for the solid surface, dissolved adsorbate, and liquid solvent respectively, t_A and t_D for the total moles of acceptor and donor sites per unit area on the surface, and x and y for the moles of these sites occupied by the adsorbate. The first two terms in the equation represent the adsorption of the dissolved adsorbate species, the second two terms the competitive adsorption of the solvent, the third term the desolvation of the adsorbate, and the fourth term the changes in solvent self-association due either to desolvation of the adsorbate or removal of solvent molecules for adsoption from the bulk solvent.

Actual quantitative use of either equation (33) or (34) is severely hampered by the necessity of not only having to know the donor and acceptor numbers of all of the species but the number of moles per unit area of each kind of interaction site [29]. Nevertheless, as will be seen, a knowledge of just the donor and acceptor numbers alone is still sufficient to qualitatively predict conditions that will maximize or minimize the occupation of certain kinds of sites and with them trends in the degrees of adsorption, contact angles, work of adhesion, and spreading coefficents as a function of the Lewis acid-base properties of the interacting phases.

IMPLICATIONS FOR COLLOID CHEMISTRY

As noted in Table V, the phenomenon of ionic dissociation is one of the topics subsumed by the generalized Lewis concepts. If a species XY is dissolved in an EPD solvent, the solvent will tend to interact with the more electropositive or acidic site on the molecule, leading to a heterolytic weakening of the original X-Y bond and to incipient ion formation:

$$EPD: + X-Y \rightarrow EPD--X^{\delta+}--Y^{\delta-} \qquad (35)$$

If the solvent is a sufficiently strong donor, this process will proceed to completion, giving rise to a contact ion pair consisting of a solvated cation and a naked anion, and these may, in turn, dissociate into independent ions capable of contributing to the solution's conductivity provided that the solvent's dielectric constant is high enough to separate them [30,31]:

$$EPD--X^{\delta+}--Y^{\delta-} \rightarrow EPD-X^{+} + Y^{-} \qquad (36)$$

The mechanism outlined in equation (2) for hydrogen bonding and proton transfer is a special case of this process, where X^{+} represents the proton and Y^{-} the conjugate Brønsted base. Within the context of the Lewis concepts, however, all possible cationic species X^{+} are inherently acidic.

If a strong EPA solvent is used instead, the complementary process will occur, giving rise to a solvated anion and a naked cation:

$$EPA + X-Y \rightarrow X^{\delta+}--Y^{\delta-}--EPA \rightarrow X^{+} + Y-EPA^{-} \qquad (37)$$

Obviously a liquid which is both a strong EPD and a strong EPA species (i.e., strongly amphoteric), and which also has a high dielectric constant, can combine both of these processes and will be a strongly ionizing solvent - a set properties corresponding almost exactly to those of water.

There is no reason, however, that X-Y must be a dissolved molecular fragment of some sort. It could just as well represent a molecule or a pair of adjacent atoms on the surface of an immiscible phase in contact with the liquid. In this case process (36) would lead to the transfer of X^{+} from the surface to the liquid, leaving the former with a net negative charge and the latter with a net positive charge. Likewise, process (37) would lead to the reverse transfer and to an opposite separation of charges between the surface and the liquid. A similar reversal could also occur for process (36) if the surface should prove to be a stronger donor than the liquid, as this would reverse their roles and X-Y in equation (36) would then represent the liquid and EPD the surface of the other phase. Likewise, a similar charge reversal could occur for process (37), if the surface proved to be a stronger EPA agent than the liquid.

Combining these various possiblities, one would expect to observe a sign reversal in the surface charge if a solid is immersed in a series of liquids of relatively constant EPA strength but gradually increasing EPD strength, the surface being positive when in contact with liquids that are weaker donors than the surface and negative when in contact with those that are stronger donors. Similarly, immersing a solid in a series of liquids of relatively constant EPD but gradually increasing EPA strength should give a negative surface charge for those that are weaker EPA agents than the surface and a positive charge for those that are stronger. What is being discussed here, of course, is really a generalization of a molecular mechanism for generating zeta potentials, first suggested by Fowkes [32] nearly 20 years ago for the more limited case of proton transfer, and the importance of such potentals in stabilizing colloidal dispersions, ranging from paints [33] and inks to fillers in polymers, need hardly be stressed.

Quantifying these considerations is, however, more difficult as most species are to some extent amphoteric and their EPA and EPD properties tend to work in opposite directions as far as the sign and magnitude of the zeta potential are concerned. Perhaps the simplest approach to this problem would be to use an empirical linear correlation similar to that proposed by Koppel and Pal'm in 1971 to analyze the solvent dependency of various physico-chemical properties [34]:

$$P = P_0 + \alpha \cdot DN + \beta \cdot AN + \xi \cdot \delta_d \qquad (38)$$

where P is the value of the property in the solvent of interest, P_0 is the value of the property in some reference state (preferably, but not necessarily, the gas phase or some inert solvent), α describes the sensitivity of the property to solvent basicity, β its sensitivity to solvent acidity, ξ its sensitivity to nonspecific dispersion forces, and the property in question may be a spectral transition, the logarithm of a rate constant or an equilibrium constant, a reaction enthalpy, an NMR shift, or, in our case, the zeta potential of a colloidal particle. Generally if the EPA - EPD character of the solvent is significant, it tends to swamp the last term and a simpler three-term correlation proposed by Krygowski and Fawcett [35-37] in 1974 can be used instead :

$$P = P_0 + \alpha \cdot DN + \beta \cdot AN \qquad\qquad (39)$$

QUANTITATIVE MEASURES OF LEWIS ACID – BASE STRENGTHS

In order to implement equations (25-26), (33-34), and (38-39) one requires, of course, some method of assigning DN and AN values to each species. This can be done by designing a probe which is selectively sensitive to either Lewis acidity or basicity alone and by applying equation (39) in reverse:

$$DN = (P - P_0)/\alpha \quad \text{(base-only sensitive probe, } \beta = 0) \quad (40)$$

$$AN = (P - P_0)/\beta \quad \text{(acid-only sensitive probe, } \alpha = 0) \quad (41)$$

Assigning arbitrary values to α, β and P_0 and the monitoring the changes in the selected properities P as a function of the species interacting with the probes, then serves to define DN and AN scales for the interacting species. Usually the probes are single chemical species whose properties are altered via their selective interaction with either the EPD or EPA functions of the species being characterized – in which case the probes are called reference acids and bases respectively. In principle, however, this need not necessarily be the case and the probes could, for example, be model displacement or elimination reactions whose rate or equilibrium constants are selectively sensitive to the Lewis acidity or basicity of the surrounding solvent.

As originally defined by Gutmann and coworkers [31,38] the DN is based on $-\Delta H_{rx}$ of the standard EPA probe antimony(V) chloride with the species of interest in a dilute 1,2 dichloroethane solution:

$$D: + SbCl_5 \rightarrow D\text{-}SbCl_5 \qquad DN = -\Delta H_{rx} \qquad (42)$$

where $P = -\Delta H_{rx}$, $P_0 = 0$, $\alpha = 1$ in equation (40). The AN, on the other hand, is based on the volume corrected difference at infinite dilution in the ^{31}P NMR shift induced in the standard EPD probe triethylphosphine oxide by the species of interest and that induced by n-hexane, scaled, in turn, relative to the shift induced by $SbCl_5$ in a dilute 1,2 dichloroethane solution [31,39]:

$$Et_3P\text{-}O + A \rightarrow Et_3P^{\delta+}\text{-}O^{\delta-}\text{-}A \qquad\qquad (43)$$

where $P = 100(\delta_\infty[A] - \delta_\infty[HX])_{corr}/(\delta_\infty[SbCl_5] - \delta_\infty[HX])_{corr}$, $P_0 = 0$, and $\beta = 1$ in equation (41). The factor of 100 essentially converts the AN of a species into a percentage of that shown by $SbCl_5$. Consequently calculation of $\Delta H_{rx}[DA]$ for an interaction via equation (19) requires a reconversion to the fraction, giving:

$$-\Delta H_{rx}[DA] = AN_A \cdot DN_D / 100 \qquad \text{(in kcal mol}^{-1}\text{)} \qquad (44)$$

DN values have been experimentally determined for about 53 liquids and AN values for about 34. Regrettably, however, these two sets of liquids are not identical, and their overlap allows for the complete characterization of the EPA - EPD properties of only about 25 different species. As shown in Figure 5, these can, in turn, be qualitatively classified as being primarily donor solvents (low AN - high DN), acceptor solvents (high AN - low DN), amphoteric solvents (high AN - high DN) and nonspecific or dispersion solvents (low AN - low DN) [40,41].

Luckily there are measurements of many other probes and properties available which are also selectively sensitive to either Lewis acidity or basicity. Some examples are listed in Tables VIII and IX and many additional correlations can be found in either the review of Griffiths and Pugh [42] or in the book [43] and reviews [44-46] of Reichardt. Again, by use of equation (39) and the resulting linear correlations, these are easily converted into equivalent DN and AN values:

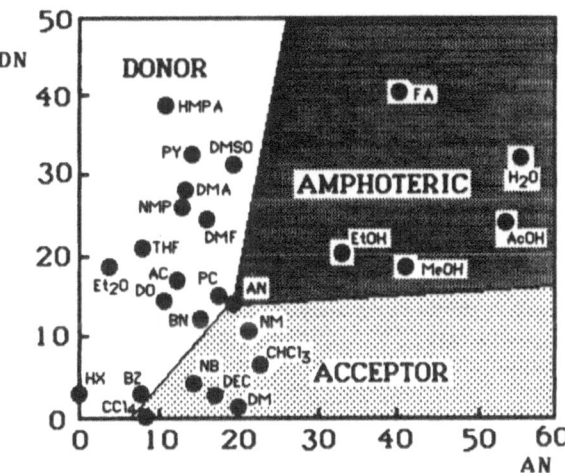

Figure 5. Plot of the donor number (DN) versus the acceptor number (AN) for a variety of solvents. Key: HX = n-hexane, Et$_2$O = diethyl ether, THF = tetrahydrofuran, BZ = benzene, HMPA = hexamethylphosphoramide, DO = dioxane, AC = acetone, NMP = N-methyl-2-pyrrolidinone, DMA = N,N-dimethylacetamide, Py = pyridine, NB = nitrobenzene, BN =benzonitrile, DMF = dimethylformamide, DEC = dichloroethylene carbonate, PC = propylene carbonate, AN = acetonitrile, DMSO = dimethylsulfoxide, DM = dichloromethane, NM = nitromethane, EtOH = ethanol, MeOH = methanol, AcOH = acetic acid.

Table VIII. Other Properties Selectively Sensitive to Lewis Basicity

SYMBOL	PROPERTY	STANDARD PROBE	AUTHORS
Q_m	$-\Delta H_{mix}[D-HCCl_3]$	$HCCl_3$	Marvel *et al.* (1940) and Searles *et al.* (1951)
$-\Delta H°_{BF_3}$	$-\Delta H_{rx}[D-BF_3]$	BF_3	Maria and Gal (1985)
PA	$-\Delta H_{rx}[D-H^+]$	H^+	Long and Munson (1973)
HCl_{sol}	Solubility of HCl in D	HCl	Gerrard *et al.* (1960)
K_a	$K_{diss}[HD]_{(aq)}$	H_3O^+	Brønsted (1929.)
$\Delta \nu_D$	IR shift for O–D bond	CH_3OD	Kagiya *et al.* (1968)
D[II,I]	Visible–UV band shifts for $VO(acac)_2$ –D	$VO(acac)_2$	Selbin and Ortolano (1964)
δ_0 ppm	Relative ^{23}Na NMR shift for Na–D	Na^+	Bloor and Kidd (1968) Popov *et al* (1970)
---	Relative ^{19}F NMR shift for F_3Cl–D	CF_3I	Gutmann *et al.* (1971)
Δppm P	Relative ^{19}F NMR shift for F–R–D	p–fluorophenol p–nitroso fluorobenzene	Taft *et al.* (1961, 1963)

$$DN = P/\alpha - P_0/\alpha \qquad \text{(base sensitive)} \qquad (45)$$

$$AN = P/\beta - P_0/\beta \qquad \text{(acid sensitive)} \qquad (46)$$

$$y = ax + b \qquad \text{(in general)} \qquad (47)$$

and in cases where either the DN or the AN alone is known, a property sensitive to both can be used to find the missing member of the pair:

$$DN = (P - P_0 - \beta AN)/\alpha \qquad (48)$$

$$AN = (P - P_0 - \alpha DN)/\beta \qquad (49)$$

Table IX. Other Properties Selectively Sensitive to Lewis Acidity

SYMBOL	PROPERTY	STANDARD PROBE	AUTHORS
Δv_A	IR shift in C=O bond for R_2C=O-A	acetophenone	Kagiya *et al.* (1968)
G	IR shift in C=O bond or S=O bond for R_2C=O-A or R_2S=O-A	benzophenone, dimethyl-formamide, dimethyl-sulfoxide	Allerhand and Schleyer (1963)
Z	Solvatochromic shift	1-ethyl-4-carbomethoxy pyridinium iodide	Kosower (1958)
E_T	Solvatochromic shift	pyridinium-N-phenol betaines	Dimroth *et al.* (1963)
x_B, x_R	Solvatochromic shift	various merocyanines	Brooker *et al.* (1951,1965)
Y	Relative rate constant	S_N1 solvolysis of t-butyl chloride	Winstein (1948)
X	Relative rate constant	S_E2 reaction of tetramethyltin with bromine	Gielen and Nasielski (1964)

By use of such linear relations for the properties summarized in Table X, our group has recently compiled a list of best averaged DN and AN values for about 150 different solvents, including water, allowing for a much more extensive semiquantitative application of the Lewis concepts than previously.

An example of this in the case of solubility is shown in Figure 6, which shows a qualitative sorting map [47,48], based on equation (26), for the miscibility of 72 binary liquid mixtures involving 21 different liquids at 25 °C. As expected, the miscible systems tend to cluster in the lower left hand corner, reflecting both favorable donor - acceptor and dispersion interactions upon mixing, whereas the immiscible systems show the opposite behavior. Once established, this map can be used, via interpolation, to predict the miscibility of other systems for which the DN, AN and δ_d parameters of the component liquids are known [49].

An example involving equation (34) and adsorption from solution is shown in Figure 7. This is based on the data of Fowkes and Mostafa [27] for the

Table X. Some Linear Relations for Determining Additional DN and AN Values

EQUATION	CORRELATION COEFFICIENT	REFERENCE
$DN = .20(\Delta \nu_D) + 3.03$	.984	42
$DN = 10.11(D[II,I]) - 12.17$	.995	42
$DN = 1.09(-\Delta HBF_3) - 4.83$	.963	60
$AN = 1.598(E_T) - 50.69$	.956	46
$\log \varepsilon = .0711(AN) + .0054(DN) + .2581$	.957	61

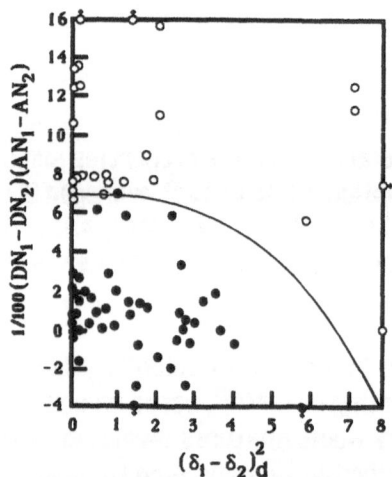

Figure 6. Sorting map for miscible (solid circles) and immiscible (open circles) binary liquid systems at 25°C.

adsorption of a basic adsorbate (poly(methyl methacrylate) or PMMA) onto an acidic surface (SiO$_2$ – acidity due to surface –OH groups) from solvents with varying EPA – EPD properties. Assuming that AN$_a$ = 0 and DN$_s$ = 0, equation (34) reduces to:

$$-xAN_s \cdot DN_a \ - (t_A- x)AN_s \cdot DN_1 \ + (x + y)AN_1 \cdot DN_a + (t_A+ t_D-2x-2y)AN_1 \cdot DN_1 \quad (50)$$

Use of solvents of approximately constant AN but variable DN makes the second term in equation (50) the controlling factor in optimizing ΔH_{ads}[DA] and this, in turn, leads to the prediction that x, the number of surface sites occupied by the PMMA, will decrease as the DN of the solvent increases. In other words, as the EPD properties of the solvent increase it will successfully compete with the basic PMMA for the acidic surface sites of the SiO$_2$. Conversely, use of solvents of relatively constant DN but variable AN will make the third term the controlling factor in optimizing ΔH_{ads}[DA], again leading to a decrease in x as the acidity of the solvent increases – this time due to the competitive solvation of the basic PMMA by the solvent. As Figure 7 shows, these predictions are confirmed by experiment.

A reversed system consisting of adsorption of an acidic adsorbate (post-chlorinated poly(vinylchloride) or Cl–PVC) onto a basic surface (CaCO$_3$) from solvents of varying EPD – EPA properties was also studied by Fowkes

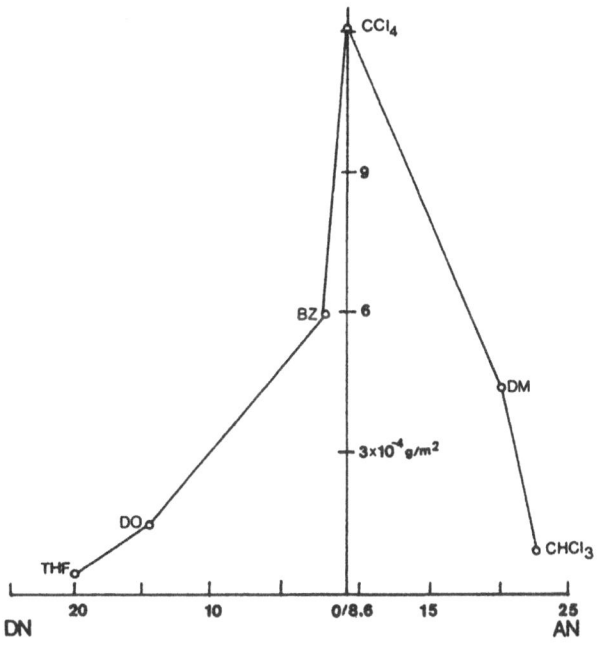

Figure 7. Adsorption of basic PMMA onto an acidic SiO$_2$ surface from solvents of variable basicity (left) and variable acidity (right)(From ref. 40).

and Mostafa. Again, by assuming that $DN_s = 0$ and $AN_s = 0$, a limiting case of equation (34) can be used which predicts that solvents of increasing AN values will decrease adsorption of Cl-PVC due to competition for the basic surface sites and solvents of increasing DN values will decrease it due to more effective solvation of the adsorbate species (Figure 8).

Williams [50] has explored the development of contact charges between a flowing liquid and a solid surface as a function of the liquid's DN. Though the results (Figure 9) were originally rationalized using an electron transfer model, they are consistent with the ion transfer model discussed earlier and show that liquids of low donicity acquire a negative charge due to transfer of X^+ to the surface of the solid, whereas liquids of high donicity acquire a positive charge due to transfer of X^+ in the opposite direction. This cross-over point is presumably an indication of the DN value of the surface. Labib and Williams [51] have also extended this work to the measurement of the zeta potentials of the suspended particles of a variety of inorganic solids as a function of the donicity of the surrounding liquid, again obtaining results (Table XI) largely in keeping with the ion transfer model, though some inconsistencies were observed. However, since not all of the liquids used had similar AN values and, as pointed out earlier, variations in this parameter can have an opposite effect on the zeta potential, these data should really be reexamined from the standpoint of equations (38) and (39).

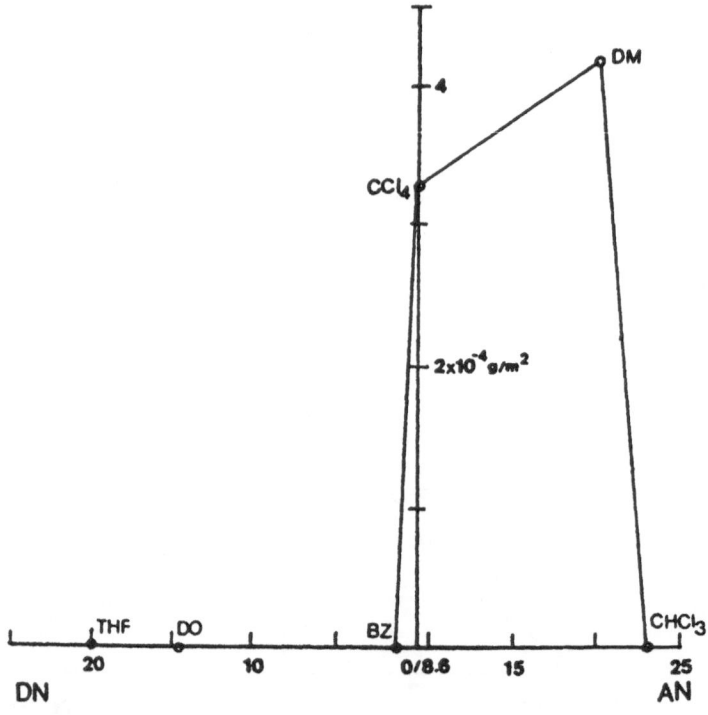

Figure 8. Adsorption of acidic Cl-PVC onto a basic $CaCO_3$ surface from solvents of variable basicity (left) and variable acidity (right)(From ref. 40).

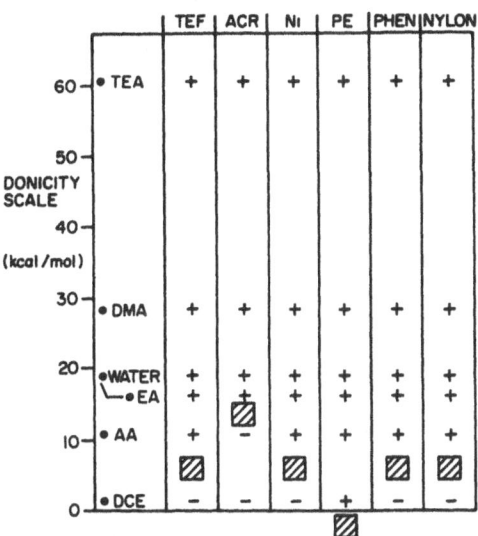

Figure 9. Charge acquired by a flowing liquid in contact with various solid surfaces. Key: TEA = triethylamine, DMA = N,N-dimethylacetamide, EA = ethyl acetate, AA = acetic anhydride, DCE = 1,2 dichloroethane, TEF = teflon, ACR = methylmethacrylate, Ni = nickel, PE = polyethylene, PHEN = phenolic polymer. Shaded areas indicate the approximate donicities of the surfaces (From ref. 50).

Use of equation (44) and the other Small-like expressions for calculating the enthalpies of EPA - EPD interactions is equivalent to the assumption that a single universal scale or order of Lewis acid - base strengths exists. Experimental evidence, however, shows that this is not rigorously true. For example, the EPD species in Table XII show a different order of base strengths depending on whether they are measured relative to their interaction with the Lewis acid Me_3Ga or the Lewis acid I_2 [52]. Even more striking examples are found for ionic Lewis acids and bases. Thus in aqueous solution the halide ions give the EPD order $F^- \gg Cl^- > Br^- > I^-$ when measured relative to the Lewis acid H^+ but the reverse order when measured relative to the Lewis acid Hg^{2+}. The qualitative interpretation of these inversions is the basis of the so-called hard and soft acid-base (HSAB) principle [53] and their quantitative treatment is to be found in the E & C equation of Drago and coworkers [54,55]:

$$-\Delta H_{AB} = E_A E_B + C_A C_B \qquad (51)$$

where the E parameters measure a species' susceptibility to electrostatic or ionic interactions and the C parameters its susceptibility to covalent or orbital perturbation interactions. Since in principle these two factors can vary independently of one another, equation (51) is capable of yielding a different order of ΔH_{AB} values for a series of bases when the reference acid is changed and vice versa.

Table XI. Zeta Potentials of Various Solids as a Function of the Donicity of the Surrounding Liquid (51)

SOLID	LIQUIDS [a]/DN								
	DCE 0.0	NM 2.7	NB 4.4	AA 10.5	EA 17.1	THF 20.0	DMA 27.8	DMSO 29.8	EDA 55.0
$CaCO_3$	+59	-104	-71	—	-92	—	-17	—	-22
TiO_2	+48	-58	—	-77	-44	—	-10	-34	—
Diamond	+53	-9	—	-42	-47	—	-70	-44	-17
ZnS	+61	-53	—	—	-92	—	-43	-45	-45
Gold	+172	+126	—	-133	-94	—	-100	-136	-205
$Na_2WO_4 \cdot 2H_2O$	-36	-21	—	—	-50	—	—	-62	-62
$Al_2(SO_4)_3 \cdot xH_2O$	+50	—	—	-114					
$Al_2(SO_4)_3 \cdot xH_2O$ (heated at 250°C for 18 hrs)	—	+72	—	—	+100	+35	—	+66	-44
Mica	+28	-52	—	-78	-40	—	-45	—	—
Mica (heated at 250°C for 18 hrs)	+39	+61	—	—	+59	-63	-23	-23	—
CaO	-72	—	—	—	—	—	-27	—	—

[a] Abbreviations the same as those used in earlier tables.

Table XII. Gas Phase Basicities as a Function of the Reference Lewis Acid

ACID	ORDER OF BASE STRENGTH [a]
Me_3Ga	$Me_3N > Et_2O > Me_2S$
I_2	$Me_3N > Me_2S > Et_2O$

[a] Abbreviations the same as those used in earlier tables.

Like the Small approach the E & C equation uses two parameters to characterize each species (E and C versus AN and DN) and four to calculate each interaction. But unlike the Small approach, it does not recognize the inherently amphoteric nature of most species and treats each of them as being either exclusively an acid or a base. Consequently the approach cannot be used to determine the contribution of EPA – EPD interactions to the self-association of a species. Thus, in choosing an quantitative approach to Lewis acid – base phenomena, we must decide between a method that treats amphoteric interactions but naively assumes universal orders of strength, on the one hand, and a method which deals with variations in orders of strength but ignores amphoteric interactions, on the other. Which is the correct choice? As might be expected with such simple models, the answer is that neither method is completely correct and the choice largely depends on the dictates of the systems one is dealing with. For example, a crude quantum mechanical treatment of EPA – EPD interactions by Klopman and Hudson [56,57], using second-order perturbation theory, shows that a minimum of three terms is necessary to approximately describe the interaction:

$$\Delta E_{AD} = \sum_{rs} \frac{Q_r Q_s}{R_{rs}} + \sum_{\substack{m \\ \text{of D}}}^{occ} \sum_{\substack{n \\ \text{of A}}}^{unocc} \frac{2(\Sigma_{rs} c_r^m c_s^n \beta_{rs})^2}{E_m - E_n} + \sum_{\substack{n \\ \text{of A}}}^{occ} \sum_{\substack{m \\ \text{of D}}}^{unocc} \frac{2(\Sigma_{rs} c_r^m c_s^n \beta_{rs})^2}{E_n - E_m} \tag{52}$$

The first of these represents a classical electrostatic or charge-control term, whereas the second and third represent covalent or orbital control terms, one in which species D acts as the EPD agent and the other in which species A does. The importance of these orbital control-terms depends largely on the size of their denominators and these, in turn, reflect the difference in the energies of the donor and acceptor orbitals and especially those of the HOMOs and LUMOs. The smaller these energy differences the larger the respective orbital-controlled interactions and vice versa. It is reasonable to view the $AN_A \cdot DN_D$ and $AN_D \cdot DN_A$ terms in the Small equation as empirical approximations of these orbital control-terms and, indeed, Paoloni et al.[58] have actually established approximate correlations between the DN and AN values of a species and the energies of the species' HOMO and LUMO orbitals. Although the present writer is not aware that any similar correlations have been established for the E & C equation, it is still not unreasonable to view the $E_A E_B$ term as an empirical approximation of the charge-control term and the $C_A C_B$ term as an approximation of the first of the two orbital-control terms. In short, the Small approach views the independent variation of the last two terms in equation (52) as being the most important, whereas the E & C equation views the independent variation of the first two terms as being the most important.

Actually the relative importance of the terms varies with the systems

Table XIII. Comparison of the DN-AN Approach and the E & C Equation

	E & C EQUATION	DONOR AND ACCEPTOR APPROACH
PARAMETERS	E_A, E_B, C_A, C_B	DN_D, AN_A, DN_A, AN_D
SYSTEMS	Nonassociated and Dilute	Associated and Condensed
SPECIES	Variable Hardness and Nonamphoteric	Similar Hardness and Amphoteric

being studied. The donor and acceptor sites on a species are usually located on different atoms and in dilute systems, and particularly in the gas phase, the species, for stereochemical reasons, will probably select one of the two modes in its interactions. In condensed systems, however, where the species is surrounded on all sides, it is likely to excercise both modes of interaction simultaneously, though usually with two different partners. Thus amphoteric properties become decidedly more important in the liquid and solid phases. Likewise, if one is dealing with systems in which the only donor atoms are N and O and most of the acceptor sites are H-bonding, then inversions in strength due to variations in hardness and softness are at a minimum, in constrast to systems also having S and P donor sites, for example, or Hg and I_2 as acceptors. These conclusions are summarized in Table XIII and it is on this basis - namely the fact that in the systems which we have been describing, we are working with relatively hard species in the liquid and solid states and with processes in which competition with the self-association of the pure components is important, that we have chosen to work with the Small equation and with the donor and acceptor number scales. Fowkes, on the other hand, has successfully applied the E & C equation to some of the same systems [27-28,59]. The ideal solution, of course, would be to have an E and C parameter for both the donor and acceptor ability of each species, though this would result in a total of four parameters per species and eight per interaction, and perhaps rather stretches the usefulness of an empirical approach to Lewis acid-base interactions.

CONCLUSIONS

As the above examples illustrate, the Lewis concepts have an enormous potential for clarifying many phenomena of interest to chemists working in the fields of surface chemistry, colloid chemistry, and adhesion and, with the very notable exception of the pioneering work of Fowkes and his coworkers [27-28,59], this potential is still largely untapped. At present most of this work is at a semi-quantitative empirical level. But even at a purely qualitative level, the Lewis concepts are of value, if for no other reason than the fact that they refocus thinking on the *complementary matching* of properties for favorable interactions rather than on the *similarity matching* of properties which still pervades many of these areas as a result of an inappropriate extension of the dispersion-only arguments used in the original theory of regular solutions.

REFERENCES

1. G. N. Lewis "Valence and the Structure of Atoms and Molecules", pp. 141-142, The Chemical Catalog Company, NY, 1923.
2. G. N. Lewis, *J. Franklin Inst* **226**, 293 (1938).
3. K-H Sun and A. Silverman, *J. Am. Ceram. Soc.* **28**, 8 (1945).
4. K-H Sun, *Glass Ind* **29**, 73 (1948).
5. R. S. Mulliken and W. B. Person, "Molecular Complexes: A Lecture and Reprint Volume", Wiley-Interscience, NY, 1969.
6. W. B. Jensen, "The Lewis Acid-Base Concepts: An Overview", Wiley-Interscience, NY, 1980.
7. H. Burgi, *Angew. Chem. Int. Ed. Engl* **14**, 460 (1975).
8. J. D. Dunitz, "X-Ray Analysis and the Structure of Organic Molecules", Cornell University Press, Ithaca, NY, 1979.
9. V. F. Traven and R. West, *J. Am. Chem. Soc* **95**, 6824 (1973).
10. H. C. Brown, *J. Phys. Chem.* **56**, 821 (1952).
11. H. Ratajczak and W. J. Orville-Thomas, "Molecular Interactions", Vol. 1, Chap.1, Wiley-Interscience, NY, 1980.
12. P. Dirac, *Proc. Royal Soc. London, Ser. A* **123**, 714 (1929).
13. J. H. Hildebrand, *J. Am. Chem. Soc*. **38**, 1452 (1916).
14. J. H. Hildebrand and R. L. Scott, "The Solubility of Nonelectrolytes", Reinhold, NY, 1950.
15. J. H. Hildebrand, M. M. Prausnitz, and R. L. Scott, "Regular and Related Solutions", Van Nostrand - Reinhold, NY, 1970.
16. A. F. M. Barton, "Handbook of Solubility Parameters and Other Cohesion Parameters", CRC Press, Boca Raton, FL, 1983.
17. C. M. Hansen, *Ind. Eng. Chem. Prod. Res. Dev.* **8**, 2 (1969).
18. C. M. Hansen and A. Beerbower in "Kirk Othmer Encyclopedia of Chemical Technology", A. Standen, Editor, Suppl. Vol., 2nd ed., p .889, Wiley-Interscience NY, 1971.
19. D. L. Wernick, *Ind. Eng. Chem. Prod. Res. Dev.* **23**, 240 (1984).
20. A. W. Francis, "Critical Solution Temperatures", Adv. Chem. Series No. 31, American Chemical Society, Washington, D.C., 1961.
21. P. A. Small, *J. Appl. Chem.* **3**, 71 (1953).
22. R. A. Keller, B. L. Karger, and L. R. Synder in "Gas Chromatography 1970", R. Stock and S. F. Perry, Editors, p. 125, Institute of Petroleum Research, London, 1971. The original equation also contained an induced dipole term, though the work of Martin *et al*[23] suggests that it can be neglected.

23. A. Beerbower, P. L. Wu, and A. Martin, *J. Pharm. Sci.* **73**, 179, 188 (1984).
24. L. A. Girifalco and R. J. Good, *J. Phys. Chem.* **61**, 904 (1957).
25. F. M. Fowkes, *J. Phys. Chem.* **66**, 382 (1962).
26. F. M. Fowkes, *Ind. Eng. Chem.* **56**(12), 40 (1964).
27. F. M. Fowkes and M. A. Mostafa, *Ind. Eng. Chem. Prod. Res. Dev.* **17**, 3 (1978).
28. F. M. Fowkes in "Physicochemical Aspects of Polymer Surfaces", K. L. Mittal, Editor, Vol. 2, p. 583, Plenum, NY, 1983.
29. K. Tanabe, "Solid Acids and Bases", Academic Press, NY, 1970.
30. V. Gutmann, *Angew. Chem. Int. Ed. Engl.* **9**, 843 (1970).
31. V. Gutmann, "The Donor-Acceptor Approach to Molecular Interactions", Plenum, NY, 1978.
32. F. M. Fowkes, *Disc. Faraday Soc.* **42**, 246 (1966).
33. P. Sørensen, *J. Paint Tech.* **47**, 31 (1975).
34. I. A. Koppel and V. A. Pal'm in "Advances in Linear Free-Energy Relationships", J. Shorter, Editor, Chap. 5, Plenum, NY, 1972. The original equation also contained a dipole term. We have also substituted a more logical choice of symbols.
35. T. M. Krygowski and W. R. Fawcett, *J. Am. Chem. Soc.* **97**, 2143 (1975).
36. W. R. Fawcett and T. M. Krygowski, *Aust. J. Chem.* **28**, 2115 (1975).
37. W. R. Fawcett and T. M. Krygowski, *Can. J. Chem.* **54**, 3283 (1976).
38. V. Gutmann and E. Wychera, *Inorg. Nucl. Chem. Lett.* **2**, 257 (1966).
39. U. Mayer, V. Gutmann, and W. Gerger, *Monatsh. Chem.* **106**, 1235 (1975).
40. W. B. Jensen, *Rubber Chem. Techn.* **55**, 881 (1982).
41. W. B. Jensen, *Chemtech.* **12**, 755 (1982).
42. T. R. Griffiths and D. C. Pugh, *Coord. Chem. Rev.* **29**, 129 (1979).
43. C. Reichardt, "Solvent Effects in Organic Chemistry", Verlag Chemie, NY, 1979.
44. C. Reichardt, *Angew. Chem. Int. Ed. Engl.* **18**, 98 (1979).
45. C. Reichardt, *Pure Appl. Chem.* **54**, 1867 (1982).
46. C. Reichardt in "Molecular Interactions", H. Ratajczak and W. J. Orville-Thomas, Editors, Vol. 3, Chap. 5, Wiley-Interscience, NY, 1982.
47. For background on the philosophy of sorting maps see A. Beerbower and W. B. Jensen, *Inorg. Chim. Acta* **75**, 193 (1983) and Wold [48].
48. S. Wold and M. Sjöström in "Chemometrics: Theory and Applications", R. F. Gould, Editor, ACS Symposium Series No. 52, Chap. 12, American Chemical Society, Washington, D. C., 1977.
49. W. B. Jensen in "Proceedings Symposium on Migration of Gases, Liquids, and Solids in Elastomers", D. Hertz, Editor, 126th National Meeting of the Rubber Division, American Chemical Society, Denver, CO, 1984.
50. R. Williams, *J. Colloid Interface. Sci.* **88**, 530 (1982).
51. M. E. Labib and R. Williams, *J. Colloid Interface. Sci.* **97**, 356 (1984).
52. K. F. Purcell and J. C. Kotz, "Inorganic Chemistry", p. 218, Saunders, Philadelphia, PA, 1977.

53. R. G. Pearson, Editor,"Hard and Soft Acids and Bases", Dowden, Hutchinson and Ross, Stroudsburg, PA, 1973.

54. R. S. Drago and B. Wayland, *J. Am. Chem. Soc.* **87**, 3571 (1965).

55. R. S. Drago, *Struct. Bonding (Berlin)* **15**, 73 (1973).

56. R. F. Hudson and G. Klopman, *Theor. Chim. Acta* **8**, 165 (1967).

57. G. Klopman, Editor, "Reactivity and Reaction Paths", Wiley-Interscience, NY, 1974.

58. A. Sabatino, G. La Manna, and L. Paoloni, *J. Phys. Chem.* **84**, 2641 (1980).

59. F. M. Fowkes, D. O. Tischler, J. A. Wolfe, L. A. Lannigan, C. M. Ademu-John, and M. J. Halliwell, *J. Polymer Sci.* **22**, 547 (1984).

60. P. Maria and J. Gal, *J. Phys. Chem.* **89**, 1296 (1985).

61. R. Schmid, *J. Solution Chem.* **12**, 135 (1983).

SURFACE CHARGE CHARACTERISTICS OF CHROMIUM (IV) DIOXIDE IN TETRAHYDROFURAN WITH SPECIAL REFERENCE TO MAGNETIC-INK DISPERSIONS

G. F. Hudson, S. Raghavan, and D. A. Roylance[*]

Department of Materials Science and Engineering
University of Arizona
Tucson, Arizona 85721

[*]IBM Corporation, General Products Division
Tucson, Arizona 85744

Electrophoretic mobilities of acicular chromium dioxide (CrO_2) particles have been measured in tetrahydrofuran containing sodium perchlorate, phosphatidylcholine, and cetyl trimethyl ammonium bromide (CTAB). In sodium perchlorate solutions, CrO_2 is positively charged; in phosphatidylcholine and CTAB solutions, the charge on the CrO_2 particles changes from positive to negative as the solute concentration is increased. There seems to exist a good correlation between the zeta potential values and the stability of CrO_2 dispersions indicating the presence of electrostatic stabilization in these systems. Application of Derjaguin, Landau, Verwey, and Overbeek (DLVO) theory to the dispersions shows that energy barriers as high as 24 kT can exist when particles have a zeta potential of approximately 70 mV.

INTRODUCTION

Stabilization of dispersions of magnetic oxides in nonaqueous liquids of low dielectric constant is an important requirement for the manufacture of magnetic recording tape for high-density recording. Any clumping or agglomeration of particles in the magnetic oxide dispersion, referred to as magnetic ink, will result in poor quality tape. Efficient and optimum dispersion of the oxide particles is usually achieved through mixing and grinding in the presence of dispersants. It is not clearly understood whether commercially used dispersants provide electrostatic stabilization, steric stabilization, or both. Particularly, there is skepticism concerning the importance of electrostatic stabilization, mainly because of difficulties in understanding and establishing charging mechanisms and charge-carrying species in

61

nonaqueous liquids. Recent work by Fowkes et al.[1] with polymeric dispersants has shown that stabilization of dispersions in organic liquids can be achieved by sufficiently strong acid-base interactions between the solid and dispersant, and that the zeta potential is a measure of the strength of the acid-base interaction. In fact, it was clearly shown that the stabilization mode can switch from steric to electrostatic as the dispersant concentration is increased.

In the preparation of magnetic inks, dispersants are added to ensure optimal dispersion of the magnetic oxide particles in the organic carrier solvent. Among the myriad of compounds that have been found and patented as good dispersants, lecithin has been the most popular in the manufacture of magnetic inks. In spite of its widespread use, little is known regarding the mechanism by which lecithin functions as a dispersant. Part of the reason for this is the fact that lecithin is a natural product. As such, its composition varies from supplier to supplier and can even change from lot to lot. Because the constituents of lecithin interact with a particular magnetic oxide to different extents, changes in the ratio of these constituents may affect some properties of the magnetic oxide dispersion, such as rheology and stability.

The origin of surface charge on oxide particles dispersed in nonpolar liquids is not as readily understood as in aqueous systems. Excellent reviews of proposed charging mechanisms in nonaqueous liquids can be found in articles by Parfitt and Peacock[2] and Fowkes et al.[1], and in a recent book by Kitahara.[3] For oxide particles dispersed in nonaqueous liquid, two charging mechanisms are currently favored: (a) preferential adsorption of ions, and (b) proton transfer between the solid and solvent or dispersant components. In the first mechanism, the surface charge is postulated to arise from the preferential adsorption of cationic or anionic species, based on such factors as ion size and specific interaction capacity. The second mechanism has been invoked to explain the charge on titanium dioxide particles dispersed in polar and nonpolar liquids,[4] and on alumina dispersed in dichloromethane.[5] Fowkes et al.[1] propose that in liquids of low dielectric constant dispersants tend not to form ionic species in solution, but rather form ions in adsorbed films on particle surfaces. When this occurs, acid-base interactions and proton transfer take place between the particle surface and the dispersant. By this mechanism, an acidic dispersant should produce a positive zeta potential on basic particle surfaces, while a basic dispersant should produce a negative zeta potential on acidic particle surfaces.

The objective of the research reported here was to study the surface charge characteristics of chromium dioxide (CrO_2) in tetrahydrofuran with special reference to the stability of CrO_2 dispersions. The zeta potential of CrO_2 particles has been measured in THF solutions containing certain inorganic and organic salts, and in a THF solution containing a pure lecithin component, while the interrelation between zeta potential and stability has been established through particle growth studies.

EXPERIMENTAL

Materials   The CrO_2 particles in the sample used were mostly submicron in size and acicular in shape, with an aspect ratio (length to diameter) of approximately 10:1. The surface area of the sample was 22 m^2/g as determined by the B.E.T. method using nitrogen gas as the adsorbate.

Tetrahydrofuran (THF: 99.5% min. THF, 0.05% max. Furan, 0.03% max. moisture, 0.015% max. peroxide, 0.025% inhibitor (butylated hydroxytoluene)) was used as the solvent for these investigations. Other chemicals used include sodium perchlorate, cetyl trimethyl ammonium bromide, and phosphatidylcholine.

Methods   Moisture content of the THF was determined using Karl Fischer titration. The conductivity of solutions was measured using an electrophoresis cell and a high-voltage booster accessory for an electrophoresis unit. The electrophoretic mobility measurements were made with the same electrophoresis unit using a glass-Teflon[*] cell of a 10-cm path length. Potential gradients of 40 to 100 V/cm were applied across the cell using the high-voltage booster. For the electrophoretic mobility measurements, 0.01 g of CrO_2 was dispersed in 100 ml of solution of a predetermined composition contained in a 125 ml Erlenmeyer flask. Dispersions were produced by placing the flask in an ultrasonic bath for 15 seconds, followed by conditioning on a wrist-action shaker for the required time period. Zeta potentials were calculated from mobility values, assuming the particles to be cylinders with a radius smaller than the double-layer thickness by using the following equation:

$$\zeta = \frac{8 \pi \eta}{\varepsilon_r \varepsilon_0 E} u, \qquad (1)$$

where ζ is the zeta potential (mV), η is the viscosity of the suspending medium, ε_r is the relative dielectric constant (7.8) of the medium, ε_0 is the permittivity of free space (1.1×10^{-12} C/(V cm)), E is the potential gradient (V/cm), and u is the electrophoretic mobility. When u is expressed as $\mu m/s$ per V/cm, the value of zeta potential (in mV) can easily be shown to be

$$\zeta = 1.46 \times U \times 10^6 \text{ mV}. \qquad (2)$$

The stability of dispersions was evaluated with a submicron-particle-size analyzer. This instrument uses photon correlation spectroscopy to determine the size of dispersed particles. Particle growth with respect to time was monitored for samples containing 0.01 g of CrO_2 per 100 ml of solution.

RESULTS

Effect of $NaClO_4$ Initially, we desired to determine the zeta potential of bare CrO_2 particles in THF. Because of problems of thermal overturn in the electrophoresis cell resulting from the low conductivity of THF, such measurements were not possible. Therefore, sodium perchlorate was added to increase the conductivity of THF to allow mobility measurements. The preliminary tests conducted to evaluate

---

[*] Trademark of E.I. DuPont de Nemours, Inc.

the effect of conditioning time on electrophoretic characteristics of CrO_2 in $NaClO_4$ solutions indicated no significant variation with conditioning times up to 8 hours. Hence,
a conditioning time of 20 minutes was chosen for all subsequent tests. Electrophoresis results obtained by varying
$NaClO_4$ concentration in THF (containing approximately 1200
ppm H_2O) are presented in Table 1. The zeta potential of
CrO_2 in sodium perchlorate solution is positive and increases
as the concentration of $NaClO_4$ decreases. Note that the decrease in zeta potential with increasing electrolyte concentration is similar to the trend that we would expect in aqueous systems.

Table 1.   Zeta Potential of CrO_2 in $NaClO_4$ Solutions in THF.
($\approx$ 1200 ppm water)

Concentration of $NaClO_4$, M	Zeta potential, mV
10^{-2}	48.0 ± 1.0
10^{-3}	49.9 ± 1.3
10^{-4}	54.1 ± 1.8
10^{-5}	67.4 ± 3.0

Figure 1 shows the results of stability tests conducted
on CrO_2 dispersions in THF solutions containing $NaClO_4$. Note
that the dispersion is more stable (i.e., the growth rate is
less) at lower $NaClO_4$ concentrations. This result correlates
well with the electrophoresis data in that a high zeta potential at the solid-liquid interface imparts more stability to
the dispersion.

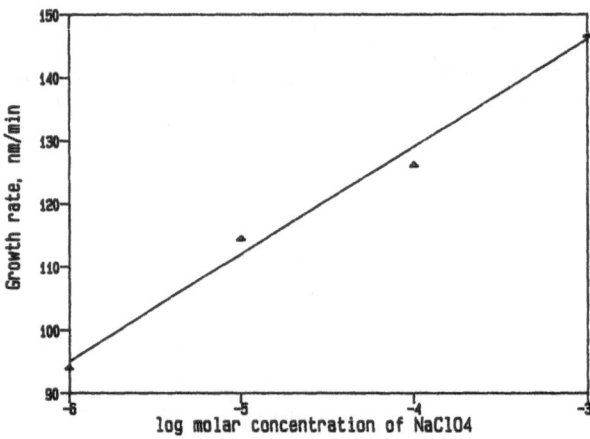

Figure 1.   Stability of CrO_2 suspensions in THF as a function
of $NaClO_4$ concentration.

Results of conductivity measurements made by varying the
concentration of $NaClO_4$ in THF are shown in Figure 2.   At
approximately 1200 ppm water, the conductivity of these solutions increases as $NaClO_4$ concentration increases. This indicates that more $NaClO_4$ is dissociated at higher concentrations.

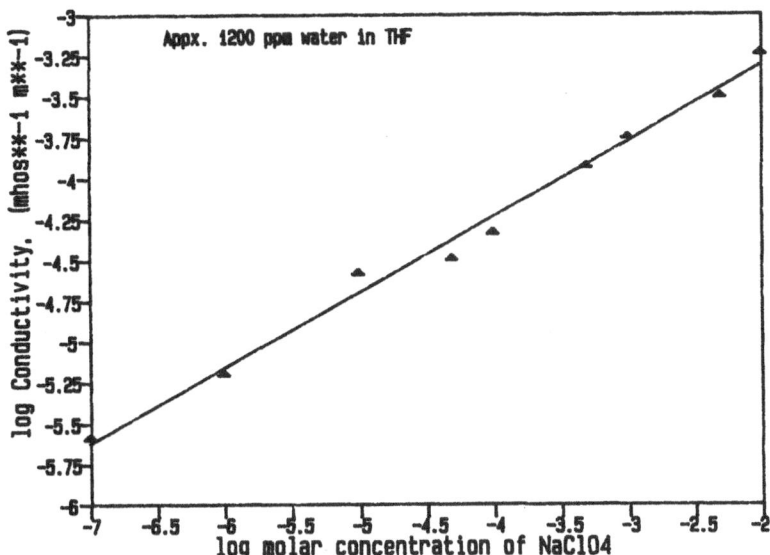

Figure 2.   Conductivity of THF as a function of NaClO₄ concentration.

Effect of phosphatidylcholine   Lecithin, a natural product, is used as a dispersant in the production of magnetic
inks.   The main ingredients of lecithin are phosphatides:
phosphatidylcholine, phosphatidyl ethanolamine, and phosphatidyl inositol.   As a first step towards understanding the
dispersing power of lecithin, the electrokinetic characteristics of CrO_2 in THF solutions containing phosphatidylcholine
(PC) were investigated.   The molecular structure of PC
appears in Figure 3.   Results of electrophoretic measurements
in PC solutions presented in Figure 4 show that CrO_2 has a
positive zeta potential at low PC concentrations (< 0.0002
g/l), and that a reversal in zeta potential occurs between
0.0002 and 0.0003 g/l PC.

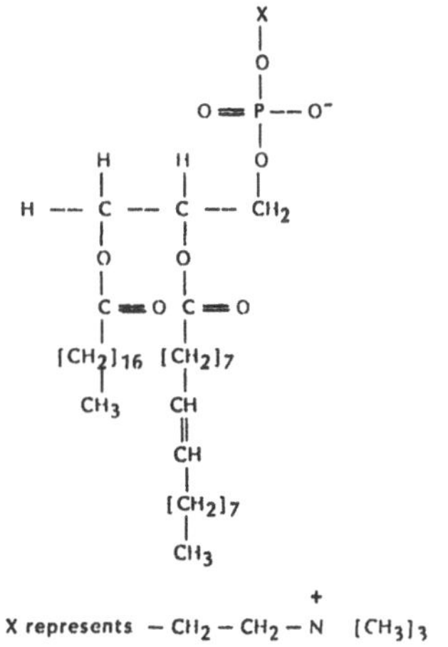

$$X \; represents \; -CH_2-CH_2-N^+ \; [CH_3]_3$$

Figure 3.   Structure of Phosphatidylcholine.

65

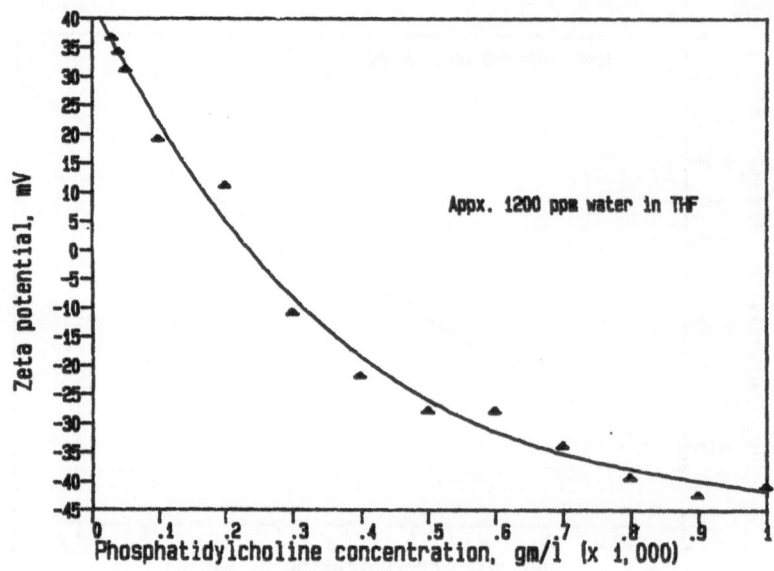

Figure 4.   Zeta potential of CrO_2 in THF as a function of phosphatidylcholine concentration.

    Because the zeta potential gradually changes from +38 mV to -42 mV as PC concentration is increased, the stability of CrO_2 dispersions was measured at different PC concentrations to correlate dispersion stability with zeta potential values. Stability should vary with the change in zeta potential, and exhibit a minimum near the PZR (point of zeta reversal) if electrostatic forces are operative. The results of the stability measurements carried out at three PC concentrations, viz. 0.00002, 0.0002, and 0.001 g/l, are presented in Figure 5. The rate of particle growth at a PC concentration of 0.0002 g/l is more than twice that at 0.00002 and 0.001 g/l PC. These results correlate quite well with the measured zeta potential values.

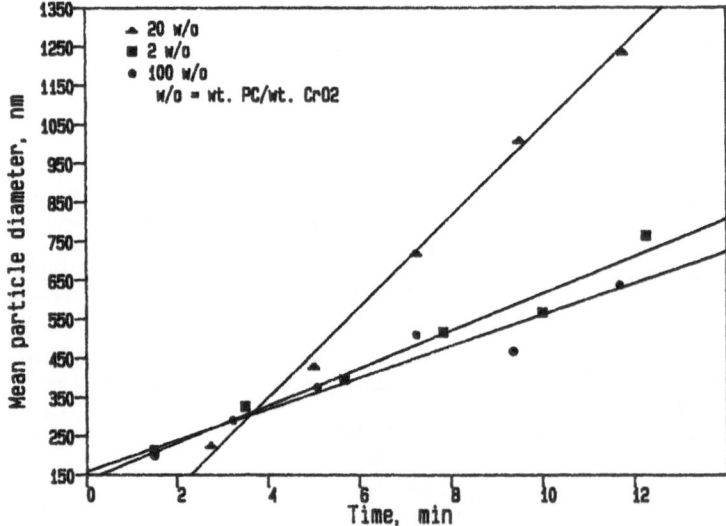

Figure 5.   Stability of dilute suspensions of CrO_2 in solutions of phosphatidylcholine in THF.

<u>Effect of CTAB</u>   The anchoring head in a phosphatidyl-
choline molecule consists of two functional groups: a phosphate group and a quarternary amine group.   To better understand the interaction of the quarternary amine group, cetyl
trimethyl ammonium bromide (CTAB) was chosen as a model surfactant and the electrokinetic characteristics of CrO_2 in
CTAB solutions were studied. Figure 6 shows that as CTAB concentration is increased, the zeta potential of the CrO_2 particles changes sign from positive to negative.   Conditioning
time appears to have some influence on the CTAB concentration
at which charge reversal occurs.

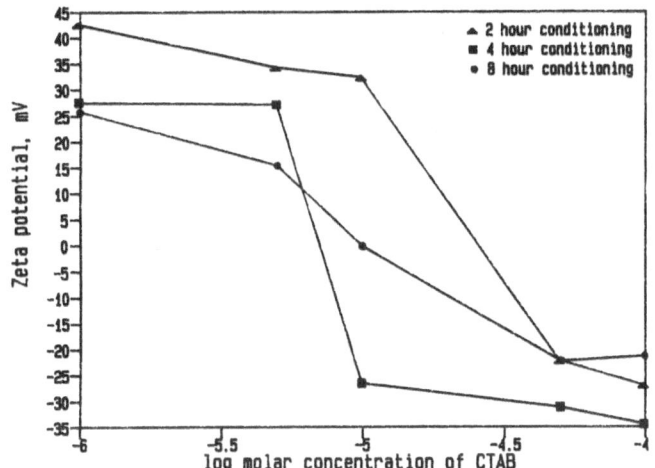

Figure 6.   Zeta potential of CrO_2 in THF as a function of
cetyl trimethyl ammonium bromide concentration.

The effect of $NaClO_4$ on the zeta potential of CrO_2 in
CTAB is presented in Figure 7.   In these tests, $NaClO_4$ was
held constant at 10^{-4} <u>M</u> while CTAB concentration was varied.

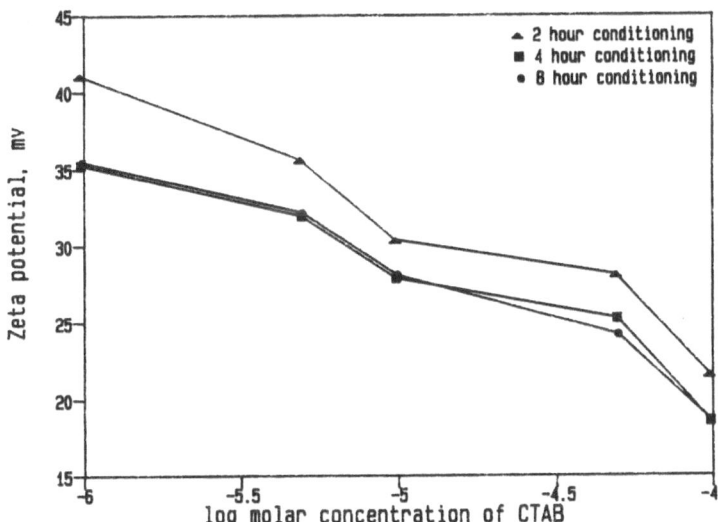

Figure 7.   Zeta potential of CrO_2 in THF as a function of
cetyl trimethyl ammonium bromide (CTAB) concentration in the
presence of 10^{-4} <u>M</u> $NaClO_4$.

From Figure 7 note that, in the presence of sodium perchlorate, CrO_2 particles are positively charged over the entire CTAB concentration range studied and that charge reversal does not occur.

DISCUSSION

Effect of $NaClO_4$ The preferential adsorption of Na^+ ion appears to be the mechanism responsible for the observed positive charge of CrO_2 in sodium perchlorate solutions in THF. It has been documented through nuclear magnetic resonance (NMR) studies[6] that sodium ion forms a very stable 1:1 complex with THF. Such a formation implies that THF can behave as a Lewis base and promote the dissociation of $NaClO_4$, as follows:

$$(THF) + NaClO_4 \;\rightleftharpoons\; (THFNa)^+ + ClO_4^-. \tag{3}$$

The $(THFNa)^+$ ion can be further complexed by THF as $[Na(THF)_4]^+$. As it was observed that the conductivity of THF increases as $NaClO_4$ concentration increases, increased sodium perchlorate concentration would shift the equilibrium in (3) to the right and promote the complexing reaction to occur. The preferential adsorption of $(NaTHF)^+$ or $[Na(THF)_4]^+$ can account for the positive charge on CrO_2 particles in $NaClO_4$ solutions. Therefore, as $NaClO_4$ concentration is increased, two competing effects should occur: (1) an increase in the adsorption of Na^+ ions, which should increase the surface potential, and (2) compression of the double layer surrounding the CrO_2 particle, which would lead to a decrease in the zeta potential. Therefore, we would ideally expect to find a maximum in the zeta potential as $NaClO_4$ concentration is increased. From the data in Table 1, we observed no such maximum. Experiments at $NaClO_4$ concentrations less than 10^{-5} $\underline{M}$ would have probably revealed such a maximum.

Effect of Phosphatidylcholine   The observed reversal of the zeta potential (from positive to negative) in the presence of phosphatidylcholine, which is a zwitterion, can be explained as follows. If it is assumed that the pH of the water in THF is greater than the pKa of both the "phosphoric acid" group and the quarternary amine group, the phosphatidylcholine molecule will possess a net negative charge. Specific adsorption of this negatively charged phosphatidylcholine molecule (via the phosphate group) can explain the zeta potential reversal.

The concentration of phosphatidylcholine has a marked effect on the stability of the CrO_2 suspensions. At a PC concentration of 0.0002 g/l, where the zeta potential is almost zero, the dispersion exhibits a minimum. At 0.00002 and 0.001 g/l PC, the growth rate is greatly reduced, hence a more stable dispersion. This implies that electrostatic stabilization forces are operative when PC is used as a dispersant.

Effect of CTAB   The sign of the zeta potential of CrO_2 particles in CTAB solutions is a function of CTAB concentration. At CTAB concentrations in excess of 10^{-5} $\underline{M}$, CrO_2 is characterized by a negative charge. The negative zeta potential is most likely caused by the preferential adsorption of bromide ions. Bromide ion, being smaller and more easily hydrated than the large cetyl trimethyl ion, should show more affinity to the surface of CrO_2.

68

In the presence of NaClO$_4$, CTAB cannot reverse the zeta potential of the CrO$_2$ particles. The zeta potential remains positive over the entire range of CTAB concentrations tested. This indicates that sodium ions have a high affinity for the CrO$_2$ surface, and adsorb in preference to bromide ions. The decrease in zeta potential with increasing CTAB concentration can be attributed to compression of the double layer by the presence of ClO$_4^-$ and Br$^-$ counterions.

Computation of Interaction Energy between CrO$_2$ Particles

For the system of CrO$_2$ particles dispersed in THF, four potential energies of interaction between particles should be considered. These are (a) electrostatic, (b) steric, (c) Van der Waals, and (d) magnetic. The first two are repulsive (positive) energies, while the latter are attractive (negative) energies.

The potential energy of repulsion caused by electrical double layer interactions was evaluated following the method of Kandori et al.[7] by considering the particles to be cylindrical in shape. Assuming constant potential surfaces and using the Debye-Huckel approximation, the potential energy of interaction of two cylindrical double layers, Φ_{el}, can be expressed as follows:

$$\Phi_{el} = \frac{\sqrt{a} \, L \, \epsilon_r \, \epsilon_0 \, \kappa \, \Psi_0^2}{\pi} \int_0^\infty \left[\frac{dx}{\exp(\kappa x^2 + \kappa H_0) + 1} \right]. \tag{4}$$

In the above equation, H_0 is the shortest distance between the surfaces of the cylindrical particles, H is the distance between the particle surfaces, x^2 is $H - H_0$, a is the particle radius, L is the particle length, and Ψ_0 is the surface potential. A Romberg numerical integration procedure was used to evaluate the integral in (4).

Stabilization by steric repulsion is not likely to be a major factor in the cases reported here, as both phosphatidylcholine and CTAB are relatively short-chained surfactants (i.e. < 5 nm). It is believed[8] that a surfactant or polymer chain length of at least 7 nm is required to provide steric stabilization.

The potential energy of Van der Waals interactions of two cylindrical particles can be expressed as follows:[9]

$$\Phi_{VDW} = -\frac{A_{212}}{48} \left[\frac{1}{\left[\frac{H_0}{2}\right]^2} + \frac{1}{\left[\frac{H_0}{2} + a\right]^2} + \frac{1}{\left[\frac{H_0}{2} + \frac{a}{2}\right]^2} \right]. \tag{5}$$

In the above equation, A_{212} is the effective Hamaker constant, and is related to the Hamaker constant of CrO$_2$ (A_{22})

and THF (A_{11}) by the expression

$$A_{212} = (\sqrt{A_{22}} - \sqrt{A_{11}})^2.$$ (6)

A_{212} values of 1×10^{-20} and 3×10^{-20} Joules were used for these calculations.

In two recent papers, Kandori et al.[7,10] have calculated the total energy of interaction of acicular, magnetic, iron oxide particles suspended in cyclohexane by taking into account magnetic interactions. Their calculations, based on parallel interactions between cylindrical particles, show that the magnetic attractive energy can be represented by

$$\Phi_{mag} = -\left[\frac{M}{4\pi}\right]^2 \frac{v^2}{[2a + H]^3} ,$$ (7)

where M is the magnetic moment of the particle and v is the particle volume. Kandori and coworkers used the saturation magnetic moment as the value for M in their calculations, which is actually quite debatable. The particles should be randomly oriented while dispersed, and the net magnetic moment should therefore be much smaller than the saturation magnetic moment. While we are aware that magnetic interactions could be quite important, the contribution of magnetic attraction has been omitted in this analysis because of the lack of data on the magnetic behavior of CrO_2 particles while dispersed in THF. This area merits detailed consideration for future research.

Equations (4) and (5) were combined in a computer routine to obtain the total energy of interaction as a function of separation distance for a surface potential of 70 mV and double-layer thickness of 45 nm. This double-layer thickness

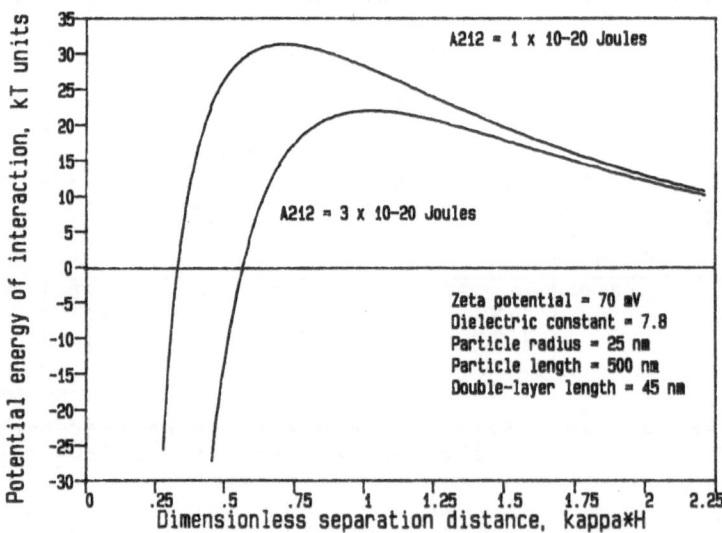

Figure 8. Calculated parallel interaction energy between CrO_2 particles in THF according to equations (4) and (5).

was calculated using the method of Fowkes et al.[1] The calculated results using Hamaker constant values of 1×10^{-20} and 3×10^{-20} J are displayed in Figure 8, from which you can discern that a potential energy barrier of about 22 to 31 kT may exist in the system, depending on the value of the Hamaker constant. Thus, electrostatic stabilization of a dispersion of bare CrO_2 particles is possible (in the presence of $NaClO_4$). Because the adsorption of lecithin has been found to decrease or reverse the zeta potential to a value less than 70 mV, the energy barrier for the interaction of lecithin-coated particles will be less than 20 kT.

SUMMARY AND CONCLUSIONS

To better understand the stability behavior of CrO_2 magnetic inks, the nature of the surface charge on CrO_2 has been studied in the presence of $NaClO_4$, CTAB, and phosphatidylcholine. Our investigations show that the magnitude of the zeta potential at the CrO_2/solution interface can be as high as 70 mV in $NaClO_4$ mixtures. In the case of $NaClO_4$ and CTAB, preferential adsorption of ionic species seems to be the charging mechanism involved. Application of DLVO theory to the CrO_2 suspensions shows that a potential energy barrier of 30 kT may be present when the zeta potential is 70 mV.

ACKNOWLEDGMENTS

The authors wish to gratefully acknowledge the finacial support provided by the IBM Corporation, General Products Divsion, Tucson, Arizona, to carry out the reasearch reported here. Special thanks are extended to R. D. Stacy, R. J. Madeya, R. M. Phelan, and M. A. Mathur, employees of the IBM Corporation, General Products Division, Tucson, Arizona, for their valuable suggestions and comments during the course of this research.

REFERENCES

1. F. Fowkes, H. Jinnai, M. Mostafa, F. Anderson, and R. Moore, in "Colloids and Surfaces in Reprographic Technology," M. Hair and M. Croucher, Editors, American Chemical Society Symposium Series No. 200, p. 307, ACS, Washington, DC, 1982.
2. G. D. Parfitt, and J. Peacock, in "Surfaces and Colloid Science," E. Matijevic, Editor, Vol. 10, p. 153, Plenum Press, New York, 1978.
3. A. Kitahara, in "Electrical Phenomena at Interfaces," A. Kitahara, Editor, p. 119, Marcel Dekker, New York, 1984.
4. K. Tamaribechi and M. Smith, J. Colloid Sci., 22, p. 404 (1966).
5. A. Foissy and G. Robert, Ceramic Bull., 61, (2), p. 251 (1982).
6. C. Hammonds and M. Day, J. Phys. Chem., 73, (4), p. 1151 (1969).
7. K. Kandori, A. Kazama, K. Kon-no, and A. Kitahara, Bull. Chem. Soc. Jpn., 57, p. 1777 (1984).
8. T. Sato and R. Ruch, "Stabilization of Colloidal Dispersions by Polymer Adsorption," Marcel Dekker, New York, 1980.
9. E. Verwey and J. Th. G. Overbeek, "Theory of Stability of Lyophobic Colloids," Elsevier, Amsterdam, 1948.
10. K. Kandori, K. Kon-no, and A. Kitahara, Nip. Kag. Kaishi, (7), p. 963 (1983).

INTERACTIONS BETWEEN POLYELECTROLYTES

AND OXIDES IN AQUEOUS SUSPENSIONS

Joseph Cesarano III and Ilhan A. Aksay

Department of Materials Science and Engineering
University of Washington
Seattle, Washington 98195

Colloidal stability of aqueous α-Al_2O_3 suspensions with polymethacrylic acid (PMAA) was studied in the pH range of 3 to 10. Stability is related to the adsorption of PMAA on α-Al_2O_3 as controlled by the chemistries of both the α-Al_2O_3 surface and the PMAA. The adsorption behavior of PMAA on α-Al_2O_3 is basically "high affinity" type near and below the zero point of charge of α-Al_2O_3 (pH 8.7). A stability map was determined which outlines the critical amount of adsorbed PMAA required to achieve dispersion.

INTRODUCTION

Polyelectrolytes are widely used in industrial applications to prepare highly concentrated (> 50 v/o) ceramic suspensions which are subsequently fabricated into dense components by sintering. Although a variety of polyelectrolytes are commercially available and are used effectively in the preparation of suspensions, their role in colloidal stabilization is not clearly understood.

In this paper, we aim to provide a clearer understanding of particle dispersion with adsorbed polyelectrolytes and relate it to the surface chemistries of the suspended particles and the polyelectrolyte. Aqueous oxide suspensions will be considered as a general example. The particular emphasis will be on the dispersion of α-Al_2O_3 suspensions with polymethacrylic acid (PMAA).

BACKGROUND INFORMATION

(1) Polyelectrolytes

A polyelectrolyte is a polymer that contains ionizable functional groups which are capable of dissociating into ionized charged sites. A typical polyelectrolyte structure is shown in Fig. 1. In general, there are three types of polyelectrolytes. Anionic polyelectrolytes contain acid groups which can ionize to form negative sites, e.g., $-COO^-$ or $-OSO_3^-$. Cationic polyelectrolytes contain basic groups that form positive sites, e.g., $-NH_3^+$. Polyelectrolytes which have both acid and basic groups are termed polyampholytes and can be negatively or positively charged. Since polyelectrolytes are ionic in nature they are generally water soluble.

In aqueous systems, the solution behavior and the extent to which the functional groups dissociate and the overall charge of the polymer are dependent on the surrounding pH and ionic strength. Using polymethacrylic acid (PMAA) as an example (Fig. 1), each acid group on the polymer chain will have its own effective dissociation constant (pQ); therefore, the fraction of the functional groups which are dissociated (α) will be dependent on the pQ values and the pH.

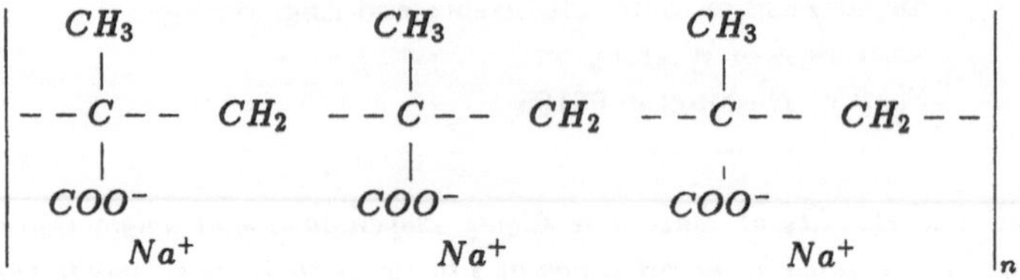

$$
\begin{array}{|cccccc}
CH_3 & & CH_3 & & CH_3 & \\
| & & | & & | & \\
--C-- & CH_2 & --C-- & CH_2 & --C-- & CH_2-- \\
| & & | & & | & \\
COO^- & & COO^- & & COO^- & \\
Na^+ & & Na^+ & & Na^+ &
\end{array}_n
$$

For each acid group:

$$R\text{-}COOH + H_2O \Leftrightarrow R\text{-}COO^- + H_3O^+$$

$$Q = \frac{(R\text{-}COO^-)(H_3O^+)}{(R\text{-}COOH)}$$

$$pQ = -logQ = pH - log\left(\frac{\alpha}{1-\alpha}\right)$$

Fig. 1: Structure and dissociation reaction for polymethacrylic acid – Na salt.

(2) Surface Charging of Oxides

When a metal oxide powder (i.e., Al_2O_3) is placed in an aqueous medium, surface reactions take place until equilibrium is established between the solid and the aqueous medium. This can be viewed as a two step process: formation of surface hydroxyls due to surface hydration, followed by ionization of these hydroxyl groups to yield a positively or negatively charged surface depending on the pH of the solution.[1]

Many authors have modeled the dissociation of the hydroxide surface to yield a positively or negatively charged surface. The common explanation is that the surface sites ionize to form positive and negative sites and the resulting sites then react with counterions in solution.[1-6] In general, the dissociation of the surface hydroxyl groups is represented as:

74

$$MOH_2^+ = MOH^o + H_s^+ \qquad K_1 = \frac{[MOH^o][H_s^+]}{[MOH_2^+]} \quad \text{, and} \qquad (1)$$

$$MOH^o = MO^- + H_s^+ \qquad K_2 = \frac{[MO^-][H_s^+]}{[MOH^o]} \qquad (2)$$

where,

M is a surface metal ion site
MOH^o is a neutral, hydroxylated surface site
MOH_2^+ is a positive surface group
MO^- is a negative surface group
H_s^+ is a proton located at the surface, and
$K_{1,2}$ are the respective equilibrium reactions constants.

The reactions above show that at low pH values the reactions will proceed to the left and the formation of positive MOH_2^+ surface sites will be dominant. Similarly, at high pH values (low concentration of H_s^+) the reactions will proceed to the right and the formation of negative MO^- surface sites will be dominant. It should be noted that at any given pH there are equilibrium concentrations of all of the surface species (i.e., MOH_2^+, MOH^o, MO^-). For example, at high pH values there are mostly MO^- sites, but depending on the reaction constants there can still be appreciable concentrations of MOH^o and MOH_2^+ sites. At a characteristic pH, the concentrations of MOH_2^+ and MO^- sites will be equal and the overall net surface charge will be zero. This characteristic pH for a given material is termed the zero point of charge (zpc) for that material. For pH values below the zpc, the net surface charge is positive and above the zpc the net surface charge is negative.

In general, the characteristic pH for the zpc for a given oxide is strongly dependent on the relative basic and acidic properties of the solid. The crystalline form, material preparation, and the degree of surface hydroxylation will also affect the zpc. For example, the zpc for Al_2O_3 has been reported to be from 6.5 to 9.8.[1,7,8]

(3) Polyelectrolyte Adsorption Behavior

When charged polymers are adsorbed onto oxides to cause flocculation or stabilization, it is generally true that electrostatic interaction is the primary adsorption mechanism. In other words, polyelectrolytes adsorb much more appreciably when oppositely charged to the solid adsorbent. When the polyelectrolyte and solid are similarly charged, some adsorption can still occur, but this adsorption is comparatively low.

Some typical examples of polyelectrolyte adsorption on oxides are given below. In the flotation industry, the percent recovery of a mineral is related to how efficiently the collector (anionic or cationic surfactant) is adsorbed. As shown in Fig. 2, the percent recovery (and adsorption of anionic and cationic surfactants) is mainly determined by the surface charge on goethite ($HFeO_2$). Therefore, pH is the most important variable. Below pH 6.7 goethite is positively charged and above pH 6.7, it is negatively charged. Therefore, it is clearly shown that the collector must be anionic when the solid is positively charged and cationic when the solid is negatively charged. Flotation ceases at pH 12.3 as a result of hydrolysis of the cationic surfactant at such a high pH.[10] Therefore, it can be concluded that the basic adsorption is electrostatic in nature and involves the ionized form of the surfactant. Modi and Fuerstenau[11] have shown that these phenomena hold for corundum (Al_2O_3), and Iwasaki, Cooke, and Choi[12] have shown that the flotation of hematite (Fe_2O_3) can be explained by these principles.

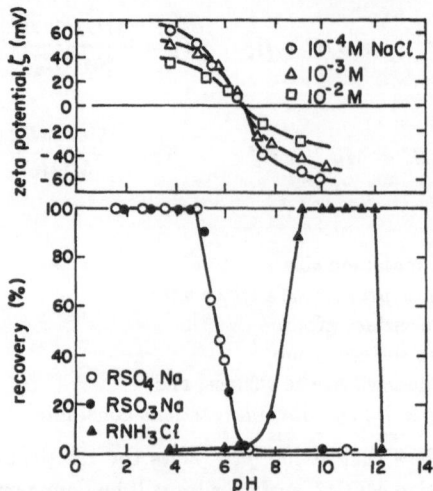

Fig. 2: The dependence of the flotation properties of goethite on surface charge. Upper curves are zeta potential as a function of pH at different concentrations of sodium chloride; lower curves are the flotation recovery in $10^{-3}M$ solutions of dodecylammonium chloride, sodium dodecyl sulfate, or sodium dodecyl sulfonate. From ref. 9 .

Gebhardt and Fuerstenau[13] studied the adsorption of anionic polyacrylic acid (PAA) on rutile (TiO_2), hematite, and silica as a function of pH. The amount of PAA adsorbed on Fe_2O_3 is shown in Fig. 3. The zpc of Fe_2O_3 occurs at pH 8.3 and the adsorption decreases to near-zero values at pH values above 8.3. Similar adsorption behavior was found for TiO_2 which has a zpc of 6.3. No adsorption in the pH range 3-9 was observed for silica which has a zpc at approximately pH 2.5. In conclusion, for PAA adsorption on oxides it was determined that the PAA adsorption plateau level decreases with increasing pH until near-zero values of adsorption occur near the zpc of the oxide but some specific adsorption still occurs at the zpc. It was also determined that positive zeta potentials are reversed to negative as PAA is adsorbed.

Lopatin[14] found similar behavior when polymethacrylic acid (PMAA) was adsorbed onto anatase (TiO_2). It was suggested that the adsorbed molecules form a monolayer of interpenetrating coils and that electrostatic bonding to anatase, which is positively charged on the acid side of an isoelectric zone between pH 4 and 6, is the primary mechanism for adsorption at pH values from 2 to 4. Below pH 2, entropy effects and hydrogen bonding may be the dominant factors; above pH 4, adsorption falls rapidly to zero.

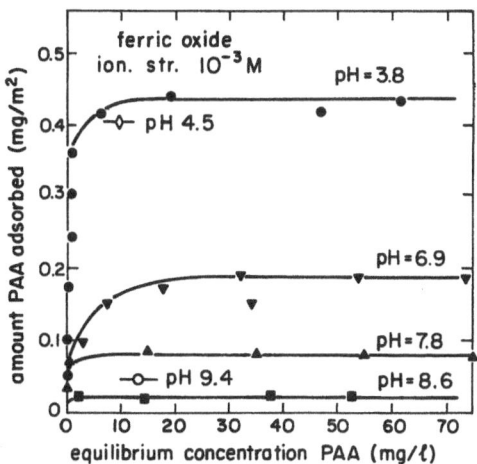

Fig. 3: Amount of PAA (MW 2×10^6) adsorbed on hematite as a function of equilibrium PAA concentration for various pH values. From ref. 13 .

EXPERIMENTAL MATERIALS AND METHODS

In order to more clearly unify the concepts previously discussed for a working system, some experiments were conducted using polymethacrylic acid as the polyelectrolyte and α-Al_2O_3 as the colloidal powder.

(1) Materials and Chemicals

The Na^+ salt of polymethacrylic acid (PMAA-Na)* used had an average molecular weight of 15,000 (Fig. 1). The alumina powder used was a very high purity α-Al_2O_3.[†] The average particle diameter was 0.37 μm and the surface area was 5.9 m^2/g.

The water used was distilled and deionized. Adjustments of pH were completed with analytical grade HCl and $NaOH$ solutions and ionic strength was adjusted with $NaCl$ as desired.

(2) Potentiometric Titrations

To measure the fraction of dissociated $COOH$ groups (α) versus pH for PMAA-Na and the surface charge (σ) versus pH for Al_2O_3, potentiometric titrations of the control electrolytes were compared to titrations of the samples.[‡]

---

*Polysciences, Inc., Pittsburgh, PA.

[†]Sumitomo Chemical America, Inc., New York, NY (AKP-30; > 99.99% pure).

[‡]Radiometer TRS822 automatic titration unit. Radiometer, Copenhagen, Denmark.

This procedure of using potentiometric titrations to determine surface chemistries is clearly outlined by Hunter.[15] Titrations are completed on a blank electrolyte solution and electrolyte solution with a known amount of sample. The difference between the amounts of titrant added to obtain a certain pH is the amount of titrant that reacted with the sample. With this information both the fraction of functional groups dissociated on polymers and the surface charge on ceramic powders can be easily calculated.

The titrations were completed in a nitrogen atmosphere and in each case 40 ml samples had known but small volumes of PMAA-Na (< 0.1 g) and Al_2O_3 (< 2 vol %) and when necessary NaCl solutions were used to adjust ionic strength. Prior to titration, the Al_2O_3 powder was cleaned to remove soluble ions using soxhlet extraction. With this technique the powder is continuously washed with freshly distilled water.

(3) Adsorption Isotherms

The adsorption of PMAA-Na on α-alumina was also determined using the titration unit. Suspensions of 20 vol % Al_2O_3 were prepared with various amounts of PMAA at various pH values. The samples were then put into a gentle mechanical shaker for approximately 24 hours and then centrifuged for 45 minutes at 2000 RPM. A known amount of supernatant was then analyzed by titration to determine the amount of PMAA left in solution.

(4) Settling and Zeta Potential Experiments

Settling experiments were completed with 2 vol % Al_2O_3 suspensions. Various amounts of PMAA-Na were added to the suspensions and were then ultrasonicated and magnetically stirred for at least 4 hours. Suspensions were then poured into graduated cylinders and after several days the final sedimentation cake heights were recorded.

Zeta potential measurements of each sample described above were completed with a Micro-Electrophoresis Apparatus.* These measurements can only be completed on very dilute suspensions. Therefore, it was very important that the 2 vol % Al_2O_3 samples were centrifuged and the supernatant carefully decanted into a beaker. Then a portion of the sediment was remixed with the supernatant. Prior to taking the measurements the new dilute suspensions were very briefly ultrasonicated and magnetically stirred for 15 minutes to ensure that only singlet particles were measured. At least ten measurements were completed for each sample to ensure accuracy.

(5) Viscosity Measurements

Viscosity measurements of the suspensions were done with a rotary viscometer.†

---

* Rank Brothers, Bottisham, Cambridge, England.

†Brookfield Engineering Laboratories, Inc., Stoughton, MA.

EXPERIMENTAL RESULTS AND DISCUSSION

(1) Surface Chemistry of PMAA-Na and α-Al_2O_3

The polyelectrolyte polymethacrylic acid-Na^+ salt (PMAA-Na) (Fig. 1), with an average molecular weight of 15,000, has approximately 138 available carboxylic acid sites or functional groups per molecule. Depending on the solvent conditions (i.e., pH and ionic strength), the fraction of functional groups which are dissociated (i.e., COO^-) and those which are non-dissociated (i.e., $COOH$) will vary. As the fraction dissociated (α) increases from $\sim$ 0 to $\sim$ 1.0, the polymer surface charge varies from relatively neutral to highly negative. Therefore, the behavior of the polymer in solution is also dependent on the solvent conditions. Fig. 4 shows the fraction of dissociated acid groups as a function of pH and background $NaCl$ concentration. As the pH and salt concentration increase, the dissociation and negative charge characteristics of the polymer increases.

Fig. 4 shows that at pH values of $\geq$ 8.5 the PMAA is effectively totally negative with $\alpha \sim 1$. In this condition, experimental evidence shows that the PMAA molecules are in the form of relatively large expanded random coils in solution.[16] This results from electrostatic repulsion between the negatively charged surface sites. As pH is decreased the number of negatively charged sites also continually decreases until the PMAA is effectively neutral near pH 3.4 and $\alpha \rightarrow 0$. In this condition the PMAA chains approach insolubility and form relatively small coils or clumps.[16]

A plausible explanation for the role of salt concentration on the degree of dissociation is that the presence of salt results in electrostatic shielding between negatively charged sites on the PMAA and this causes a decrease in the probability of having non-dissociated acid groups (i.e., $COOH$) as explained below. At a given pH, the system is in a dynamic equilibrium state where any given acid group is part of the time dissociated and part of the time non-dissociated but the overall fraction dissociated remains constant. Upon an increase in salt concentration, the probability of COO^- groups being stable with surrounding Na^+ ions in close proximity increases and therefore the probability of H_3O^+ ions reacting with available COO^- groups to form $COOH$ decreases. Therefore, upon dissociation, it is more difficult for COO^- groups to reform to $COOH$ and the dynamic equilibrium is shifted to a higher fraction of dissociation. Similarly, salt decreases the activity of H_3O^+ which shifts the equilibrium towards more dissociation as shown in Fig. 1.

Fig. 5 shows a measure of the relative charge density (σ) on the surface of the Al_2O_3 particles as a function of pH. At every pH, there is a large number of positive, neutral, and negative sites. The σ value gives the overall net charge density. At the zpc, the number of positive sites equals the number of negative sites and the net charge equals zero. For α-Al_2O_3, the zpc is at approximately pH 8.7. Both σ values and the zpc are in very close agreement to that of Hasz.[17]

In Fig. 5 a comparison with Fig. 4 indicates that there should be a great deal of electrostatic attraction between the negatively charged polymer and the positively charged Al_2O_3 particularly in the pH range from 3.5 to 8.7. This will be related to adsorption below.

(2) Adsorption of PMAA-Na on α-Al_2O_3

Fig. 6 shows the resulting adsorption for various pH values plotted as mg PMAA adsorbed per m^2 surface area of Al_2O_3 versus the initial amount of PMAA-Na added (on a dry weight basis of Al_2O_3). The solid diagonal line represents the adsorption behavior that would occur if 100% of the PMAA added were to adsorb.

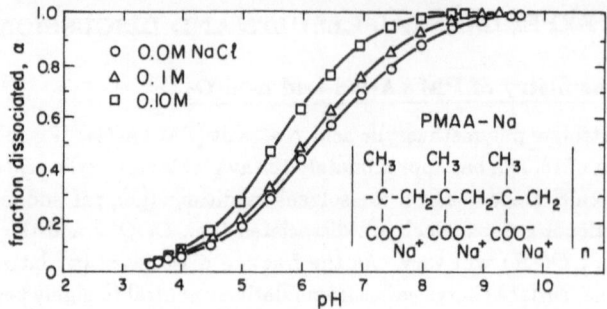

Fig. 4: Fraction of acid groups dissociated versus *pH* as a function of salt concentration for polymethacrylic acid-Na salt, MW 15000.

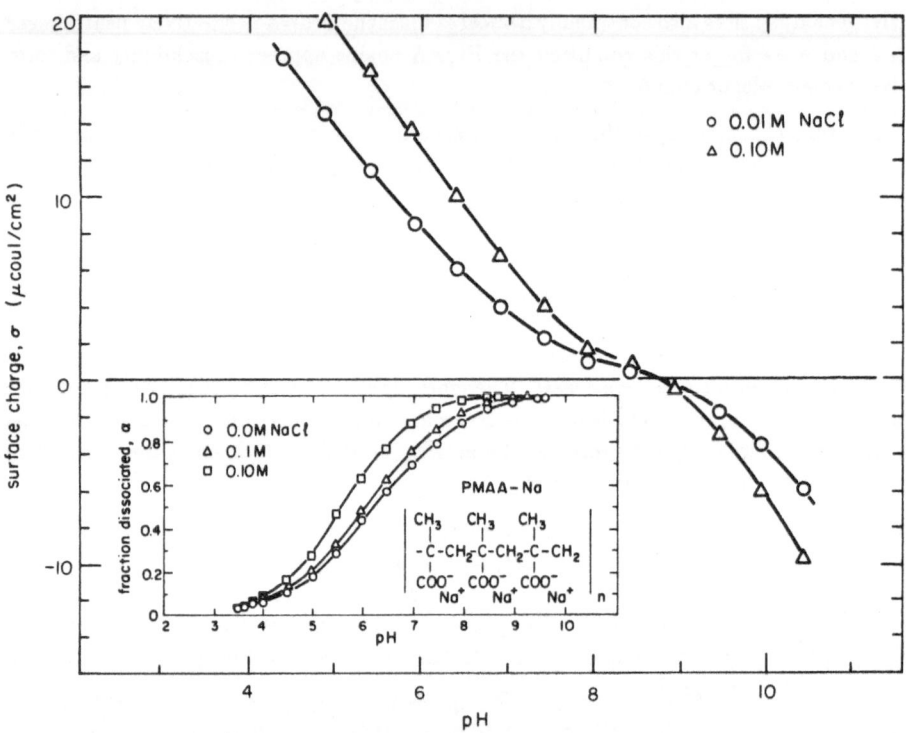

Fig. 5: The relative surface charge versus pH for Sumitomo AKP-30 α-Al_2O_3 with an insert of Fig. 4.

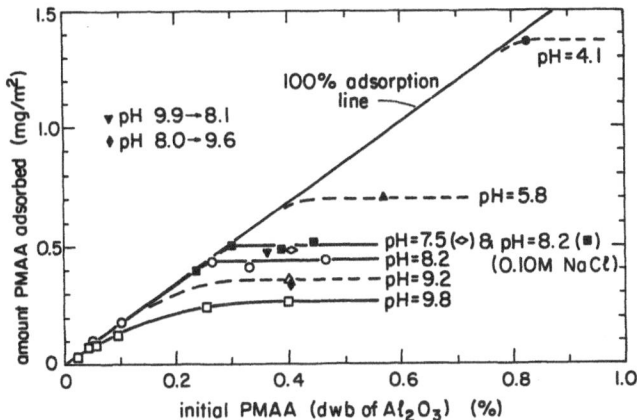

Fig. 6: The amount of polymethacrylic acid-Na salt adsorbed on Sumitomo AKP 30 α-Al_2O_3 as a function of initial PMAA-Na added.

It is clearly shown that the amount adsorbed increases greatly with decreasing pH. This is in agreement with recent polyelectrolyte adsorption theory developed by Van der Schee et al.[18,19] as explained below. For pH values above the zpc, α is approximately 1 and the negatively charged sites on the polyelectrolyte chains tend to repel each other. This repulsion suppresses the formation of loops in the adsorbed conformation. Consequently, the polyelectrolyte chains adsorb in a relatively flat conformation with each chain covering a relatively large amount of surface area. It should be noted that there still can be polyelectrolyte tails extending into solution even though the formation of loops is inhibited. The determined values for polyelectrolyte adsorption in this configuration were on the order of a few tenths of a mg/m^2. As the pH is decreased and α approaches 0 (Fig. 4), the polyelectrolyte chains become uncharged and the formation of loops in the adsorbed configurations is enhanced. Thus, the covered surface area per adsorbed chain is relatively small, and it takes more adsorbed chains to form a saturated monolayer. Commonly measured values of this type of adsorption are a few mg/m^2 which is in agreement with the experimental results.[20]

Fig. 6 also shows that there is a difference between the adsorption behavior above the zpc and below the zpc. For pH 9.8, there is a gradual attainment to the saturated adsorption plateau; but, at pH 8.2 and below plateau adsorption is reached without having any appreciable PMAA left in solution. Therefore, it is concluded that at pH values near and below the zpc, the adsorption behavior is basically "high affinity" type adsorption where practically all of the PMAA which is added adsorbs on the surface and follows the 100% line until a saturation plateau level is reached. Only after that limit does non-adsorbed PMAA become appreciably present in solution. This also agrees with adsorption theory when there is an added electrostatic influence for adsorption and a negative or partly negative polyelectrolyte adsorbs on a positively charged surface.[19] Above the zpc, due to similar net charges, a barrier for adsorption exists, so that in order to adsorb appreciable amounts of PMAA there has to be a corresponding equilibrium concentration of PMAA in solution. Adsorption in this region is due to the presence of positive surface sites even though the net surface charge is negative.

Fig. 6 also shows that the presence of background salt can slightly increase the adsorption. This occurs because the background salt has an electrostatic shielding effect between negatively charged sites on the PMAA thereby causing the chains to behave more like uncharged polymers and enhancing the development of loops. This type of behavior is commonly observed in many systems and is in agreement with adsorption theory.[18,19] It should also be noted that even with 0.1 M $NaCl$ concentrations stable suspensions are achieved after saturation adsorption occurred.

The degree of reversibility of this system is indicated by the ▼ and ♦ points in Fig. 6. At these points, the suspensions were made at the initial pH values indicated and shaken for 4 hours. The suspensions were then adjusted to the indicated pH values and shaken for 20 or more hours before an analysis was completed. The system which is initially at pH 9.9 and then adjusted to pH 8.1 adsorbs a final amount which is consistent with what is expected if the initial pH is already 8.1. This is expected since at pH 9.9 there is appreciable non-adsorbed PMAA in solution which could be easily adsorbed once the pH is lowered. A change in the configuration of the adsorbed PMAA may occur but there are no observable changes in the suspension behavior. On the other hand, when the initial pH is 8.0 and then adjusted to 9.6, total reversibility is not observed but only part of the expected PMAA desorbs. This shows that the PMAA is relatively strongly held on the Al_2O_3 surface. It also shows that with time most of the PMAA can desorb, at least for pH values near and above the zpc.

(3) Consolidated State

The effect of incomplete adsorption and flocculation on the consolidated state was determined by sedimentation and centrifugation experiments. Fig. 7(a) shows that trace amounts of PMAA-Na (< 0.25 %) induces flocculation and large sedimentation volumes. This results because the binding energy between the particles is high and the particle clusters that form during consolidation behave as rigid flow units and do not pack densely.[21] In contrast, at concentration levels of approximately 0.35-0.5% PMAA-Na, the binding energy between particles is low and the particle clusters display relatively denser packing structure.[21]

These data are also correlated with the zeta potential measurements for the same systems. Fig. 7(b) shows that with increasing polymer content, the zeta potential decreases to zero and then reverses sign. Above 0.5% polymer, the zeta potential approaches a nearly constant value. This is due to saturated monolayer adsorption of the anionic polyelectrolyte.

If one correlates the data from Figs. 7(a) and (b) it can be concluded that with small additions of PMAA-Na (i.e., < 0.1%) there is charge neutralization and subsequent flocculation. In this region flocculation is mainly due to electrostatic patch model flocculation whereby flocculation takes place because negatively charged patches due to the adsorbed polyelectrolyte are attracted to positively charged patches of surface on other particles. When the molecular weight is relatively low then maximum flocculation occurs when the zeta potential is approximately zero. This type of behavior is discussed by Bleier and Goddard[22]. In the

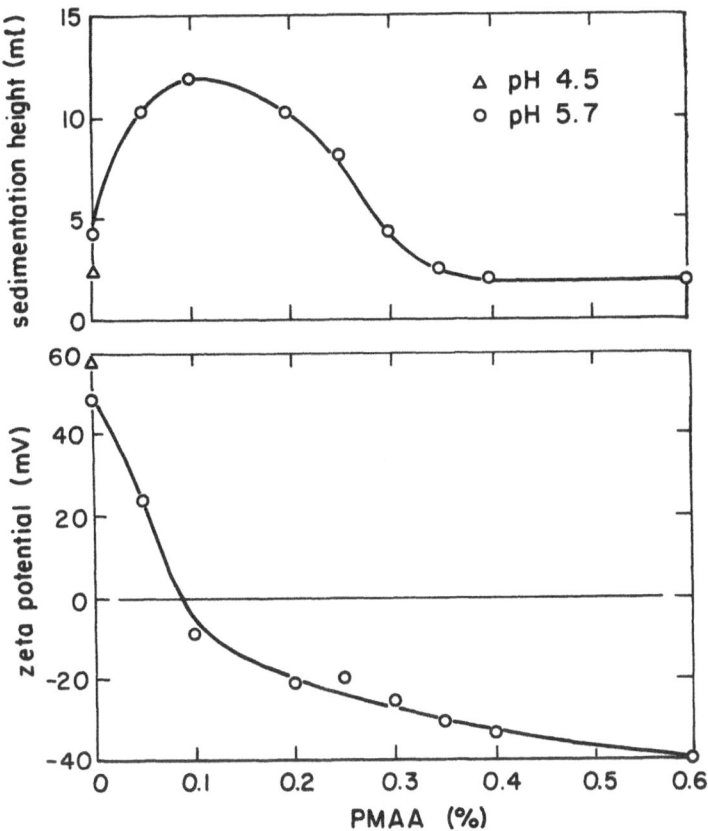

Fig. 7: (a) Sedimentation height and (b) zeta potential versus percent poly-methacrylic acid-Na (MW $\sim$ 15,000) for 2 vol % suspensions of Sumitomo AKP 30 α-Al_2O_3.

range 0.1 to 0.3% PMAA-Na, flocculation is due to a combination of two effects: (1) the zeta potential is relatively low in magnitude, and (2) incomplete adsorption results in polymer bridging where two or more particles can be mutually adsorbed by polymer chains.

It can also be concluded that the PMAA results in more efficient consolidation. Upon settling, the pure Al_2O_3 case (0% polymer) shows a larger sediment volume than the system with 0.4% polymer added even though the zeta potentials are +48 mV and -33 mV, respectively. The stability determined by settling rates and sedimentation heights do not appear to be equal until zeta potentials of $\geq$ 58 mV are reached for the pure Al_2O_3 suspensions. This means that the increased stability and dispersion of the 0.4% polymer system may be due to an enthalpic steric stabilization interaction in combination with the repulsion effect due to electrostatics. These mechanisms together can be termed electrosteric.

From these results some preliminary conclusions can be made: (1) only very minute amounts of PMAA-Na are necessary to flocculate Al_2O_3 suspensions, (2) a critical amount of PMAA-Na is needed before stabilization occurs, and (3) PMAA-Na does indeed provide an enhanced stabilization effect.

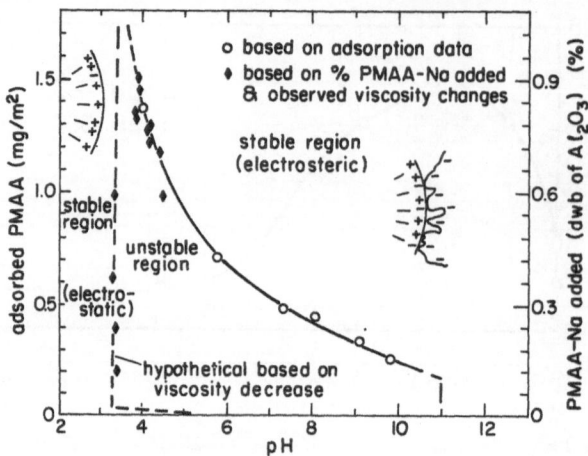

Fig. 8: A stability map showing the amount of adsorbed PMAA-Na required to form stable suspensions of Sumitomo AKP 30 α-Al_2O_3 as a function of pH.

(4) Stability Map for PMAA-Na/Al_2O_3 System

With the above information and the adsorption data in Fig. 6 a stability map was constructed and is shown in Fig. 8.

Fig. 8 is a plot of the amount of PMAA adsorbed at the plateau levels versus pH. In other words, for any pH near and below the zpc, it is a measure of the amount of PMAA which must be adsorbed to achieve a stable suspension. Regions above the curve are stable and dispersed. Regions below the curve are unstable and show an onset to flocculation. The reason for the assumed steep decrease in adsorption at pH 3.3 is because of observed viscosity decreases. For example, as pH is decreased on samples which are initially stable, the viscosity noticeably increases when crossing the stability/instability boundary region. This is expected since at those lower pH values, it takes more PMAA to maintain a stable suspension and these suspensions are below that critical level. This is the same as moving to the left from a stable region on Fig. 8 and entering the unstable region. Once the pH is decreased further to pH 3.3, deflocculation reoccurs and viscosity sharply decreases and approaches that of a system stabilized at pH 3.3 without any PMAA present. Since at pH 3.3 Al_2O_3 is stable without the presence of PMAA due solely to electrostatic interactions, it is believed that there are a combination of two possible mechanisms for the observed restabilization of Al_2O_3 below pH 3.3 in the presence of PMAA. One explanation is that below pH 3.3 the PMAA sufficiently desorbs since this is also the pH at which the PMAA loses its negative charge and approaches that of a neutral polymer thereby losing a strong driving force for adsorption (Fig. 4) and losing its solubility. Concurrently, some PMAA may still be adsorbed but the effectively neutral polymer layer may be small enough so that the electric double layer

Table 1: Rheology data at various pH's for 20 v/o AKP-30 suspensions with 0.312%(dwb) of PMAA-Na MW 15,000.

pH	$\eta(cp)$ at 10 RPM
5.2	1650
7.4	< 10
8.7	< 10
9.3	< 10
9.8	< 10
10.2	< 10

(formed by the highly positive Al_2O_3 surface) may extend past the polymer layer. In this way, electrostatic stabilization due to electric double layer interactions can still occur.

Table 1 illustrates the transition from stability to instability based on viscosity measurements. A series of samples with 0.312% PMAA were prepared at various pH values. Viscosities above pH 7 are < 10 cp while the viscosity at pH 5.2 is more than two orders of magnitude greater, indicating flocculation. This is in agreement with Fig. 8 which shows a stability/instability transition at $\sim pH$ 7 for 0.312% PMAA.

CONCLUSIONS

1. The adsorption behavior of polymethacrylic acid on alumina is very much dependent on solvent conditions and the surface charge characteristics of the PMAA and Al_2O_3: as pH is decreased the adsorption of PMAA increases until insolubility and charge neutralization of the PMAA is approached; for pH values near and below the zero point of charge for Al_2O_3, there is an added electrostatic attractive potential for adsorption which results in a high affinity type adsorption behavior.

2. For pH values near and below the zpc, PMAA induces flocculation until an adsorption saturation limit is reached and the binding energy between particles is reduced below a critical level.

3. Once saturation adsorption occurs and stability is achieved, the PMAA induced stabilization shows an enhanced stability as compared to suspensions stabilized electrostatically without the presence of polyelectrolytes.

4. The stability map introduced here can be a useful processing tool for tailor-making suspensions with varying amounts of polymer from low to high pH's depending on the desired properties.

ACKNOWLEDGEMENTS

The authors gratefully acknowledge the contributions of Dr. Alan Bleier who directed part of the work at the Oak Ridge National Laboratory and Prof. Hans Lyklema whose insight and discussion were very helpful. This research was sponsored by the Advanced Research Projects Agency of the Department of Defense and was monitored by the Air Force Office of Scientific Research under Grant No. AFOSR-83-0375.

REFERENCES

1. P. L. de Bruyn and G. E. Agar, "Surface Chemistry of Flotation," in *Froth Flotation* 50th Anniversary Volume, D. W. Fuerstenau, ed., The American Institute of Mining Metallurgical, and Petroleum Engineers, Inc., (1962) p. 91.

2. S. Voyutsky, *Colloid Chemistry* (Translated from Russian by N. Bobrov), MIR Publishers, Moscow (1978).

3. W. J. Wnek and R. Davies, J. Colloid Interface Sci., **60** 361 (1977).

4. R. J. Hunter, J. Colloid Interface Sci., **37** 564 (1971).

5. D. E. Yates, S. Levine, and T. W. Healy, Trans. Faraday Soc., **70** 1807 (1974).

6. G. A. Parks and P. L. de Bruyn, J. Phys. Coll. Chem., **66** 967 (1962).

7. G. A. Parks, Chem. Rev., **65** 177 (1965).

8. M. Robinson, J. A. Pask, and D. W. Fuerstenau, J. Am. Ceram. Soc., **47** 516 (1964).

9. I. Awasaki, S. R. B. Cooke, and A. F. Colombo, "Flotation Characteristics of Goethite," U.S. Bur. Mines, Rep. Invest. 5593 (1960).

10. F. F. Aplan and D. W. Fuerstenau, "Principles of Nonmetallic Mineral Flotation," in *Froth Flotation* 50th Anniversary Volume, D.W. Fuerstenau, ed., The American Institute of Mining, Metallurgical, and Petroleum Engineers, Inc., (1962) p. 170.

11. H. J. Modi and D. W. Fuerstenau, AIME Trans., **217** 381 (1960).

12. I. Iwasaki, S. R. B. Cooke, and H. S. Choi, AIME Trans., **217** 237 (1960).

13. J. E. Gebhardt and D. W. Fuerstenau, "Adsorption of Polyacrylic Acid at Oxide/Water Interfaces," in *Colloids and Surfaces*, Elsevier Scientific Publishing Co., Amsterdam, (1983) p. 221.

14. G. Lopatin, "The Adsorption of Polymethacrylic Acid from Solution," Ph.D. Thesis, Polytechnic Institute of Brooklyn (1961).

15. R. J. Hunter, *Zeta Potential in Colloid Science - Principles and Applications*, Academic Press, New York (1981).

16. R. Arnold and J. Th. G. Overbeek, Recueil, **69** 192 (1950).

17. W. C. Hasz, "Surface Reactions and Electrical Double Layer Properties of Ceramic Oxides in Aqueous Solution," M.S. Thesis, Massachusetts Institute of Technology (1983).

18. H. A. Van der Schee and J. Lyklema, J. Phys. Chem. **88** 6661 (1984).

19. J. Papenhuijzen, H. A. Van der Schee, and G. J. Fleer, J. Colloid Interface Sci., **104** 540 (1985).

20. G. J. Fleer and J. Lyklema, in *Adsorption from Solution at the Solid/Liquid Interface* G. D. Parfitt and C. H. Rochester, eds., p. 153, Academic Press, New York (1983).

21. I. A. Aksay and R. Kikuchi, "Structures of Colloidal Solids," in *Ultrastructure Processing of Ceramics, Glasses, and Composites II*, L. L. Hench and D. R. Ulrich, eds., Wiley & Sons, New York (1986).

22. A. Bleier and E. D. Goddard, *Colloids and Surfaces*, 1 407 (1980).

PRESSURE FILTRATION OF MONOSIZED COLLOIDAL SILICA

M. Velazquez[*] and S. C. Danforth

Department of Ceramics
Rutgers University
P.O. Box 909,
Piscataway, N.J.   08854

The research described herein is directed at establishing an enhanced understanding of the basic processing behavior of uniform fine ceramic powders. Monosized colloidal silica has been used as a model system to determine the effects of the processing conditions on consolidation behavior (particle packing) during pressure filtration from aqueous suspensions. In this study, 0.3 and 0.6 μm SiO_2 powders were pressure filtered (20-40 psi) to determine the effects of the solids loading and state of dispersion on the green density, cake permeability, specific cake resistance, microstructural uniformity, and pore size distribution of the SiO_2 green bodies. Pressure filtration from aqueous dispersions with increased solids loading resulted in lowered green densities, decreased specific cake resistance (to fluid flow), α_c, increased permeability, K_c, and increased median pore size. Similar effects were observed when green bodies were formed from low pH (agglomerated) as opposed to high pH (dispersed) aqueous suspensions. Pressure filtration of monosized colloidal silica produced green bodies with uniform (non-ordered) particle arrangements.

I.   INTRODUCTION

The current level of research and development of high technology ceramics for structural, electronic, optical applications etc., is at a feverish pitch.[1] Many advances have been made in the areas of new ceramic materials, composite ceramics, and fabrication methods. While the great potential of ceramics has been realized in numerous areas, there are still many applications for which poor component property reliability and uniformity are key limiting factors. It is increasingly evident that the starting powder characteristics and the details of the consolidation process (prior to densification) are the origins of many of the limiting defects (impurities, microstructural heterogeneities, etc), that are found in these technologically important ceramic materials.

---

[*]Corning Electronics, Raleigh, N.C.

The ability to form ceramics with the required level of reliability requires extreme levels of microstructural control over the powder characteristics, particle packing, and pore size uniformity.[2]

Recent studies[3-14] have shown that there are numerous potential advantages associated with the processing of an idealized ceramic powder. There can be no one universal definition of an ideal ceramic powder for all materials and processes, yet the following is an extremely useful set of guidelines. As a goal, an ideal ceramic powder has the following general characteristics:[2] 1) a small average particle size, typically ≤ 1 μm, 2) a narrow particle size distribution, with a ratio of the maximum to average particle diameter of < 1.5, 3) a regular particle morphology, tending toward spherical, 4) a high degree of chemical and phase purity, and 5) freedom from particle agglomeration.

There are numerous advantages associated with processing of such idealized ceramic powders. The ability to accurately analyze and describe the characteristics of idealized powders is enhanced. Use of such powders has led to the potential to model the behavior of forming processes.[10,11,12,16,17] Narrow particle size distributions result in increased colloidal stability against agglomeration.[16] The small particle size and narrow size distribution, regular shape, and lack of agglomeration allow the formation of green microstructures with extremely uniform particle packing.[4-14]

Particle packing of monosized spheres can range from: 1) poor, i.e., low green densities (<40%) from dry packing of highly agglomerated powders, to 2) fair, i.e. (40-60%) for packing of agglomerated powders via filtration or sedimentation, to 3) very good, for random ideal packing (~63%) of fully dispersed powders via colloidal or pressure filtration, sedimentation etc., from aqueous or non-aqueous suspensions, to 4) extremely good (~65%-74%), for ordered packing of fully dispersed monosized powders via sedimentation or filtration from liquid suspensions. Uniform packing, both random ideal and/or ordered, has been demonstrated for a number of systems: ZrO_2,[8] Si,[18,19] SiO_2,[11,13,14,20,23] Al_2O_3[21,22] TiO_2,[3,4] B_2O_3-SiO_2,[24] among others. It has also been shown that such uniformly packed green bodies can be sintered to theoretical density at significantly lower temperatures, for shorter times (than with conventional powders), with minimized grain growth. This has resulted in more uniform microstructures, reduced defect sizes, narrower defect size distributions, and, therefore, should yield enhanced property uniformity etc.

While it is clear that the goal of producing the ultimate in particle packing uniformity (ideal ordered packing, Figure 1), should be studied and attempted, it is currently unclear to the authors if some of the types of defects present in these ordered arrays might potentially act as property limiting flaws in the sintered ceramic components. These defects range from "vacancies", "interstitials", "dislocations", "low angle grain boundaries", and "high angle grain boundaries." It is likely that "vacancies", "interstitials", and "dislocations" will not present significant problems. In contrast, "the high angle grain boundaries", or the randomly packed "interdomain" regions[25] may very well act as critical flaws. These flaws will be the last to be eliminated during sintering (if then) due to their very high pore coordination numbers,[26] R, (R, the number of particles surrounding and defining the pore in three dimensions). It has been shown[26] that pores with less than the critical pore coordination number ($R<R_c$) will be eliminated very early on (at low T, and short t) in the sintering process. Those pores with $R<R_c$ are associated with regions of high local packing densities, i.e, regions where

the particle coordination is high, ~8-12. In these regions, the diffusion distances needed for densification are short, and local densification can occur without the need for coarsening via grain growth.

Sintering of green bodies with non-homogeneous particle packing proceeds by:[11,26,27,28] rapid densification of densely packed agglomerates or domains which may then pull away from the surrounding matrix. This results in an increase in the size of inter-agglomerate or inter-domain pores which can then only be eliminated by significant grain growth during sintering at high temperatures or for long times.

It is a combination of both the narrow particle size and the narrow pore size distribution (i.e. uniform particle packing) that results in the enhanced sintering behavior of fine monosized powders in random ideal or ideal-ordered packing arrangements. In view of the defects present in ideal-ordered (polycrystalline or polydomain) green bodies, and their potential to act as critical flaws, it may be more effective (as well as more feasible) to strive for maximum packing density and uniformity of packing in random-ideal particle arrangements.[6,19]

Of the various methods for producing green bodies, only techniques involving powders incorporated in liquids (aqueous, non-aqueous, or liquid polymer) offer the potential for complete dispersion (elimination of agglomerates) and, in turn, the opportunity to achieve maximum packing density and uniformity of packing on the scale of the particle diameter (< 1 μm). There are numerous references detailing the forces and mechanisms involved in powder dispersions.[29-33] Due to space limitations, a detailed discussion will not be presented here.

There are several different methods of consolidating fine ceramic powders into ceramic components using dispersed powders: slip casting,[34-36] tape casting,[37-39] injection molding,[40,41] colloidal filtration,[10,11,12,25,42,44] sedimentation,[4,13,14,43] and centrifugal casting.[25] Of interest here is colloidal filtration and the effects of the dispersion parameters on the nature of the particle packing and uniformity in the green body.

The mechanisms of filtration have been detailed recently by Kim.[22] Some key points made are: 1) that filtration should not be treated as a diffusional process (as in some earlier studies), 2) that the treatment presented by the Kozeny-Carman model makes the assumption of uniform flow throughout the porous particle bed, an inappropriate assumption for colloidal filtration, 3) that the Adcock-McDowall[45] treatment is appropriate when the resistance of the mold or filtration media is accounted for, and 4) that the more general Dal and Deen model reduces to the Adcock-McDowall treatment when the mold resistance can be ignored.

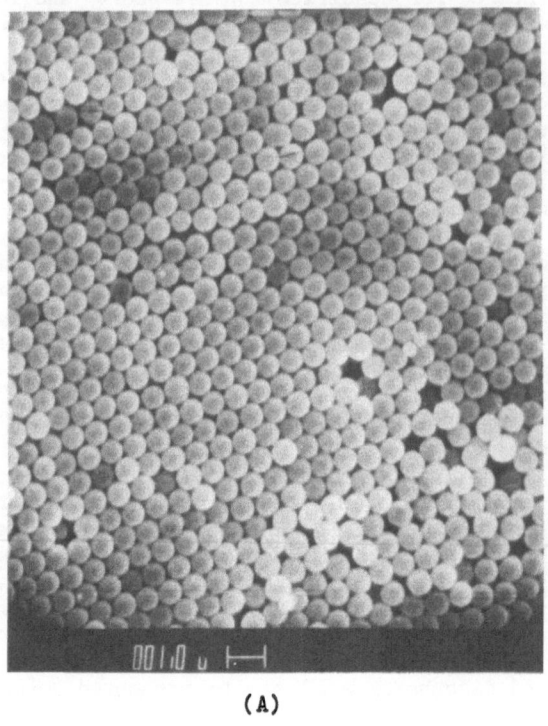

(A)

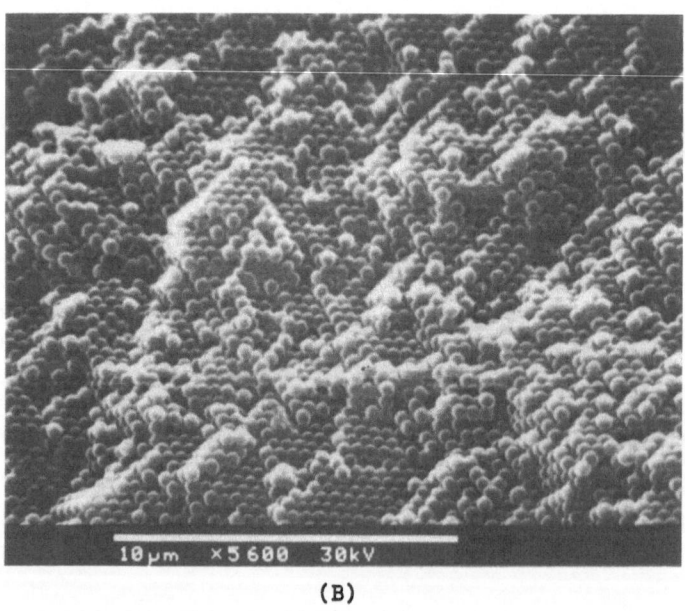

(B)

Figure 1.   SEM micrographs of ordered packing for monosized SiO_2 powders, A) top surface of settled body; B) fracture surface of settled body.

Using Darcy's[46] law, Kim[22] utilized conservation of matter to evaluate the rate of filtration of a colloidal suspension using a porous plaster mold as the medium. The final result was that the linear pressure drop across the consolidated layer is given by the expression:

$$p_t - p_0 = (\frac{\xi_c^2}{t})(\frac{\eta n \alpha_c}{2} + \frac{\eta n^2 \alpha_m}{2\epsilon_m}) \qquad (1)$$

where: p_t = the applied pressure
p_0 = ambient pressure
ξ_c = location of the suspension-consolidated
        layer interface (related to layer thickness)
t = time
η = liquid filtrate viscosity
α_c = specific resistance of consolidated layer
α_m = specific resistance of mold
ϵ_m = volume fraction of voids in mold
n = system parameter = $1 - X_p - \epsilon_c/X_p$
X_p = volume fraction of solids in the suspension
ϵ_c = volume fraction of voids in consolidated layer

This analysis indicates that the specific cake resistance, α_c, increases rapidly with reductions in particle size (for monosized spheres), and void fraction in the consolidated layer. This analysis[22] has shown that the casting rate and uniformity for colloidal powders can be substantially enhanced by the use of increased filtration pressure $(p_t - p_0)$.

Using a different analysis, Bridger et al,[42] treated the filtration behavior of uniformly sized particles using traditional flow equations in resistance form. In this approach, the resistance to fluid flow is made up of two components, the cake resistance and the medium resistance. The result, assuming a constant cake and mold resistance, and where $Rm \ll \alpha_{ave}$, was:

$$pt/ \eta v = \frac{\alpha_{ave}}{2} W_c + R_m \qquad (2)$$

where: p = applied pressure
t = time
v = volume of filtrate per unit area of cake
q = dv/dt
η = filtrate viscosity
W_c = mass of cake per unit area
α_{ave} = average specific cake resistance
R_m = filter medium resistance

By recording the filtrate flow rate, they were able to evaluate α_{ave} for filter cakes formed from a variety of different aqueous polystyrene latex dispersions. The key results of this study will be presented below in Section III.B.

The authors, in an approach similar to Aksay's[10], treated the permeability of a porous material to the flow of an incompressible fluid according to Darcy's law[51] as:

$$K = \frac{Q\eta}{A(\Delta P/L)} \qquad (3)$$

where: K = permeability
Q = fluid flow rate
η = fluid viscosity
ΔP = pressure difference across an area A of a specimen of thickness L

Because solids are being deposited during the process of filter pressing, a moving boundary condition exists. Darcy's law is still obeyed in general, yet certain corrections must be made to model the behavior. The law of deposition describes the change in thickness of the formed cake with time as:[51]

$$\frac{\partial x_c}{\partial t} = \frac{-f_s}{\phi(1-\phi_c)(1-f_s)} (V_n) \qquad (4)$$

where: x_c = increase in cake thickness over time t
f_s = volume fraction of solids
ϕ = porosity of the filter medium
ϕ_c = porosity of the powder compact
V_n = volume flow rate normal to the powder compact

Using Darcy's law with appropriate boundary conditions, and with use of the law of deposition, the permeability, K_c, of the filter-pressed powder compact can be expressed as[51]:

$$K_c = \frac{x_c^2 \eta K}{2(P_a K w_e t - \eta x_c L)} \qquad (5)$$

where: x_c = filter cake thickness
η = viscosity of filtrate
K = permeability of filter to filtrate from Equation (3)
P_a = applied pressure
t = time of filter pressing
L = thickness of filter
$w_e = [(1 - f_s)(1 - \phi_c)]/f_s$

This expression indicates that the permeability of a powder compact can be experimentally determined by using a filter-pressing apparatus. It will be possible then to evaluate the influence of processing parameters on the permeability of the filter-pressed powder compacts of mono-sized SiO_2 particles (as they are consolidated) in relation to the state of dispersion and the nature of the particle packing.

In this investigation, monosized colloidal SiO_2[47] was used as a model material to study the effects of the dispersion and filtration parameters on the nature of the particle packing in the resulting green bodies. The dispersion parameters varied were pH and solids content, while filtration was carried out over a small range of pressures. The green compacts were evaluated to determine: green density, specific cake resistance, α_{ave}, cake permeability, K_c, pore size distribution, and microstructure.

II. EXPERIMENTAL PROCEDURES

Monosized colloidal SiO_2 particles were prepared using a technique similar to that reported by Stober et al.[47] for the hydrolysis and polymerization of amorphous SiO_2 from tetraethyl orthosilicate (TEOS – $Si(C_2H_5O)_4$)[+].

Table 1 shows typical reactant concentrations for the synthesis of 0.6 µm and 0.3 µm powders. Distilled TEOS and 3:1 n-propanol:methanol were mixed in one flask, with alcohol, ammonia and water in a second flask. After mixing the contents of the two flasks, powders were observed to form after ~2 min (by the onset of turbidity). After 24 hours in the growth solution, powders were cleaned by repeated centrifugal settling[++] (3000 g for 1 hour) and redispersion cycles (high speed mixer) in distilled H_2O.

Once cleaning was completed, small powder samples were set aside for analysis as follows. Particle size and microstructural analysis was conducted using SEM[+++,++++] on samples coated with a thermally evaporated Au-Pd conductive layer. Size and size distribution analysis was conducted by both SEM and photon correlation spectroscopy[*]. Surface area was measured by BET method of N_2 gas adsorption – desorption[**]. Powder densities were measured using He pycnometry[***] on dried powders. Dispersions were prepared by additions of 2M HCl[****] to the SiO_2 suspensions to lower the pH. Solids loadings were varied by controlled additions of distilled water to concentrated aqueous suspensions. At this stage, the suspensions were dispersed using ultrasonic agitation[!] for 3 min. at a 50% power setting.

Pressure casting of the aqueous silica suspensions was carried out in a stainless steel filter press[!!] using high purity He gas to apply the overpressure. The filter media used was a 0.2 µm pore size cellulose acetate material.[!!!] Before filtration commenced, the filter media was premoistened with distilled H_2O. Filtration pressures and times ranged from 5–40 psi and 0–60 min, respectively.

Measurements of cake permeability (K_c) involved the following sequence of steps: 1) casting for a given time period (to achieve > 3 mm cake thickness), 2) decanting the excess dispersion (some must be present), 3) repressurisation to drive out excess water, 4) cake removal, 5) measuring the weight and thickness of the wet cake, 6) drying the cake for $\geq$ 48 hours at 80°C, and 7) measuring the dry cake weight and thickness.

[+]Fisher Scientific, Fairlawn, NJ
[++]DPR, 6000, I.E.C., Needham Hts., MA
[+++]Etec Autoscan, Haywood, CA
[++++]AMRAY 1200, Bedford, MA
[*]Coulter N-4, Langley Ford Inst., Amherst, MA
[**]Quantasorb 05-10, Quantachrome Corp., Syosset, NY
[***]AutoPycnometer 1320, Micromeritics, Norcross, GA
[****]Fisher Scientific, Fair Lawn, NJ
[!]Heat Systems Ultrasonics, Plainview, NY
[!!]Seitz Model 5 (modified), Seitz Filters, Milldale, CT
[!!!]Versapore 200, Gelman Sci., Ann Arbor, MI
[@]Pore Sizer 9305, Micromeritics, Norcross, GA

Evaluation of the specific cake resistance, α_c, was performed using the same filter press. The press was suspended above an electronic balance in order to continuously monitor the collected filtrate. The filtrate weight was recorded every 10 seconds for the initial 5 min and every minute thereafter for $\geq$ 40 min. Wet and dried cakes were characterized as above.

Pore (capillary) size distributions of pieces of dried filter pressed SiO_2 compacts were analyzed using scanning Hg porosimetry[e] in a 5 cc sample holder using pressures up to 30,000 psi which corresponds to measurement of capillaires down to 6 nm.

III. RESULTS AND DISCUSSION

A. Powder Characteristics

The particular characteristics of the powders used in this study are presented in Table I. Two lots of powders were used having 0.6 μm and 0.3 μm average sizes. Figures 1 and 2 show the powders (size and

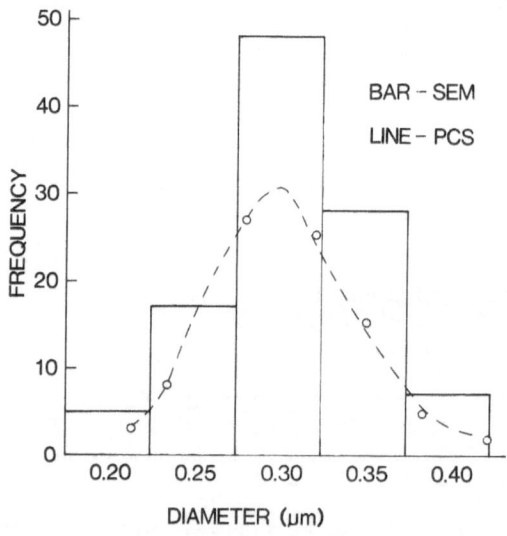

Figure 2. Particle size distribution for 0.3 μm SiO_2 powders used in this investigation (SEM and photon correlation spectroscopy).

spherical shape) and the particle size distribution for the 0.3 μm powders via SEM (100 particles) and photon correlation spectros

Table I SiO$_2$ Synthesis Parameters

0.6 μm SiO$_2$

TEOS mol/l	H$_2$O mol/l	NH$_3$ mol/l	3:1 n-Propanol/Methanol mol/l
0.28	3.9	0.92	8.45

0.3 μm SiO$_2$

TEOS mol/l	H$_2$O mol/l	NH$_4$OH$_{aq}$ mol/l	3:1 n-Propanol/Methanol mol/l
0.16	0.67	0.45	6.34

copy. Note the narrowness of the size distribution. The standard deviation for the 0.3 μm powders was 0.05 μm. Both powders had measured He pycnometric densities of 2.08 g/cc which is very close to values reported in the literature for amorphous SiO$_2$[14]. The surface areas for the 0.6 and 0.3 μm SiO$_2$ were 10.3 m^2/g and 18.53 m^2/g respectively. These correspond to equivalent spherical diameters of 0.26 μm and 0.15 μm respectively, both smaller than the measured sizes. This result indicates that the particles have rough or microporous surfaces which result in higher surface areas and correspondingly lower equivalent spherical diameter. A comprehensive analysis of similar SiO$_2$ powders can be found in the papers by Sacks et al.[13,14]

B. **Pressure Filtration**

The results of the filtration studies are presented in Figures 3,4 and 5 and Table II. Figure 3 shows a typical plot of the raw data (for α_{ave} determination) plotted as time per unit volume vs. W_c. The slope of the linear portion of the curve, utilizing Equation 2, yields values for the specific cake resistance, α_{ave}. The slope is taken after an initial non-linear region, usually after W_c= 0.1 g/cm. It has been reported[42] that this initial region corresponds to a structural instability in the initial filtration layer that is collapsed or relaxed after sufficient layer thickness has developed. The exact cause and nature of this phenomenon is not understood at this time.[42,48]

Figures 4 and 5 show the observed dependence of both K_c and α_{ave} on the solids loading of the filtered suspension for the 0.6 and 0.3 μm powders respectively (K_c for 0.6 μm only). It can be seen that K_c increases rapidly with increased f_s for both 0.6 and 0.3 μm SiO$_2$ suspensions. The rate of increase in both cases is qualitatively similar. It should be noted that the values of K_c for the 0.3 μm SiO$_2$ are a factor of approximatively 10^1-10^2 smaller than for the 0.6 μm SiO$_2$ compacts .

The corresponding dependence of α_{ave} on f_s is shown in Figure 5. The specific cake resistance drops by more than a factor of 2 with an increase in f_s from 0.03 to 0.63. Table II shows the results of the effects of pH on the specific cake resistance and permeability.

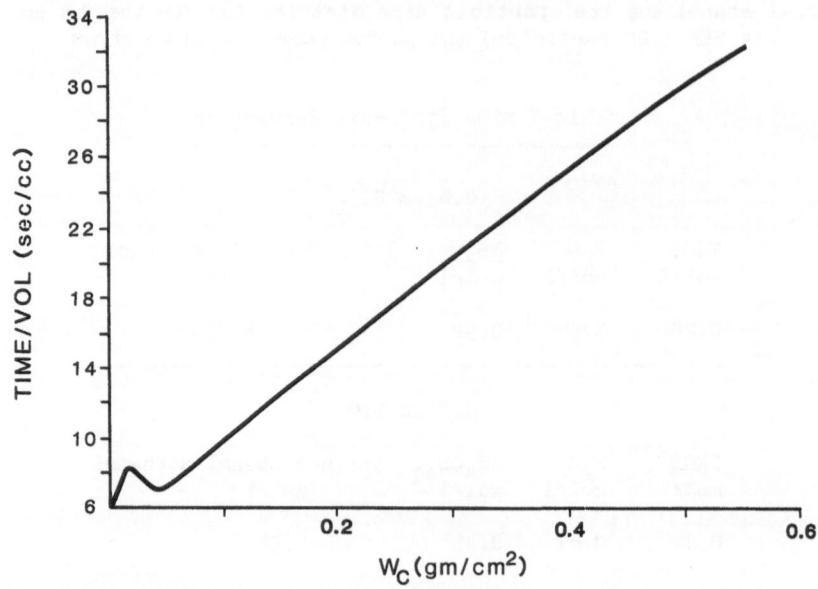

Figure 3. Plot of time/volume vs. W_C from filtration experiments.

The results in Figures 4 and 5 and Table II show a clear increase in α_{ave} and a decrease in K_c as each system is modified to enhance particle dispersion stability. For the silica, increased pH corresponds to increased negative surface charge (zeta potential)[13,14] and interparticle repulsive forces. This results in reduced agglomeration and more uniform, dense particle packing. In a similar fashion, the

Table II Filtration Results

Dispersion pH	Cake Porosity	$\alpha_{ave}(\times 10^{11}$ cm/gm$)$	K_c(millidarcy)
9.8	48%	11.4	0.0042
3.6	51%	6.8	0.0090
2.9	55%	5.7	0.0097

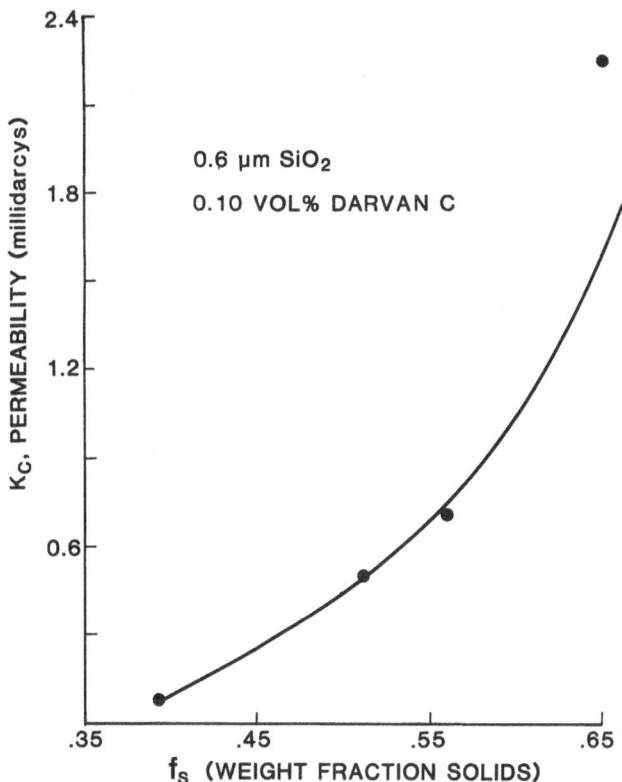

Figure 4. Plot of permeability, K_c, vs. solids loading, f_s, from filtration studies (0.6 μm SiO_2).

reduction of the ionic strength for the polystyrene latex dispersions[42] resulted in increased interparticle repulsion, enhanced particle packing density and uniformity, and increased α_{ave} valves.

Density measurements were made on all of the SiO_2 green bodies. Results showed that there was a decrease in green density from 52% to 42% as the pH of the suspension was lowered from 9.7-2.9. No significant changes in density were measured for samples cast from different solids content dispersions. Samples cast from high pH suspensions (9.6-9.8) had essentially constant green densities between 51-54% regardless of f_s (which ranged from 2.2% to 63%).

C. Pore (Capillary) Size Distribution

In addition to the above analysis, green bodies were evaluated for their pore (capillary) size distribution using Hg intrusion porosimetry. The results of Hg porosimetry are presented in Figure 6, where the data for the following samples are presented: 1) 0.3 μm SiO_2, f_s = 0.17, pH = 9.8, P = 40 psi; 2) 0.3 μm SiO_2, f_s = 0.17, pH = 2.8, P = 40 psi; 3) 0.3 μm SiO_2, dry pressed at 5,000 psi. Figure 6 shows a

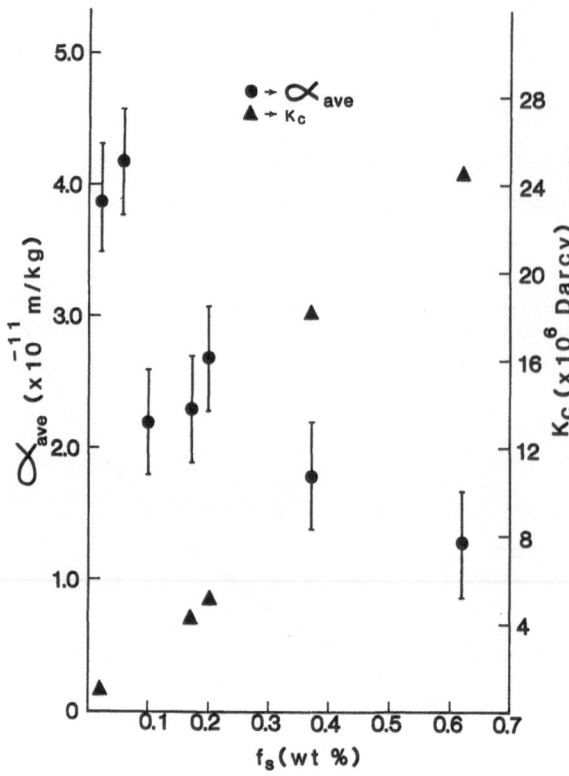

Figure 5. Plot of specific cake resistance, α_{ave}, and permeability, K_c, vs. solids loading, f_s, (0.3 µm SiO$_2$).

typical plot of the pore (capillary) size distribution of a green body pressure cast from monosized 0.3 µm SiO$_2$ powder. This sample (3A) was filter pressed at a pH of 9.7, from a 17 wt % SiO$_2$ suspension. The pore fraction was 0.48 and the specific cake resistance for this sample was high (2.3 x 10^{+12} cm/gm).

For this sample (and the majority of samples cast from aqueous dispersed SiO$_2$ suspensions), the largest pore (capillary) size measured was < 10 µm. In addition, most samples had few pores (capillary) above approximatively 1-5 µm in size. The median pore (capillary) size ranged from approximatively 0.05 µm to approximatively 0.07 µm for samples cast from these aqueous suspensions, depending on the solids content and pH of the suspension. This size agrees reasonably well with that cited for the relationship between the median pore size and the particle size.[50] This predicts a 0.043 µm median pore size for the 0.3 µm particles. Table III shows the measured median pore (capillary) sizes for samples of 0.3 µm

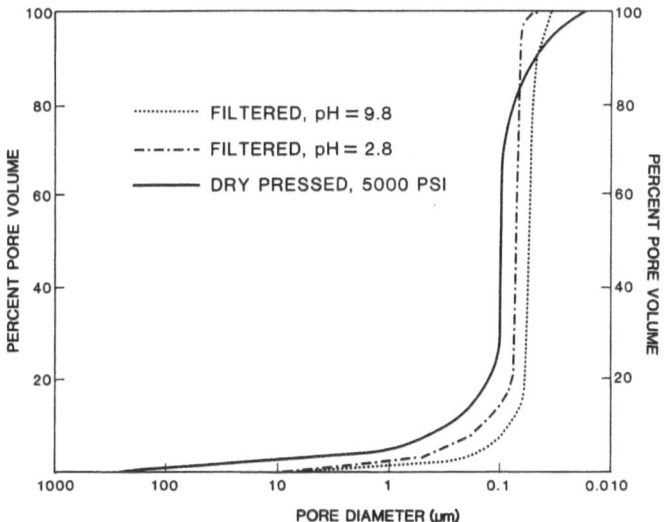

Figure 6. Pore (capillary) size distributions for green bodies filtered
and pressed using 0.3 μm SiO₂.

Table III Consolidation Parameters and Pore (Capillary) Size

Sample No.	Vol. Fr.	pH	Pressure (psi)	Vol. Fr. Poros.	Particle Size μm	Median Pore Size μm	Maximum Pore Size μ m
3A	0.17	9.7	40	0.48	0.30	0.055	9.5
5K	0.022	9.6	40	0.48	0.30	0.058	9.5
4K	0.62	9.6	40	0.47	0.30	0.063	9.2
3D	0.17	2.8	40	0.55	0.30	0.068	9.5
2L	0.2	9.7	20	0.48	0.30	0.058	9.5
3J	0.2	9.6	40	0.46	0.30	0.058	8.8
G-1	---	---	5.000[#]	0.53	0.30	0.097	360

[#]Die Pressed (Dry)

99

SiO_2 cast under different conditions. The median pore (capillary) size is seen to increase from 0.055 μm to 0.068 μm when the pH of the casting dispersion is lowered from 9.7 to 2.8, for a 17 wt % solids loading (samples 3A and 3D).

The median pore (capillary) size was also observed to increase from 0.058 μm to 0.063 μm for an increase in the solids content of (pH 9.6-9.8) slips from 2.2 wt % to 62 wt % (samples 5-K and 4-K). It was determined that the filtration pressure (in the range of 20 to 40 psi) had no significant effect on the median pore (capillary) size (0.058 μm) or the overall nature of the pore size distribution (samples 2L and 3J).

In addition to evaluating the filter pressed compacts discussed above, we have prepared green samples by dry pressing the 0.3 μm SiO_2 powders discussed above. These were utilized for comparative purposes only. Samples were dry pressed in a 0.5" steel die at 5000 psi without binder.

The dry pressed 0.3 μm SiO_2 green body showed a significant increase in the median pore (capillary) size (from approximately 0.06 μm for the cast 0.3 μm SiO_2 samples) to 0.1 μm. More significantly, the percentage of pores larger than 1 μm is much greater for the dry pressed sample than the cast samples. In addition, the dry pressed sample has a significant distribution of pores (capillaries) between 10 and 400 μm! In contrast, cast pieces had maximum pore (capillary) sizes of approximately 10 μm. This result is dramatic evidence of the advantages of wet methods over dry methods of consolidating fine ceramic powders.

Figures 7 and 8 show different aspects of the microstructural features of various samples prepared in this study. Figure 7 shows the as cast top surface and Figure 8 shows the fracture surface of the 0.3 μm, f_s = 0.17, pH 9.7 sample. This shows the uniform nature of the particle packing and the relatively narrow distribution of pore sizes. SEM micrographs of the dry pressed powders showed increased agglomeration, decreased packing density, decreased uniformity of packing, and corroborated the increase in the width of the pore size distribution (from the Hg porosimetry) which results from dry pressing the 0.3 μm powder. The small changes in the pore size distribution between samples filter pressed at pH 9.7 vs. 2.8 or between samples filter pressed from 2.2 vs. 62 volume % solids were not observable in the SEM.

From these results, several very important points can be made. The maximum pore size, the median pore size, and the width of the pore size distribution can be reduced: a) by reducing the particle size, b) by replacing dry pressing by colloidal filtration/slip casting, and c) by forming the cast body from a more fully dispersed (less agglomerated) suspension, i.e. a high pH and a low solids content suspension for SiO_2 in H_2O. The effects of the pH or the solids content (of the suspension) on the nature of the particle packing is discussed below. Under low solids content or high pH conditions, the particles arriving at the interface between the suspension and the filter cake will have sufficient mobility (due to the large repulsive interparticle forces >10kT) or time (due to the relatively low rates of deposition at the interface) to arrange themselves in a densely packed (random) structure. This will yield a green body with: a smaller pore size, increased green density, a narrower pore size distribution,[13,14] an increased specific cake resistance, and a decreased permeability.

Figure 7. SEM of the top, as cast surface of filtered green body from the 0.3 µm SiO_2.

Figure 8. SEM of the fracture surface of filtered green body from 0.3 µm SiO_2.

In the case of a low pH or a high solids content, the inter-particle repulsive forces will be low or the critical solids content will have been reached so that particles will flocculate prior to consolidation (either while still in the suspension, or as the solids content rises as the particles approach the suspension-body interface). The packing of these flocs or "domains"[10,11,25] will therefore result in a more open structure, with a lower density, a larger median pore size, and increased permeability and decreased specific cake resistance. While this type of packing will no doubt lead to faster casting rates, the detrimental effects on the microstructure and its uniformity are evident.

IV. SUMMARY AND CONCLUSIONS

Monosized colloidal silica has been used as a model system to investigate the effects of dispersion conditions and consolidation parameters on the nature and uniformity of the green microstructure of a ceramic body. Pressure filtration was chosen as the consolidation technique for its similarities to conventional slip casting.

Monosized SiO_2 powders, 0.6 and 0.3 μ m, were suspended in aqueous media of different pH and solids loadings. These suspensions were filter pressed to investigate the effects of solids loading, state of dispersion and the filtration pressure on the green density, the pore size and distribution, the cake permeability and specific cake resistance and the nature of the particle packing.

The results showed that under filtration conditions which promoted full dispersion, i.e. high pH or low solids loadings, the particle packing was enhanced with resultant: increased specific cake resistance (decreased permeability), increased green density, decreased median pore size, and more uniform particle packing.

Under conditions where there was an increased tendency for agglomeration in the suspension (prior to the particles reaching the suspension-body interface), the particle packing was poorer, with a resultant: lower specific cake resistance (increased permeability), lowered green density, increased median pore size, and less uniform packing.

These results are in qualitative agreement with recent work of other investigators,[10,11,13,14,16,19,25,42] showing that with proper control over interparticle forces and consolidation conditions, green bodies with very uniform microstructures (and fine defect sizes) can be formed with uniform fine (ideal) ceramic powders.

Key issues that still remain are: 1) how to optimize the level of microstructural uniformity, i.e., particle packing uniformity,in the green body (using these ideal powders), and 2) how to develop consolidation methods or techniques which will allow the production of large, three dimensional, complex shaped green bodies at a rate, and in such a fashion as to allow the levels of microstructural uniformity generated in the laboratory to be duplicated in industry. Work in these areas is ongoing in this and in other laboratories.

V. ACKNOWLEDGEMENTS

The financial support for this work from the Center for Ceramics Research at Rutgers University is greatfully acknowledged. The authors would also like to express their deep thanks to Gene Krug for the Hg porosimetry work and help in many other areas. The assistance of Mary Jane Smyth and Tom O'Leary is also acknowledged.

VI. REFERENCES

1) J. B. Wachtman, "Ceramic Fever -- Advanced Ceramics in Japan", Ceramic Industry, 121, 6 (1983).

2) H. K. Bowen, "Basic Research Needs on High Temperature Ceramics for Energy Applications," Mater. Sci. and Eng., 44, 1-56 (1980).

3) E. A. Barringer and H. K. Bowen, "Formation, Packing, and Sintering of Monodisperse TiO_2 Powders," J. Am. Ceram. Soc., 65, C-199 (1982).

4) E. A. Barringer, "Monodisperse Colloidal TiO_2," Ph.D. Thesis, M.I.T., (Sept. 1983).

5) J. S. Haggerty, "Growth of Precisely Controlled Powders from Laser Heated Gasses," in Ultrastructure Processing of Ceramics, Glasses, and Composites, pp. 353-366. Edited by L. L. Hench, D. R. Ulrich, J. Wiley and Sons, N.Y. 1984.

6) S. C. Danforth and J. S. Haggerty, "Mechanical Properties of Sintered and Nitrided Laser Synthesized Silicon Powder," J. Am. Ceram. Soc., Communications, 66, 4, C-58-59, April (1983).

7) M. F. Yan, et al., "Effect of Impurities and Pores on Grain Boundary Mobility," Bull. Am. Ceram. Soc., 56, 291 (1977).

8) W. H. Rhodes, "Agglomerates and Particle Size Effects on Sintering Yttria Stabilized Zirconia," J. Am. Ceram. Soc., 64, 19 (1981).

9) H. K. Bowen, R. L. Pober, E. A. Barringer, M. V. Parrish, and N. Levoy, "Dispersion and Packing of Narrow Size Distribution Powders," Report of the Committee on the Workshop on the Reliability of Multi-layer Ceramic Capacitors, Publication NMAB-400, National Academy Press, Washington, D.C. (1983).

10) I. A. Aksay and C. H. Schilling, "Mechanics of Colloidal Filtration," Advances in Ceramics, Forming of Ceramics, V. 9, Ed. J. A. Mangels, pp. 85-93 (1984).

11) I. A. Aksay, "Microstructure Control Through Colloidal Filtration," Advances in Ceramics, Forming of Ceramics, V.9, Ed. J. A. Mangels, pp. 94-104 (1984).

12) M. Velazquez, S. C. Danforth, "Casting of Monodisperse Colloidal Silica," Advances in Ceramics, Forming of Ceramics, V.9, Ed. J. A. Mangels, pp. 105-114 (1984).

13) M. D. Sacks, T. Y. Tseng, "Preparation of SiO_2 from Model Powder Compacts: I, Formation and Characterization of Powders, Suspensions, and Green Compacts," J. Am. Ceram. Soc., 67, 8, 526-532 (1984).

14) M. D. Sacks, T. Y. Tseng, "Preparation of SiO_2 Glass from Model Powder Compacts: II, Sintering," J. Am. Ceram. Soc., 67, 8, 532-537 (1984).

15) F. F. Lange, "Sinterability of Agglomerated Powders," J. Am. Ceram. Soc., 67, 2, 83-89 (1984).

16) M. T. Strauss, T. A. Ring, H. K. Bowen, "Interaction Energies and Pressures in Concentrated, Acqueous Suspensions of Polydisperse Ceramic Powders," Am. Ceram. Soc., Fall Mtg., 101-B-84P, San Francisco, CA, Oct. (1984).

17) W. B. Russel, private communication (1985).

18) S. C. Danforth, "Synthesis and Processing of Uniform Fine Ceramic Powders, <u>Proceedings of the Lecture Meeting on Advanced Ceramics,</u> Tokyo Inst. of Techn., Tokyo, Japan, Oct. (1984), Ed. S. Somiya et al., Terra Scientific Publishing Co., Tokyo, Japan (1986).

19) G. J. Garvey, J. M. Lihrmann, J. S. Haggerty, "Dispersion, Shaping, and Reaction-Bonding of Laser Synthesized Silicon Powders," Am. Cer. Soc., Fall Mtg., 103-B-84P, San Fransisco, CA, October (1984).

20) T. C. Huynh, "Synthesis of Monosized SiO_2 Powder," Ph.D. Thesis, MIT, June (1985).

21) H. K. Bowen, private communication (1982).

22) S. I. Kim, I. A. Aksay, "Pressure Filtration of Colloidal Systems," Am. Ceram. Soc., Fall Mtg., 102-B-84P, San Fransisco, CA, October, (1984).

23) C. Han, I. A. Aksay, "Uniform Densification of Bimodal Particle Systems," Am. Ceram. Soc., Fall Mtg., 11-B-84P, San Francisco, CA, October (1984).

24) E. A. Barringer, N. Jubb, B. Fegley, R. L.+ Pober, and H. K. Bowen, "Processing Monosized Powders," in <u>Ultrastructure Processing of Ceramics, Glasses, and Composites,</u> pp. 315-333. Ed. L. L. Hench, D. R. Ulrich, J. Wiley and Sons, N.Y. (1984).

25) I. A. Aksay and C. H. Schilling, "Colloidal Filtration Route to Uniform Microstructures," <u>Ultrastructure Processing of Ceramics, Glasses, and Composites,</u> pp. 439-447. Ed. L. L. Hench, D. R. Ulrich, J. Wiley and Sons, N.Y. (1984).

26) F. F. Lange, "Sinterability of Agglomerated Powders," <u>J. Am. Ceram. Soc., 67,</u> 2, 83-89 (1984).

27) S. C. Danforth, unpublished work (1982).

28) F. F. Lange, B. J. Kellett, "Grain Growth Driven by External Surface Curvature," Am. Ceram. Soc., Fall Mtg., 37-B-84P, San Fransisco, CA, Oct. (1984).

29) E. Matijevic, "Colloidal Stability and Complex Chemistry," <u>J. Colloid Interface Sci., 43,</u> 217 (1973).

30) J. Th. G. Overbeek, "Recent Developments in the Understanding of Colloidal Stability," <u>J. Colloid Interface Sci., 58,</u> 357 (1977).

31) D. J. Shaw, <u>Introduction to Colloid and Surface Chemistry,</u> Third Ed., Butterworths and Co., London, England (1980).

32) R. H. Ottewill, "Stability and Instability in Disperse Systems," <u>J. Colloid Interface Sci., 58,</u> 408 (1977).

33) P. J. Anderson, "Characteristics of Zeta Potential Against Concentration Relations," <u>Trans. Faraday Soc., 54,</u> 562 (1958).

34) G. W. Phelps, S. G. Maguire, W. J. Kelley, R. K. Wood, <u>Rheology and Rheometry of Clay-Water Systems,</u> Cyprus Industrial Minerals Co. (1982).

35) J. E. Funk, "Slip Casting and Casters," <u>Advances in Ceramics, Forming of Ceramics,</u> V. 9, Ed. J. A. Mangels, pp. 76-84 (1984).

36) D. S. Adcock and I. C. McDowall, "Mechanism of Filter Pressing and Slip Casting," <u>J. Am. Ceram. Soc., 40,</u> 355 (1957).

37) E. S. Tormey, R. L. Pober, H. K. Bowen, P. D. Calvert, "Tape Casting Future Developments," <u>Advances in Ceramics, Forming of Ceramics,</u> V. 9, Ed. J. A. Mangels, pp. 140-149 (1984).

38) R. J. MacKinnon, J. B. Blum, "Particle Size Distribution Effects on Tape Casting Barium Titanate," <u>Advances in Ceramics, Forming of Ceramics,</u> V. 9, Ed. J. A. Mangels, pp. 150-159 (1984).

39) R. E. Mistler, D. J. Shanefield, R. B. Runk, "Tape Casting of Ceramics," <u>Ceramics Processing Before Firing,</u> Ed. by G. Y. Onoda, Jr., and L. L. Hench, John Wiley and Sons, N.Y., N.Y., pp. 411-448 (1978).

40) J. A. Mangels, W. Trela, "Ceramic Components by Injection Molding," <u>Advances in Ceramics, Forming of Ceramics,</u> V. 9, Ed. J. A. Mangels, pp. 220-233 (1984).

41) C. L. Quackenbusch, K. French, J. T. Neil, "Fabrication of Sinterable Silicon Nitride by Injection Molding," Ceram. Eng. Sci. Proc., 3, 1-2, 20-34 (1982).

42) K. Bridger, M. Tadros, W. Leu, F. Tiller, "Filtration Behavior of Suspensions of Uniform Polystyrene Particles in Aqueous Media," Separation Sci. Technol., 18, 12-13, 1417-1438 (1983).

43) R. Allman III, "Polycrystalline Colloids and Their Implications on Microstructure Evolution," M.S. Thesis, Univ. of California, Los Angeles, CA (1983).

44) W. B. Russel, private communication (1984).

45) D. S. Adcock and I. C. McDowall, "Mechanism of Filter Pressing and Slip Casting," J. Am. Ceram. Soc., 40, 355 (1957).

46) A. E. Scheidegger, The Physics of Flow Through Porous Media, 3rd Ed., Univ. of Toronto Press, Toronto (1974).

47) W. Stober, A. Fink, and E. Bohn, "Controlled Growth of Monodisperse Silica Spheres in the Micron Size Range," J. Colloid Interface Sci., 26 62 (1968).

48) K. Bridger, private communication (1984).

49) G. W. Phelps, J. S. Dennis, "Particle Size of Feldspar and Flint as a Factor in Slip Casting," J. Am. Ceram. Soc., 44, 4, 149-156 (1961).

50) O. J. Whittemore, "Mercury Porosimetry of Ceramics," Powder Technology, 29, 167-175 (1981).

51) R. E. Collins, Flow of Fluids Through Porous Media, Reinhold Publ. Co., N.Y. (1961).

SOL-GEL ROUTES TO CERAMICS AND GLASSES   I. GELS

Francis J. Bonner

Department of Chemical Engineering
Vanderbilt University
Nashville, TN 37235

This work describes some significant differences between alkoxide gels and colloidal particulate gels which are important for fundamental understanding and for the manufacture of ceramic and glass products. The experimentally observed differences in rheology between the two gel types are rationalized using fundamental theories advanced by Iler. The most significant difference is the irreversible nature of the gel formed from ethanolic solution of tetraethyl orthosilicate (TEOS) and the highly reversible nature of the fumed silica gels in water and in alcohol. Two possible reasons for this difference are advanced: variation in the fundamental type of dominant network bond, viz. primary chemical bonds for the alkoxide gels versus secondary physical type bonds for the fumed silica gels, and in the amount of adsorbed liquid that could function as an internal lubricant. Other important observations relating to the fundamental structure of the alkoxide gels involve the duration of highly exothermic reactions and the formation of colloidal particles that appeared to promote gellation. These observations support the conclusion that the alkoxide gels could be formed by a network of colloidal particles or by precursors to these solid particles, i.e., by aggregates of hydrolyzed TEOS oligomers.

INTRODUCTION

Graham[1] extensively investigated silica gels over 100 years ago. The Brintzingers[2] mixed tetraethyl orthosilicate (also termed tetraethoxysilane and abbreviated TEOS) and water to synthesize and to polymerize silicic acid over 50 years ago. Kistler[3] prepared silica aerogels, a highly porous glass-like material, via a sol-gel route using sodium silicate solutions also over 50 years ago. Several decades later, similar materials were being made by reacting tetramethyl orthosilicate and water in methanol[4]. Ethanolic solutions of tetraethyl orthosilicate and water became a more popular sol-gel route, probably because this system is less toxic and generally easier to handle than its methyl homologe[5,6].

A surge of inquiry into sol-gel routes to ceramics and glasses is now occurring. This increased interest stems from demands for more efficient manufacturing, especially with regard to energy use, and, perhaps more importantly, for both new and improved materials, especially for electronic and optical applications. Sol-gel processes and materials are increasingly important subjects for international meetings, e.g., at Padua[7] in 1981, Gainesville[8] in 1983, Albuquerque[9] and Nashville[10] in 1984. Sakka[11], Klein[12], and Brinker and Scherer[13], among others, have provided useful reviews.

Many of the perceived advantages of the sol-gel route stem directly and indirectly from reducing the maximum processing temperature to one-half or one-fourth that for melt glasses, ie., from 2000 - 3000 $^{\circ}$C to 700 - 1400 $^{\circ}$C. Reagents and composite materials of a larger variety and increased purity can be used and the rheological properties of the sol-gels can be exploited to make materials and end-products with better purity and homogeneity as well as with other improved and novel properties. Disadvantages probably are mostly related to the relatively high cost of raw materials, large shrinkage and long processing times. Mackenzie[14] assesses the advantages of sol-gel processes relative to conventional glass and ceramic processes in the context of reviewing proposed and attempted applications.

The alkoxide route in terms of the tetraethoxy orthosilicate, water, ethanol system remains the chief sol-gel route of scientific and of practical interest. Within the last couple of years, however, research works have discussed sol-gel routes based on less expensive pre-formed colloidal solids, such as fumed silica, in water[15] and in non-aqueous liquids such as low molecular weight alcohols[16, 17]. Figure 1 outlines these sol-gel routes of current interest and identifies important processing steps.

Iler[18] marshalls an impressive amount of experimental evidence to support his view that at least in aqueous systems the polymerization of silicic acid should not be considered completely analogous to classical carbon-based polymerizations but rather that silicic acid forms oligomers and then aggregates of these oligomers coalesce into colloidal solid particles which, in turn, chain and cross-link to form gels. In

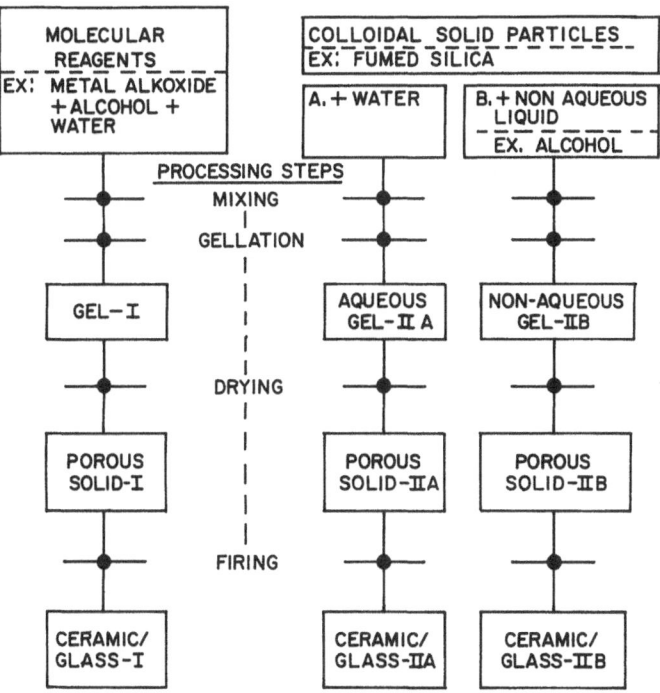

Figure 1. Sol-gel routes of current interest.

contrast, work on the alkoxide sol-gel route to ceramics and glasses is generally conceived and explained in terms of classical condensation polymerization to form macromolecules which cross-link to form gels, although gellation by cross-linking of solid particles is sometimes considered to occur under restricted conditions. The recent review by Brinker and Scherer[13] is an example of this latter viewpoint, with some consideration of Iler's work. Further resolution of this dichotomy in viewpoint is important for fundamental understanding and for controlling the properties of the gels and their final products.

ALKOXIDE GELS

Background

While recognizing that the relative proportions of alkoxide, water, and alcohol along with acidity and processing variables such as order and degree of mixing significantly affect gel properties[11,19], our present discussion limits itself to the formulation described by Klein and Garvey[20] as typical: 43 vol.% TEOS, 43 vol.% ethanol, and 14 vol.% water. This gives (moles H_2O/(mole TEOS) approximately 4/1 or (moles H_2O/(mole reactive ethoxy group) approximately 1/1. The stoichiometry of this formulation is especially interesting. If the hydrolysis went to completion, all the TEOS would become silicic acid and all the water initially present would be replaced by an equal molar amount of ethanol:

$$Si(OC_2H_5)_4 + 4H_2O \rightarrow Si(OH)_4 + 4C_2H_5OH. \tag{1}$$

If the silicic acid subsequently polymerized, water would reform in the system, e.g., in terms of a first step dimerization suggested by Brintzingers' data[2]:

$$2Si(OH)_4 \rightarrow (HO)_2 \overset{O}{Si-O-Si}(OH)_2 + 2H_2O. \tag{2}$$

With hydrolysis and polymerization occuring simultaneously, water content would depend on the relative reaction rates. Assink and Kay[21] recently concluded from their high resolution 'H NMR measurements that hydrolysis and polymerization occur at widely different rates such that the just mentioned 4/1 stoichiometry can produce under sufficiently acidic conditions essentially complete hydrolysis of the TEOS before gellation, whereas changing from acidic to more basic conditions after 40 minutes can produce incomplete hydrolysis in the gel state.

Experimental Observations and Discussion

The following schematic summerizes a set of experiments we performed to demonstrate the effects of acid and base on gellation of the just-described formulation.

A	+	B	→	C
7.0 ml H_2O		21.5 ml TEOS		21.5 ml TEOS
+	+	+		21.5 ml C_2H_5OH
10.75 ml C_2H_5OH		10.75 ml C_2H_5OH	→	7.0 ml H_2O

  Note: A stirred into B

1) A + B $\longrightarrow$ C_1 (pH 5-6) $\xrightarrow{\text{1 day}}$ Clear | All

2) A + B $\xrightarrow[\text{conc. NH}_4\text{OH}]{\text{2 drops}}$ C_2 (pH 8-9) $\xrightarrow{\text{1 day}}$ Cloudy | Very Fluid

3) A + B $\xrightarrow[\text{conc. HCl}]{\text{2 drops}}$ C_3 (pH4) $\xrightarrow{\text{1 day}}$ Clear |

After 10 days:
1) Relatively few rather large particles $\qquad$ C_1 + 1 drop conc.
    suspended and settled out $\qquad\qquad\qquad$ NH_4 OH = Gel
    $\qquad\qquad\qquad\qquad\qquad\qquad\qquad\qquad\qquad$ within 30 min.

Note: Repeated: after only five days, C_1 was clear and the addition of the concentrated NH_4OH produced no change.

2) C_2 + 1 drop conc. HCl :   No change.

3)   Clear C_3 + 1 drop conc. NH$_4$OH → Local cloud of ppt. + 1 drop

       conc. NH$_4$OH   Local cloud of ppt. → gel

Note: While some precipitate settled out as large white aggregates, a gel appeared to grow outward from clouds of particles. Within 60 minutes, a transparent gel mass, with some opaque occlusions at the bottom, had formed in a pool of liquid.

Shifting to acidic pH shortened the aging time required before the base addition caused rapid gellation. Basic conditions at the outset caused the rapid formation of colloidal particles but not gellation. These results are consistent with hydrolysis of the TEOS being catalyzed by both acid and base if both the gel and the colloidal suspension were the results of processes involving Si-OH groups.

Dropwise addition of base to the 10-day aged acidic solution produced colloidial particles that appeared to promote gellation. This is completely consistent with Iler's description of basic conditions causing silicic acid to form colloidal solid particles which then under more acidic conditions cross-link into a gel[18], recognizing that the drop of concentrated NH$_4$OH probably caused a local shift to basic pH which was then gradually shifted back to acid pH by the surrounding liquid.

The addition of acid followed shortly by the addition of base to the TEOS system produced upon each addition a highly exothermic reaction lasting several minutes. Figure 2 understates these heat effects since the reagents were stirred in an uninsulated glass jar at ambient temperatures circa 20°C. The first rise in temperature can be associated with the hydrolysis of the TEOS into silicic acid. The second rise in temperature can be associated with the polyermization of the silicic acid that forms solid particles. In terms of the observed temperature rise both reactions took several minutes. The absence of any observed continuous increase in temperature and in viscosity up to gellation suggests that there were no significant progressive molecular polymerization and macro-molecular cross-linking occurring throughout the bulk of the liquid that lead directly to gellation. Again Iler's description of gellation seems more appropriate; nevertheless, the possibility of gellation by progressive polymerization and macromolecular cross-linking should be checked further by very precise calorimetric and viscometric measurements.

The TEOS gels formed as transparent masses in a pool of low viscosity liquid. They seem to have an open structure so that liquid appears to flow through them. The gels can be described as hard-elastic and somewhat brittle. They do not easily deform and are essentially

111

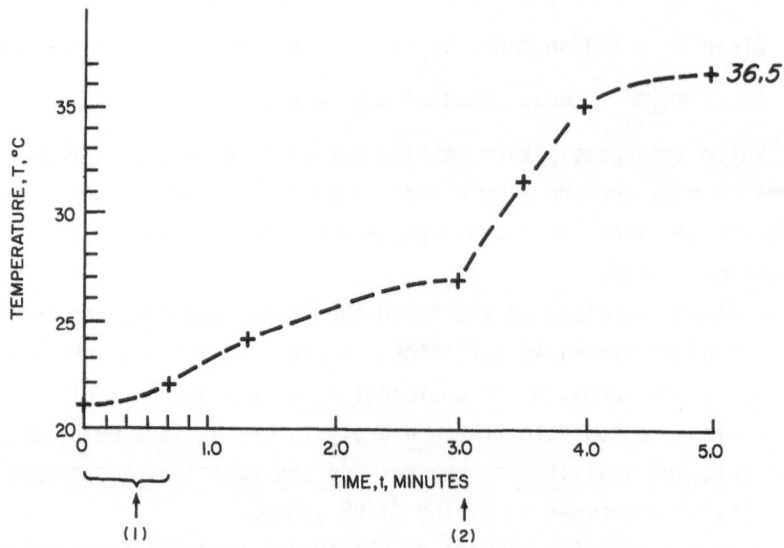

Figure 2. Exothermic reactions. Experimental conditions: (1) stirred A into B. A: 7.0 ml H_2O + 10.7 ml C_2H_5OH + 1 drop concentrated HCl. B: 21.5 ml TEOS + 10.7 ml C_2H_5OH. Continued stirring slowly for a total of 5 minutes. (2) Added 1 drop concentrated NH_4OH after 3 minutes. Additional experimental observations: (1) After the NH_4OH addition, much white precipitate quickly formed and settled out as large aggregates. Smaller particles remained suspended. (2) No apparent viscosity change within the first hour. (3) Eventually gelled.

irreversible. They are rather easily broken-down with a stirring rod or spatula into sharp-edged particles that form a slush in the liquid. This irreversibility suggests that primary chemical bonds are an important factor in gellation.

COLLODIAL PARTICULATE GELS

Background

The gels made by mixing fumed silica into water and into non-aqueous liquids, e.g., the alcohols (methanol, ethanol, propanol), are markedly different from the gels made from alcoholic solutions of TEOS and water. The following physical-chemical description of Cab-O-Sil fumed silica, extracted from the manufacturer's (Cabot Corp.) technical literature, is pertinenet to understanding the gellation process and the properties of the resultant gels.

Burning silicon tetrachloride in a hydrogen-oxygen flame produces fumed silica. Molten spheres of silica form. These primary particles

112

fuse irreversibly while still semi-molten into irregularly extended structures. These aggregates become entangled into larger structures during
further cooling and handling. Thse agglomerates can be disentangled and
dispersed in a suitable liquid. Silicon and oxygen atoms at the surface
of the solid silica particles can each form hydrogen-type bonds, as
bridges between the solid particles and with entities in the surrounding
liquid medium, e.g. to participate in the chaining and cross-linking of
the silica particles into a gel and to adsorb water reversibly. Interparticle siloxane bonds (Si-O-Si), i.e., primary chemical bonds, could
also form to chain and cross-link the silica particles. The 1.46 refractive index of Cab-O-Sil is sufficiently close to that of some solvents,
e.g., glycerine and butyl alcohol, that Cab-O-Sil can be added to these
liquids without loss of clarity. We have made some transparent gels on
this basis.

The rheological properties of liquid systems containing fumed
silica are strongly dependent on such factors as solids loading, degree
of dispersion, and the hydrogen bonding nature of the liquid. The solids
content required for gellation can be minimized or maximized and the drying characteristics of the resultant gel can be significantly changed by
suitable choice of liquid.

Our discussion here will now be limited to 30 wt.% Cab-O-Sil Grade
M-5 in water and in methanol, as representative. The gels are formed by
mixing the colloidal silica into the liquid with a Waring blender. Further procedural detail and experimental results with solids loadings from
20 to 50 wt.% in these and other liquids are available.[16]

Cab-O-Sil Grade M-5 has the following properties: surface area of
200 ± 25 m^2/g, primary particle diameter of 14×10^{-9} m, a bulk density
of 32 kg/m^3, maximum loss in weight on heating at $105^{\circ}C$ of 1.5%. This
weight loss is reversible and represents loosely bound water. The actual
weight loss of a particular sample will depend on its humidity exposure
history. We have dried samples at $110^{\circ}C$ and $600^{\circ}C$ and blended them
into non-aqueous liquids in a glove bag under a dry nitrogen blanket.
These extra drying precautions did not affect the observations of this
present work. The $600^{\circ}C$ dried sample, however, was somewhat more
difficult to mix into the methanol.

Experimental Observations and Discussions

The water and methanol based fumed silica gels are translucent,
with little or no liquid rejection. They can be dried at room temperature
to coherent solid masses. The gels are relatively strong in compression

but weak in shear. Slight hand-stirring is sufficient to re-liquefy the gels, which readily re-gel when the stirring stops. The reversible nature of these gels is also apparent in Figure 3, which gives the rate of viscosity increase after blending in terms of measurements with a Brookfield rotational viscometer. After gellation a concentric ring of gel surrounds

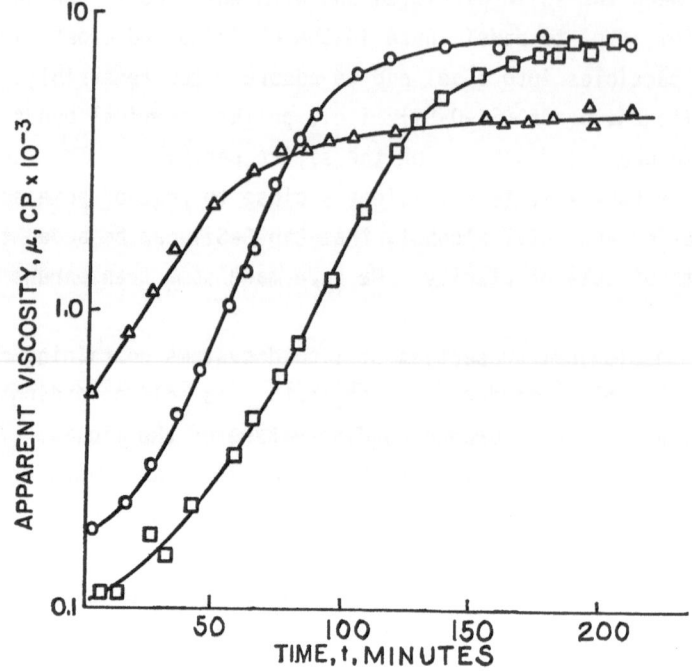

Figure 3. Rheology of fumed silica gellation. Brookfield viscosities at 25.0°C using a No. 4 spindle at 30 rpm. Cab-O-Sil Grade M-5 at 30 wt.% in methanol (Δ) and in deionized water (0) and after reliquefication of the water based gel in a Waring blender (□).

a pool of liquid centered on the rotating spindle. The yield stress of the gel should determine the diameter of this liquid pool. An intercept formed by extrapolating the rising and plateau portions of the curve might be used to define an apparent gellation time. The viscometer was calibrated with sugar solutions, standardizing on container, spindle, and rotational speed. The term "apparent viscosity" is used since this calibration does not fully account for the just described non-Newtonian behavior. The water based gel was reliquefied in the Waring blender. No attempt was made to optimize the conditions of reblending. Differences in the rising portions of the curve could have been caused by different particle morphology and by different thermal history resulting from different frictional resistance or blending. It is noteworthy that the re-gel reached the same viscosity plateau as the original gel. Fumed silica gels over a year old were similarly reliquefied and regelled. In contrast, similarly trying to reliquefy the alkoxide gel resulted in a dry, coarse, granular powder, which turned into a slush upon adding ethanol. This slush did not gel.

SUMMARY AND CONCLUSIONS

The differences observed with the different sol-gel routes are summarized in Table I.

Table I. Comparison of gellation and gels.

I TEOS System	IIA & IIB Fumed Silica System
Overall Process:	
1. Highly Exothermic	Relatively small or nil heat effect
After gellation:	
2. Excess fluid	Little or no fluid
Gel Features:	
3. Hard-Elastic	Soft-Elastic
4. Brittle	Easily deformed
5. Irreversible	Reversible

This present work draws the following conclusions. The fumed silica gels are composed of a network of solid particles. The TEOS gels could also be composed of a network of solid particles or composed of a network of precursors to solid particles, i.e., aggregates of hydrolyzed TEOS oligomers. Experimental evidence in this work supporting this possibility regarding the TEOS gels includes: (1) the short duration of observed exothermic reactions relative to gellation, and (2) the formation of solid particles that seemed required for gellation and that seemed to promote the gellation, under conditions similar to those described by Iler[18] for the gellation of silicic acid via the networking of solid particles. Two possible explanations for the differences in reversibility, i.e., the highly reversibly nature of the fumed silica gels versus the irreversibility of the TEOS gels, are: (1) The fumed silica gels physically adsorb most of the liquid medium to provide an internal lubricant for reliquefication and regelling; (2) The network bonds are substantially different: secondary, hydrogen-type physical bonds for the fumed silica gels and primary chemical bonds for the TEOS gels.

ACKNOWLEDGEMENTS

The author acknowledges with much appreciation the generous encouragement and help extended by Dr. Donald L. Kinser, Professor of Materials Science, and the laboratory work of Vanderbilt student Charles P. Cooper.

REFERENCES

1. T. Graham, J. Chem. Soc., 17, 318 (1894).

2. H. and W. Brintzinger, Z. Anorg. Allg. Chem., 196, 44 (1931).

3. S. S. Kistler, J. Phys. Chem., 36, 52 (1932).

4. S. J. Teichner, G. A. Nicolaon, M. A. Vicarini, G. E. E. Gardes, Adv. Colloid Interface Sci., 5, 245 (1976).

5. H. D. Cogan and C. A. Setterstrom, Chem. Engr. News, 24, No. 18, 2499 (1946).

6. R. Aelion, A. Loebel, F. Eirich, J. Am. Chem. Soc., 72, 5705 (1950).

7. V. Gottardi, Editor, J. Non-Cryst. Solids, 48, 1 (1982).

8. L. L. Hench and D. R. Ulrich, Editors, "Ultrastructure Processing of Ceramics, Glasses, and Composites," John Wiley, New York, 1984.

9. C. J. Brinker, D. E. Clark, D. R. Ulrich, Editors, "Better Ceramics Through Chemistry," Materials Research Society Symposia Proceedings, 32, North Holland, New York, 1984.

10. R. A. Weeks, D. L. Kinser, and G. Kordas, Editors, J. Non-Cryst. Solids, 71, 1 (1985).

11. S. Sakka, in "Treatise on Materials Science and Technology," M. Tomozawa and R. H. Doremus, Editors, Vol. 22, pp. 129-167, Academic Press, New York, 1982.

12. L. C. Klein, The Glass Industry, p. 27, May 1982, and p. 14, January, 1981.

13. C. J. Brinker and G. W. Scherer, J. Non-Cryst. Solids, 70, 301 (1985).

14. J. D. Mackenzie, in reference 8, pp. 15-26.

15. E. M. Rabinovich, D. W. Johnson, Jr., J. B. MacChesney, E. M. Vogel, D. L. Wood, J. Am. Ceramic Soc., 66, 683, 688, 693 (1983).

16. F. J. Bonner, G. Kordas, D. L. Kinser, J. Non-Cryst. Solids, 71, 361 (1985).

17. F. Ehrburger, V. Guérin and J. Lahaye, Colloids and Surfaces, 9, 371 (1984).

18. R. K. Iler, "The Chemistry of Silica," Chapter 3, John Wiley, New York, 1979.

19. S. Sakka and K. Kamiya, J. Non-Cryst. Solids, 48, 31 (1982).

20. L. C. Klein and G. J. Garvey in "Soluble Silicates," J. S. Falcone, Jr., Editor, ACS Symposium Series 194, pp. 293-304, American Chemical Society, Washington, DC, 1982.

21. R. A. Assink and B. D. Kay, in reference 9, pp. 301-312.

PART II. ADHESION OF FILMS AND COATINGS INCLUDING RESISTS

IMPORTANCE OF INTERFACES IN MICROELECTRONIC DEVICE FABRICATION

J.N. Helbert, F.Y. Robb,
B.R. Svechovsky, and N.C. Saha

Motorola SPS
Process Technology Lab and Semiconductor Analytical Lab/SRDL
5005 E. McDowell Rd.
Phoenix, AZ 85008

Microelectronic device fabrication currently relies totally upon photoresist processing for pattern delineation in integrated circuit fabrication. Adhesion of polymeric photoresist patterns, especially those of submicron dimensions, to the various substrates encountered is therefore of paramount importance. Poor photoresist adhesion has been cured by a variety of processes involving (1) double silane promoter applications, (2) plasma treatments, and (3) vapor phase HMDS treatment. Overall, the vapor phase HMDS treatment appears to be the most efficient, and excellent photoresist adhesion test results have been obtained.

INTRODUCTION

Resist adhesion is a major concern in integrated circuit fabrication production, and failures can be intermittent, ambient dependent, and reoccuring in nature. In fact, resist image adhesion failure is a major contributor to production wafer rework rates.

Integrated electronic circuits are built vertically layer by layer[1] on silicon wafers utilizing chemically different layer materials. Obviously, the attainment of patterned masking resist layers on these different substrates requires that the resist images adhere well at the chemically different surfaces. These surfaces include single crystal silicon (both doped and undoped), polycrystalline silicon, silicon oxide of several stoichiometric forms, silicon nitride, metals like aluminum or gold, and high temperature organic polymers like polyimides.

For the devices of these circuits to function, these layers must be patterned into specific geometries such as metal conduction lines, silicon active areas, or polysilicon gate lines.[2] This patterning is accomplished by using e-beam or photosensitive materials,[1] because the substrates themselves are either not directly patternable or they are not economically patterned directly. We[3,4] and others[5,6,7] have demonstrated that these patterns sometimes lift during development and/or wet etching of underlying materials.

In this work, we will describe several resist adhesion promotion techniques which have proven successful in device fabrication at Motorola. We will explain why these methods work by focusing on the chemical or physical nature of the substrate-resist interface.

BACKGROUND
Silane surface chemistry

One of the earliest uses of chemical adhesion promoters in the electronics industry involved the application of hexamethyldisilazane (HMDS).[7] This material is still the workhorse of the industry. HMDS is a silazane, a class of materials which is known to be monofunctionally reactive with OH containing substrate materials.[5,6] Since silicon dioxide surfaces are known to be covered with SiOH surface species called silanols,[8] it is not surprising that HMDS might react with these surface groups as shown:

$$(CH_3)_3Si \, _2 \, NH + 2 \, SiOH \rightarrow 2 \, Si-O-Si(CH_3)_3 + NH_3$$

Verification of this reaction will be provided via ESCA analysis in this work.

The question is why would this reaction render adhesion between the resist and the substrate? The surface methyl blanket resulting from the HMDS reaction above[6] should not in itself create a condition for improved photoresist adhesion. Furthermore, this methyl blanket has been proposed on the basis of surface wetting and thermodynamic studies and had not been verified directly until this work. The photoresist Novalak resin is a polar polymer soluble in aqueous bases, and should not be particularly attracted to the methylated substrate except through weak dispersion forces. This question will be addressed later in this paper.

Other classes of silanes, namely alkoxy and halogenated silanes[3,9,10] are known to react with -OH containing compounds, and therefore, should also function as adhesion promoters or surface modifiers for OH containing substrates. References 3, 9 and 10 describe many of these materials applied to silicon dioxide substrates. As for HMDS treatment, ESCA evidence of these reactions to verify covalent bonding to surface silanol groups will be provided; surface coverage or thicknesses will also be provided by angle resolved ESCA analysis.

According to Plueddemann,[11] only a monolayer of organosilane is theoretically needed to achieve adhesion for glass systems being bonded to polymer resins like resists.

Plasma Treatment

Plasmas are known to affect the surfaces of wafers placed in them.[12] A plasma is a gas under low pressure that is excited by an electric field. Gas ionization and dissociation occurs in a plasma due to electron/molecule and ion/molecule reactions.[12] These reactive plasma species are capable of chemically or physically changing a wafer surface from that of the bulk material. Surface conditions are important to adhesion,[13] and it follows that a plasma treatment could affect adhesion.

Depending on where in the plasma the wafer is placed and what kind of plasma is used, the surface can be changed in three ways: elements can be deposited on the surface,[14] the surface can be cleaned of contaminants,[12] and active sites or broken bonds can be created on the surface layer.[15]

Adhesion changes due to plasma pretreatment were expected, in that both chemical and physical surface changes have been reported as a result of plasma treatments. Physical damage has been shown to be a function of the ion bombardment energy, ion dose, and plasma chemistry.[15,16] Higher ion energies, larger doses, small ions (especially hydrogen and helium) and carbon-containing gases all have been shown to produce more physical damage.

Chemical surface changes have been reported due to plasma exposure.[14,17] Deposition of fluoro- and chlorocarbon polymers is common in the CCl_x and CF_x type discharges, such as those studied here. When using a fluorocarbon plasma for etching, the etch rate depends on the ratio of F/C in the system.[14] As this ratio decreases, the system reaches a point of polymerization, where etching is slowed and polymer is deposited on surfaces.

The composition of a plasma changes throughout the reactor. The electrodes, in an rf discharge, develop a negative charge. Thus, positively charged ions are drawn to and bombard wafers placed there, physically changing their surfaces. Adhesion sites can be added and surface cleaning can occur, through the sputtering of contaminants. As the wafer is moved from the electrodes, it enters the glow region. The glow region is field free, so ions are only weakly accelerated towards the wafer surface (via the floating potential). Therefore, little ion bombardment occurs and mainly chemical processes take place there, as opposed to physical surface modifications.

ESCA Considerations

In recent years, X-ray Photoelectron Spectroscopy (XPS) or Electron Spectroscopy for Chemical Analysis (ESCA) has emerged as a very powerful spectroscopic technique to investigate the chemistry and elemental compositions of the first few atomic layers (less than 50A)   of a wide range of organic and inorganic surfaces. The basic principle of ESCA has been described in great detail in numerous publications.[19] Typically, the samples are irradiated with aluminum K-alpha or magnesium K-alpha x-rays to generate photoelectrons. These photoelectrons are energy analyzed and detected as photoelectron peaks in the ESCA spectrometer. The intensities and the binding energies of such photoelectron peaks contain information about the concentration and chemical environment of the atoms from which the photoelectrons are ejected.

Low radiation damage, minimal surface charging, and high surface sensitivity have made ESCA a particularly suitable technique to investigate polymer surfaces.[20] The surface sensitivity or the analyzing depth (x) in ESCA may be defined by the relationship: $x = 3\lambda \sin \theta$, where λ is the inelastic mean free path (IMFP), and θ is the take-off angle of the photoelectron. The value of λ is dependent upon the material and the kinetic energy of the electrons. In conventional ESCA, λ is less than 30 A. Surface sensitivity is enhanced if data collection is made at low take-off angles. In the angle resolved ESCA instrumentation, the take-off angle θ can be varied from grazing angles of 2° to 90° by tilting the sample with respect to the analyzer.

The wafer surfaces, treated with silane coupling agents like HMDS, may be thought of as substrates with a uniform thin overlayer (0) of thickness d. The photoelectron intensities from the overlayer (I_o^d) and from the substrate (I_s^d) are given by, ($I_o^d = I_o^\alpha [i-e(-d/\lambda \sin \theta)]$ and $I_s^d = I_s^\alpha [e (-d/\lambda \sin \theta)]$, where I_o^o and I_s^o are the respective intensities from the overlayer and substrate material of infinite thickness.

High angular resolution, coupled with the ability of ESCA to distinguish different chemical environments of atoms, provide an excellent opportunity for delineating the chemical nature of any treated wafer surface in a nondestructive way.

Effect of Moisture and Surface Cleanliness

Relative humidity is one of the principal factors controlling the quality of adhesive bonds.[21] Furthermore, Mittal has pointed out that adsorbed water on SiO_2 surfaces is undesirable for photoresist adhesion.[6] SiO_2 surfaces can be dehydrated at temperatures from 200 to $800°C.$[8,21] This is a reversible process, and these surfaces will rehydrate in air. Therefore, dehydrated wafer surfaces should not be exposed to air before applying an adhesion promoter, which is not the case for conventional liquid track priming systems. The Yield Engineering Systems LP-III and the Imtec Star 2000 reactors prevent air contact to the wafer through in-situ wafer dehydration followed by vapor priming, and as a result they should lead to improved adhesion promoter processing. Results for Star 2000 treated wafers will be reported in this work.

Of equal importance to adhesion, is cleanliness of the substrate.[22] Cleanliness cannot be expressed in absolute terms, but what is really important is reproducibility of wafer surface preparation. Surface adhesion processes must create stable and reproducible surfaces to be successful. In general, physical cleaning techniques, such as plasma treatments with a wafer on the cathode, will also make the substrate more reactive or adhering.

Adhesion Definition

What is meant in this work by adhesion? The term is not used here in a structural adhesion sense, but in the practical sense.[6] For resist images, adhesion is merely whether the resist image sticks or "lifts" (also termed "floating image")[6] during development or etch. If it lifts, adhesion failure occurs and a pattern defect will result, and in turn the desired circuit will most probably not function in the desired manner. This practical definition of adhesion represents the force or the work required to effect separation of the adhering patterned layer, and depends upon the molecular interactions at the interface.[23] Since future IC fabrication technology will be concerned with dry etching technology due to design rule shrinkage, only the events at the photoresist/substrate interface and how they are affected by development action will be considered in this work.

Adhesion can be the result of secondary forces such as mechanical, electrostatic (either induced image or dipole), van der Waals (dispersion) or chemical (chemisorption) between the photoresist film and wafer substrate to be patterned. These forces run from weak for van der Waals and possibly mechanical to stronger for chemical or image forces: they range from 2-4 kcal/mole to as high as 10 kcal/mole, respectively. Dipole-dipole interactions may extend 2 - 3 atomic layers from the interface, while dispersion forces and mechanical forces are thought to be of very short range.[21] Mechanical forces can be thought of as those due to increased surface area. Chemical or chemisorption forces are short range and directional in nature and are usually the strongest. In reality, solid layers must be thought of as being held together by a combination of these forces. Recently, van der Waals and acid-base interactions have been hypothesized as the only interactions which are important to adhesion.[23b,c]

EXPERIMENTAL
Wafer Testing

The resists used in this work are all conventional positive photoresists.[1,13] They are all proprietary formulations, but generically they are composed of a mixture of (1) Novalak resins, (2) photoactive components of the diazoquinone type, (3) leveling agents and/or surfactants, and (4) glycol-based spinning solvents. The specific resists tested were Polychrome 129, Hunt 204, KTI-II, Dynachem OFPR 800, and AZ 1350. They were all applied by conventional spinning technique by either an SVG track, or by Headway or Solitec manual spinner systems.

Prior to resist spinning by conventional techniques, the substrates were water rinsed and dried 30 minutes to one hour at 200°C. Wafers utilized for simulated wafer rework testing were cleaned with acetone, followed by 10-30 minutes of O_2 plasma cleaning at 300 watts. The commercially available promoters (Petrarch Systems Inc.) were applied as dilute solutions 0.3 - 7% by weight in acetone or xylene. The photoresists were all cured 30 minutes at 90°C in a filtered dry air convection oven.

A practical experimental approach for testing polymeric resist adhesion during pattern development has been adopted, where the test pattern is one that yields a series of unexposed resist islands of varying dimensions of 0.5 to 5 microns wide and 4 to 20 microns long, surrounded by larger and exposed/developed areas. After development, the "lifted" island images were simply counted under a Zeiss Axiomat microscope. The final relative adhesion improvement was determined as a "normalized" Delta Lift calculated by:

$$\text{Normalized Lift} = \frac{(\# \text{ Lift controls} - \# \text{ Lift treated})}{\# \text{ Lift controls}}$$

Control substrates were not treated with promoters, and usually exhibited 50 - 90% island image "lifting" when adhesion failure occurred (see Figure 1). In some control tests, no lifting occurred, but inferior adhesion was observed at pattern image edges. Adhesion failure in control samples occurred when the lithographic exposure was e-beam (Cambridge EBMF2) or optical (Cobilt 2020H or Ultratech 1000). Simulated rework samples usually exhibited adhesion failure by the second or third rework cycle.

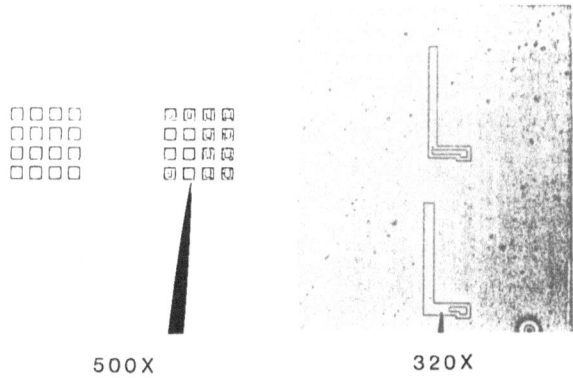

500X   320X

Figure 1. Optical micrographs of "lifted" photoresist test images. The pointers mark areas where island images or parts of them are missing due to adhesion failure following exposure and development.

Test wafers treated with most of the adhesion promoters used in this work were capable of withstanding 30-40% overdevelopment without incurring any test pattern "lifting", and adhesion promotion on both new and reworked substrates was achieved. Control samples, on the other hand, often failed even when the control substrates were underdeveloped, as is the case for the substrate pictured in Figure 1. Underdevelopment is indicated by resist nonclearing around the edges of the written outer squares.

A lot of testing dealt with vapor application of HMDS. These tests were carried out in a Star 2000 marketed by Imtec in Santa Clara, CA. This machine provides an in-situ dehydration bake at $150^{o}C$ in dry nitrogen prior to HMDS vapor priming also at 150°. Thus, contact to air before adhesion promoter application was eliminated.

The Star 2000 is a microprocessor controlled all stainless (316) steel HMDS vapor reaction chamber. This second generation system automatically sequences through the in-situ dehydration bake and vapor HMDS prime cycle; it can also be operated manually or easily programmed with a different process sequence. The Star 2000 gas distribution and vacuum plumbing system is all welded fitting (VCR) construction with air operated Nupro bellows-sealed valves.

ESCA Analysis

ESCA measurements were carried out in a PHI Model 5300 ESCA spectrometer using magnesium K-alpha x-rays. The base pressure in the analyzing chamber was 1×10^{-9} torr or better during analysis. In the PHI Model 5300 system, the variable take-off angle measurements are performed by rotating the sample on an axis through the sample surface. The measurements can be made from grazing angles of 0° to 10° past normal.

In the present investigation, the data were collected normally at 45° and 5° take-off angles. Typically for each sample, a survey scan (1000 - 0) eV was recorded to identify the elements present on the surface. Then high resolution multiplex data were collected over a narrow energy range for individual elements to determine the elemental compositions as well as the binding energies (BE) of the detected atoms. All data collection and subsequent treatment were accomplished with a Perkin-Elmer Series 7000 dedicated computer system with Version D software. The software package contains all essential sub-routines for data treatments such as smoothing, deconvolution, and curvefitting.

The BE values of photoelectron peaks were corrected by using different references for different substrates. For Y58 silicon wafers, binding energies were referenced to the Si2p peak at 99.2 eV, which in turn was referenced to the implanted Ar2p peak at 241.9 eV. For SiO_2 and Si_3N_4 samples, all BE values were referenced to the Ar2p peak at 241.9 eV. To accomplish this, argon was implanted intentionally in selected SiO_2 and Si_3N_4 samples. Thus all binding energies are referenced indirectly to the Ar2p signal at 241.9 eV. The atomic compositions were calculated from peak areas using Phi elemental sensitivity factors.

Plasma Treatments

The equipment used to plasma pretreat surfaces varied. The systems and conditions utilized (power, pressure, gases, and flows) are summarized in Table I. Argon, nitrogen, oxygen, and fluorine based chemistries were studied in the Applied Materials 8110 Hex reactor. A one to two hour O_2 clean, followed by 30 minute inert gas pretreatment, preceded all the inert gas surface treatments.

For comparison, surface pretreatments in low ion-energy CF_4 and argon plasmas (Tegal 400 barrel reactor) and very high ion-energy argon ion-milling (Veeco Micro etch system) were also evaluated.

Table I. Reactor Conditions used and Etch Rates of Polysilicon, Nitride, and Plasma Oxide Substrates. The Etch Rates were Calculated from the Amount of Film Loss in 1-2 min. Runs

Reactor	Power(W)	Press(mTorr)	Gas	Flow(sccm)	Etch Rate (A/min) Poly	P.Oxide	Nitride
Hex	1100	65	CHF_3	75	68	382	0
			O_2	12	110	525	150
			CF_4	75	199	448	650
			N_2	75	0	15	0
			O_2	75	12	7.5	0
			Ar	108	0	12	0
Barrel	100	500	CF_4	20	2.5	6.5	0
Single	100	500	CCl_4	8			
wafer			N_2	8	439	403	713
(13.56MHz)			H_2	2.7			
			Ar	7			
			CCl_4	8			
			N_2	8	1689	596	633
			Ar	7			
Single Wafer	60	800	Cl_2	100	5949	565	1375
(200 kHZ)			Cl_2	65			
			H_2	35	3222	65	835

Arcing was observed for the argon plasma in the hex reactor with the conditions used in Table I. Later optimization studies produced an alternate set of conditions (75 sccm argon, 25 mtorr and 800 Watts), which yielded equivalent adhesion results without arcing.[24] Chlorine chemistry was studied in two single wafer reactors: (1) a laboratory-built parallel plate reactor, powered at 13.56 MHz, and (2) a lower frequency (200 kHz) Perkin-Elmer Cornerstone system.

RESULTS AND DISCUSSION
Single Crystal Silicon

The reason for studying this substrate is one of ease of ESCA analysis and detection sensitivity considerations. Although the IC is built upon a single crystal silicon wafer, this substrate and interface are usually not seen by the patterning photoresist in new MOS fabrication processes; this may not be the case for older product process flows.

A substrate was needed, where detection of carbon and silicon-containing surface species would not be over-shadowed by signals from the bulk species, even at low angles.

A system whose oxide was grown at low temperature, where a high silanol group surface population is known to exist, was also desirable. Oxides grown at higher temperatures, such as thermal oxide, are known to have lower populations of silanols as well as lower silanol group densities.[8]

ESCA survey scans for these samples show the presence of Si, O, and C. The elemental compositions for the samples are shown in Table II. The results do not indicate any change in C/Si ratios following HMDS treatment, liquid or vapor phase. O/Si ratios of HMDS treated samples, particularly after vapor phase treatment, do indicate a detectable lowering of the relative oxygen surface concentration compared to blank wafers.

Figure 2a shows a typical high resolution Cls XPS spectrum recorded for a blank Y58 silicon wafer at 5° take-off angle. Deconvolution after background subtraction, gives three well defined peaks at 285.5 ± 0.1 eV, 287.0 ± 0.1 eV and 289.7 ± 0.1 eV. Cls spectra from all Y58 blank wafers show the presence of these three types of carbon, although relative intensities are found to vary slightly. The peak at 285.5 eV is assigned to a CH_x (hydrocarbon) species, the 287.1 eV peak to $-CH_2O$ (e.g. alcohol or hydroperoxide) species, and the 289.7 eV peak to a $-CO_2$ (carboxylic acid or ester) adsorbed species. These carbon compounds are found on nearly all substrates, and are thought to be adsorbed from the processing ambient.

Table II.   XPS Elemental Compositions from Different Y58 Samples

Sample	Take-off angle	Si	Composition (at %) C	O	C/S	O/Si
Blank Y58 (1)	5°	25.28	21.24	53.48	0.84	2.11
Blank Y58 (2)	5°	21.20	30.49	48.31	1.43	2.28
Blank " (3)	5°	25.21	23.11	51.68	0.92	2.05
Blank 58(cured)	5°	19.78	28.25	51.97	1.43	2.62
HMDS (SVG)Y58 cured (1)	5°	24.65	29.27	46.08	1.18	1.87
" (2)	5°	27.88	23.02	49.15	0.82	1.76
HMDS (SVG)/Y58 No cure	5°	22.49	31.58	45.93	1.40	2.04
HMDS (*2000)/Y58 cured, (1)	5°	28.53	27.42	44.05	0.96	1.54
HMDS (*2000)Y58 cured (2)	5°	27.30	23.43	49.26	0.85	1.80
HMDS (*2000)/Y58 cured, (3)*	5°	31.16	21.88	46.96	0.70	1.50

*sample stored in air for 7 days.

Figs. 2 (b & c) show Cls XPS spectra from HMDS (SVG)/Y58 and HMDS (*2000) treated Y58 silicon wafers. Interestingly, the Cls spectrum for the HMDS (*2000)/Y58 sample consists mainly of a single peak with only a small shoulder at higher binding energy. The B.E. of the main Cls peak is 284.5± 0.1 eV, which is 1.0 eV lower than the adsorbed hydrocarbon peak in the blank wafers. This spectrum is attributed to the $-CH_3$ surface groups resulting from the HMDS treatment, covalently anchored to the surface. We favor covalent bond formation, because overnight storage of the sample in the high vacuum (2 x 10^{-10} torr) of the ESCA analyzer chamber did <u>not</u> reduce the Cls peak intensity. This would not have occurred if the signal was due to just adsorbed HMDS; the signal would not have been as stable.

The binding energies of the two small peaks deconvoluted from Figure 2c correspond to those of adsorbed hydrocarbon, CH_x, and the $-CH_2O-$ type of species found in the blank wafers (Figure 2b). The Cls spectral changes (Figs. 2a and 2c) strongly suggest that the vapor phase HMDS treatment cleans the wafer very efficiently by removing all three types of adsorbed carbon compounds, followed by the spreading of the methyl blanket from the surface HMDS reaction across the substrate. The small amounts of carbon at 285.5 eV and 286.8 eV are probably residues of the original contamination or that of small amounts of readsorbed species. Since the samples were handled in air and spectra recorded at least 24 hours after vapor priming, this process must be considered to be a very good surface stabilizing treatment. Furthermore, the liquid phase reaction of HMDS from the wafer track could not remove the adsorbed species as completely.

Figures 3(a–c) show the XPS spectra resulting from Si2p transitions from the (a) Y58 blank, (b) HMDS (SVG track) and (c) HMDS (*2000)/Y58 wafer, respectively. All spectra were recorded at 5° take–off angle. These spectra clearly indicate the evolution of a new peak between the elemental silicon and the SiO_2 peaks. The growth is very pronounced in the case of the HMDS (*2000)/Y58 treated wafer. The new Si2p peak, arising from HMDS treatment of the wafer, is centered at 101.8 eV (see

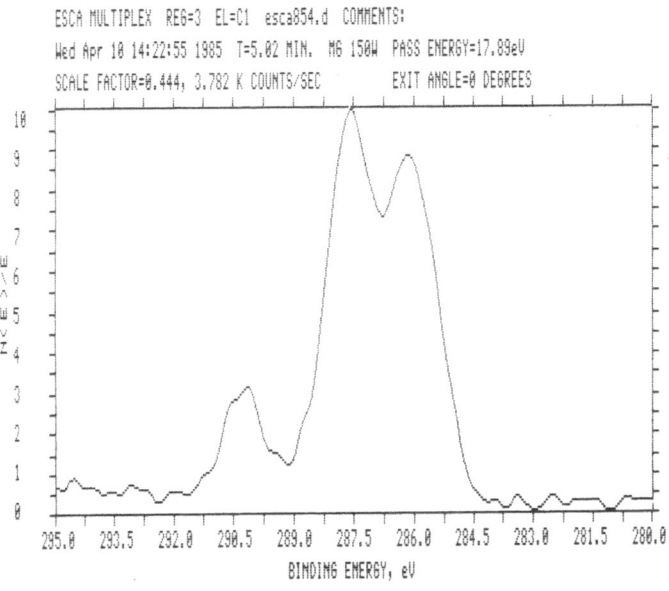

Figure 2a.   Cls XPS spectrum for blank silicon wafer (Y58).

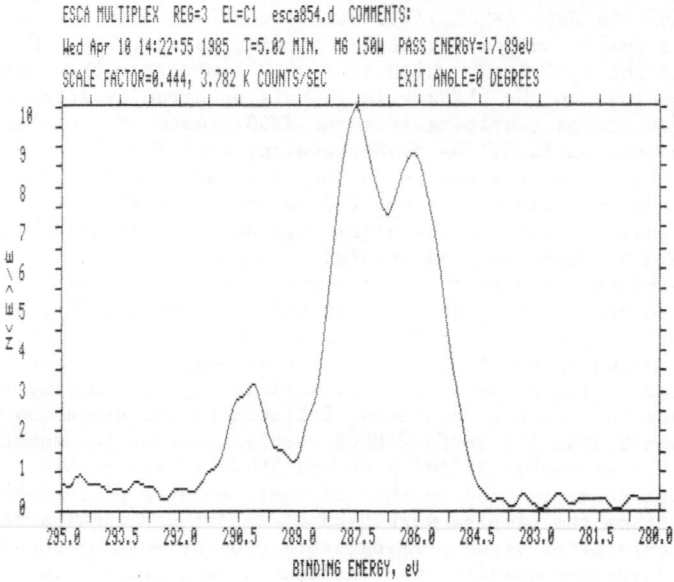

Figure 2b. C1s XPS spectrum for SVG conventional liquid HMDS treated silicon wafer.

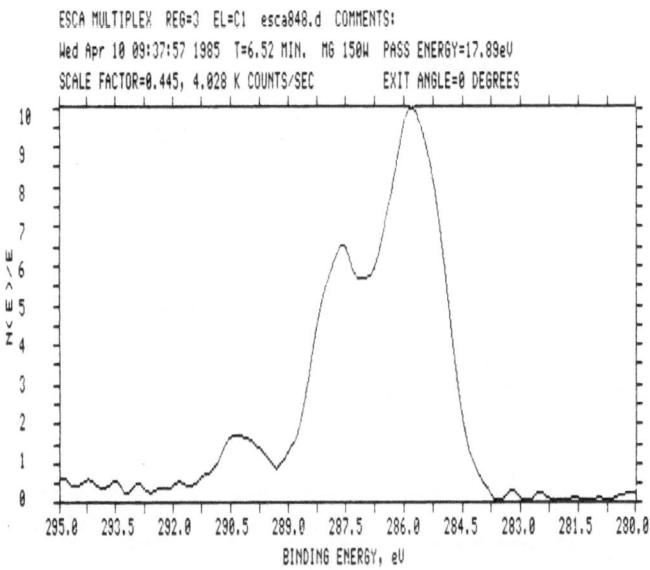

Figure 2c. C1s spectrum for Star 2000 vapor HMDS treated silicon wafer.

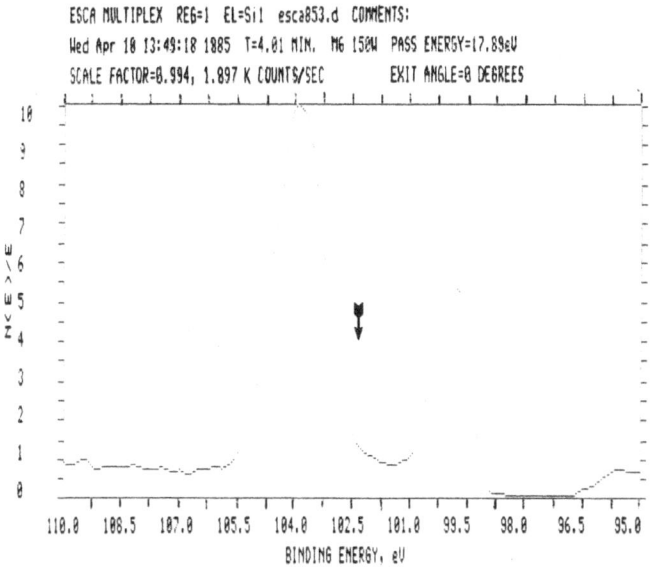

Figure 3a. XPS spectra for Si2p transitions from Y58 blank.

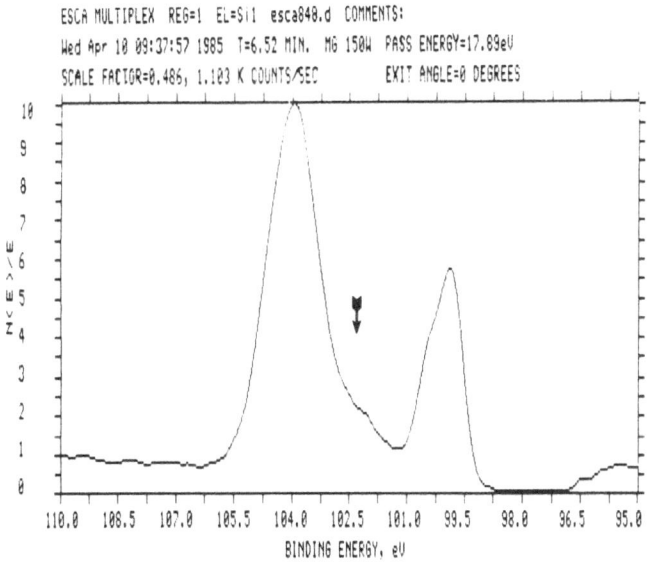

Figure 3b. XPS spectra for Si2p transitions from HMDS liquid treated on SVG Track Y58.

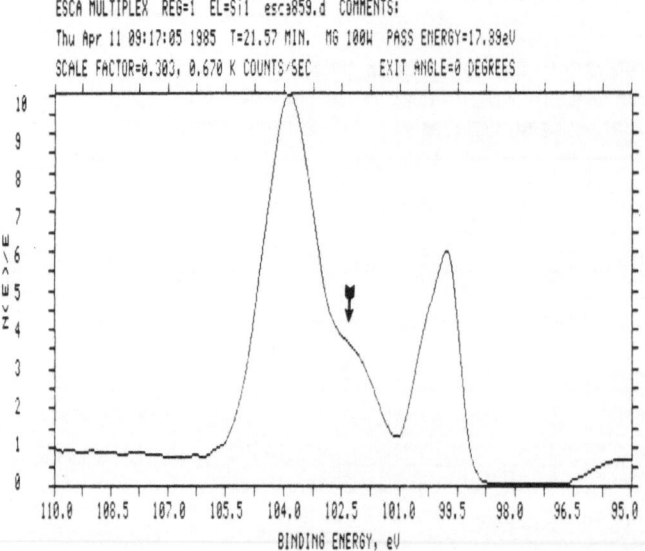

Figure 3c.   XPS spectra for Si2p transitions from Star 2000 vapor primed Y58 wafer.

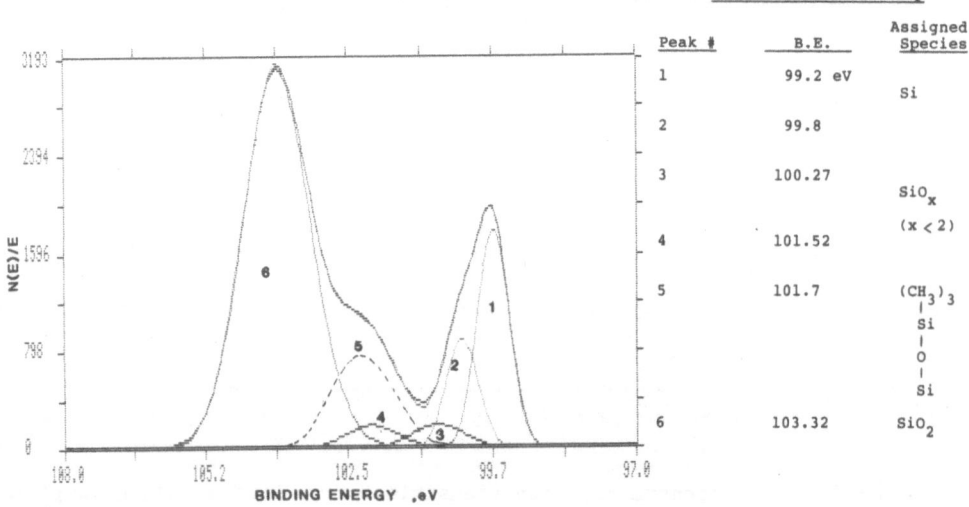

Peak #	B.E.	Assigned Species
1	99.2 eV	Si
2	99.8	
3	100.27	SiO_x
4	101.52	$(x < 2)$
5	101.7	$(CH_3)_3$ Si O Si
6	103.32	SiO_2

Corrected B.E. of Si2p

Figure 4.   Deconvoluted Si2p XPS spectrum for Star 2000 treated Y58 silicon wafer.

Figure 4). The other five peaks present in the spectrum of the blank wafer are assigned to Si^o, SiO, Si_2O_3 and SiO_2 species.[25] The new peak at 101.8 ± 0.1 eV is assigned to $(CH_3)_3$Si-O- type of Si species formed on the surface due to the HMDS reaction. The presence of this new Si2p peak is consistent with the earlier interpretation of the C1s XPS data.

XPS spectra resulting from O1s transitions are relatively less informative, and are very similar in all three types of samples. Each spectrum is fitted with two peaks, a main peak at 532.7 ± 0.1 eV assigned to O1s from SiO_2 and a smaller peak centered at 533.5 ± 0.1 eV. This is not surprising because most oxygen functional groups have B.E. close to 532.0 eV, except the carboxylic compounds, where the B.E. is centered at 533.6 eV. The area under the peak at 533.6 eV is 22.0% of the main peak in the case of the Y58 blanks and HMDS (SVG)/Y58 treated wafers, where the adhesion promotion reaction is known to be less efficient. In the case of HMDS (*2000)/Y58 treated wafer, it is only $\approx$ 10.0%. The peak at 533.5 eV may be assigned to the adsorbed carboxyl species. This assignment, although speculative, supports the proposed surface modification scheme based upon the carbon and silicon ESCA data. More importantly, the *2000 treated wafer is less susceptible to surface recontamination than the liquid promoted and blank wafers. In addition, these treated substrates should also be less susceptible to water adsorption due to the increased hydrophobicity of the wafer.

SiO_2 Substrates

Since SiO_2 substrates appear frequently during IC fabrication, the adhesion test results for this substrate are important. Four types of oxides have been extensively tested. They are (1) thermal oxide grown at 1100^oC, (2) softer oxides processed by conventional spin-on-glass technology, (3) phosphorus doped LPCVD oxide, and (4) low temperature (200^oC) plasma deposited oxide.

Adhesion has been achieved on these oxides through a variety of processes. Conventional liquid phase application of HMDS, however, was not adequate for the latter three substrates listed above; it did provide adequate photoresist adhesion for thermal oxides, however. For the last three substrates, a double adhesion promoter process was needed and developed. This process has been incorporated into actual device fabrication processing.

The double promoter process involves the successive application of liquid promoter solutions of vinyltrichlorosilane (VTS) and 3-chloropropyltrimethoxysilane, followed by successive cure cycles in dry N_2 at 90^oC before photoresist application. The double promoter process evolved, because complete labelling of SiOH surface groups of low temperature oxides was felt not to occur in a single promoter application. Interestingly, a double HMDS liquid promoter process failed to yield adequate adhesion as well.

Later in time, the successful but somewhat complex double promoter process was replaced by the "vapor phase" HMDS process in the Star 2000; then superior resist image adhesion was obtained on all four oxide substrates with all the photoresists tested.

The double promoter process is felt to approach the same quality of surface preparation as that for the vapor HMDS treatment through improved surface cleaning capability. Furthermore, it must be concluded that the in-situ dehydration baking and vapor phase processing is of great importance. If this were not so, then suitable adhesion should have been achieved on SiO_2 substrates by a simple double HMDS (liquid) treatment.

What are the mechanisms of these two successful promoter processes? To answer this question, ESCA analysis was employed.

ESCA survey scans recorded for thermal oxide surfaces indicate the presence of Si, C and O. The elemental compositions for those oxides are similar to those for the Y58 blank system of Table II. Again, the simple elemental compositions do not show any dramatic change due to the surface chemical modifications by the HMDS treatment, except for the decrease in O/Si ratio. But, the ESCA data do show that the chemical nature of these elements changes significantly, i.e., C/O containing molecular impurities are being replaced by a covalently bonded Si/O/C containing stabilization layer. As seen for Y58 wafers, XPS spectra resulting from Si2p transitions for HMDS treated oxide wafers also depict the evolution of a new peak at 101.8 ± 0.1 eV. Again, this peak is most prominent in the Star 2000 HMDS treated oxide sample, and is assigned to the Si2p peak present on the oxide surface due to $(CH_3)_3$ Si-O- surface species.

C1s XPS spectra for the treated surfaces are not well resolved. From the deconvoluted spectra, the decreases in the main contamination peak at 284.8 eV, and the other two peaks at 1.7 eV and 4.0 eV higher B.E. can be followed. The intensities of these peaks are notably much lower in the oxide samples compared to those of Y58 wafers, consistent with the lower density of surface silanols or contamination adsorption sites between the two surfaces. After vapor phase HMDS treatment, the contribution of these peaks is greatly reduced and a new main C1s peak centered at 284.6 eV appears as for the Y58 samples, which is assigned to the $-CH_3$ group, due to the HMDS stabilization reaction.

When conventionally applied VTS is used, adhesion is increased to SiO_2 substrates, and no trace of Cl remains at the surface. ESCA results for VTS treated PSG oxides show reproducible increases in carbon concentration as expected from $-Si-O-Si-(CH=CH_2)$ surface reaction products. Consistent with this hypothesis, broadening of the ESCA Si2p peak at lower B.E. is also observed. The broadened Si2p spectrum can be simulated by two Gaussian curves with 1.8 eV FWHM centered at 103.0 and 103.7 BE. The 103.7 eV peak would be the same as that observed for the blank with the second peak at 103.0 eV attributed to the Si product of the VTS surface reaction.

ESCA results for double promoted oxides are less definitive than those previously described, but definitely show differences in the Si2p spectra, hence, indicating that again surface changes are occurring due to the irreversible surface chemical reactions. At the same time, these oxide surfaces are being at least partially cleaned of C/O containing atmospheric contaminants.

Polysilicon Substrates

This substrate is usually encountered during MOS device fabrication for defining gate structures. For submicron CMOS processing, adhesion problems encountered with this substrate are even more severe.

Here again, standard liquid HMDS promoter processing fails to be adequate. Historically, double promoter processes were needed as found for the troublesome low temperature oxides. Great care must be exhibited here, however, because oleophobicity is often created if the wrong two promoters are selected resulting in dewetting of the treated substrate by the photoresist spinning solution. This often occurs, even though photoresist solutions contain surfactants to aid in wetting. This more "black art" processing involved the use of VTS again but followed this time by the use of nitrogen-containing silanes, such as 1, 3-divinyltetramethyldisilazane (6208) or HMDS- and in this order.

Again, double liquid treatments of HMDS did not prove adequate for poly substrates. As for the troublesome oxide surfaces, vapor phase HMDS also works for the poly substrate. Thus the double promoter process utilizing somewhat "poor handling" promoters for poly can be replaced with the vapor phase HMDS process.

ESCA results of treated poly substrates indicate that the poly surface, like that of the blank single crystalline Si wafer, is covered with a thin layer of SiO_2 (<20A). For this substrate, surface layer changes are detected by observing the SiO_x/Si ratio versus that of the blank. VTS/HMDS and VTS/6208 treated surfaces have increased SiO_x/Si ratios of 6.8 - 7.0, while the blank exhibits values of 2.5 - 2.6. Increases also concomitantly occur in the carbon concentration. The increases in SiO_x/Si ratio and carbon content after surface treatment is interpreted as due to the formation of an additional surface layer through the reaction of the organosilanes and surface-SiOH groups. If a thicker layer had resulted, no Si^0 would have been detected, because it would have been out of the ESCA detection depth, therefore, the reactions are restricted primarily to the first few surface layers.

It is not too surprising that vapor phase HMDS also improves adhesion to this substrate. The native oxide has been shown on Y58 wafers to be easily treated and/or passivated. Actual resist image "lift" testing on vapor promoted poly wafers produced superior results and no image lifting occurred even for first generation resists known to be susceptible to "lifting."

Silicon Nitride Substrates

Since this substrate is also used extensively in device fabrication, it was also tested for resist adhesion. Photoresist adhesion to silicon nitride was found to be generally poorer than that for oxides. In fact, IBM researchers[26] developed a O_2 plasma process specifically to combat adhesion problems associated with that substrate. This O_2 plasma treatment renders the surface more SiO_x rich; then SiO_x double promoter processes and even the liquid HMDS treatments become somewhat successful at promoting photoresist adhesion. We have found that this processing can be eliminated by using just vapor phase HMDS treatment.

Unfortunately, the interpretation of ESCA data for silicon nitride surfaces is rather complex due to overlapping Si2p signals. They do indicate, however, that the Si_3N_4 wafer surface is contaminated with C and SiO_x and SiO_xN_y compounds. Cls ESCA data show that vapor phase HMDS treatment clearly removes carbon contamination as in the case of Y58 substrates, although it is not as efficient in this case. Liquid phase track HMDS treatments, on the other hand, were ineffective in cleaning nitride surfaces, and this is probably why "lifting" was observed so frequently for this substrate.

Gold and Aluminum Substrates

These substrates are used in conventional metallization fabrication processing. For aluminum and gold, adhesion failure is somewhat less frequent than for the other substrates; actually, adhesion failure could not be achieved for aluminum testing. In fact, no promoter is usually required for these layers even though they are frequently used. "Lifting" with these substrates usually occurs when the device wafers are reworked due to photoresist or other lithographic problems. Reworked wafers exhibit statistically poorer adhesion,[5] and a promoter is usually needed to insure good adhesion confidence.

Good adhesion has been achieved for gold substrates,[4] using chelating type adhesion promoters. Chemisorption of carbon monoxide(CO), adsorption with stronger and more directional bonding than that for van der Waals attraction forces, are well-known for metal surfaces like gold.[21] Sellwood has shown that the chemisorbed CO aligns on metal surfaces with the C towards the substrate and the oxygen pointing outward.[27] Therefore, trimethylsilylacetamide (T3250), a carbonyl containing silane, is very likely to adsorb and align on the Au surface similar to that observed for CO the model compound. The possibility of electrostatic adhesion forces existing between the C-O dipole and that of the polar Novolak resin is also very likely.

Although no adhesion failures were observed for aluminum substrates as found for gold with a variety of liquid silanes, VTS and vapor phase HMDS treated aluminum wafers were analyzed via ESCA to determine if surface changes occurred with these treatments. For VTS treated substrates, a dramatic 20X increase in the Si/Al ratio occurred with a concurrent decrease in carbon surface concentration. These findings support the hypothesis of a covalently bonded $-O-Si-CH=CH_2$ species 5-10A in thickness across the wafer. ESCA results for HMDS treated Al wafers were void of any trace of surface reaction product. Therefore, HMDS is a less effective surface modifier than VTS.

Gold coated substrates treated with these silanes were analyzed. First of all, the three promoters yielded differing layer thicknesses on the wafers. For T3250, the promoted layer was <30A or 1-2 layers in thickness. For 2-(diphenylphosphino)ethyltriethoxysilane (6110) and 3-mercaptopropyltriethoxysilane (8502), layers >>50A to <1000A were observed. All three promoters were successful in promoting resist adhesion, but the latter promoters are not recommended unless R.I.E. or ion-milling etch processes are to be used. Due to their thickness, these layers may create a masking problem for plating or wet etching processing applications.

For T3250 processed wafers, ESCA shows an increase in Si and oxygen surface concentrations with a concurrent loss in carbon concentration. Therefore, carbon surface contamination on the gold is being replaced with the chemisorbed primer species. HMDS treated Au surfaces exhibited no such change in surface chemical nature, consistent with negative adhesion "lifting" test results for that promoter.

The latter thicker adhesion layers are most likely polysiloxane condensation polymer layers. Photoresist adhesion to these layers may be electrostatic in nature, while the adhesion of the promoter layers to the gold substrate may be the result of chemisorption forces between the chelating moieties of the promoter layer and the gold surface atoms.

Plasma Treatments

Improved adhesion was achieved by plasma treatment for two different substrates, LPCVD Si_3N_4 and plasma deposited SiO_2. No improvement was obtained for plasma treated polysilicon substrates, but adhesion was not adversely affected either. Plasma treatments were also found to be successful in restoring overpromoted wafers, wafers doubly promoted or overpromoted by accident to the point of photoresist solution non-wetting (i.e., oleophobicity), to their original state. This process prevents the scraping of valuable wafers.

Wafers with the fluorine chemistry based plasma treatments exhibited erratic image "lifting" results (see Figure 5). Various chemistries (CF_4, CHF_3+O_2, and CHF_3) were compared in an attempt to define the effect of the F/C ratio on adhesion. The F/C ratio has been shown to control the amount of polymer deposition.[14] CF_4, CHF_3/O_2, and CHF_3, have decreasing F/C ratios and would, therefore, produce increasing amounts of CF_x polymer on the surface. The adhesion results obtained for the three systems, however, showed no discernable trends. In addition, neither exposure time, nor substrate type (see Figure 5) nor ion bombardment energy had any reproducible effect on adhesion.

The mixed results with these fluorine chemistries could be due to competition between two mechanisms: physical damage/roughness and polymer

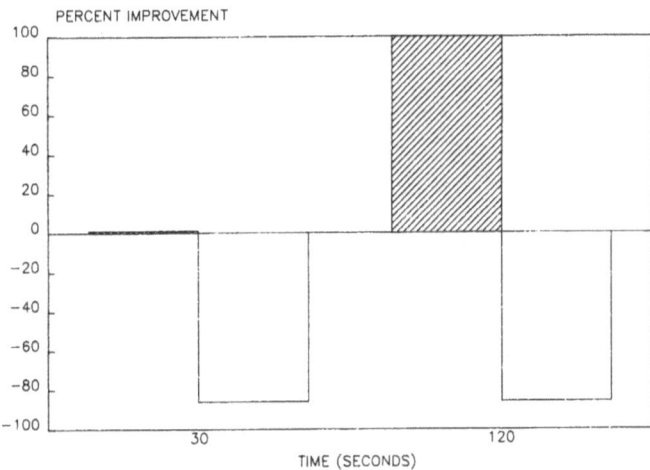

Figure 5. "Lifting" data for fluorine plasma treated wafers.

formation. Saturated CF_2 layers have been shown to decrease the surface adhesion due to decreased surface energy or low surface tension.[23] Ion bombardment, however, would break surface bonds causing an "incomplete" polymer passivation layer which could quite conceivably increase adhesion. If irreproducible CF_2 layers are formed, then adhesion "lifting" results would become mixed as was experimentally observed.

The chlorine plasma produced very little adhesion improvement (see Figure 6). This was somewhat surprising as CCl_2 layers have been shown to enhance adhesion in certain situations and to possess surface tension values of several times greater than that for CF_2 layers.[28]

Factors that may be contributing to the adhesion degradation in these chlorine-based plasmas (as compared to the other plasmas) are as follows:

(1) The ion energy in the single wafer systems may be less than with the hex reactor, although this is not clear.

(2) The Cl ion is larger in size than the F ion so that its damage may not be as deep. In addition, Si-Cl dangling surface bonds may produce more steric hindrance than Si-F bonds.

(3) CCl_x polymers may better "passivate" the underlying surfaces, minimizing damage.

The "inert" gas (N_2 and Ar) and O_2 plasma pretreatments all produced excellent adhesion, much better than either the fluorine or chlorine based plasmas (see Figure 7).

Ion-energy was shown to be a primary adhesion controlling factor via a comparison of argon plasma pretreatments in the barrel reactor, hex, and ion-miller. Exposure times in each treatment were calculated from estimations for equal ion doses. The normalized lifting was 0% (everything lifted) for the barrel and 100% (nothing lifted) for both the hex and the ion-miller. Detailed inspection of the wafers revealed, however, that although none of the islands lifted from the hex treated samples, some deformation at the pattern edges could be found as compared to the images of ion-milled treated wafers.

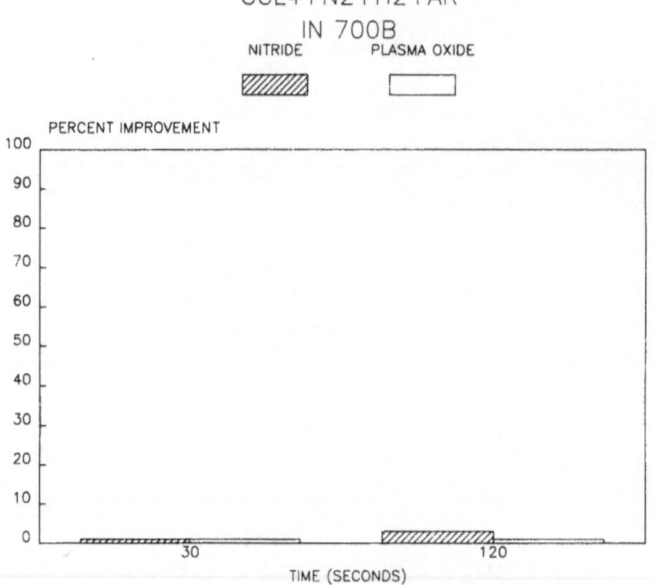

Figure 6. "Lifting" data for chlorine plasma treated wafers.

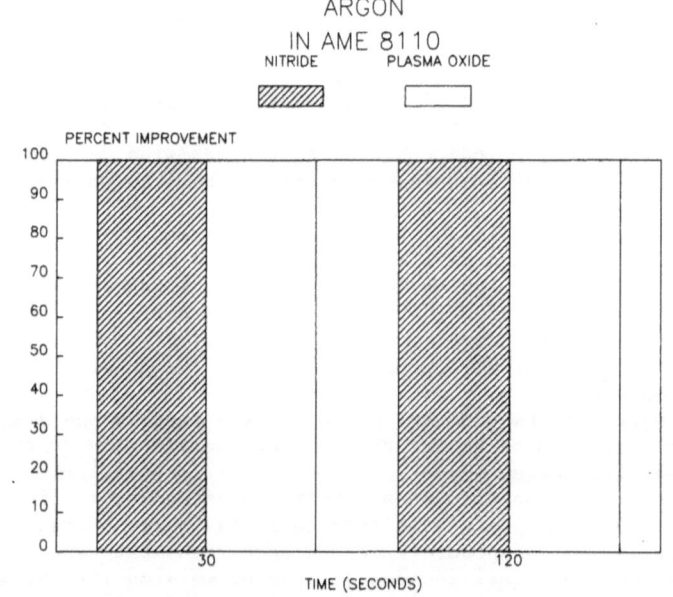

Figure 7. "Lifting" data for Argon plasma treated wafers.

The ion bombardment energy, which is proportional to purely physical damage, was not directly measurable in these systems. The ion-miller, however, would produce the highest ion-bombardment energies of the systems studied, while the barrel reactor would produce the lowest. The hex (and the single wafer reactors) would lie somewhere in between.

To help clarify the plasma treated wafer results, ESCA analysis of Ar and CHF_3/O_2 treated nitride wafers was carried out. The CHF_3/O_2 treatment left a definite polymer layer (CF_x), while the Ar RIE treatment created a surface with increased carbon and oxygen concentrations vs. untreated control wafers. The higher carbon surface concentration is most likely due to sputtered material from the hex reactor cathode trays. The higher oxygen concentrations may result from the hex reactor or may be the product of resputtered C and O containing surface impurities, which have been previously identified for nearly every substrate. Surface scanning electron micrographs of Ar treated wafers revealed no signficant differences in the surface roughness of treated and untreated surfaces, meaning any roughness must be on the order of 200 A or less. A very small amount of F was also found on the argon treated surface and attributed to the fact that the hex system was routinely used for oxide etching in a fluorine chemistry.

These results suggested that the mechanism for plasma adhesion promotion was different than for conventionally used methods, but further ESCA work on Ar treated substrates was felt to be needed.

ESCA analysis of Ar treated Y58 substrates provides an interesting perspective on Ar plasma treatments. It is a substrate where a lot of data exist, and the individual Si2p and C1s signals are usually well resolved. Ar RIE treated samples are striking in that they have a thicker than normal (i.e.>>50A) native oxide layer. The oxide layer is also chemically different from native oxide layers with a higher than normal concentration of intermediate SiO_x compounds. As found for Ar treated Si_3N_4 wafers, a higher concentration of surface carbon occurs with Y58 samples as it should since the source of carbon is from the reactor. Interestingly, no detectable Ar concentration is found in these wafers; this is not the case for Ar ion-milled samples where concentrations as high as 3% can be detected.

The Ar RIE treatment is creating an oxide layer thicker than the native oxide on the Y58 and most likely on the Si_3N_4 wafers as well. This is a similar result to that of the IBM O_2 plasma treatment of Si_3N_4 wafers, i.e., both provide a more promotable oxide layer upon the difficult nitride coated wafer. To the resist, the nitride wafer now looks like an oxide wafer. The fact that SiO_2 promoter processes would work for the Ar treated surfaces is therefore not too surprising.

If longer range dipole (electrostatic) forces contribute to the overall adhesion forces as hypothesized, then an oxide wafer would have the greater force of adhesion due to the greater difference in electronegativities between oxygen and nitrogen. Thus, the polar resin of the photoresist would prefer to adhere to promoted or stabilized oxide surfaces than those of Si_3N_4, consistent with experimental results. Remember that the carbon surface concentration is also higher after Ar RIE treatment, but instead of this layer being molecular in nature as found for atmospheric impurities, it is a polymerized layer capable of acting as a passivating or stabilizing layer as hypothesized for other substrate systems. This interpretation is consistent with the dramatic improvements in adhesion that occurred for Ar treated nitride and oxide substrates (see Figure 7).

Photoresist and Processing Effects

Although the photoresists employed were generically similar in formulation composition, they exhibited significantly different adhesion image "lifting" results. The number of resists tested had to be restricted due to the large numbers involved, but data for a representative number of first and second generation resists was assembled. A large amount of data for adhesion failures observed in actual device fabrication was also assembled to create the whole data base.

The majority of the results are for standard track applied liquid promoter processing. "Lifting" was observed for first generation resists, like PC 129 and KTI II, on Si_3N_4 and SiO_2 substrates. On polysilicon, resist image lifting for resists like PC 129 and AZ 1450 was observed. Hunt 204 images have lifted on oxide surfaces, and in one case, not even double promoter processing could solve the problem. The representative second generation resist tested, OFPR-800, did not suffer from "lifting" even on blank wafers, although image edge "lifting" or poor image edge acuity did occur for images on control wafers.

When vapor priming was used on the Star 2000, no "lifting" occurred for any of the resists on the "lifting-susceptible" substrates tested. This dramatic result must be attributed to the process of that system. The in-situ dehydration bake of that system is far superior to that of older processing. There, the wafer was (1) dehydration baked in dry-N_2 convection ovens, (2) cooled in air, and (3) track adhesion promoted and resist coated in fab area ambient. Obviously, the wafers could be rehydrated or surface contaminated in the older processing scheme. The plasma treated or Star 2000 treated wafers, could be stored in ambient for 1-3 days without image "lifting" occurring, therefore, these processes stabilized the wafer surfaces.

CONCLUSIONS

Superior resist adhesion or resistance to resist image "lifting" has been achieved by (1) conventional double chemical treatments, (2) plasma treatments, and (3) vapor phase HMDS treatments. The different and successful adhesion processes have one point of commonality. They all dehydrate, clean, and chemically stabilize or passivate the surface requiring photoresist adhesion in some way. Furthermore, the improved adhesion observed must be the result of stronger forces than simple short range van der Waals forces. Electrostatic forces extending farther than the first monolayer at the interface must be hypothesized to be occurring for these systems as well.[29] Of course, acid-base forces of attraction cannot be ruled out either.

REFERENCES

1. D.W. Hess, Chemtech, 432 (July 1979).

2. R.S. Muller and T.I. Kamins, "Device Electronics for Integrated Circuits," John Wiley & Sons, New York, 1977.

3. J.N. Helbert and H.G. Hughes, in "Adhesion Aspects of Polymeric Coatings," K.L. Mittal, editor, 499, Plenum Press, New York, 1983.

4. J.N. Helbert, J. Electrochem. Soc., 131, 451 (Feb. 1984).

5. C.A. Deckert and D.A. Peters, in "Adhesion Aspects of Polymeric Coatings," K.L. Mittal, editor, 469, Plenum Press, New York, 1983; in "Proceedings of the 1977 Kodak Microelectronics Seminar," 13 (1977); Circuits Manuf. April, 1979.

6. K.L. Mittal, Solid State Technol., 89 (May 1979).

7. R.H. Collins and F.T. Deverse, U.S. Patent 3,549,368 (1970).

8. B.E. Wagner, J.N. Helbert, E.H. Poindexter and R.D. Bates, Surface Sci., 67, 251 (1977), and references therein.

9. C.G. Armistead, A.J. Tyler, F.H. Hambleton, S.A. Mitchell, and J.A. Hochey, J. Phys. Chem., 73, 3947 (1969).

10. R.L. Kaas and J.L. Kardos, Polymer Eng. Sci., 11, 11 (1971).

11. E.P. Plueddemann, J. Adhesion, 2, 184 (July 1970).

12. B. Chapman, "Glow Discharge Processes," John Wiley and Sons, New York, 1980.

13. W.S. DeForest, "Photoresist Materials and Processing," McGraw-Hill, New York, 1975.

14. J.W. Coburn and H.F. Winters, J. Vac. Sci. Technol., 16, 394 (March 1979).

15. S.W. Pang, D.D. Rathman, D.J. Silversmith, R.W. Mountain, and P.D. DeGraff, J. Appl. Phys., 54, 3272 (1983).

16. R.G. Frieser, F.J. Montillo, N.B. Zingerman, W.K. Chu, and S.R. Mader, J. Electrochem. Soc., 130, 2237 (1983).

17. A.G. Nagy and D.W. Hess, J. Electrochem. Soc., 129, 2530 (1982).

18. C.B. Zarowin, Solid State Technol., 1148 (May 1983).

19. K. Siegbahn, C. Nordling, G. Johansson, J. Hedman, P.F. Heden, K. Hamrin, V. Gelius, T. Bergmark, L.O. Werme, R. Manne, and Y. Baer, "ESCA Applied to Free Molecules "(North Holland, Amsterdam, 1969), T.A. Carlson, "Photoelectron and Auger Spectroscopy," Plenum Press, New York, 1975.

20. D.T. Clark, in "Advances in Polymer Science," H.J. Cantow, editor, Springer Verlag, Berlin, 24, 125 (1977); D.T. Clark, in "Physicochemical Aspects of Polymer Surfaces," K.L. Mittal, editor, Vol. 1,p. 3, Plenum Press, New York, 1983.

21. N.A. Debruyne, in "Adhesion and Adhesives," R. Houwink and G. Salomon, editors, Chap. 1, Elsevier, Amsterdam, 1965.

22. K.L. Mittal, in "Surface Contamination: Genesis, Detection and Control", K.L. Mittal, editor, Vol. 1,p. 3, Plenum Press New York, 1979.

23a. K.L. Mittal, J. Vac. Sci. Technol., 13, 19 (Jan. 1976).

   b. K.L. Mittal, Pure Appl. Chem., 52, 1295 (1980).

   c. F.M. Fowkes, in "Physicochemical Aspects of Polymer Surfaces," K.L. Mittal, editor, Vol. 2,p.583,Plenum Press, New York, 1983.

24. F.Y. Robb, in "Proc. 5th Symp. Plasma Processing," Vol. 85-1,p. 1, The Electrochem. Soc., (1985).

25. F.J. Grunthaner, P.J. Grunthaner, R.P. Vasquez, B.F. Lewis, and J. Maserjian, J. Vac. Sci. Technol., 16 (5), 1443 (1979).

26. M.R. Gulett, M.L. Trudel, and K. Stewart, US Patent 4,330,569, 1982.

27. P.W. Sellwood, "Adsorption and Collective Paramagnetism," Academic Press, New York, 1962.

28. Kirk-Othmer Encyclopedia of Chemical Technology, 2nd edn.,p. 707, Interscience, New York, 1966.

29. H. Yanazawa, Colloids Surfaces, 9, No. 2, 33 (March 1984).

ROLE OF SILANES IN POLYMER-POLYMER ADHESION

Edwin P. Plueddemann

Dow Corning Corporation
Midland, Michigan 48640-0995

Organofunctional silanes are generally recommended as
"coupling agents" in bonding organic polymers to mineral
substrates. Certain silane primers that are effective in
bonding two dissimilar polymers to glass are also effective in
bonding the two polymers to each other. The higher melting
polymer may be coated with a silane primer and contacted with
molten lower-melting polymer. In some cases the silane
adhesion promoter may be added to the lower melting polymer to
obtain unprimed adhesion when fused against the other polymer.
Two very dissimilar polymers (polyethylene and polybutylene
terephthalate) may each be primed with appropriate silane
primers and bonded by fusing a third polymer between them.
Higher melting polymers may be fibers, films, or vulcanized
elastomers including silicones. The mechanism of adhesion is
believed to be formation of interpenetrating polymer networks
between the primer and the other polymer phases.

I. INTRODUCTION

Silane coupling agents are very effective in bonding organic
polymers to mineral fillers and reinforcements[1]. By selecting an
organofunctional silane for optimum bonding of a given polymer it is now
possible to bond virtually any thermosetting or thermoplastic polymer to
glass and other minerals. Some typical silane coupling agents that may
also be useful in polymer to polymer bonding are shown in Table I.

TABLE I. Typical Organofunctional Silanes for Organic to Organic
Bonding

No.	Organofunctional Group	Chemical Structure	Trade Name
A.	Epoxy	$CH_2\text{-}CHCH_2OCH_2CH_2CH_2Si(OCH_3)_3$ (with epoxide O)	DC® Z-6040
B.	Methacrylate	$CH_2\text{=}\overset{CH_3}{C}\text{-}CO\text{-}O\text{-}CH_2CH_2CH_2Si(OCH_3)_3$	DC® Z-6030
C.	Diamine	$H_2NCH_2CH_2NHCH_2CH_2CH_2Si(OCH_3)_3$	DC® Z-6020
D.	Vinylbenzyl cationic	$V.B.\overset{HCl}{NHCH_2}CH_2NHCH_2CH_2CH_2Si(OCH_3)_3$	DC® Z-6032

In the fiberglass industry, silane coupling agents are
pre-hydrolyzed and applied to glass from dilute aqueous solutions.
Upon drying, the silanols of the coupling agent condense with silanols
of the glass to form covalent siloxane bonds with the surface as
outlined in Figure 1.

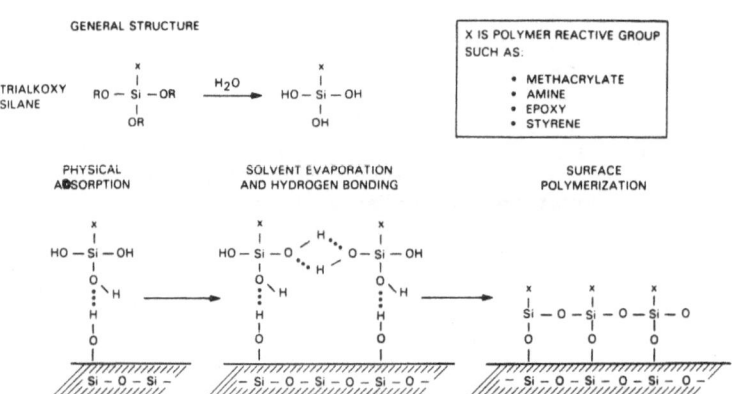

Figure 1. Bonding of Silane Coupling Agents to Glass.

When the organofunctional group of silicon copolymerizes with a
polymer, the polymer becomes chemically bonded to the treated glass.
In practice, the silane coupling agent primer is generally not a simple
monolayer, but is a multilayer of oligomeric siloxanes. The oligomers
are originally soluble, and fusible and may bond with polymers by at
least three mechanisms besides simple copolymerization (Figure 2).

144

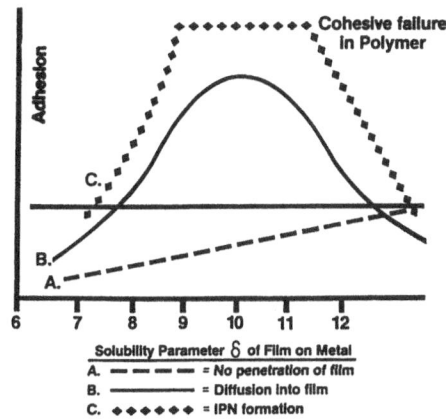

Figure 2. Three Mechanisms of Adhesion of a Polymer δ = 10 to Primer
Film.

1. If the films are baked excessively on a surface, they become
insoluble and infusible as crosslinked functional silicones. Organic
polymers generally do not form strong bonds to such films. The degree
of adhesion is determined by the surface energy of cured primer films,
and can be predicted roughly from chemical composition of the primer.
Initial adhesion may even be less than to an unmodified mineral surface,
but the bond is resistant to water since cured silicone films are
generally hydrophobic.

2. If a liquid polymer or polymer solution contacts the primer
while it is soluble and fusible, the primer may inter-diffuse into the
polymer as predicted from solubility parameters of the two phases.
Polystyrene composites made with non-reactive silane finishes on glass
showed maximum properties with siloxanes having a solubility parameter
of about 10, Figure 3. Each composite was given a 'total rating' that
was derived from the sum of dry and wet strengths of 4-ply laminates.

3. If the fusible siloxane films include organofunctional groups
that can react chemically with the polymer, the inter-diffused layer
may become an interpenetrating polymer network (IPN) which provides
even better adhesion across the interface. Polystyrene laminates with
glass sized with a vinyl silane or a methoxytolylsilane appear to form
such an IPN structure at high temperatures, while a phenylsiloxane
layer forms a simple inter-diffused layer, Figure 4.

Inter-diffusion of primer and polymer may be modeled as in
Figure 5. If the inter-diffused phases crosslink with a minimum of
co-reaction the model becomes an interpenetrating polymer network
(IPN).

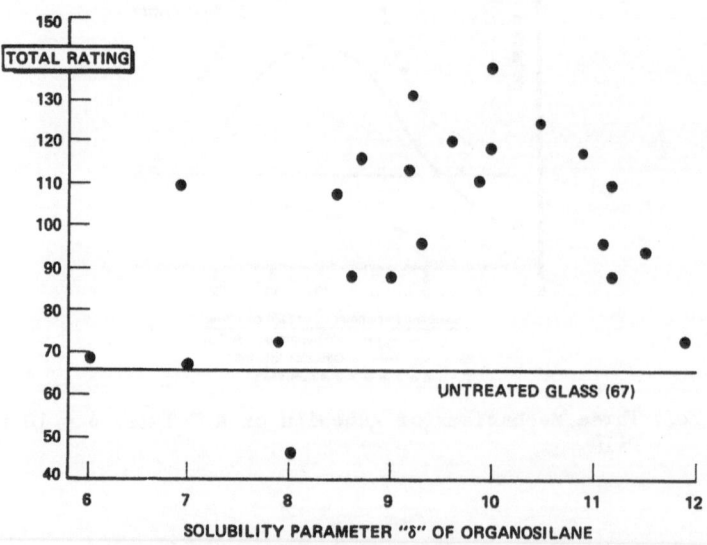

Figure 3. Relation of Solubility Parameter of Primer with
Performance of Polystyrene–Glass Composites (from Ref. 1)
Courtesy of Plenum Press.

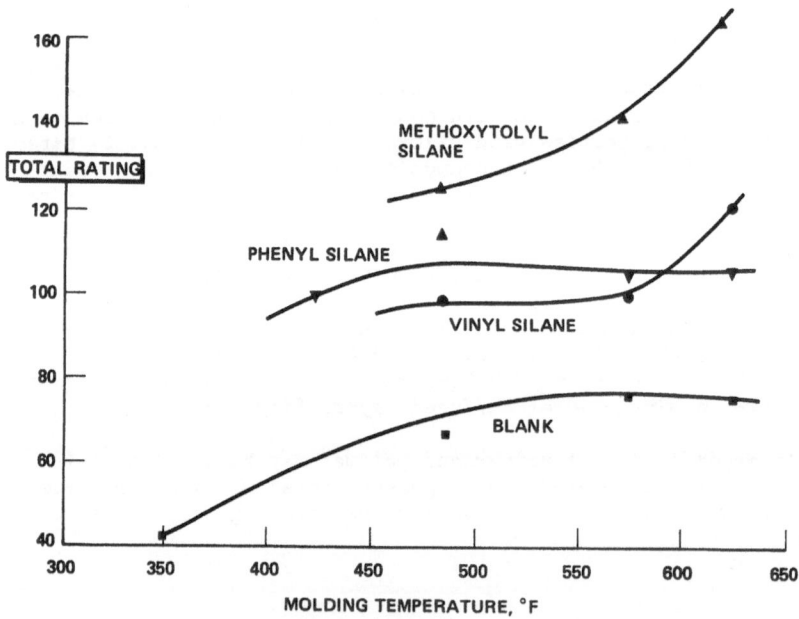

Figure 4. Effect of Molding Temperature on Properties of
Silane–Modified Polystyrene–Glass Composites(from Ref. 1.)
Courtesy of Plenum Press.

146

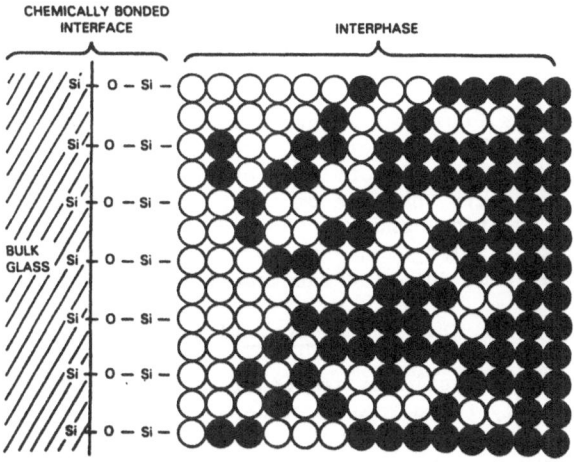

Open circles indicate regions of coupling agent.
Closed circles indicate regions of polymer.

Figure 5.   Inter-diffusion Model for Adhesion of a Thermoplastic
to Silane-Primed Glass.

Adhesion of a PVC plastisol to an amine-functional siloxane layer
shows all three forms of adhesion,[2] depending upon the temperature at
which the siloxane layer was dried, Figure 6. Primer films baked
above 200°C were impervious to the plastisol such that only surface
bonding was possible. Peel strength of the PVC to this film was only
about 4 Ncm^{-1}, but there was no loss in adhesion after 1 day in 50°C
water. When the primer was baked for 15 minutes at temperatures to
150°C, the initial adhesion of PVC was perfect (cohesive failure), but
adhesion dropped to below 4 Ncm^{-1} after 1 day in 50°C water. When the
primer film was baked at 150-200°C, the PVC had good initial adhesion
and retained a peel strength of over 15 Ncm^{-1} after 1 day in 50°C
water. All peel strengths were measured at 90 degrees with a static
load.

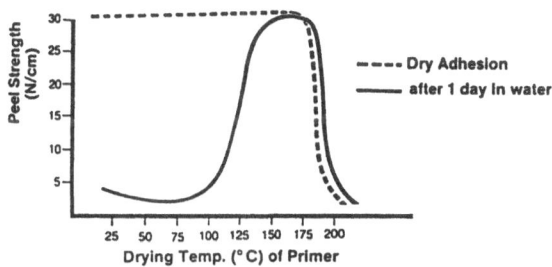

Figure 6.   Adhesion of a PVC Plastisol to Oligomeric Siloxane
Primers.

The three areas of Figure 6 were considered to illustrate surface
bonding, inter-diffusion, and IPN bonding, respectively. IPN formation
at the interface is preferred for maximum bonding, and requires a
proper matching of solubility parameters between the primer and
polymer, and a controlled reactivity of the primer film.

II. BONDING POLYMERS TO POLYMERS

A. Mechanism

A silane primer film that interpenetrates different polymers and bonds them to glass should interpenetrate such polymers brought into intimate contact and bond them to each other. In order to get intimate contact, it is necessary that one of the polymers be in a liquid state (melt or solvent solution) and wet the other surface. The silane primer is not a gap-filling adhesive, but is a "zipper" to bond the two surfaces when they come into intimate contact.

B. Primer Compositions

Monomeric trialkoxysilanes are generally not good film formers when applied directly to polymer surfaces. Several suggested formulations of modified silane primers are proposed that are known to be effective in polymer to polymer adhesion.

1. Partially Prehydrolyzed Silane in a Solvent

50 parts silane   50 parts methanol
5 parts water   1 part dicumyl peroxide (with unsaturated silanes)

The mixture should be allowed to stand at room temperature for a few hours to equilibrate the oligomeric siloxane structure. The concentrated primer may then be diluted to about 10% solids in any convenient solvent (e.g., isopropanol). These primers are very effective in bonding epoxy, urethane or phenolic adhesives, melamine-alkyd coatings, PVC varnishes and plastisols, and thermoplastics in general (with silane D).

2. Silane-Modified Tackifying Resins

Cold blends of commercial tackifying resins and silane C in suitable solvents are effective primers for adhesion of hot melt thermoplastic elastomers. Primers are made by mixing 2 parts silane C with 100 parts of the tackifying resin in any suitable solvent. Promising combinations are then optimized for a particular polymer by varying the silane content from 1 to 10 parts per 100 parts of resin.

3. Silane-Modified Melamine Resins

High solids etherified melamine resins offered under such trade names as Cymel® (American Cyanamid) and Resimene® (Monsanto) are ideal materials for preparing silane-modified primers for engineering thermoplastics and certain condensation polymers. Controlled resin reactivity based on a stable triazine ring, along with a high solubility parameter, provides a material well adapted for forming IPNs with resins having a comparable solubility parameter range. Silane modifiers may be epoxy, methacrylate, or amine-functional. For most engineering thermoplastics, the preferred primer comprises about 10 parts of epoxy-functional silane (A) in 90 parts of melamine resin (e.g., Cymel® 303). Acid catalysts for melamine resin reactions generally are not used in such primer applications.

III. SPECIFIC SYSTEMS

A. Thermoplastic Elastomers to Thermoplastic Elastomers

Cold blends of tackifying resins with small proportions of diamino-functional silane (C) are very effective as primers or additives for improving the adhesion of thermoplastic elastomers[3]. More polar elastomers (urethanes, polyesters) are best modified with a blend of the epoxysilane (A) in a melamine resin such as Cymel® 303[4]. Silane-modified tackifier primers are ineffective with vulcanized rubbers, or with rigid thermoplastics like polystyrene, rigid polyacrylates, poly-propylene or polyethylene.

It is not possible to predict the best tackifying resin, and the optimum proportion of silane C in a primer for elastomers, but a few simple microscope slide tests can help to optimize the system. The resin is dissolved in toluene/isopropanol and mixed with 1 to 10 parts of silane C, based on 100 parts of resin. Thermoplastic elastomers are hot-pressed against microscope slides coated with these primers and observed for initial adhesion and retention of adhesion in water. With optimum primer composition, the elastomer cannot be pulled from glass and retains adhesion after soaking in water for a week (rating = 4).

Rating 3 indicates good initial adhesion, but loss of much adhesion in water. Fair initial adhesion (several lbs/in pull strength, but interfacial failure) is rated a 2. Poor initial adhesion is rated a 1. Several ratios of silane C (Z-6020) in a terpene phenolic resin LTP-135 (this resin is discontinued, but Hercules Piccofyn-100 gives similar results) with several elastomers are shown in Table II.

Table II. Adhesion of Rubbers to Picco LTP-135 Primers on Glass

%Z-6020 Added to LTP-135 in Primers	Adhesion* of Rubbers Pressed 150°C, 1 min.				
	Kraton 1102	SBR 1006	Butyl Enjay 035	Isoprene Shell 305	EPT Enjay 3509
0	1	1	1	1	1
0.01	2	3	1	1	1
0.05	3	4	1	1	1
0.10	4	4	2	2	1
0.5	4	4	2	3	1
1.0	4	4	3	3	1
2	4	3	4	4	2
5	4	1	3	3	3
10	3	1	3	2	4

* 1 = poorest
  4 = best

The optimum proportion of silane C in the primer varied from 0.1 to 10%, with best across the board response at 2% silane. Results similar to adhesion on glass were observed on ceramic, wood, and common metals. Other resins might have different optimum concentrations of silane C for adhesion of the same elastomers.

A primer that bonds two or more polymers to glass should bond the polymers to each other. This concept was tested by fusing the lower-melting polymer against the primed higher-melting polymer as shown in Tables III and IV.

TABLE III. Adhesion of Elvax® 150 to Primed Surfaces.
Peel strength (N/cm) of films fused at indicated temp. (°C)

Primer - Resin with 2% Silane C	Elvax® 150 to Glass[1]	Elvax® 150 to Kraton® 1102[1]	Kraton® 1102[3]	
			Dry	Wet
No Primer	2	3	20	1
Piccofyn-100	19	8	40(c)	40(c)
Piccotex-100	4	12	40(c)	40(c)
" (no silane)	2	8	23	2
Piccovar L-60	23(c)	19	40(c)	40(c)
Staybelite Resin	5	8	40(c)	25
Hercolyn-D	15	13	30	14
Chlorez-700	6	12	40(c)	40(c)
Cymel® 303 (10% silane A)	5	7	13	8

(c) = Cohesive failure in polymer
[1] Polymer fused at 125°C
[2] Wet - 1 hr. in 50°C water
[3] To glass - fused at 225°C

TABLE IV. Adhesion of Estane® 5701 to Primed Surfaces.
Peel strength (N/cm) of films fused at indicated temp. (°C)

Primer - Resin with 2% Silane C	To Glass (200°C)		To Kraton® 1102 (175°C)
	Dry	Wet	
No Primer	1.1	-	3.5
Piccofyn-100	40(c)	40(c)	20
Piccotex-100	40(c)	40(c)	5.8
Piccotex-100 (no silane)	8.5	-	3.0
Piccovar L-60	40(c)	17.3	4.0
Staybelite Resin	7.7	-	15.8
Hercolyn-D	40(c)	17.3	4.6
Chlorez-700	16.2	3.9	1.6
Cymel® 303 (10% silane A)	40(c)	40(c)	2.0

(c) = Cohesive failure in polymer
[1] Wet = 2 hrs. in 80°C water

B. Bonding to Silicones

It is generally difficult to bond organic polymers to vulcanized silicone elastomers. Primer II-B-1 (based on partially prehydrolyzed silane C), and primer II-B-2 (2% silane C in Piccotex®-75 tackifying resin) are fairly good primers on some room temperature vulcanized (RTV) silicone elastomers, but for most silicones it is necessary to activate the surface by flame or plasma treatment. Contacting the silicone surface with a blue gas flame for about one second is sufficient to activate it. Organic polymers such as urethanes, EVA copolymers, styrene-butadiene block copolymers and others that bond to silane primers on glass will bond to silane primers on flame-treated silicone as shown in Table V. Dow Corning also supplies a curable tie-coat adhesive (SYLGARD® 577) that will bond vulcanized silicone elastomers to many organic surfaces, including polyester (Mylar®), nylon, polyimide (Kapton®), as well as to mineral surfaces.

TABLE V.   Bonding to Cured Silicone Elastomers (N/cm peel).

Silicone	Primer on Silicone	Elvax® 150 at 150°C		Kraton® 1102 at 210°C	
		Untreated	Flame Treat	Untreated	Flame Treat
DC 9596	None	0.6	0.6	0.2	1.0
DC 9596	B-1	1.5	23.1	1.5	c
DC 9596	B-2	0.8	10.4	0.3	c
DC 3140	None	0.2	0.5	0.3	-
DC 3140	B-1	0.3	c	0.2	5.0
DC 3140	B-2	0.4	13.5	0.3	9.6

B-1 = Partially prehydrolyzed silane C cf. II-B-1
B-2 = 2% silane C in Piccotex® 75   cf. II-B-2
c   = Cohesive failure in polymer at >30 N/cm.

C. PVC Plastisols to Various Surfaces

The partially prehydrolyzed amino-functional silane C is an excellent primer for adhesion of PVC plastisols to mineral surfaces[5]. Silane-modified melamine resins are generally effective in bonding to engineering thermoplastics. Various Weight Ratios of partially hydrolyzed silane C (Z-6020-W) in Cymel® 303 were tested in bonding a PVC plastisol to various plastic surfaces (Table VI). Mixtures were generally better primers on most organic surfaces than the melamine or Z-6020-W alone. Mixtures of equal parts by weight of Cymel® 303 and monomeric silane C were relatively ineffective in this application.

TABLE VI.   Peel Strength of PVC Plastisols on Organic Polymers (N/cm)
(Fuse 5 min. at 175°C).

Polymer	Primer (25% in IPA) on Polymer Surface				
	None	C-W[1]	Cymel[2]	CW/Cymel[3]	C/Cymel[4]
Glass	0.7	c	6.9	c	c
Plexiglas®	c	1.1	19.0	17	16
Polyester (Mylar®)	1.9	1.9	2.3	c	1.9
Polycarbonate (Lexan®)	1.9	17.0	10.4	14	4
Polysulfone	1.2	9.6	4.2	c	8
Nylon	1.5	8.2	10.4	c	c
Kapton® Film	0.7	9.6	c	c	c
Epoxy Laminate	0.7	9.4	5.8	c	17
Phenolic Laminate	0.7	7.7	7.7	c	12
Polyester Laminate	0.7	15.4	7.7	c	19
Dacron® Fabric	5.8	c	-	27	-

[1] Partially prehydrolyzed silane C (Dow Corning® Z-6020 silane)
[2] Cymel® 303 product of American Cyanamid
[3] 1/1 by weight mixture of (1) and (2)
[4] 1/1 by weight mixture of (2) with unhydrolyzed silane C
(Dow Corning® Z-6020 silane)
c = Cohesive failure in PVC at >30 N/cm

D. Bonding Polymers of Widely Differing Solubility Parameters

Partially prehydrolyzed silane D with 1% added Dicumyl peroxide is a good primer for bonding most thermoplastic polymers to mineral surfaces, but is especially effective with hydrocarbon polymers such as the polyolefins. A melamine resin (Cymel® 303) modified with silane C (Z-6040) is a good primer for bonding engineering thermoplastics to minerals but is almost completely ineffective with polyolefins. These two primers were used to bond a slab of polyethylene to a polyester (Mylar®) film through hot-melt adhesives. Either an EVA terpolymer (DuPont® CXA 1025) or an elastomeric polyester (Goodyear® Vitel 5571) were good hot-melt adhesives to the primed surfaces. Adhesion to unprimed polyethylene and polyester were relatively poor (Table VII).

TABLE VII.   Bonding Polyethylene to Mylar® Through "Hot Melts" at 150°C.

"Hot Melt" Polymer[1]	Primer on PE	Peel Str. (N/cm) PE	Primer on Mylar®	Peel Str. (N/cm) to Mylar®
CXA 1025[1]	None	1.8	None	0.8
CXA 1025	A	*	B	*
Vitel 5571[2]	None	3.5	None	11.5
Vitel 5571	A	*	B	*

* Cohesive failure in adhesive at greater than 30 N/cm
[1] EVA terpolymer, product of DuPont
[2] Elastomeric polyester, product of Goodyear
Primer A = 1% dicumyl peroxide in silane H, (10% in methanol)
Primer B = 10% silane C in Cymel® 303, (10% in isopropanol)

E.   Adhesion of Crosslinkable EVA to Various Plastics.

A peroxide crosslinked ethylene vinylacetate copolymer (EVA) has been formulated by Springborn Labs, Enfield, CT, for encapsulating solar cells in modules. Twenty-two different primers were identified that were useful in improving adhesion of the encapsulant to the many surfaces encountered in solar cell modules[6]. It was later observed that a silane-modified melamine primer gave essentially perfect adhesion of ethylene copolymers to all plastic surfaces considered for solar cell module construction. A mixture of 10 parts silane A, 10 parts Silane B, in 80 parts melamine resin (Resimene® 740) diluted to 10% solids in isopropanol was used as a primer on the plastic surfaces. Both formulated EVA (A-9918) and (15295) gave excellent adhesion when cured 15 min. at 125°C against primed plastic surfaces. Primed plastics included:

Polyimide	Kapton® film	(DuPont)	
Polyester	Mylar® sheet	(DuPont)	
Acrylic	Korad® 6300	(Xcel)	white
	Scotchpar® 20 CP	(3M)	white
	Korad® 212	(Xcel)	clear
	Acrylar X-22417	(3M)	clear
Fluorocarbon	Tedlar® 200-BG-30-UT	(DuPont)	clear
	Tedlar® 150-BL-30-WH	(DuPont)	white
	Chem C-20® FEP	(treated side)	
Polycarbonate	Lexan®	(GE)	
Nylon sheet	generic		

IV. CONCLUSION

A general concept has been demonstrated: silane primers that bond two different polymers to glass will also bond the polymers to each other. Since silanes are available that will bond virtually any polymer to glass, it should be possible to bond all polymers to each other. Many practical applications of inter-coat adhesion, adhesive bonding of composites, and coated film adhesion should result from applying this concept.

In general, the primer will not be a silane monolayer, but will probably be an oligomeric siloxane, or a silane-modified polymer precursor applied in sufficient thickness to have film-forming properties. The mechanism of adhesion to polymers involves mutual inter-diffusion often with interpenetrating polymer network (IPN) formation.

Silane-modified polymers are better adhesion promoters for organic-organic adhesion than the silane or the polymer alone, indicating that the silane modifies the polymer more than merely providing a mechanism for bonding to glass. A challenge remains to identify the complete mechanism of adhesion. This will allow more rapid development of polymer-to-polymer adhesives, as well as indicating methods of improving the adhesion of polymers to mineral surfaces.

REFERENCES

1. E. P. Plueddemann, "Silane Coupling Agents," pp. 134-136, Plenum Press, New York (1982).
2. E. P. Plueddemann, 39th Ann. Conf. Reinforced Plastics, SPI 4-c, 1984.
3. E. P. Plueddemann, Applied Polymer Symposium, No. 19, 75-90 (1980).
4. E. P. Plueddemann, US Patent 4,231,910 (to Dow Corning), Nov. 4, 1980.
5. E. P. Plueddemann, US Patent 4,228,061 (to Dow Corning), Oct. 14, 1980.
6. D. R. Coulter, E. F. Cuddihy, and E. P. Plueddemann, "Chemical Bonding Technology for Terrestrial Photovoltaic Modules," DOE/JPL-1012-91) JPL Publication 83-86, Nov. 15, 1983.

INFRARED AND X-RAY PHOTOELECTRON SPECTROSCOPY OF THIN SILANE FILMS ON IRON AND TITANIUM

F. J. Boerio and D. J. Ondrus

Department of Materials Science and Engineering
University of Cincinnati
Cincinnati, Ohio   45221

The properties of primer films formed by γ-aminopropyl-triethoxysilane (γ-APS) adsorbed from dilute aqueous solutions onto metal adherends depend on the pH of the solutions. Reflection-absorption infrared (RAIR) and x-ray photoelectron (XPS) spectroscopy were used to determine the structure of thin γ-APS films on iron and titanium. Results from RAIR showed that films formed by adsorption at pH 10.4 were composed of partially polymerized polysiloxanes containing significant amounts of absorbed carbon dioxide. Such films appeared to react with epoxy resins. Films formed at pH 8.0 were much more polymerized and reacted with epoxy resins to only a very limited extent. These results could be related to the dry strength of iron/epoxy lap joints. Joints prepared from adherends primed with γ-APS at pH 10.4 consistently had higher dry strengths than joints prepared from adherends primed with γ-APS at pH 8.0 because of the greater primer/adhesive reaction in the former case. Results obtained from XPS showed that films formed on iron at pH 8.0 contained more oxygen than films formed at pH 10.4. This may indicate that pH affects the orientation of the molecules on the surface. However, no chemical shifts were observed that could be related to the adsorption of γ-APS onto the oxidized surface of iron.

INTRODUCTION

Organofunctional silanes such as γ-aminopropyltriethoxysilane (γ-APS) have been used to improve the hydrothermal stability of glass fiber reinforced composites for many years. More recently it has been shown that silanes can be used to formulate primers for improving the wet strength of adhesive bonds to metals. For example, Schrader and Cardamone[1] found that γ-APS was a useful primer for enhancing the wet strength of adhesive bonds between anhydride-cured epoxies and titanium. Boerio and Dillingham[2] reported similar results for bonds formed between titanium and an epoxy cured with a tertiary amine. γ-APS has also been shown to be an effective primer for iron/epoxy[3] and aluminum/epoxy[4] adhesive bonds.

155

γ-APS primers are usually applied to metal adherends by adsorption from dilute aqueous solutions and several investigators have reported that the performance of the primers depends strongly on the pH of the solutions. Boerio and Williams[3] showed that iron/epoxy lap joints prepared from adherends pretreated with dilute aqueous solutions of γ-APS at pH 8.0 retained about 75% of their initial strength after immersion in water at 60°C for 60 days. Joints prepared from adherends pretreated with solutions of γ-APS at pH 10.4 retained about 50% of their strength after similar exposure to water but joints prepared from unprimed adherends retained only about 25% of their strength.

Boerio and Dillingham[2] reported that γ-APS primers were about equally effective when applied to titanium/epoxy adherends from aqueous solutions at pH values of 10.4 and 8.0. However, the primers were less effective when applied to the adherends at a pH value of 5.5.

Mittal[5] investigated the dry strength of bonds between polyimide coatings and oxidized silicon wafers that were pretreated with dilute (0.01%) aqueous solution of γ-APS and reported interesting pH effects. The peel strength was a maximum when the primer was applied at the natural pH of the solutions (8.0). When the pH of the primer solutions was raised or lowered the peel strength decreased.

We have been interested in relating the performance of silane primers to their molecular structure and have already described the use of infrared spectroscopy to determine some of the structural features of γ-APS primer films applied to metal substrates. The results obtained indicated that the structure of the films was a complex function of variables such as pH, drying time and temperature, and substrate. Primer films formed on iron and titanium at pH 10.4 were composed of siloxane polymers containing a great deal of absorbed carbon dioxide[2]. Somewhat similar siloxane polymers were formed at pH 8.0, but the amino groups were protonated and there was no absorbed carbon dioxide[2].

γ-APS primer films formed on copper were also composed of siloxane polymers but the oxidized surface of the copper was etched during deposition of films from aqueous solutions at pH 10.4, and films formed at such pH values contained copper atoms coordinated to the amino groups[6]. The oxidized surface of aluminum alloys was also etched during deposition of γ-APS primer films at high pH values. Films formed on alloys containing copper also contained copper coordinated to the amino groups on the γ-APS molecules[7].

The purpose of this paper is to describe some additional results that have been obtained using infrared and x-ray photoelectron spectroscopy to determine the structure and properties of γ-APS primer films on iron and titanium. The use of XPS to determine the effect of pH on the locus of failure in iron/epoxy and titanium/epoxy lap joints as a function of immersion time in water at elevated temperatures will be described in a subsequent publication[8].

EXPERIMENTAL

Substrates for infrared spectroscopy and x-ray photoelectron spectroscopy were prepared by mechanically polishing interstitial-free iron and titanium-6Al, 4V coupons using alumina polishing compounds on polishing wheels covered with billiard cloth. Samples were prepared for infrared spectroscopy by immersing substrates into 1% aqueous solutions of γ-APS (A-1100, Union Carbide Corp., as received) for a few minutes and then blowing the mirrors dry with nitrogen. Results obtained previously using ellipsometry indicated that this procedure resulted in primer films that were about 100 Å in thickness. Similar techniques were used to prepare samples for XPS except that the sample mirrors were rinsed in water after being removed from the γ-APS solutions and before being dried in nitrogen. Considering the nature of the XPS spectra obtained, the thickness of the rinsed films was close to that of a monolayer.

Infrared spectra were obtained from the primer films on the metal mirrors using a computer-controlled Perkin-Elmer 180 infrared spectrophotometer and external reflection accessories provided by Harrick Scientific Co. The infrared spectra of the primer films were always obtained by recording the spectrum of a film-covered mirror and then subtracting the spectrum of a film-free mirror. XPS spectra were obtained from rinsed primer films on metal mirrors using a Perkin-Elmer 5300 x-ray photoelectron spectrometer and Mg K_α radiation. The spectrometer was equipped with a sample stage that could be tilted to enable non-destructive depth profiling by varying the exit angle of the photoelectrons accepted by the electron energy analyzer.

RESULTS AND DISCUSSION

The infrared spectrum obtained from a primer film deposited on an iron mirror from a 1% aqueous solution of γ-APS at pH 10.4 is shown in Figure 1A. The spectrum was characterized by a strong, broad band near 1100 cm^{-1} and by weaker bands near 1640, 1570, 1470, 1310, and 930 cm^{-1}. Infrared spectra of γ-APS monomer are dominated by strong bands near 1100, 960, and 800 cm^{-1} that are related to the epoxy groups[9]. The absence of these bands from the spectrum shown in Figure 1A indicates that the γ-APS was hydrolyzed in the aqueous solution. However, the band near 1100 cm^{-1} in Figure 1A is assigned to a stretching mode of siloxane bonds, indicating that the hydrolyzed γ-APS polymerized on the surface of the iron mirrors to form a siloxane polymer. The weak band near 930 cm^{-1} is assigned to residual, unreacted silanol groups.

The origin of the bands near 1640, 1570, 1470, and 1310 cm^{-1} has been the subject of considerable discussion. Boerio and Williams[10], and Ishida[11] assigned these bands to a bicarbonate salt formed between the amino groups of γ-APS and carbon dioxide absorbed from the atmosphere. Dreyfuss and Eckstein[12] suggested that γ-APS formed a carbamate with carbon dioxide. The important point is that the bands near 1640, 1570, 1470, and 1310 cm^{-1} in the infrared spectra of films formed by γ-APS adsorbed onto iron mirrors from aqueous solutions at pH 10.4 are related to absorbed carbon dioxide and not to some structural feature of the γ-APS molecules.

After the sample mirrors used to obtain Figure 1A were heated in an oven at 100°C for twenty minutes, the spectrum shown in Figure 1B was obtained. The strong, broad band near 1100 cm^{-1} split into components near 1150 and 1040 cm^{-1} and the weak band near 930 cm^{-1} decreased in intensity, indicating additional crosslinking in the siloxane polymers formed on the surface of the iron mirrors. At the same time, the bands near 1640, 1570, 1470, and 1310 cm^{-1} also decreased in intensity, demonstrating that the absorbed carbon dioxide was easily driven off.

We were interested in determining the effect that the absorbed carbon dioxide has on the crosslinking in films formed by γ-APS deposited from aqueous solutions at pH 10.4. Fresh films were deposited on iron mirrors in a nitrogen atmosphere that was reasonably free of carbon dioxide. Only very weak bands were observed near 1640, 1570,

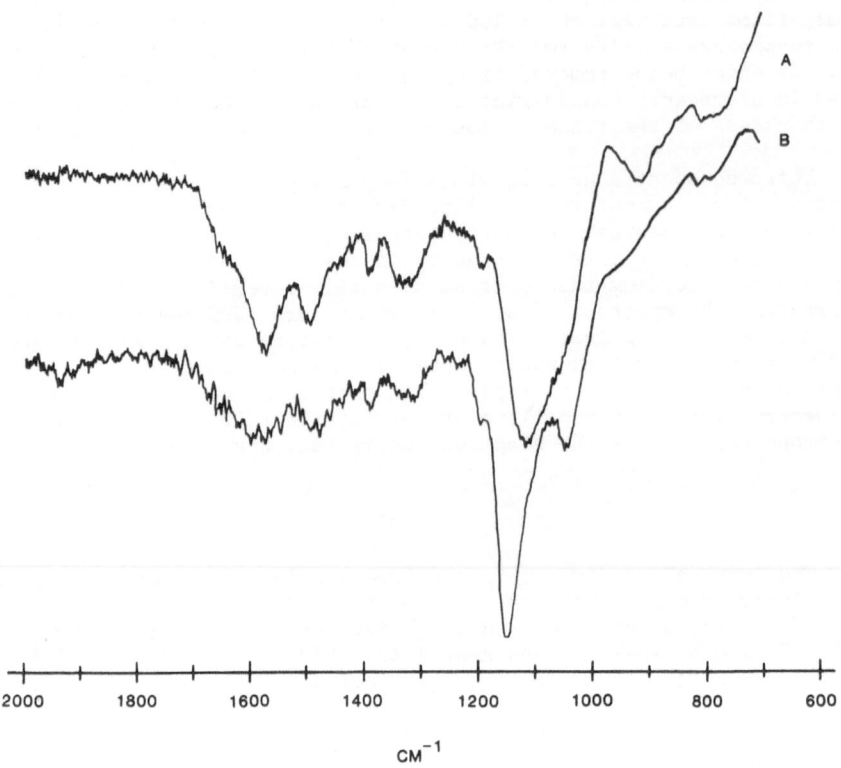

Figure 1. Infrared spectra of thin films formed by γ-APS adsorbed onto
iron mirrors from 1% aqueous solutions at pH 10.4 (A) – before and (B) – after heat treating at 100°C for 20 minutes.

1470, and 1310 cm^{-1} in the infrared spectra of these films, confirming
the assignment of these bands to absorbed carbon dioxide (see Figure 2A).
A pair of strong, well resolved bands assigned to siloxane bonds was
observed near 1150 and 1040 cm^{-1}, implying that the as-formed films prepared in the absence of carbon dioxide were highly crosslinked and that
atmospheric carbon dioxide inhibited polymerization in γ-APS films
formed by adsorption from aqueous solutions at pH 10.4. A weak band
due to residual silanol groups was again observed near 930 cm^{-1}.

   After the mirrors used to obtain the spectrum shown in Figure 2A
were heated in an oven at 100°C for twenty minutes, the spectrum shown
in Figure 2B was obtained. Only a small increase in the resolution of
the bands near 1150 and 1040 cm^{-1} was observed, confirming that the as-
formed films prepared in the absence of carbon dioxide were highly
crosslinked.

   In another series of experiments the effect of pH on the structure
of films formed by γ-APS adsorbed onto iron mirrors from dilute aqueous
solutions was determined by adjusting the pH of the solutions to 8.0 by
the addition of hydrochloric acid. Infrared spectra of such films were
characterized by a strong band near 1150 cm^{-1} and by weak bands near
1600, 1500, 1230, 925, and 900 cm^{-1} (see Figure 3A). The bands near

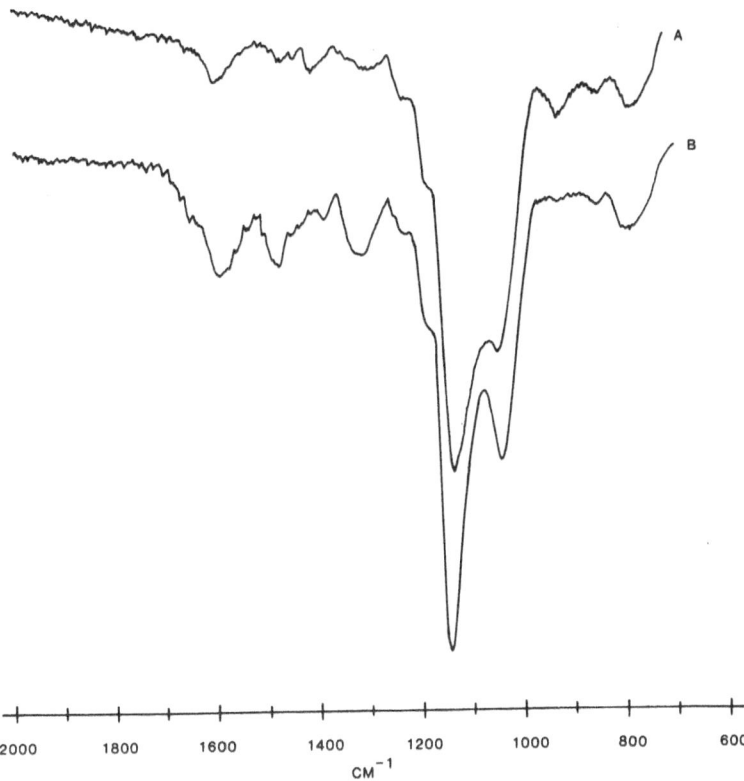

Figure 2.   Infrared spectra of γ-APS adsorbed onto Iron from 1% aqueous solution pH 10.4:   (A) Film formed in the absence of CO_2 and (B) same film heat treated at 100°C for 20 minutes in the presence of CO_2.

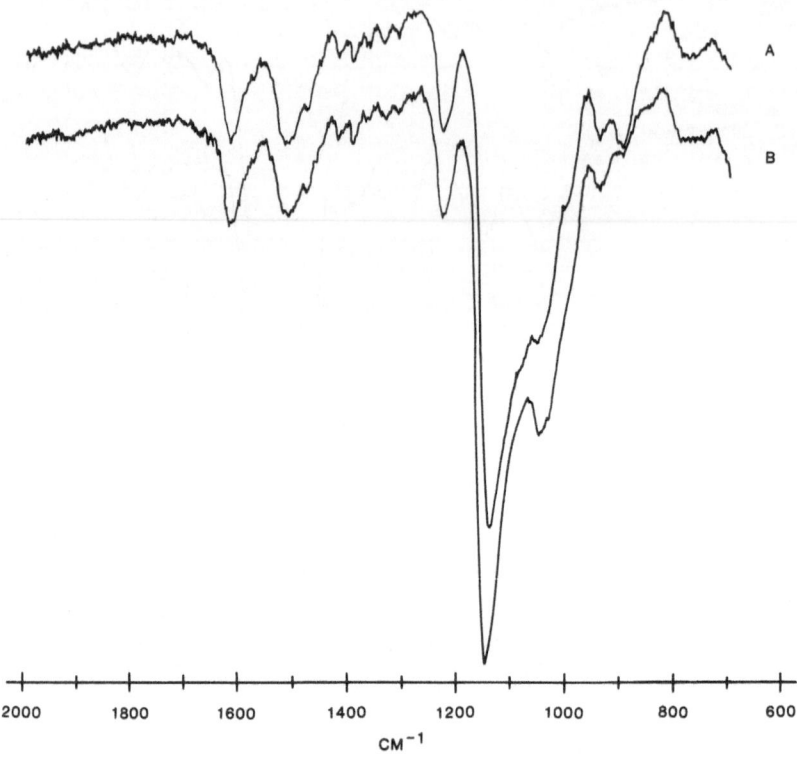

Figure 3. Infrared spectra of thin films formed by γ-APS adsorbed onto
iron mirrors from 1% aqueous solution at pH 8.0 (A) - before
and (B) - after heat treating at 100°C for 20 minutes.

1600, 1500, and 1230 cm^{-1} were assigned to the bending and rocking vibrations of protonated amino groups, implying that the amino groups formed hydrochlorides. The strong band near 1150 cm^{-1} and the shoulder near 1040 cm^{-1} were again assigned to the stretching modes of siloxane bonds and the presence of this doublet was considered to indicate that the films formed at pH 8.0 were highly polymerized. The weak band near 900 cm^{-1} was assigned to residual silanol groups but the origin of the band near 930 cm^{-1} was not determined.

After the mirrors used to obtain the spectrum shown in Figure 3A were heated in an oven at 100°C for twenty minutes, the spectrum shown in Figure 3B was obtained. The resolution of the doublet near 1150 and 1040 cm^{-1} increased somewhat and the intensity of the band near 900 cm^{-1} decreased somewhat due to a small increase in the crosslinking. However, the bands near 1600, 1500, and 1230 cm^{-1} were not changed at all and it was concluded that heat treating the films did not result in the dissociation of the amine hydrochlorides.

The nature of the reaction between epoxy resins and γ-APS primer films was also investigated. In one case a primer film was deposited on an iron mirror by adsorption from an aqueous solution at pH 8.0. The infrared spectrum obtained from that film is shown in Figure 4A. A thin film of an epoxy polymer was then applied over the primer and the iron mirror was heated at 100°C for 20 minutes. The spectrum shown in Figure 4B was then obtained. Bands due to the epoxy were clearly observed near 1510 and 1260 cm^{-1}. The mirrors were then rinsed in methylethylketone to remove unreacted epoxy and the spectrum shown in Figure 4C was obtained. The bands near 1510 and 1260 cm^{-1} were no longer observed, indicating that there was very little retained epoxy and that there was very little reaction between the epoxy and the primer films formed at pH 8.0.

In another case, a primer film was deposited on an iron mirror by adsorption from an aqueous solution at pH 10.4. The infrared spectrum obtained from that film is shown in Figure 5A. A thin film of epoxy resin was applied to the mirror and the mirror was heated in an oven at 100°C for twenty minutes. The mirror was rinsed with methylethylketone to remove unreacted epoxy and the spectrum shown in Figure 5B was then obtained. Bands due to retained epoxy were clearly observed near 1510 and 1260 cm^{-1} showing that there was some reaction between the epoxy and the primer films formed at pH 10.4.

The molecular structure of the primer films was related to the dry strength of iron/epoxy lap joints prepared using a tertiary amine curing agent[8]. The breaking strength of joints prepared from adherends primed with γ-APS at pH 10.4 was always greater than that of joints prepared from joints primed at pH 8.0. Moreover, the locus of failure was always well within the epoxy resin when the adherends were primed at pH 10.4 but the locus of failure was closer to the epoxy/primer interface when the adherends were primed at pH 8.0.

These differences in the dry strength and the locus of failure of the iron/epoxy lap joints were related to the structure of the adhesive/primer interface. As indicated above, there is little reaction between epoxy resins and γ-APS primer films deposited from aqueous solutions at pH 8.0. Joints prepared from adherends primed with γ-APS at pH 8.0 tend to have relatively low dry strengths and to fail near the adhesive/primer interface. There is more reaction between epoxy resins and γ-APS primer films deposited at pH 10.4. Joints prepared from adherends primed with γ-APS at pH 10.4 have relatively high dry strengths and tend to fail well within the adhesive.

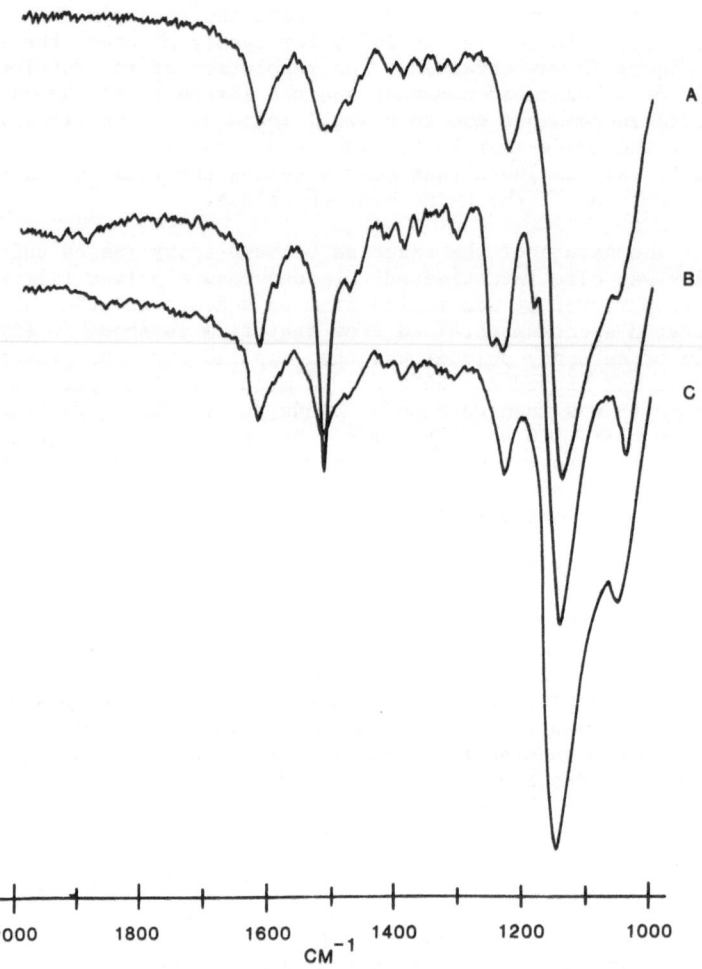

Figure 4. Infrared spectra of reaction between air dried γ-APS film and
Epon 828:   (A) γ-APS film as applied at pH 8.0 on Iron.
(B) Same film after applying a thin film of Epon 828 (spin
coating) and heating to 100°C for 20 minutes.   (C) Same
film rinsed thoroughly in MEK.

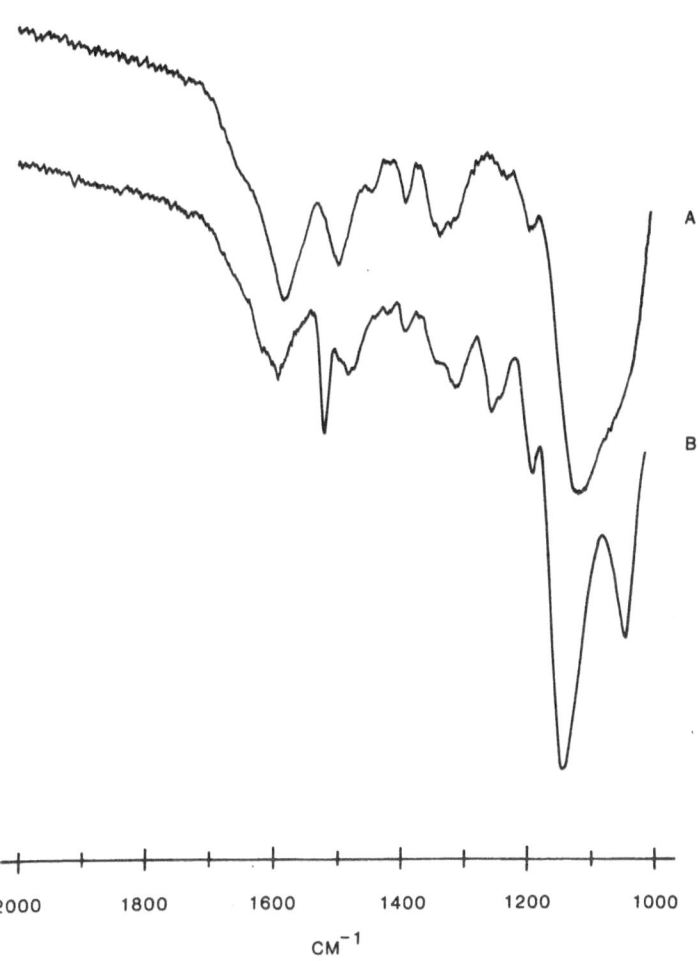

Figure 5.   Infrared spectra of reaction between air dried γ–APS film and
            Epon 828:   (A) γ–APS film as applied at pH 10.4 on Iron.
            (B) Same film after applying a thin film of Epon 828 (spin
            coating), heating to 100°C for 20 minutes and rinsing tho-
            roughly in MEK.

However, it is interesting to note that the wet strength of iron/ epoxy lap joints shows the opposite behavior. That is, joints prepared from adherends primed with γ-APS at pH 8.0 have higher breaking strengths than joints prepared from adherends primed with γ-APS at pH 10.4 after lengthy immersion in warm water[2,8]. We have previously suggested that this tendency might be related to the structure of the primer/oxide interface[2]. That is, primer films deposited at different pH values might have different orientations on the oxide surface. As a result, we have used x-ray photoelectron spectroscopy (XPS) to investigate the effect of pH on the structure of the primer/oxide interface. Some preliminary results are described below.

When XPS survey spectra were obtained from mechanically polished but unsilanated iron mirrors, the only elements detected were iron, oxygen, and carbon. The carbon was mostly attributed to adsorbed hydrocarbons. The Fe(2p) and O(1s) spectra are shown in Figures 6 and 7, respectively. The Fe($2p_{3/2}$) peak was observed near 711.0 eV, indicating that mostly Fe(III) was present in the oxide and that the oxide was mostly Fe_2O_3[13]. When the exit angle was 90° and photoelectrons from relatively deep (~50Å) within the sample were detected, an additional Fe($2p_{3/2}$) peak characteristic of Fe(0) was observed near 706.8 eV. This peak was not observed when the exit angle was 10° and only photoelectrons from the outermost atomic layers (~10Å) were detected.

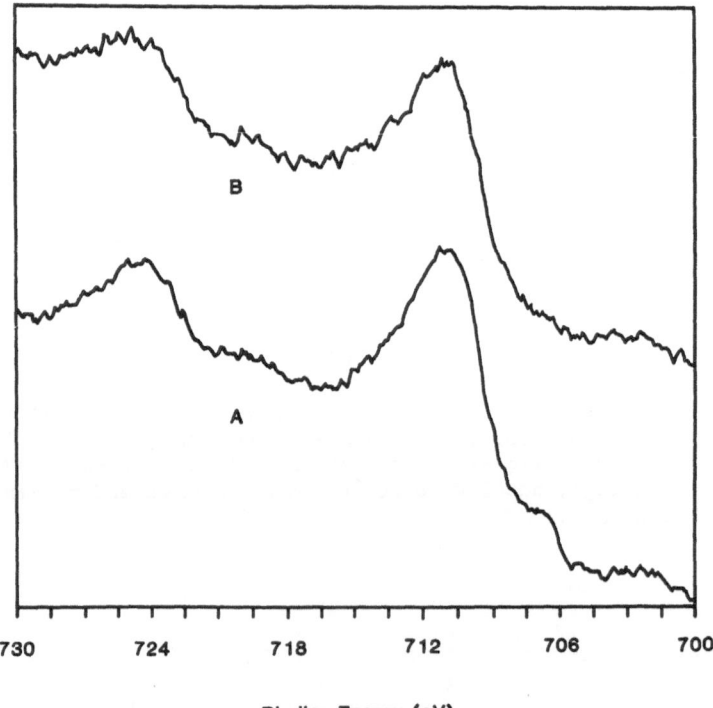

730   724   718   712   706   700

Binding Energy (eV)

Figure 6. Fe(2p) XPS spectra obtained from mechanically polished iron mirrors at exit angles of (A) - 90° and (B) - 10°.

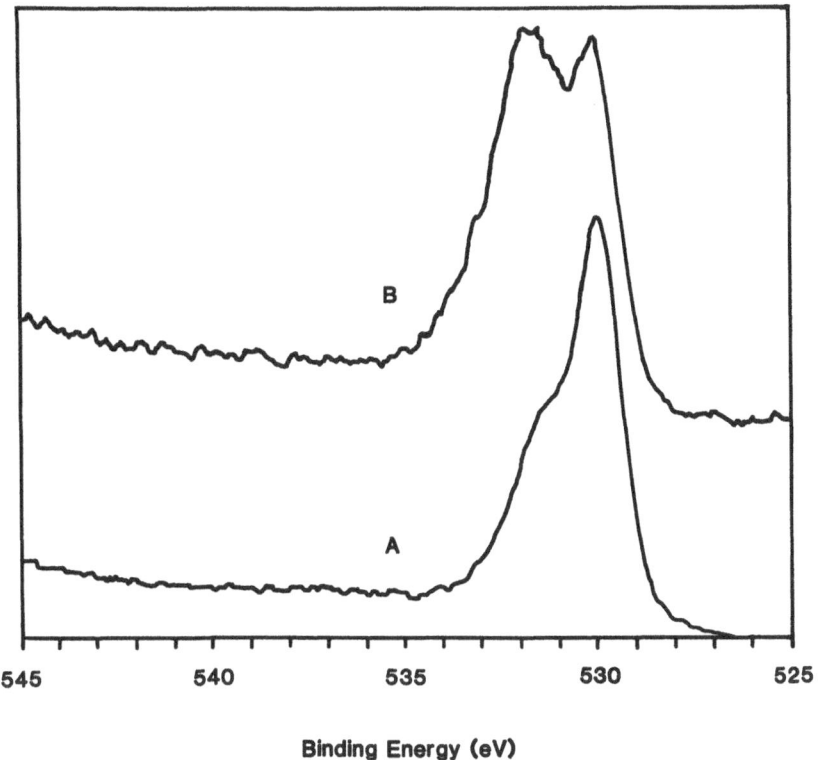

Binding Energy (eV)

Figure 7. O(1s) XPS spectra obtained from mechanically polished iron
mirrors at exit angles of (A) - 90° and (B) - 10°.

The structure of the O(1s) spectra for the polished iron was complex
(see Figure 7). When the exit angle was 90°, a sharp peak was observed
near 530.0 eV with a broad shoulder at higher binding energies. When the
exit angle was 10°, the high energy shoulder became a well resolved band
near 531.8 eV. In fact, as shown in Figure 8, the O(1s) spectra obtained
from iron mirrors at an exit angle of 10° could be resolved into three
components near 530.0, 531.8, and 533.3 eV. The peaks near 530.0 and
531.8 eV were assigned to the oxide and to a surface hydroxide, respectively[13]. The high energy component near 533.3 eV may be related to
adsorbed oxygen.

Next , an approximately monomolecular layer of γ-APS was adsorbed
onto a polished iron mirror by immersing the mirror in a 1% aqueous
solution of γ-APS at p^H 10.4 for 30 minutes and then rinsing the mirror
in water. The elements iron, carbon, oxygen, silicon, and nitrogen were
detected in survey spectra obtained from such samples.

The Fe(2p) spectra obtained from these mirrors were essentially the
same as those obtained from bare iron mirrors (see Figure 6). The
Si(2p) spectra showed an interesting dependence on exit angle (see
Figure 9). When the exit angle was 90°, the Si(2p) peak had nearly the
same intensity as the Fe(3s) peak near 94.0 eV. However, when the exit
angle was 10°, the Si(2p) peak was much stronger. In all cases the
Si(2p) peak was near 102.5 eV, a position characteristic of thick γ-APS
films. No chemical shifts that could be related to adsorption on the
oxidized iron surface were observed.

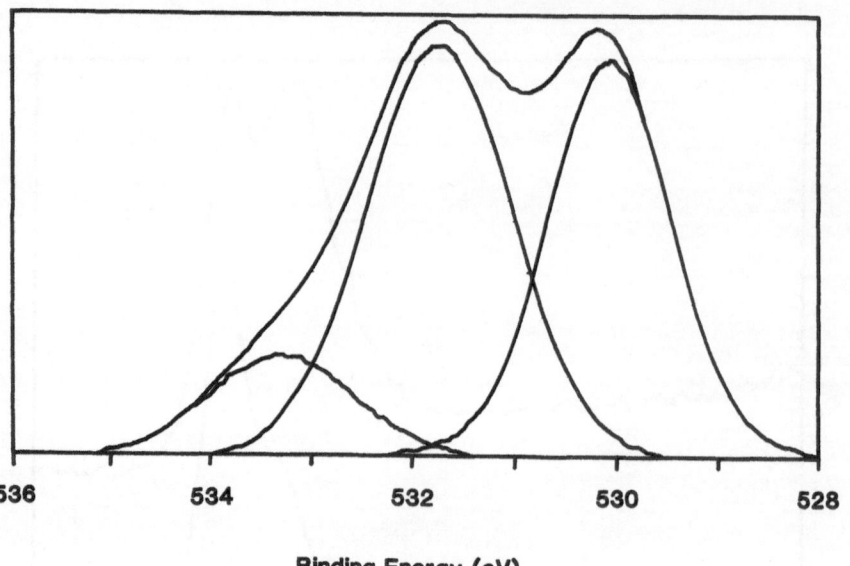

Binding Energy (eV)

Figure 8. Deconvolution of O(1s) XPS spectra from mechanically polished
iron mirrors at an exit angle of 10°.

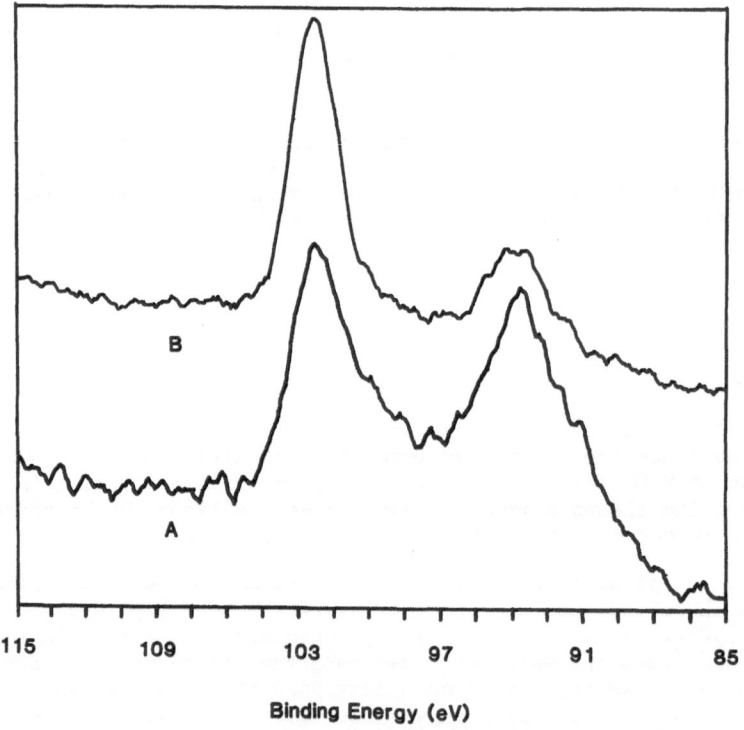

Binding Energy (eV)

Figure 9. Si(2p) XPS spectra from γ–APS films that were adsorbed onto
iron mirrors from 1% aqueous solutions at pH 10.4 and then
rinsed. The exit angles were (A) – 90° and (B) – 10°.

Similar results were obtained for the N(1s) spectra. As shown in Figure 10, the N(1s) spectra were observed near 400.0 eV, the same position that is observed for thick films of γ-APS. Once again, no chemical shifts related to adsorption on the oxide were observed.

The O(1s) spectra for films formed by the adsorption of γ-APS onto iron mirrors from aqueous solutions at pH 10.4 are shown in Figure 11. For an exit angle of 90°, a peak characteristic of the oxide was again observed near 530.0 eV and a broad shoulder was observed at higher binding energies. When the exit angle was 10°, a broad band was observed near 531.8 eV and the oxide band was observed as a shoulder. The O(1s) spectra for γ-APS adsorbed onto iron at pH 10.4 could not be adequately explained by the same three bands (near 530.0, 531.8, and 533.3 eV) that explained the O(1s) spectra of the unsilanated iron substrates. At least one additional band, near 532.2 eV, was required to fit the O(1s) spectra of iron mirrors that were treated in aqueous solutions of γ-APS at pH 10.4 and then rinsed in water. The band near 532.2 eV is considered to be characteristic of siloxane bonds.

Finally, an approximately monomolecular film of γ-APS was adsorbed onto a polished iron mirror by immersing the mirror in a 1% aqueous solution of γ-APS at pH 8.0 for 30 minutes and then rinsing the mirror in water. Once again the elements iron, carbon, oxygen, nitrogen, and

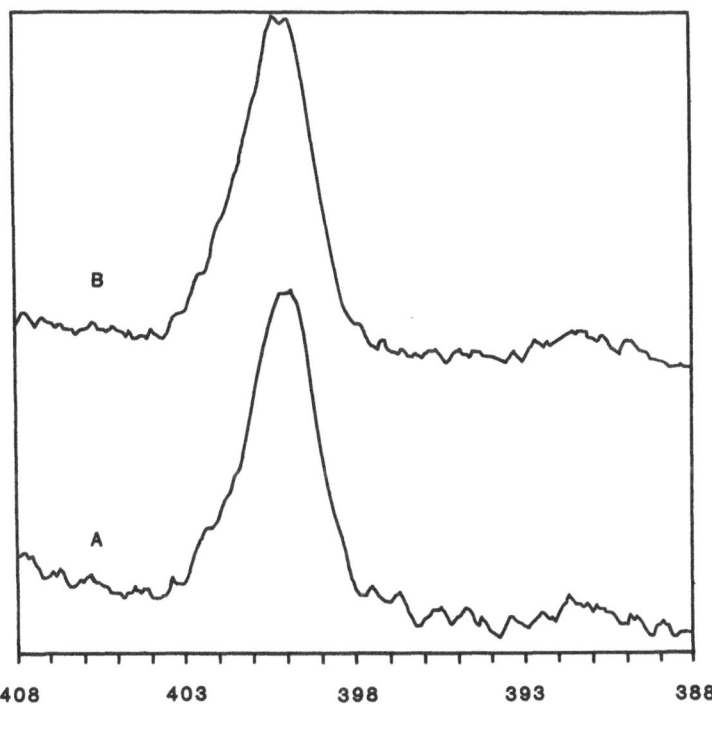

Binding Energy (eV)

Figure 10. N(1s) XPS spectra from γ-APS films that were adsorbed onto iron mirrors from 1% aqueous solutions at pH 10.4 and then rinsed. The exit angles were (A)-90° and (B)-10°.

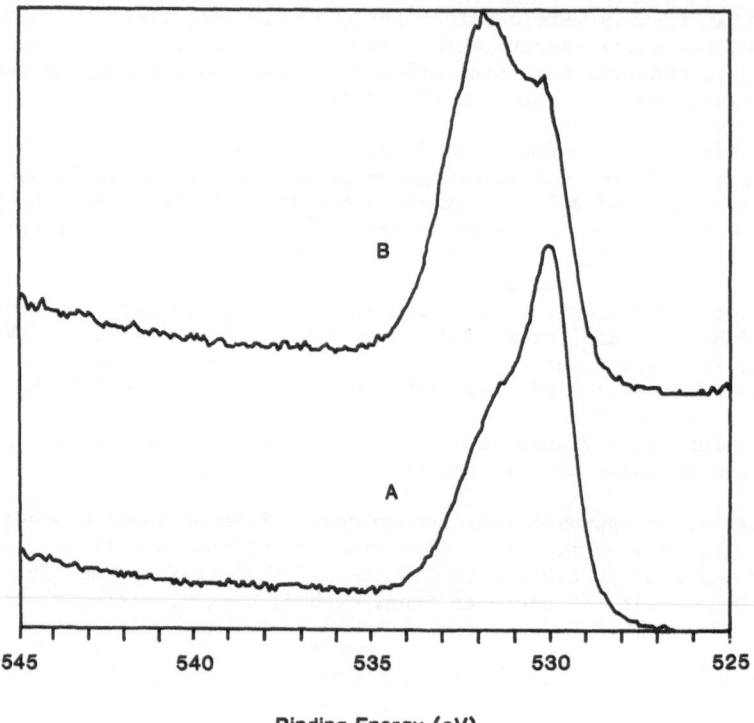

Figure 11. O(1s) XPS spectra from γ-APS films that were adsorbed onto iron mirrors from 1% aqueous solution at pH 10.4 and then rinsed. The exit angles were (A)-90° and (B)-10°.

silicon were detected from the survey spectra. The Fe(2p) spectra were very similar to the Fe(2p) spectra of bare iron and the N(1s) and Si(2p) spectra were very similar to the corresponding spectra for iron mirrors treated with γ-APS at pH 10.4. The most important feature in the XPS spectra of iron mirrors treated with γ-APS at pH 8.0 concerned the nature of the O(1s) peak (see Figure 12). The O(1s) peak was more intense for films formed on iron at pH 8.0 than at pH 10.4, indicating that pH may affect the orientation of the γ-APS molecules. When the exit angle was 90°, a sharp band assigned to oxygen in the oxide lattice was observed near 530.1 eV with a broad, unresolved shoulder extending to higher binding energies. When the exit angle was 10°, a strong, broad band was observed near 532.2 eV with a moderately strong shoulder near 530.1 eV. Once again the band near 532.2 eV was assigned to O(1s) electrons from oxygen atoms in siloxane bonds.

The XPS results can be summarized as follows. No significant chemical shifts have been observed in the spectra of γ-APS adsorbed onto iron mirrors from aqueous solutions at pH values of 8.0 or 10.4. The most interesting features of the spectra of γ-APS on iron concern the nature of the O(1s) peak. The O(1s) spectra of the bare substrate are composed of components near 530.0, 531.8, and 533.3 eV. O(1s) spectra of the silanated substrates have at least one additional component, near

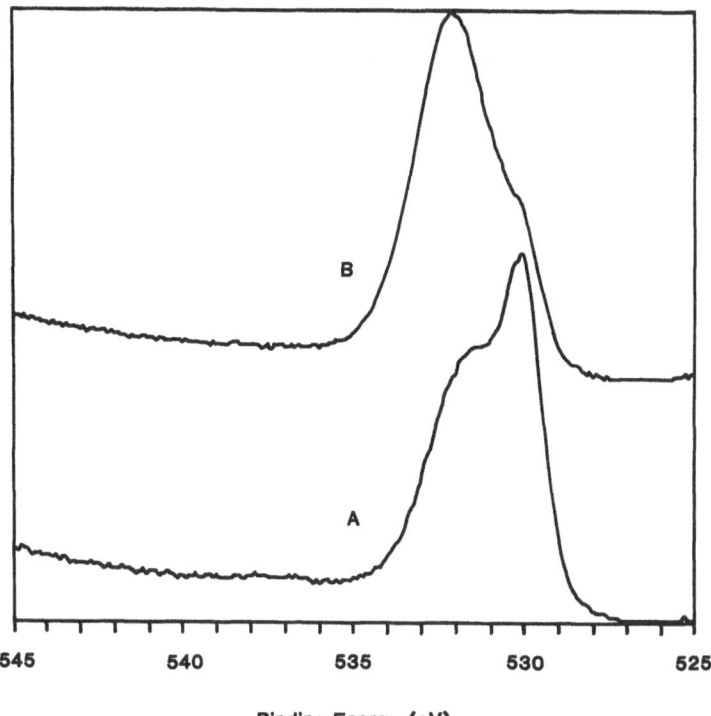

Binding Energy (eV)

Figure 12. O(1s) XPS spectra from γ-APS films that were adsorbed onto
iron mirrors from 1% aqueous solutions at pH 8.0 and then
rinsed. The exit angles were (A) - 90° and (B) - 10°.

532.2 eV, that is related to siloxane bonds. More careful analysis of
the O(1s) spectra of γ-APS films on iron mirrors may yield additional
information about the mechanisms by which γ-APS is adsorbed onto iron
and the reasons why iron/epoxy lap joints prepared from adherends primed
with γ-APS at pH 8.0 have higher wet strengths than joints prepared from
adherends primed with γ-APS at pH 10.4. Such an analysis is presently
underway.

SUMMARY

The dry strength of iron/epoxy lap joints prepared from adherends
primed with γ-APS correlates with the extent of reaction between the
epoxy and the primer. There is considerable reaction between epoxies
and γ-APS films deposited on iron substrates from aqueous solutions at
pH 10.4 but less with films deposited at pH 8.0. The wet strength
of iron/epoxy lap joints seems to depend on the characteristics of the
primer/oxide interface but a preliminary investigation of the interface
using XPS did not reveal significant differences for films deposited
pH 10.4 and those deposited at pH 8.0.

ACKNOWLEDGEMENTS

This research was supported in part by a grant from the Office of Naval Research. The assistance of Kristen A. Boerio in preparing the figures is gratefully acknowledged.

REFERENCES

1. M. E. Schrader and J. A. Cardamone, J. Adhesion 9, 305 (1978).
2. F. J. Boerio and R. G. Dillingham, in "Adhesive Joints: Formation, Characteristics, and Testing, "K.L. Mittal, ed., p. 541, Plenum Press, New York, 1984.
3. F. J. Boerio and J. W. Williams, Appl. Surf. Sci. 7, 19 (1981).
4. F. J. Boerio and C. A. Gosselin, "Proc. 36th Ann. Conf., SPI Reinf. Plastics/Composites Inst.", Sec. 2G, 1981.
5. D. Suryanarayana and K. L. Mittal, J. Appl. Polym. Sci. 29, 2039 (1984).
6. F. J. Boerio, J. W. Williams and J. M. Burkstrand, J. Colloid Interface Sci. 91, 485 (1983).
7. F. J. Boerio, C. A. Gosselin, R. G. Dillingham and J. M. Burkstrand, "Proc. 15th Natl. SAMPE Tech. Conf." 15, 212 (1983).
8. F. J. Boerio and D. J. Ondrus, J. Adhesion, accepted for publication, 1986.
9. F. J. Boerio, L. H. Schoenlein and J. E. Grievenkamp, J. Appl. Polym. Sci. 22, 203 (1978).
10. F. J. Boerio and J. W. Williams, "Proc. 36th Ann. Conf., SPI Reinf. Plastics/Composites Inst.", Sec. 2F, 1981.
11. S. Naviroj, J. L. Koenig and H. Ishida, "Proc. 37th Ann. Conf., SPI Reinf. Plastics/Composites Inst.", Sec. 2C, 1982.
12. P. Dreyfuss and Y. Eckstein, J. Adhesion 15, 163 (1983).
13. C. R. Brundle, T. J. Chuang and K. Wendelt, Surf.Sci., 68, 459 (1977).

ZIRCOALUMINATE COUPLING AGENTS AS HIGH PERFORMANCE ADHESION PROMOTERS

L. B. Cohen

Cavedon Chemical Co., Inc.
Woonsocket, Rhode Island 02895

Adhesion problems encompass such a varied range of
adherends that no single adhesion promoter possesses suitable chemistry to be the optimum choice in all systems.
Organofunctional zircoaluminates are a unique family of
compositions which afford optimum performance where coatings
(and other adherends) must be bonded to a metallic substrate. Reactivity and synthesis considerations are explored to provide an understanding for the mandatory preparation in solvent at <30 weight percent active matter
The direct incorporation of zircoaluminates in coatings
or adhesives without the use of priming or other precursor
steps results in chemical attachment of the inorganic portion
of the molecule to metal substrates by oxo, hydroxy bridges
while the organofunctional portion of the molecule reacts
with the resin upon curing. Improvements in salt spray and
humidity resistance (coatings) are reported. Reaction mechanisms are explored which contribute to basic adhesion in
systems where zircoaluminates are used to promote intercoat
adhesion of acrylic to phenolic, improved durability of a
bonded steel joint, and improved performance of metal conversion chemicals (phosphates). Joint strength improvement is
shown quantitatively by improved T-peel strengths on an EPDM
system bonded together by an elastomeric zircoaluminate containing adhesive.

1. INTRODUCTION

Adhesion promoters are a chemically diverse group of additives
which share a single performance characteristic; that is, when added to
an adhesive in small quantity, i.e. <2 percent based on resin solids,
the work required to separate the adherends may be shown to increase
dramatically. This macroscopically observed increase is what has been
referred to as "practical adhesion" by Mittal[1] and is readily documented
using a variety of quantitative ASTM test procedures. Of a more elusive
and esoteric nature is what Mittal refers to as "basic adhesion", by
which is meant the cumulative effect of the various intermolecular
interactions, which collectively reveal themselves as "practical adhesion".[2]

There are an infinite variety of adhesion problems which can be identified in terms of their specific compositional characteristics. Generically, these problems may be classified into eight common groups:

1. Coating (plastic) to metal
2. Coating (plastic) to primer (plastic)
3. Rubber to metal
4. Rubber to rubber
5. Metal to glass
6. Metal to metal
7. Metal to plastic
8. Plastic to glass

The first two groups are unique in having a single interface and a single bulk phase; the remaining classes having two interfaces and two bulk phases.

2. WHY USE ZIRCONIUM AND ALUMINUM TO SYNTHESIZE AN ADHESION PROMOTER?

Logically, a functional adhesion promoter should share chemical similarities (and hopefully reactivity) with each of the adherends under consideration. Presented with the breadth of chemistry represented in the eight separate cases of adherend combinations, it is unreasonable to pre-suppose that any single material will be so multifarious in its chemistry as to be the optimum promoter in each case. Hence, the market place has long recognized the supremacy of silanes as materials of choice for the adhesion of diverse resins to glass. However, the only quasiorganometallic character of silanes renders them less useful for enhancement of adhesion of polymer coatings to metallic substrates (or the bonding of two metal adherends).

What type of chemistry then would be most appropriate for this most vital task that most often parades under the banner of corrosion inhibi-tion? An organometallic compound provides the obvious answer, but what metals? What chemical, colorimetric, toxicological, and other properties must such metals have? The heightened awareness of toxicity and carcino-genicity results in immediate exclusion of chromium (Cr^{+3}, Cr^{+6}), nickel, cadmium, and mercury. The very aesthetic essence of coatings forbids color containing species such as manganese, cobalt, and copper. The real world considerations of availability and cost further eliminates ruthenium, rhodium, palladium, platinum, iridium, silver, and gold. Presumably, any performance chemical must be chemically stable; thus, the metallic center must not undergo either facile oxidation or reduction, so we must discount iron and zinc. Other metals such as yttrium, niobium, technetium, moly-bdenum, vanadium, and scandium are simply not available in useful forms as items of commerce. Non-transition metals exemplified by sodium and magnesium are too limited by valency restrictions to be of use. Titanium, although fulfilling many of the stipulated prerequisites, has an organo-metallic chemistry characterized by excessive reactivity and/or hydrolytic sensitivity. Not surprisingly, such materials are most often used as homogeneous transesterification catalysts; in essence, the chemical insta-bility providing kinetically appealing continuous access to the metal center.

3. ZIRCOALUMINATE COMPOSITIONS

When this review of elemental attrition concludes its march across the periodic table, there are but two survivors remaining: aluminum and zirconium. Aluminum offers the specific advantage of having a propensity for varied mineral surfaces. Its limited valency capacity can be compen-sated for by chemically reacting it with zirconium moieties thereby in-troducing the advantages of the polyvalent second row transition element.

This synergistic combination forms the inorganic backbone to which a variety of organofunctional moieties may be chemically tied. The organometallic compositions thus prepared offer several functionalities including:

- Amino
  - different compositions available with one or more amino group per molecule.
  - mixed amino and other functional groups available as part of a single molecular specie.
- Carboxy
- Methacryloxy
- Oleophilic
- Mercapto

It is the intrinsic chemistry of such species that as their concentration increases, the rate of inorganic polymerization increases simultaneously (Figure 1).

As graphically depicted (Figure 1) the rate dependency upon concentration is linear up to approximately 30 weight percent organometallic; thereafter, the rate of polymer formation assumes an exponential dependency. Hence, the preparation of stable compounds may be accomplished only by maintaining an active matter content of <25 weight percent. The other 75 weight percent is one of three solvents which is present as a vital part of the synthesis process throughout, specifically,

1. Lower alcohols
2. Propylene glycol
3. Methyl ether propylene glycol

The solvent plays more than a passive role in that it is responsible for the solubilization of the active matter in the adhesive resin or solvent and hence, may dramatically affect performance. For concise reference purposes, the commercial identification of the various zircoaluminates (Tradename CAVCO MODS) is included (Table I).

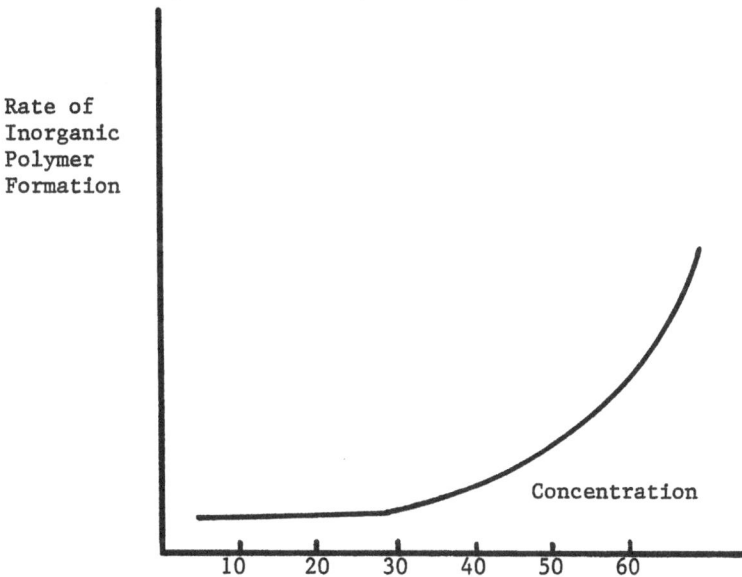

Figure 1. Zircoaluminate Polymerization as a Function of Concentration of Active Zircoaluminate.

Table I. Zircoaluminate Products

Commercial Name	Organo Functionality	Solvent
A	Amino	Lower alcohols
APG	Amino	Propylene glycol
C	Carboxy	Lower alcohols
CPM	Carboxy	Propylene glycol, methyl ether
CPG	Carboxy	Propylene glycol
C-1	Carboxy (enhanced inorganic hydroxy)	Lower alcohols
C-1PM	Carboxy (enhanced inorganic hydroxy)	Propylene glycol, methyl ether
F	Oleophilic	Lower alcohols
FPM	Oleophilic	Propylene glycol, methyl ether
M	Methacryloxy	Lower alcohols
MPM	Methacryloxy	Propylene glycol, methyl ether
MPG	Methacryloxy	Propylene glycol
M-1	Methacryloxy/ Oleophilic	Lower alcohols
M-1PM	Methacryloxy/ Oleophilic	Propylene glycol, methyl ether
S	Mercapto	Lower alcohols
SPM	Mercapto	Propylene glycol, methyl ether
APG-1	Proprietary	Propylene glycol
APG-2	Proprietary	Propylene glycol
APG-3	Proprietary	Propylene glycol

4. MOLECULAR ADHESION CONSIDERATIONS WITH ZIRCOALUMINATES IN SPECIFIC SYSTEMS

4.1 Coating (Primers) Applied to Metal Surfaces

Mechanistically, adhesion promotion with zircoaluminates is most readily understood in the specific instance of a coating (primer) applied to the metal surface. As depicted in the accompanying diagram, it is envisioned that the Al/Zr inorganic portion of the molecule will chemically attach itself to the metal by way of oxo or hydroxy bridges (Figure 2).

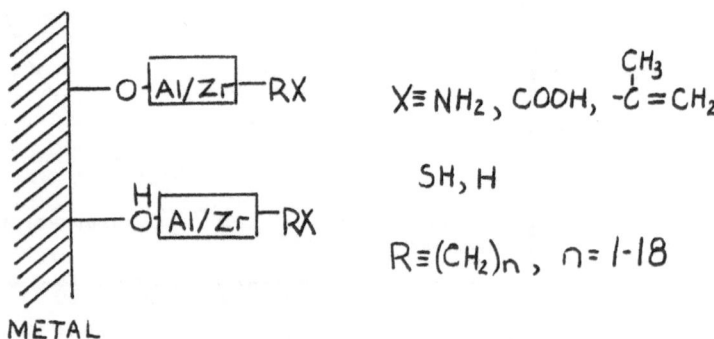

Figure 2. Zircoaluminate Attachment to Metal Substrates.

The natural abundance of various clays containing Al-O-Al linkages and the mineral Zircon containing Zr-O-Zr linkages substantiates the thermodynamic stability of such bonds. Hence, the attachment of the zircoaluminate to the metal substrate is substantially irreversible[3] resulting in long term coating durability improvements as manifested by improved salt spray and humidity resistance[4].

Preliminary results examining the presence of zircoaluminates on metal or mineral surfaces further suggests the formation of covalent bonding between zircoaluminate metal centers and the metal surface[5].

Subsequently, upon curing the coating resin (or solvent removal by ambient air drying) the organofunctional group will react with the resin backbone (Figure 3) in the case of a polyester coating.

Thus, by the combination of the zircoaluminate carboxy condensation with resin terminal OH at one end of the molecule and the aluminum and zirconium attachments by oxo and hydroxy bridges to metal substrates at the other end, the zircoaluminate becomes a highly effective agent for enhancing coating adhesion. Similar instances of coating adhesion are observed with epoxy, epoxyester, acrylic, and alkyd coating systems.

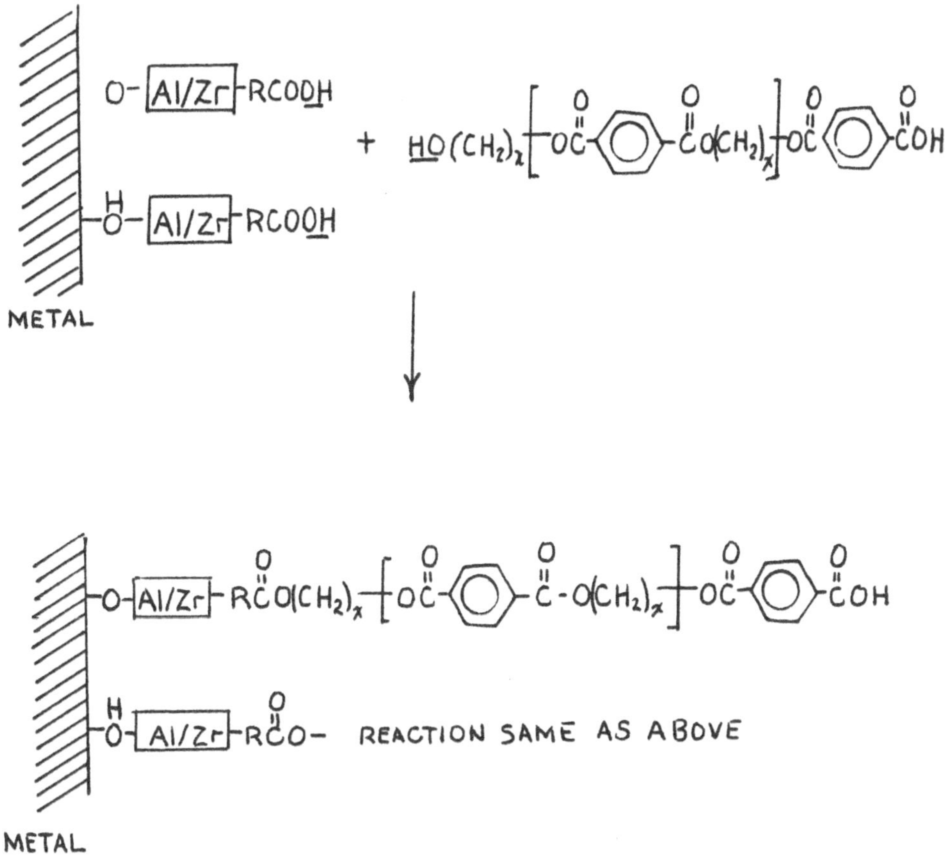

Figure 3. Zircoaluminate Modified Metal Coated with Polyester.

Not unexpectedly, zircoaluminates are finding use in the application of metal conversion chemicals, specifically, iron and zinc phosphate, which are coated as very thin films (0.005 - 0.4 mils). Such films give increased corrosion resistance and enhanced adhsion of the organic coating to be applied by the automotive or appliance manufacturer. Zircoaluminates have found use both to enhance the phosphate adhesion to the base metal and to minimize the phosphate crystal size in the deposition process. In the latter instance, the metal protection is known to increase as crystal size decreases.[6]

4.2 Intercoat Adhesion Between Phenolic and Acrylic

Zircoaluminates have also been successfully employed to enhance the adhesion of an acrylic topcoat to a phenolic undercoat. Although the ideal elements of metal and organic coating suitable for organometallic adhesion promotion are not present, it is still quite possible to propose a useful mechanism. Recognizing the phenolic base coat to be a fully cured primer at the time of application of the acrylic (which contains the zircoaluminate) the substrate may be envisioned to be a surface having a high hydroxy population providing ready sites of attachment for the hydroxy bearing metal centers (Figure 4).

Subsequent curing of the acrylic topcoat would result in the condensation of the zircoaluminate carboxy functionality with either the acrylic bearing carboxy groups or terminal hydroxy groups (Figure 5).

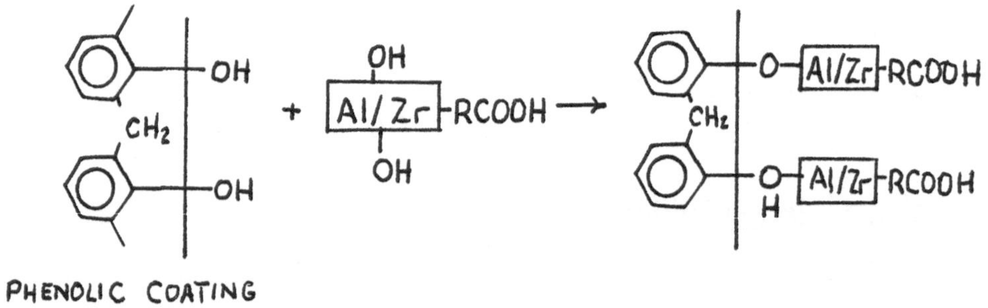

PHENOLIC COATING

Figure 4.   Phenolic Primer Coated with Acrylic Topcoat Containing Zircoaluminate.

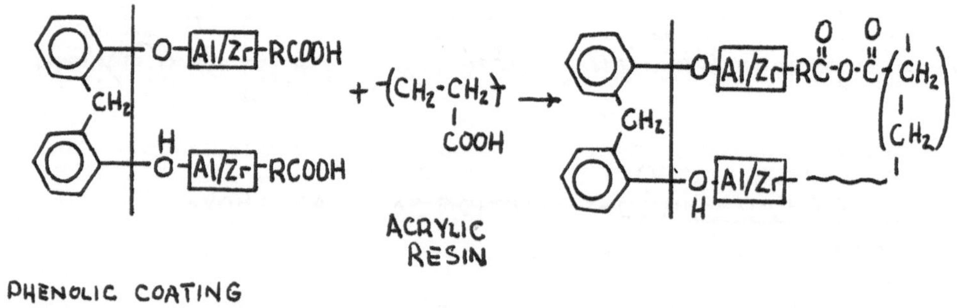

PHENOLIC COATING

Figure 5.   Cure Mechanism for Acrylic/Zircoaluminate Topcoat on Phenolic Basecoat.

176

4.3 Epoxy Structural Adhesives for Metal Bonding

Amino functional zircoaluminate has been used in mineral filled epoxy adhesives for the bonding of steel joints and has been shown to outperform glycidoxy functional silane (CH_2-CH-CH_2-O-$(CH_2)_3$-Si-$(OMe)_3$) as measured by time to failure when such a joint is placed under load and exposed to conditions of moderate temperature and high humidity. It is reasonably expected that the glycidoxy functionality will become covalently linked to the epoxy resin in the instance of the silane and that the amino functionality of the zircoaluminate will also become covalently linked to the epoxy resin. Hence, differences in performance are likely attributable to the stability of the bonds formed respectively by the silicon or aluminum/zirconium to the metal surface (Figure 6).

The precise nature and reactivity characteristics of such zircoaluminate bonds will be the subject of a study which will shortly commence.

4.4 Rubber Bonded to Rubbber

Both silanes and zircoaluminates have been successfully used to promote adhesion between two rubber adherends (to be discussed in section 5). It is exceedingly difficult to view the surface of a fully cured rubber constituted exclusively of highly stable C-H bonds and satisfactorily explain improvements in adhesion which do not derive from chemical reactivity. Section 5 specifically discusses results in an unfilled system where there are no changes in adhesive rheology or wetting characteristics induced by the addition of either adhesion promoter.

5. PRACTICAL ADHESION IMPROVEMENT ON A BONDED EPDM SUBSTRATE

EPDM has been used in diverse applications which require a highly stable non-reactive substrate in the presence of heat and humidity. Not surprisingly, one area of substantial commercial importance is the use of EPDM rubber sheet as a single ply roof membrane. The very property of environmental stability which renders the EPDM so useful in this application also results in great difficulty in bonding adjacent sheets at the seam. Nonetheless, both silanes and zircoaluminates have been used at comparable levels in elastomeric adhesives to dramatically enhance such bonding. Moreover, for reasons not clearly understood, certain zircoaluminate containing adhesives have actually yielded T-peel strengths (ASTM D-1872) 20%-50%

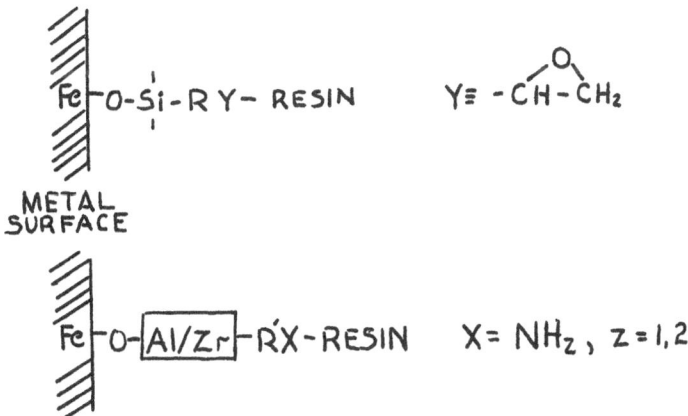

Figure 6. Silane and Zircoaluminate Use for Adhesive Bonding of Metal Joints.

greater than a similar aminofunctional silane containing composition (Table II).

The particular adhesive used contains significant toluene which evaporates in application over 15-30 minutes to leave a tacky surface. During this time, the adhesion promoter undoubtedly migrates to the adhesive substrate interface. Although molecular alignment and interaction is inexplicable, the presence of zircoaluminate (or silane) is manifested in a shift from interfacial failure to cohesive failure of the rubber.

Table II. T-Peel Strengths of EPDM Bonded with Zircoaluminate or Silane Containing Adhesive

Adhesion Promoter	T-Peel Strength pli	Standard Deviation
None	2-4	----
Amino Silane[2]	13.9	18.1
APG-2	16.5	20.1
APG-3	21.2	19.5

(APG comparable to aminosilane)

1. n=5, test performed in accordance with ASTM D-1872
2. Approximately 1 weight per cent adhesion promoter used (silane or zircoaluminate)

6. CONCLUSIONS

Zircoaluminates are not a serendipitous union of zirconium and aluminum; they are instead the result of a thoughtful and deliberate process to synergistically unite two common materials in an uncommon beneficial way. The highly reactive metal hydroxy groups are preserved for substrate reaction by synthesis of the product in sufficient solvent to largely eliminate internal inorganic polymerization. The organometallic nature of the organofunctional zircoaluminate has resulted in it being the optimum adhesion promoter for adhering primer coatings to metals and adhesive bonding of metal joints. Additionally, application has been found in topcoat adhesion and rubber/rubber bonding. Exceptional chemical and thermal stability afford applications for prolonged performance enhancement.

REFERENCES

1. K.L. Mittal, Editor, "Adhesion Measurement of Thin Films, Thick Films, and Bulk Coatings," American Society for Testing and Materials, Philadelphia, PA, 1978.
2. K. L. Mittal, Polymer Eng. Sci., 17, 467 (1977).
3. L. B. Cohen, High Solids Coatings, 9, (No. 3), 2 (1984).
4. Reichhold Bulletin, EPOTUF Epoxy Resin 38-690, Formula Recommendation W/R Mil Spec Primer MIL-P-53032, (September, 1984).
5. H. Ishida, et al, The Society of the Plastics Industry, Annual Conference Reprints, Session 2-B (1986).
6. C. H. Hare, "Corrosion and the Preparation of Metallic Surfaces for Painting", (Unit 26), Federation Series on Coatings Technology, (Copyright 1978), pp. 41-50.

STRESSES IN THIN POLYMERIC FILMS: RELEVANCE TO ADHESION AND FRACTURE

Robert H. Lacombe

I.B.M. Corporation
Rt. 52
Hopewell Junction, NY 12533

The problems of delamination and cracking in microelectronic structures are considered from a unified point
of view, i.e., in terms of the stresses built up in the
structure due to manufacturing processes and end use
conditions. We take the position that delamination and
cracking are simply different aspects of a larger problem,
which is concerned with the thermal-mechanical stability of
the structure. From this vantage point it is equally
important to understand the bulk mechanical and
thermodynamic properties of the materials used in the
structure as well as the detailed properties of the various
interfaces. These ideas are illustrated in detail for a
simple via structure, which is subjected to thermal
expansion mismatch stresses. Detailed finite element
calculations of the relevant stress distributions are
presented. The relevance of bulk mechanical properties on
stresses affecting adhesion at interfaces is emphasized
throughout.

I. INTRODUCTION

When one thinks offhand about the topic of "surface science," one's first thoughts are confined to phenomena which occur at or very close to some geometric surface. ·The adsorption of molecules onto a surface or the adhesion strength of a bond between two different materials are topics which readily come to mind when thinking of "surface science." In practical applications, however, problems arise in which properties of the bulk phases have a strong effect on what happens at an interface, and in a reciprocal way, properties of the interface can affect what happens in the bulk. This reciprocity between interfaces and the bulk is nowhere more evident than in microelectronic applications. The current technology attempts to build multilevel structures of metal, insulator and semiconductor which include all of the electrical functionality of tens of thousands of conventional circuits onto an area the size of a fingernail. The major thrust of the industry is to put more and more circuits into a smaller and smaller space and thereby improve circuit performance while reducing unit costs at the same time. The road to improved performance and lower cost is not an easy one, however, since device yields suffer as one pushes the technology to higher circuit densities. Of all the problems one faces, such as contamination control and heat build up, we shall be interested here rather in the general problems of delamination and cracking. Delamination and cracking are basically different symptoms of the same ailment: excessive stress in the structure. In a typical microelectronic device one has a sandwich of metal, insulator, and semiconductor materials each of which has a different coefficient of thermal expansion. The differences can be rather large in some cases. For example, silicon has an expansion coefficient near 5×10^{-6} K^{-1} whereas high temperature polymer insulators have an expansion coefficient near 40×10^{-6} K^{-1}. Furthermore, manufacturing processes such as metal deposition, solder bonding and brazing require a device to endure large temperature excursions. Stresses due to thermal expansion mismatch alone can get quite high. A simple example will give an idea of the order of magnitude of such stress levels. Assume a uniform polymer insulator film in contact with a silicon substrate. Let the thermal expansion difference be $\Delta\alpha$, the temperature excursion ΔT, and the modulus and Poisson ratio of the polymer insulator E and ν. The tensile stress in the polymer due to the thermal excursion ΔT will be approximated by:

$$\frac{E \times \Delta\alpha}{1 - \nu} \times \Delta T$$

For a typical high temperature polymer insulator E $\sim$3 GPa.[1] Further, it is reasonable to assume $\Delta\alpha \sim 30 \times 10^6$ inv. deg. K and a temperature excursion $\Delta T \sim 380°$ K. The stress in the polymer is then 52 MPa (megapascal)[$\sim$7400 psi (pounds per sq. inch)]. For some insulators this stress is already over half the ultimate tensile strength of the material. Furthermore, this is an average estimate of the tensile stress induced by a temperature excursion T. The stress level can get much higher locally due to discontinuities in the structure. In addition, such a stress level could easily pull apart a weak interface in the structure.

Of course, the situation is much more complicated in real microelectronic structures. In addition to stresses due to thermal expansion mismatch, further stresses can arise due to solvent swelling and shrinkage due to curing reactions or a host of other physico-chemical processes. Furthermore, the materials in the structure do not always respond in a linear reversible fashion. Thus one is confronted with complex yielding phenomena in polymers and strain hardening in metals. In these situations the stress level in a structure becomes a function of its entire thermal-mechanical and processing history. The modern day builder of microelectronic structures finds himself in much the same situation as the stone masons who built the great Gothic cathedrals. Bronowski[2] has reflected on the matter thus "...one has the sense that the men who conceived these high buildings were intoxicated by their new found command of the force in the stone. How else could they have proposed to build vaults of 125 feet and 150 feet at a time when they could not calculate any of the stresses? Well, the vault of 150 feet - at Beauvais, less than a hundred miles from Rheims collapsed. Sooner or later the builders were bound to run into some disaster: There is a physical limit to size, even in cathedrals."

The modern day microelectronic engineer finds himself in much the same kind of situation as the ancient stone mason except that instead of striving to build larger and larger structures constrained by gravitational forces, he attempts to build smaller and smaller structures which are constrained by atomic and molecular forces. There is also a second exception, in that the modern engineer does have the wherewithal to compute the stresses in his structures and therefore, does not necessarily face the same constraint which thwarted the stone masons of the Gothic period. The remainder of this work shall, therefore, be concerned with the problem of stresses in microelectronic structures. As alluded to earlier, the stress distribution in bulk phases can be strongly affected by properties of the interfaces and, vice versa, properties of the bulk phase strongly influence the stresses at the interfaces. In short, when considering the thermal-mechanical stability of microelectronic structures, one must at the same time be equally concerned with bulk as well as surface phenomena.

II. THE CASE OF A LONE VIA

The problem of determining the detailed stress distribution in an actual microelectronic device subjected to some realistic set of mechanical and thermal loads is a formidable problem beyond the scope of this work. Once can, however, learn much about the stress distribution in complex microstructures by studying relatively simple structures in detail. Thus, this section will closely treat the problem of stresses near a via hole in an insulating film coated on a massive hard substrate. The major assumptions to be made are as follows:

1. The via hole is isolated from any other structure in the sense that the stress distribution near the via is unaffected by other parts of the device.

2. The insulating film material is much softer than the underlying substrate and has a much higher coefficient of thermal expansion.

3. The loading on the structure arises solely from the thermal expansion mismatch between the insulator and the substrate materials. In fact, it will be assumed that the structure has

been cooled from 400 to 20°C. There is no stress in the
structure at 400°C.

4. There is perfect adhesion at all interfaces.

5. All materials respond in a linear elastic fashion.

All of the above assumptions with the possible exception of the
fifth tend to be reasonable approximations to reality in practical
situations. The choice of 400°C as a reference temperature from which
to cool is essentially arbitrary, but not necessarily unrealistic,
since metal deposition, soldering and brazing operations may occur in
a neighborhood of this temperature. For the sake of concreteness it
will be assumed that the insulator film has the properties of a high
temperature polyimide material and that the substrate is a silicon
wafer. Table I gives representative elastic constants for these
materials.

Table I

Property	Polyimide	Silicon
Young's modulus	3 GPa	110 GPa
Poisson ratio	0.35	0.14
Thermal Expansion Coefficient	35×10^{6} $^{o}K^{-1}$	5×10^{-6} $^{o}K^{-1}$

Summarizing our model, we have a silicon wafer with a thin
polyimide coating on it, and, at the center of the wafer, a via hole
exists in the film through its entire thickness down to the wafer
surface. This model has an obvious axial symmetry, and we will
compute the stresses using a finite element idealization as shown in
Fig. 1. The finite element model was generated using two dimensional
isoparametric solid elements as provided by the "ANSYS"[3] programming
system. Under the conditions of this model the stresses depend only
on the relative scale of the model and not on any absolute dimension.
The question of the effect of different via sizes will be discussed at
the end of this section.

Stresses Near An Isolated Via

Figure 2 gives an exaggerated representation of the deformation
which the via structure in Fig. 1 would suffer on being cooled from
400°C to room temperature. The polyimide material, having a much
higher coefficient of thermal expansion, is shrinking much more
quickly than the silicon substrate and this gives rise to the
calculated deformation. Note that those elements of the film which
border the silicon substrate have the most severe constraint placed
upon them and they thus suffer the highest strain levels. It is those
strains which give rise to the stress levels illustrated in Fig. 3.
The stress distribution in any continuous solid is represented by a
second rank tensor quantity called the stress tensor. In the general
case of an elastic solid there are six independent components to this
tensor. In the present case we have an axisymmetric solid laminate
made out of materials which are assumed to be homogeneous and
isotropic. In this case there are only 4 independent components of
the stress tensor. Figure 3 represents a contour diagram of the
"normal" stress component or that component which is directed

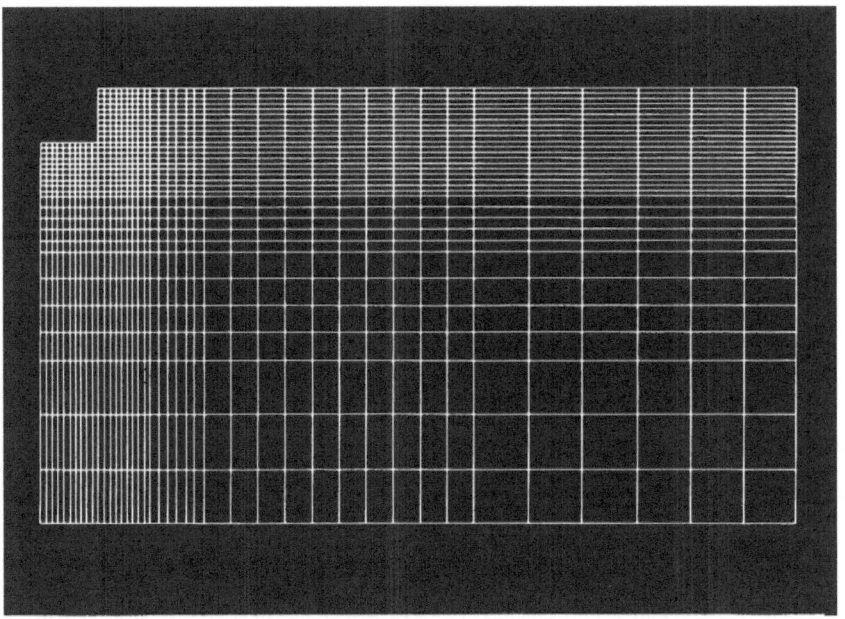

Figure 1. Finite element idealization of an isolated via. The model
has axial symmetry about the vertical axis and reflection
symmetry about the horizontal axis.

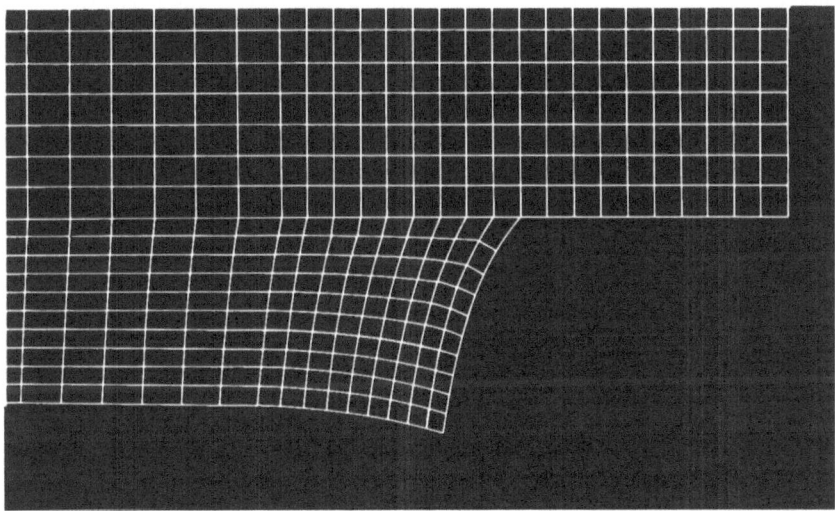

Figure 2. Via deformation on cooling from 400°C. Note that the
magnitude of the deformation is expanded for visual purposes.

perpendicular to the wafer surface. In particular this is the primary
component of the stress tensor which might cause delamination of the
film from the wafer substrate. Note the stress concentration at the
base of the via wall approaching a maximum level of 125 MPa
(megapascal) or equivalently 18,000 psi (pounds per square inch).
This level of stress is already high enough to cause delamination or
even cracking in some films.

A second tensile component of the stress tensor is the radial stress which acts in a plane parallel to the wafer surface and in a radial direction from the center of the via hole outward. Figure 4 illustrates the constant stress contours for the radial stress. Again the maximum stress level occurs at the base of the via wall and is at a very high level of 125 MPa. In a brittle coating such levels could give rise to circumferential cracks near the via rim.

The final tensile component of the stress tensor is the circumferential or "hoop" stress. This stress component also lies in a plane parallel to the wafer surface and is perpendicular to the radial stress component. Thus the circumferential stress distribution illustrated in Fig. 5 acts in a direction perpendicular to the plane of the paper. Again the maximum value occurs at the base of the via wall and is a very high 127 MPa. Such high stress levels could give rise to radial cracks. Figure 6 illustrates the three failure modes possible near a via hole. In our laboratory we have in fact observed all three modes of failure in different insulator films on hard massive substrates.

Another question is how do the various stresses vary as we vary the size of the via? This question can be answered by constructing finite element models of vias with different aspect ratios (ratio of diameter to depth). Figure 7 shows a plot of the maximum value of each of the tensile stress components as a function of via aspect ratio. Figure 7 has a number of nonintuitive features. It is surprising that all stress levels decrease as one goes to vias with smaller diameters. It is also unclear why all the stresses should reach a local maximum at an aspect ratio of 1 to 1 and then decrease in vias with larger radii only to increase again at the largest aspect ratio considered. It seems the best one can say is that one can increase or reduce via stresses by changing the via aspect ratio.

Modifying The Interphase

One question a pragmatic reader may have at this point is that, if the stresses near a via are too high, is there anything that can be done to correct the situation? The answer to that question is a qualified "yes." One of the first things that might be tried is to create a via structure with more gently sloping walls, thereby reducing the stress buildup at the base of the via. In stress analysis it is always a rule of thumb to avoid sharp edges to prevent large stress concentrations. However, electrical designers prefer vertical-walled vias to vias with gently sloping walls since the vertical walled variety consume less space and thus more circuitry can be crammed into a given area. Can we then reduce via stresses without reverting to sloping via walls? The answer is still a qualified yes if one is willing to do the extra work required to provide a soft interphase region between the hard silicon substrate and the insulator film. Figure 8 gives a finite element idealization of a possible structure. For concreteness we can let the insulator coating be two microns thick and the interphase layer 2000 $\overset{\circ}{\text{A}}$ thick. A typical polyimide insulator will have a Young's modulus on the order of 3 GPa. One would like to find a different polyimide material with a Young's modulus near 0.3GPa to serve as the interphase material. Such materials can be found among the modified polyimides such as the polyetherimides or silicone-containing polyimides. These materials can have a modulus near .1 GPa and will also tend to have a Poisson ratio near 0.4 and a thermal expansion coefficient between 100 and 200 x 10^{-6} K^{-1}. Using these hypothetical elastic constants, we can recalculate all of

184

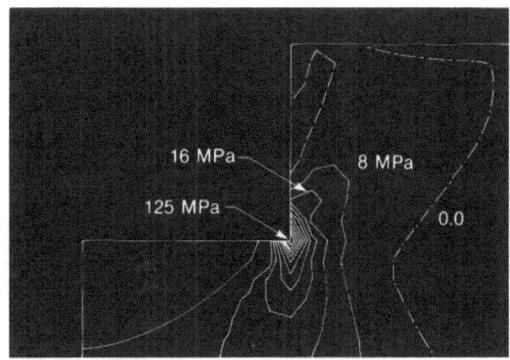

Figure 3. Normal stress distribution near via. Normal stresses act along a line perpendicular to the coating–substrate interface.

the via stresses for this new composite via structure when it is cooled from 400 to 20°C. Figure 9 gives a contour plot of the normal stress distribution generated by this process. The remarkable feature of this plot is the tremendous drop in the maximum normal stress, which still is located at the base of the via wall. The new value of 12 MPa is an order of magnitude less than the value of 125 MPa calculated in Fig. 3 for the via structure without a soft interfacial layer. This is of course very good news if one has delamination

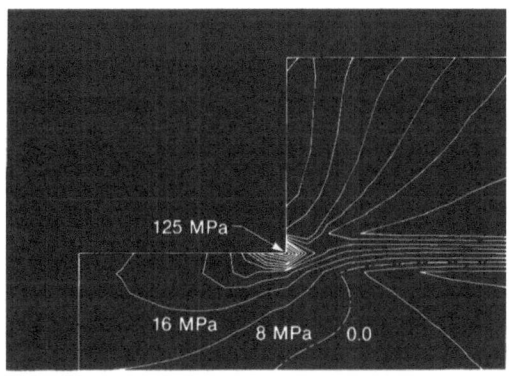

Figure 4. Radial stress distribution near via. The radial stresses act along a radial axis parallel to the coating–substrate interface.

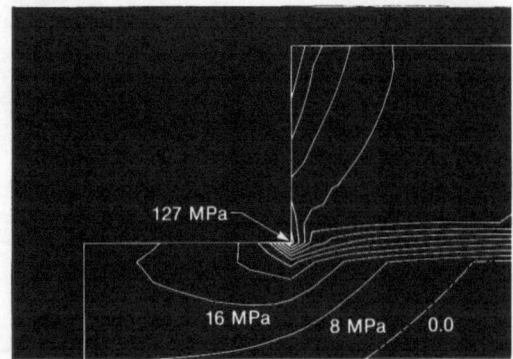

Figure 5. Circumferential stress distribution near via. The circumferential stresses act in a direction perpendicular to the plane of the paper.

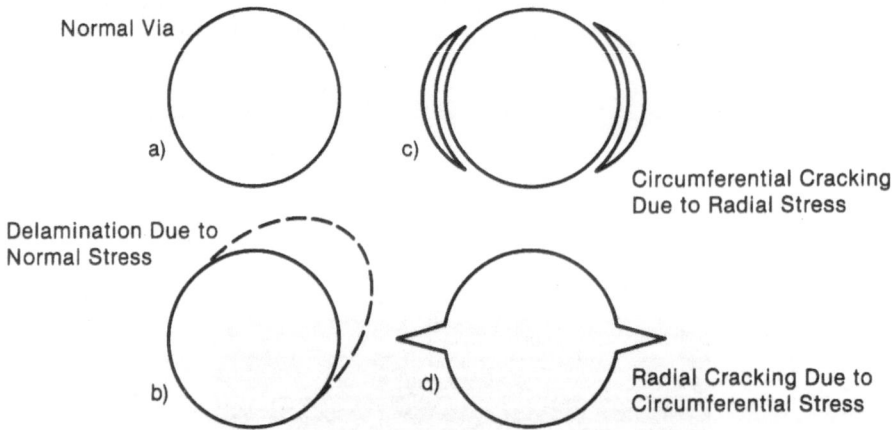

Figure 6. Modes of mechanical failure near a via.

problems but one still has to be aware of how the other stress components are behaving.

Figure 10 shows the situation for the radial stresses. Two features are of immediate interest: First, the maximum stress is now only 30 MPa as compared to 125 MPa for the unmodified via structure in Fig. 4. Second, the location of the maximum stress has moved back away from the base of the via wall into the bulk of the insulator layer. The much lower maximum stress value of 30 MPa is good news for the alleviation of circumferential cracking of the type depicted in Fig. 6c.

At this stage the introduction of a soft interfacial layer into the via structure seems to be some kind of universal panacea for relieving stress problems. However, before becoming too elated, we should examine closely the final tensile component. The circumferential stress is plotted in Fig. 11. Again there are two points to be noted. Curiously enough the location of the maximum stress has shifted from the base of the via wall to the top of the via wall. More disconcerting, however, is the high value of 103 MPa of the maximum stress, not down very much from the value of 127 MPa shown in Fig. 5 for the plane via structure. Thus if one's structure were suffering from radial cracking of the type illustrated in Fig. 6, the use of a soft interfacial layer might not bring much relief. This is a curious nonintuitive result arising from the geometrical complexity of the stress tensor.

III. CONCERNS OF MATERIALS CHARACTERIZATION

Problem of Large Deformation and Nonlinear Response

Though the stress problem discussed in the previous section is highly simplified and involves a number of idealizations, it is nonetheless instructive and gives one important clue as to what to expect under more realistic conditions. One assumption, which requires more attention, is that of assuming linear elastic response of the insulator coating when it is subjected to any finite

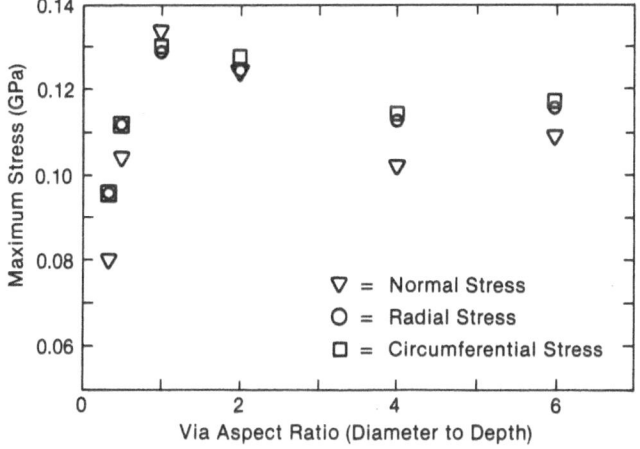

Figure 7. Plot of the maximum value of the tensile stress components versus via aspect ratio.

deformation. This assumption certainly makes the calculation of the stress distributions much easier. For instance, by referring to Fig. 7 for the linear elastic case, we find that it costs a few hundred dollars to compute all the stresses for a single via aspect ratio on a 1982 model super minicomputer. If full nonlinear response behavior must be accounted for, however, the bill for the same calculation easily runs to several thousand dollars. Thus the mechanical response properties of the materials one is working with deserve careful consideration.

If one is working with polyimide insulators, for instance, one has a class of materials which can show a wide range of thermal-mechanical response behavior. Figure 12 shows the tensile

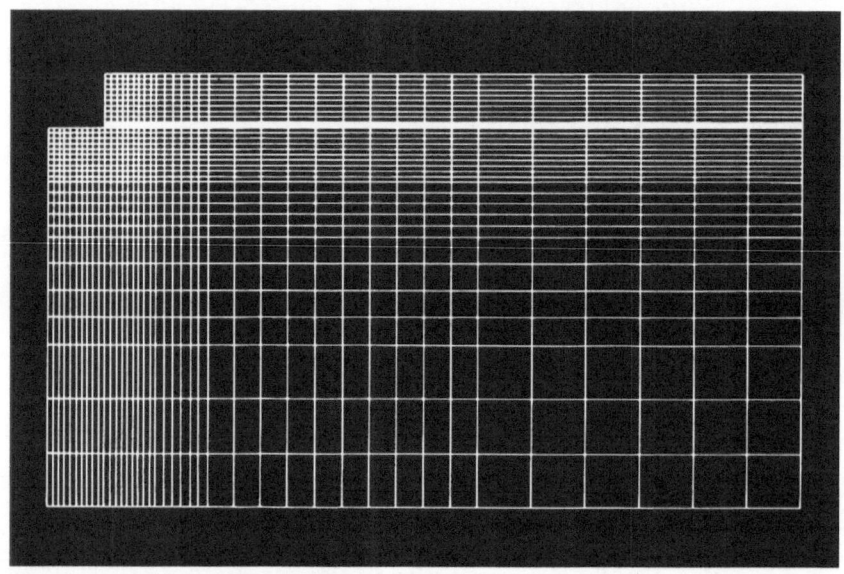

Figure 8. Composite via structure with soft interphase region between substrate and coating.

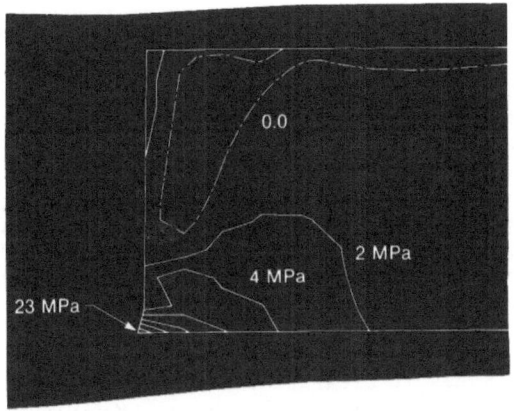

Figure 9.   Normal stress distribution for composite via structure.

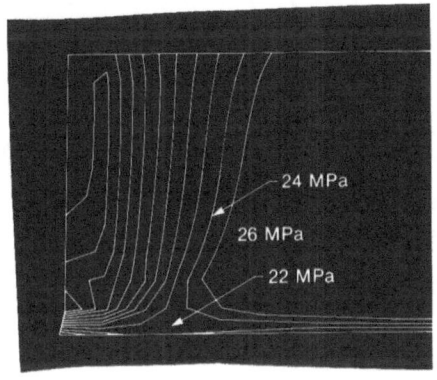

Figure 10.   Radial stress distribution for composite via structure

stress—strain behavior of three different polyimide materials. Note
the broad range of mechanical behavior ranging from a brittle
glass—like behavior of the BTDA/MDA material to a nearly rubber—like
response of the PMDA/ODA sample. The same plot shows a dashed line
that would represent the response of a perfect elastic solid, such as
was assumed for the calculations in section III. We note from those
calculations that all of the tensile stresses reached a maximum value
of 125 MPa or more in the simple via structure which was cooled from
400°C. This stress level would clearly lead to cracking in the
BTDA/MDA material. For the H—H and the PMDA/ODA materials, however,
the maximum stress would fall back to a level near 60 MPa due to
stress relief from yielding. Thus, neither of these materials would
be expected to show cracking, but delamination from the substrate is
still a possibility.

This example makes it clear that the mechanical response behavior
of the bulk layers in a multilevel laminate structure may have a very
strong effect on the level of stress, which can be built up due to
thermal cycling or other mechanical loading. In particular, the
normal stresses across an interface are affected, and thus the
durability of that interface is also strongly altered. This is a
clear and direct example of bulk properties affecting the properties
of an interface. We reach the curious conclusion that perhaps a good
way to control what is happening at an interface is to modify certain
bulk properties of the materials one is working with as opposed to
directly manipulating surface properties. For problems involving
stresses, this is especially true due to the way most solids
efficiently transfer stress loads from one phase to another across
interfaces.

Characterizing the Mechanical Properties of Thin Films

If we have established the fact that understanding the mechanical
stability of interfaces in laminated structures requires an

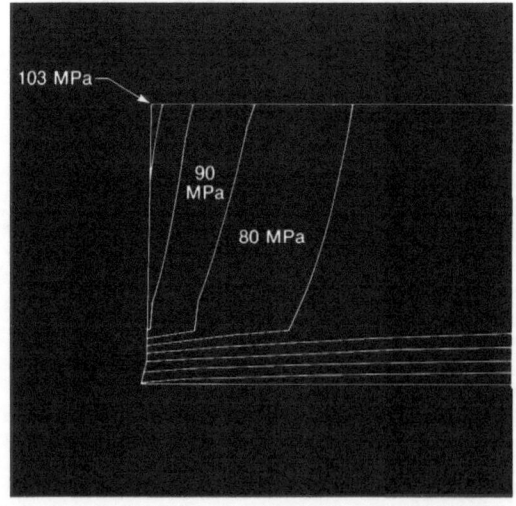

Figure 11.   Circumferential stress distribution for composite via
structure.

understanding of the mechanical properties of the bulk layers, then it becomes relevant to ask how does one characterize the mechanical properties of thin films. In the microelectronic industry, insulator films can range from a few hundred angstroms to several microns. For the present discussion we will assume we are dealing with a film thickness near 1 micron. In determining the mechanical properties of such films, one encounters a number of difficulties, a significant one being how to handle the films. With all the apparent difficulties surrounding the direct determination of thin film mechanical properties, one is tempted to ask why should we even bother with this problem in the first place? Why not just rely on the bulk mechanical properties of our materials? The answer to this question involves surface ordering effects. It is well known that the presence of a surface affects the way in which polymer chains will align themselves. Cohen and Reich,[6] for example, have studied such ordering phenomena in polystyrene films. Such ordering phenomena would be restricted to a very thin surface layer in bulk samples and thus would likely go unnoticed. However, for a 1-micron film, a layer on the order of 1000 $\overset{o}{A}$ thick is a significant fraction of the bulk material. What is more important, however, is the fact that a thin interphase layer occurring at an interface in a layered structure can dramatically alter the stress distribution in the entire structure. The example in the previous section on the modified via structure amply illustrates this point.

Polymer chain orientation at an interface might give rise to more efficient chain packing in a localized region near the surface. Such a more densely packed region would be expected to have a larger tensile modulus than the bulk phase and could thereby tend to generate higher stress levels than might normally exist otherwise. Thus, it is important to be able to directly determine the elastic behavior of thin coatings. Unfortunately, very few techniques are available for films of 1 micron and thinner, but one technique, which shows promise, involves coating a film on a cantilevered substrate. By observing the mechanical response of the composite sample, one can calculate the mechanical properties of the coating. This technique, known as the "Vibrating Reed" experiment, was first employed by Berry and Pritchet[7] in examining the thermal-mechanical properties of thin metal films.

Subsequently, Lacombe and Greenblatt[8] used the identical technique to look at the mechanical properties of 1000 $\overset{o}{A}$ PMDA/ODA polyimide coatings. The latter work, while operating at the limit of sensitivity of the technique, seemed to indicate that the thin polyimide film had a tensile modulus nearly twice that of the bulk material. A more detailed and careful study of this problem is clearly in order. Note also that the "Vibrating Reed" technique can be used to evaluate the relative toughness of polyimide films by monitoring the internal friction behavior as a function of temperature. A comparison of BTDA/MDA and PMDA/ODA type polyimides is given in reference 8.

Before leaving the topic of thermal-mechanical characterization of thin polymer films, the question of residual stress levels due to curing or manufacturing processes should be addressed. Residual stresses can be due to a variety of causes, but in polymer films there tend to be three major causes: shrinkage due to chemical reactions on curing; solvent loss on drying; and thermal expansion mismatch between film and substrate. The internal stresses in BTDA/MDA and PMDA/ODA polyimides have been discussed in detail by Geldermans et al[9] and by Goldsmith et al.[10]

Among the important results derived from these studies are the following:

1. Residual stresses in fully cured BTDA/MDA and PMDA/ODA films are independent of film thickness.

2. Film stresses in fully cured films are dominated by the effect of thermal expansion mismatch between film and substrate.
3. For cured films the state of zero stress occurs at the highest temperature to which the film was subjected.

4. The maximum stress level achieved in the BTDA/MDA films was more than twice that in the PMDA/ODA films.

Results 1 through 3 completely support the linear stress analysis of section III by assuming that the stresses are dominated by thermal expansion mismatches and that using $400^{\circ}C$ as a reference state of zero stress is perfectly valid. Furthermore, observation 1 validates the results for different film thicknesses. Observation 4 reinforces the conclusion that the linear analysis will be in substantial error for PMDA/ODA films. We already know this from the previous discussion at the beginning of this section. It is clear that an experimental study of the residual film stresses is indispensable to understanding the stress levels which can develop in actual structures.

Failure Criteria

The whole premise of this paper has been that failure phenomena such as cracking or delamination are caused by excessive stress levels. While strictly speaking this is true, in practice there are a number of difficulties which make precise prediction of failure very difficult. Not the least of these difficulties is the fact that real structures inevitably include a number of defects which cannot be accurately included in a stress analysis. Defects alter the local structural geometry in an unpredictable way and nearly always tend to give rise to regions of high local stress. The problem is further compounded by the fact that the precise conditions of local temperature and material composition can only be determined in some average sense. It is for reasons such as these that there is a very large literature on statistical failure analysis of devices.

For materials like the BTDA/MDA polyimide, however, the failure criterion for fracture is quite straight forward since the material shows simple linear stress-strain behavior right up to the failure point. For such a material, if the stress is greater than the fracture stress it fails and if less than the fracture stress then no failure. The PMDA/ODA material, on the other hand, will show yielding behavior and then enter a large region of elastic/plastic response. This is one of the most difficult types of material on which to do a failure analysis. Given the large range of strain over which the material behaves more like a rubber than a glassy solid, it is more likely that this material will exhibit delamination rather than cracking.

Though delamination and cracking are essentially alike from a stress analysis point of view, historically, the studies of delamination and cracking underwent separate developments. Detailed study in either field can easily be a life's work. Readers interested

192

in pursuing fracture failure phenomena can consult the elementary text by Broeck[11] and the many references therein. A treatment more exclusively devoted to fracture phenomena in polymers can be found in the text by Hertzberg and Manson.[12] The field of adhesion and adhesion testing is equally vast. A good overview of the field may be found in the edited compendium by Mittal.[13]

Though any detailed discussion of failure analysis and adhesion testing is clearly beyond the scope of this work, an example of a novel adhesion test for determining the adhesion of brittle films to flexible substrates will be presented to give some flavor of the subject. The test as originally conceived by Chow[14,15] is illustrated in Fig. 13a. A thin film of some brittle material is deposited onto a strip of flexible material usually by a vapor deposition technique that uses a mask. The ends of the flexible substrate are left uncoated, and the strip can thus be mounted in a tensile testing device. A uniaxial strain can now be applied to the sample until the coating material fails either through delamination or fracture. In order to be able to evaluate the interfacial free energy of adhesion, Chow[14] attempted an analytical calculation of the stresses generated during the tensile test. However, in order to keep the problem tractable, the normal stresses between the film and substrate were assumed to be negligible. This assumption makes the test of limited use for adhesion testing since normal stresses tend to be a primary cause of delamination.

In light of the above mentioned problem, we decided to recompute Chow's model by using a finite element analysis, taking into account all relevant stresses, and using a fine uniform mesh in order to ensure numerical accuracy. However, through an inadvertent oversight the actual model investigated was that in Fig. 13b instead of Chow's original model, which is shown in Fig. 13a. The model in Fig. 13b is of interest in its own right since the asymmetry gives rise to a high normal stress at the interior edge of the film coating. Figure 14 illustrates a specific example in which the flexible substrate is a PMDA/ODA polyimide and the brittle coating is vitreous silica. A tensile load of 100 g is applied, and the resulting deformation of the composite strip is illustrated. It is important to note that Fig. 14 shows only the coated half of the specimen in Fig. 13b. The edge of the coating at the middle of the strip is referred to as the interior edge. Figure 15 shows the normal stress distribution at the interior edge of the coating under the load conditions of Fig. 14. From the figure we see that the maximum stress occurs some 10 microns in from the interior edge and has a maximum value of 3.5 MPa. The edge itself is under a compressive normal stress of 2.5 MPa. The model would then predict delamination in the form of a blister very near the interior edge. In practice, however, this fine distinction might not be noticeable since any interior delamination so close to an edge could readily propagate to the edge and obscure the original locus of failure. Figure 16 illustrates the tensile stresses induced in the coating. A large uniform tensile stress exists near the interface of the polyimide and silica and has a value slightly larger than 10 MPa. A much higher level tensile stress of 27 MPa exists at the interior edge of the coating. Since the maximum tensile stress in the coating is some 5 times higher than the normal stress the sample could fail through fracture instead of delamination. However, from Fig. 16 we also see that the surface of the coating is under a tensile compression of over 4 MPa. This compressive stress will tend to suppress any crack propagation from surface defects, which is known to be one of the major modes of failure in brittle glasses. Thus, although the tensile stresses are much larger than the normal stresses at the interface, the existence of high uniform compressive stresses at the surface could easily shift the failure mode in favor of delamination from the substrate, as opposed to cracking of the film in the bulk. For these reasons the sample configuration in Fig. 13b is to be preferred over that in Fig. 13a for purposes of adhesion testing.

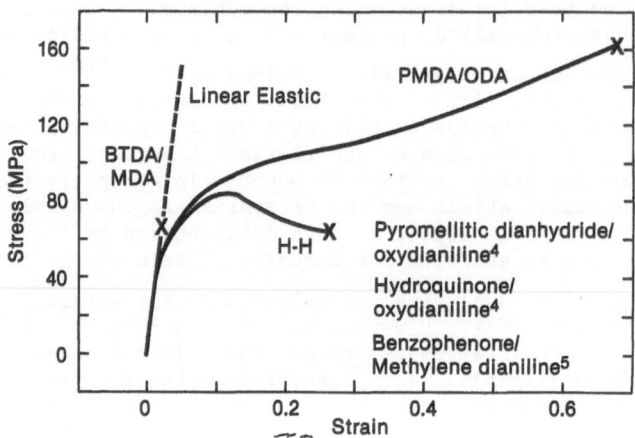

Figure 12. Tensile stress–strain behavior of three polyimides: PMDA/ODA,(Pyromellitic dianhydride/oxydianiline)4; H-H, (Hydroquinone/ oxy dianiline)[4]; BTDA/MDA, (Benzophenone/Methylene dianiline)[5].

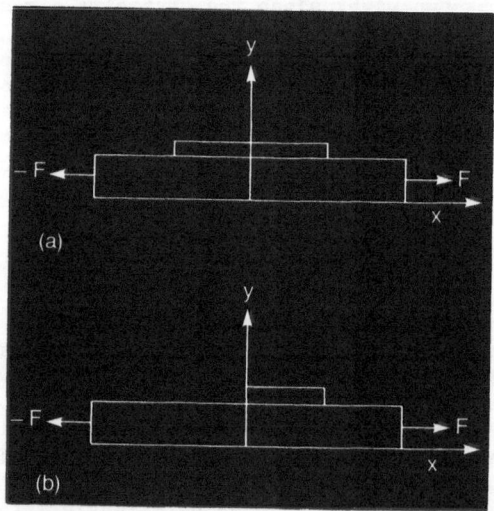

Figure 13. a) Geometry for tensile adhesion test of thin brittle coating on a thick flexible substrate. [14,15]

b) Modified geometry for tensile adhesion test of thin brittle coating on a thick flexible substrate.

IV. DISCUSSION

The analysis of failure mechanisms in microelectronic structures requires attention to both the properties of the interfaces and the bulk layers. It is futile to examine any particular interface in great detail while ignoring the mechanical properties of the bulk layers, since these properties strongly determine the state of stress

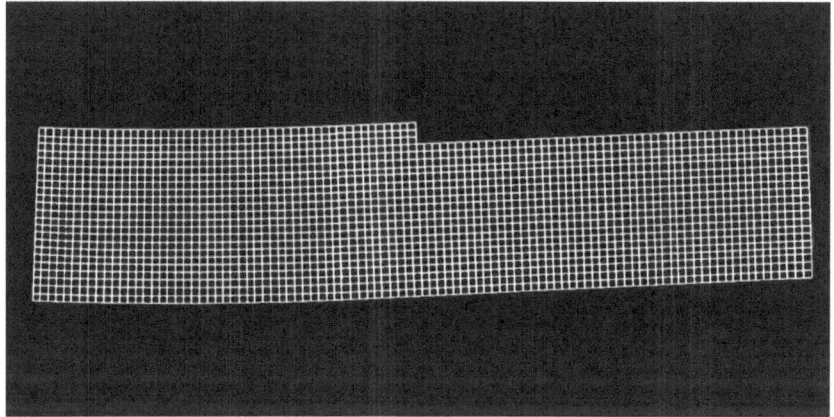

Figure 14. Deformation generated during tensile adhesion test of SiO_2 on PMDA/ODA polyimide substrate. Elastic constants are: PMDA/ODA polyimide, Young's modulus = 3 GPa, Poisson ratio = 0.35. Silica, Young's modulus = 70 GPa, Poisson ratio = 0.14. 30 micron coating on 200 micron substrate.

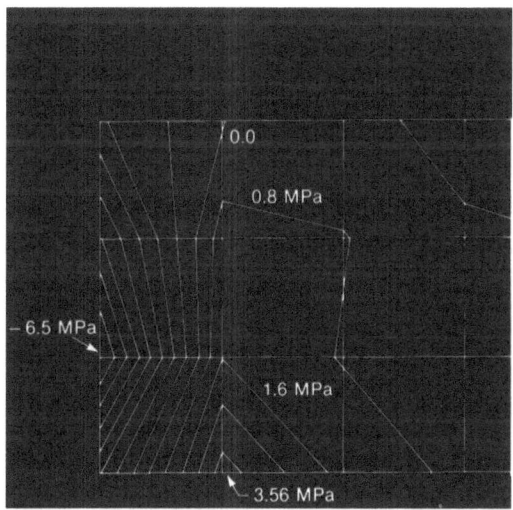

Figure 15. Normal stresses generated in tensile adhesion test specimen.

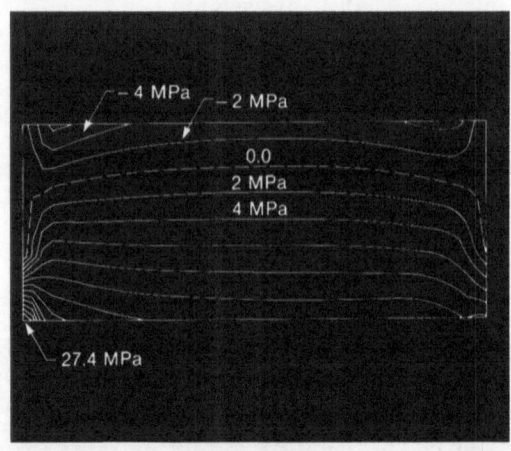

Figure 16. Tensile stresses generated in tensile adhesion test specimen.

at the interfaces. We have seen, furthermore, that thin interphase regions can radically alter the state of stress at an interface and in a large part of the bulk structure as well. Since the ultimate mechanical stability of any structure depends on the state of stress in that structure, we can only conclude that any program aimed at understanding cracking or delamination in microelectronic structures must include a joint effort to understand both the thermal-mechanical properties of the bulk layers and the physico-chemical properties of the interfaces.

Another example of interdependence of bulk and surface properties occurs whenever one attempts to quantitatively evaluate the adhesion strength of any interface. Quantitative adhesion tests invariably involve the forced separation of two different layers of a test structure in order to determine the strength of the interfacial bonds. In all cases corrections have to be made to account for the contribution of the bulk layers to the total force required to achieve separation. In the case of the Peel test, for example, Gent and coauthors[16,17] have given an account of the thermal-mechanical analysis involved. Crocombe and Adams[18] have further shown how a finite element analysis might be employed in evaluating peel test data.

In conclusion, it is fitting to speculate on what the future holds for this type of analysis. As a start, it is safe to assume that

the microelectronics industry will continue to push for ever higher
levels of integration in its microcircuit structures in order to gain
higher levels of performance, higher device reliability and lower cost
per circuit. This translates into even smaller devices, which are
more densely packed with additional levels of wiring in order to meet
input-output demands and device-to-device interconnection
requirements. Such multilevel devices will require even more
sophisticated materials analysis and stress modeling in order to
achieve a structure that will survive the rigors of the manufacturing
process and end use conditions. A comprehensive and coordinated
effort in surface and interface analysis coupled with
thermal-mechanical characterization of all the materials involved will
have to be implemented. This will provide a data base on which more
detailed stress analyses will be carried out in order to ascertain the
mechanical stability of a proposed structure before it is built. Such
calculations will become more and more critical as device structures
become more complicated and the price of mistakes becomes more
expensive.

REFERENCES

1. The official measure of stress or pressure in SI units is The
   Pascal or one Newton per meter square. The Pascal is such a small
   unit that the prefixes Giga (10^9) and Mega (10^6) are often
   affixed. Many people seem to find it impossible to think in terms
   of these units so conversions to the more popular but outdated
   British units of pounds per square inch (psi) are supplied in this
   paper. A simple approximate conversion of Pascals (Pa) to psi is
   obtained by dividing the quantity in Pascals by 7000.
2. J. Bronowski, The Ascent of Man, (Little, Brown Company, 1973),
   Chapter 3.
3. "ANSYS" is the registered trademark of Swanson Analysis Systems,
   Inc., Houston, Pennsylvania 15342.
4. A. S. Argon and M. F. Bessonov, "Plastic Deformation in
   Polyimides, with New Implications on the Theory of Plastic
   Deformation of Glassy Polymers", Phil. Mag., 35, 917, (1977).
5. M. Gupta, unpublished data.
6. Y. Cohen and S. Reich, "Ordering Phenomena in Thin Polystyrene
   Films", J. Polymer Sci: Polymer Phys. Ed., 19, 599 (1981).
7. B. S. Berry and W. C. Pritchet, "Vibrating Reed Internal Friction
   Apparatus for Films and Foils", I.B.M. J. Res. Devel., 19, 334,
   (1975).
8. R. H. Lacombe and J. Greenblatt, "Mechanical Properties of Thin
   Polyimide Films", in Polyimides, 2, K. L. Mittal Ed. (Plenum
   Publishing Corp., 1984).
9. P. Geldermans, C. Goldsmith, and F. Bedetti, "Measurement of
   Stresses Generated During Curing and in Cured Polyimide Films",
   ibid.
10. C. Goldsmith, P. Geldermans, F. Bedetti, and G. N. Walker,
    "Measurement of Stresses Generated in Cured Polyimide Films", J.
    Vac. Sci. Technol. A., 1, 407 (1983).
11. David Broeck, Elementary Engineering Fracture Mechanics, (Martinus
    Nijhoff Publishers, 1982) 3rd Ed.
12. R. W. Hertzberg and J. A. Manson, Fatigue of Engineering Plastics,
    (Academic Press, 1980).
13. Adhesion Measurement of Thin Films, Thick Films and Bulk Coatings,
    K. L. Mittal Ed. (ASTM STP 640, 1978).
14. T. S. Chow, "Adhesion of Brittle Films on a Polymeric Substrate",
    in Adhesion Science and Technology, Lieng-Huang Lee Ed. (Plenum
    Press, 1975).

15. T. S. Chow, C. A. Lon and R. C. Renwell, "Direct Determination of Interfacial Energy Between Brittle and Polymeric Films", J. Polym. Sci., Polym. Phys. Ed., 14, 1305 (1976).
16. A. N. Gent, "The Strength of Adhesive Bonds", Adhesives Age, February 1982, p. 27.
17. A. N. Gent and G. R. Hamed, "Peel Mechanics for an Elastic-Plastic Adherend", J. Appl. Polym. Sci., 21, 2817 (1977).
18. A. D. Crocombe and R. D. Adams, "Peel Analysis Using the Finite Element Method", J. Adhesion, 12, 127 (1981).

ADHESION AND CROSSLINK GRADIENT IN A PHOTORESIST

R. L. Geary[†], S. V. Babu[†], and J. Stephanie[*]

[†]Department of Chemical Engineering
 Clarkson University
 Potsdam, NY 13676
[*]IBM Corporation
 Systems Technology Division
 Endicott, NY 13760

Using chloroform and hexane as extraction solvents, we determined the fraction of reacted 'monomer as a function of exposure energy in a Riston(R) ** photoresist film. We have also measured adhesion at the resist-substrate interface and find that adhesion is inversely correlated to the crosslinking at the interface.

INTRODUCTION

Negative photoresists are used quite extensively in the fabrication of semiconductor devices and printed circuit boards. Maintaining good adhesion to the substrate and maintaining pre-defined line channel profiles are highly desirable properties of negative resist films. It is well known that both of these properties are dependent on crosslink uniformity and solvent-caused swelling throughout the resist film. However, determining the crosslink gradient is a nontrivial task[1] and no method exists to determine the gradient in the exposed negative photoresist.

In the following, we briefly consider several possible methods to determine the crosslink gradient. Three of these techniques, namely the molecular optical laser examiner (MOLE), ATR/FTIR, and Measurement and Evaluation of Surfaces by Evaporative Rate Analysis (MESERAN)(R) ***, did not possess the required sensitivity in the experimental configurations that we investigated. Our results on the extent of crosslinking were finally based on a dual-solvent extraction procedure. Furthermore, the adhesion of the resist to the substrate has been determined at different exposure energies. Finally, by correlating the adhesion of the resist to the substrate with resist crosslinking, the resist performance can be better characterized. The resist used in all of the experiments is a Du Pont product with the tradename Riston(R) .

**Riston is a registered trademark of E. I. Du Pont de Nemours and Co., Inc., Wilmington, DE.; ***Meseran is a registered trademark of the Meseran Co., a division of ERA Systems, Inc. Ooltewah, TN.

METHODS OF CROSSLINK DETERMINATION

There does exist a variety of techniques which can be used to determine the extent of polymerization that has occurred during the photocuring process. Among these are high-pressure liquid chromatography, gel permeation chromatography, Raman spectroscopy, Fourier transform infrared spectroscopy with either photoacoustic spectroscopy or attenuated total reflectance spectroscopy as sampling techniques, thermogravimetric analysis, evaporative rate analysis, solvent absorption, and solvent extraction. Four of the preceding methods were attempted in the course of this study.

In the first method a MOLE was used. The basic instrument combines a conventional optical microscope which has bright and dark field illumination, with an optical filter possessing a very low stray light level, and a multi-channel and/or mono-channel detection system[2]. A laser (argon, krypton or dye laser, etc.) is the monochromatic source used for irradiating the sample. In the laser molecular microprobe, photons generated by the laser are used to excite the sample and cause the emission of Raman lines of the various components.

The extent of crosslinking can be determined by monitoring the disappearance of 1,643 cm^{-1} peak in the Raman spectrum which is due to the C=C bond stretching mode[3,4]. In our experiments using continuous wave krypton laser, no meaningful results were obtained. The intense fluorescence caused by the initiators and dyes in the Riston Ⓡ photoresist masked the Raman signal. Since the Raman emission decays faster than the fluorescence emission, we suggest the use of a pulsed laser of the appropriate width in the MOLE system to obtain useful results.

Next, we utilized attenuated total reflectance/Fourier transform infrared spectroscopy (ATR/FTIR)[1]. The scan was done with an IBM Instruments FTIR/98. The FTIR/98 utilizes 100 scans at a resolution of 2 cm^{-1}. The 810 cm^{-1} IR band, which has been ascribed to vinyl C-H beinding, was used to monitor the degree of crosslinking. However, the technique showed no differences in the IR spectra of resist films of various thicknesses[5]. Perhaps a more sensitive ATR/FTIR system is necessary to detect these differences in the spectra.

The third method of crosslink determination tried in the course of the experiments was evaporative rate analysis (ERA). ERA had been demonstrated previously as a method for monitoring crosslinking[6]. The MESERAN Ⓡ Surface Analyzer Model 1200 with nitrogen as a sweep gas was used in all ERA experiments. The analyzer mechanically dispenses 0.02 ml of a test solution onto the sample. After the test solution has been deposited, a Geiger-Mueller detector tube begins counting the evaporation of the radioactive chemical (C^{14}) found in the test solution. The change in the evaporation rate of the C^{14} can be used to monitor crosslinking. During the experiments, it was difficult to determine the crosslinking due to the large scatter in the experimental data obtained. Interfacing the MESERAN Ⓡ apparatus with a more advanced computer would make the method more feasible.

The final method utilized in this research, which was the most successful, was solvent extraction. This method is based on the solubility differences between the crosslinked polymer, unreacted monomer, and the oligomers, and the binder and other resist components. The experimental details are described in the next section.

EXPERIMENTAL

In order to determine the crosslink gradient in the negative dry film photoresist, a series of samples of different thicknesses were required. These samples were prepared using a spin coating process. First the resist, after the removal of the Mylar®️ and the PE protective cover sheets, was dissolved in chloroform ($CHCl_3$). Approximately 4 parts by weight of $CHCl_3$ to 1 part photoresist were used. The dissolved resist was spin coated on 2.5 inch square glass plates for 30 seconds and then allowed to dry in air overnight at room temperature and ambient pressure. Chloroform, being very volatile, readily evaporated from the film. After drying, a Mylar®️ film was placed over the samples for protection.

Resist films with thickness ranging from 0.5-2.9 mils were obtained by spin coating at different speeds, typically 100-1500 rpm. The uniformity and thickness of the resist coatings were measured by a Sloan Dek Tak profilometer. The uniformity of the coated resist films was found to be excellent, with a variation of ± 0.05 mils over a scan of 1.0 inch.

The resist films were exposed in two different ways. In both methods the resist films were placed in a vacuum frame prior to and during the exposure process. In one technique the spin coated resist films were exposed through the glass face after the removal of the Mylar®️ protective cover sheet. This method was used for the determination of the crosslink gradient experiments by solvent extraction and infrared spectroscopy. In the second exposure procedure, the laminated resist was exposed through the Mylar®️ film and then used for crosslink determination by both Raman spectroscopy and evaporative rate analysis, as well as for all adhesion measurements.

For all of the photoresist samples the incident intensity of the UV light was measured at the resist surface by subtracting out the absorption of light due to either the glass plate or Mylar®️ film. The Mylar®️ film cut off all incident light of wavelengths less than 330 nm, while the glass plates allowed the transmission of light of wavelengths greater than 340 nm. All exposures were performed on an Optical Associates Inc. laboratory exposure system. The system contained a HBO 350 watt high pressure mercury lamp with a seven inch collimating lens, and an automatic shutter and timer. The lamp intensity could be changed through the variable power supply. All the experiments described here were performed at the same power setting and, hence, the same intensity.

The intensity of the UV light was measured with an International Light 700A Research Radiometer using a WBS #320 probe which scans the intensity in the range 300-400 nm. Repeated tests with the 320 probe at the same power setting gave intensity readings with an accuracy of ± 0.03 mw/cm^2. To determine the transmitted and incident intensity of the light through and at the surface of the resist film at the different wavelengths an EG & G model 550/555 Spectroradiometer was used. The system consists of a fiber optic probe which inputs data through an order sorting filter wheel assembly into the detection system. The detection system contains two parts, a holographic grating monochromator in conjunction with a harmonic rejection ordering filter. The order sorting filter assembly transmits only one or two nanometer wavelength increments of light to the detector system. The transmitted light is fed from the detection system to a radiometer/photometer which was interfaced with a Hewlett-Packard 85 desk top computer.

The exposed films were scraped off the glass plates and the extent of monomer reacted was determined by solvent extraction. In the extraction procedure, the amount of monomer reacted is determined through two successive extractions with two different solvents. The insoluble components of the first extraction with chloroform include both the crosslinked polymer and other components entangled in the polymer matrix. From a second extraction with hexane, the amount of entangled components can be determined. The weight of the entangled components can be subtracted from the weight of the insoluble material of the first extraction giving only the amount of crosslinked polymer. Dividing the amount of crosslinked polymer by the known quantity of the monomer originally present in the resist film yields the extent of monomer reacted. We will use this as a measure of the "weight percent cross-linked" too.

Since the entire resist film is dissolved in this technique, this measure is an average over the total thickness of the film. A more comprehensive description of the solvent extraction procedure can be found elsewhere[5].

To test the reproducibility of the solvent extraction technique, three separate sets of experiments at the same exposure energy (6.9 mJ/cm^2) were completed. The results are shown in Figure 1. The three lines drawn are the results of least square fits of the data. The experiments indicated a maximum error of about ±5 weight percent monomer reacted at all thicknesses. Similar findings were also seen[5] at a higher exposure energy (34.4 mJ/cm^2).

RESULTS

The results of the crosslink versus exposure energy experiments for four different resist thickness films are presented in Figure 2. Similar results were obtained for many other film thicknesses.

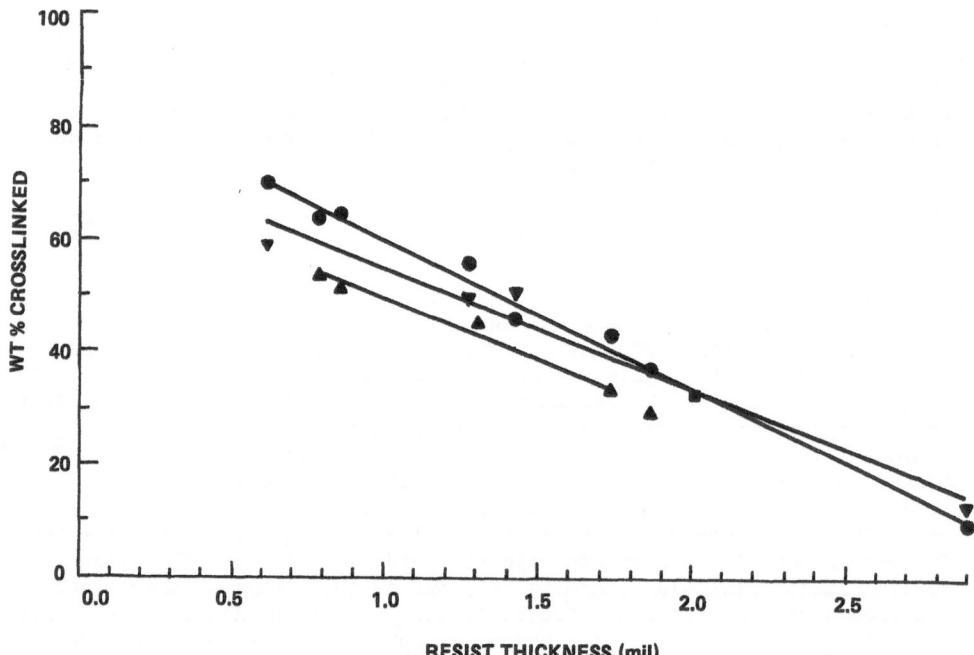

Figure 1. Reproducibility of the Riston ® photoresist extraction results (energy 6.9 mJ/cm^2). The three symbols represent three different sets of experiments conducted in three different months.

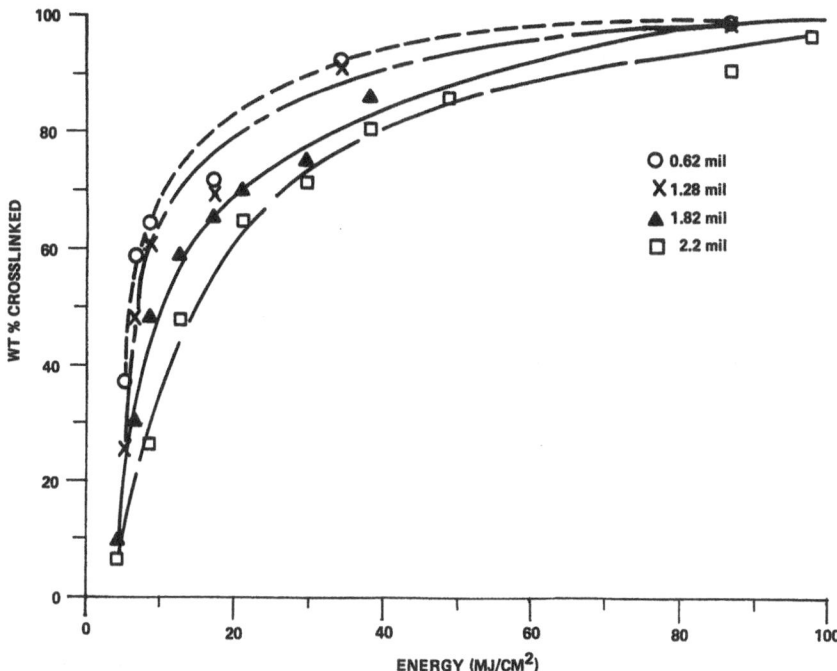

Figure 2. Weight % crosslinked as a function of exposure energy for films of varying thicknesses.

At all thicknesses, the same general trend was apparent. As the exposure energy was increased, the amount of crosslinking averaged over the entire film increased, eventually leveling off. However, it should be noted that the increase in crosslinking is much faster, as may be expected, for the thinner films. In addition, the results indicate that at lower energies (<20 mJ/cm^2) the weight percent monomer reacted increases with decreasing resist thickness. Finally, the threshold energy for the formation of insoluble material is about 4.5 mJ/cm^2. This threshold energy will depend on the solvent used.

By using the data in Figure 2 and a simple subtractive technique, a crosslink gradient can be determined. The subtractive method assumes that the spin-coated resist film thickness depicted an equal depth from the top of a thicker resist film (substrate reflectivity was minimal). Our interest here was in studying resist films of 2.2-mil total thickness. Thus, for example, a 0.5-mil-thick spin-coated resist film would simulate the top 0.5 mil of the 2.2-mil Riston® photoresist film. By knowing the total monomer reacted over different depths, the amount of crosslinking at a specific depth can be deduced. The accuracy of this procedure is improved by studying a large number of films with different thicknesses. More details are available in reference 5.

Figure 3 gives the crosslink gradients obtained by this procedure in a film of 2.2-mil thickness at various exposure energies. The crosslink gradient is definitely more evident at the lower exposure energies. For the lower exposure energies (<10 mJ/cm^2), the difference in the weight fraction of the monomer reacted at the top surface of the resist and at the resist-substrate interface approached 65%. On the other hand, at energies exceeding 34 mJ/cm^2, the variation in the crosslinking over the entire 2.2-mil resist film was only about 20%. In addition, it appears that below an exposure of 10 mJ/cm^2 polymerization at the resist-substrate interface is nonexistent. These experimental data have been

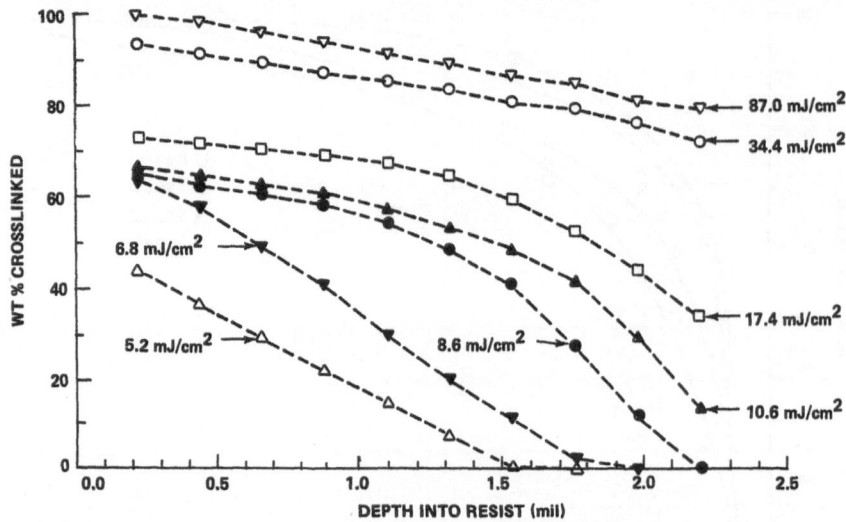

Figure 3. Crosslink gradient in Riston® photoresist at varying exposure energies.

used to test the theoretical model of Srinivasan and Babu[7]. The results are presented elsewhere[5].

ADHESION MEASUREMENT[8,9,10]

To understand the relation between adhesion and weight percent crosslinking, two separate quantitative techniques were employed to measure the resist adhesion to the substrate. The first technique, the substrate elongation test, was developed at Bell Laboratories by Parker and Ranes[11]. In this method, the resist is laminated in three 100-mil strips with 100-mil spacing onto a 0.5-inch-wide 10-oz copper substrate. Using an Instron machine with a 1,000-lb tensile load cell at a crosshead speed of 5 mm/min and gauge length of 3 inches, a load is applied to one end of the sample, holding the other end fixed. A measure of adhesion is the percent elongation of the substrate at the point of the resist failure. Through repeated trials with five samples at the same exposure energy, the substrate elongation at failure was found to vary ±0.6%.

In all samples tested, the resist failure was due to shear stresses, and the failure mode was the resist lifting from the copper substrate. A cracking of the resist or failure due to tensile stresses was not observed on any of the test samples. The results show that the adhesion of the resist to the substrate was strongest at the lower exposure energies. With increasing exposure energy, the percent elongation of the samples decreases as shown in Figure 4. This is expected since the resist becomes less pliable with exposure. If the energy range of the experiments were expanded, then one would expect the rate of decrease in the percent elongation with exposure energy to eventually level off.

A second method used to measure laminate adhesion was the cross pull test. First, the resist was laminated onto the desired substrate, then cut into pieces 11/16 inch wide by 3 inches long. After exposure, the samples were glued by a commercial structural adhesive to a second piece of substrate of the same dimensions in the shape of a cross. The samples were allowed to dry in air at room temperature. Holes were drilled in the ends of the samples, which were then placed in a fixture in an Instron Model 1122* machine. Using a 1,000-lb tensile load cell at a

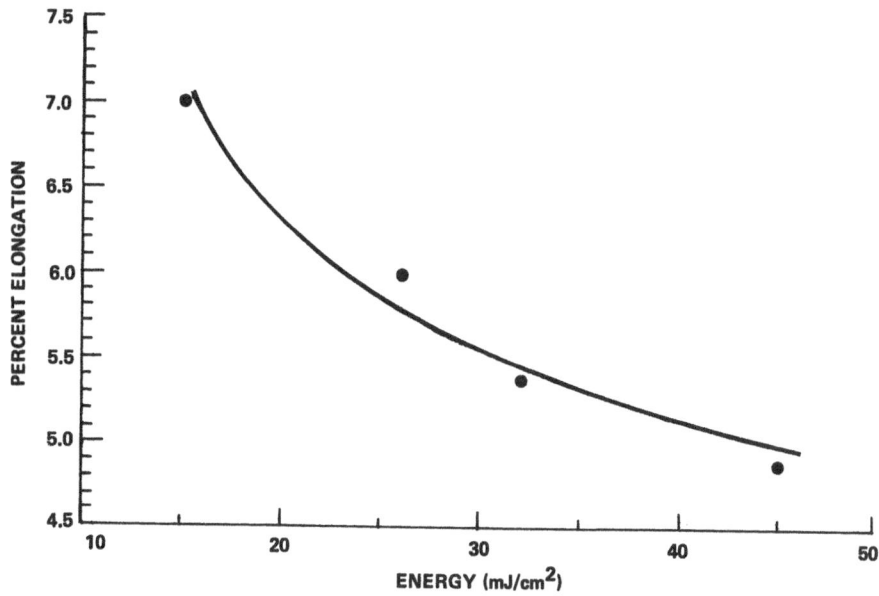

Figure 4. Adhesion of Riston® to 10-oz copper substrate, as measured by elongation at failure.

crosshead speed of 0.2 in./min, the top pieces of the samples were pulled off of the bottom pieces, with the resist laminated onto the bottom pieces.

The load at failure in pounds per square inch gives a measure of the adhesion strength of the bond between the resist and the substrate. The test was repeated for four different samples at each energy of exposure, and the results were averaged. The error of the test was found to be ± 15 lb/in^2.

The cross pull test was performed on a copper and an organic substrate. From the results (Figure 5), it is evident that the resist-organic adhesion is stronger than the resist-copper adhesion at all energies of exposure. At lower energies (<7.0 mJ/cm^2) the adhesion of the resist to both the copper and organic substrates is a constant. It is also apparent that, beyoind about 7.0 mJ/cm^2, the adhesion of the resist to the substrate is a decreasing function of exposure energy, up to an energy of about 15 mJ/cm^2. At higher energies the adhesion again remains constant. Interestingly, the energy range in which the adhesion drops coincides with the energy range in which the crosslinking at the resist-substrate interface increases. To emphasize this, we have shown the weight percent of the monomer that was crosslinked at the resist-substrate interface in the same figure.

The results of the cross pull test show that adhesion is inversely correlated to the crosslinking at the resist-substrate interface. The adhesion between the Riston® photoresist and either substrate is constant up to the onset of crosslinking at the resist-substrate interface (see Figure 5). Further, it appears that once the resist at the resist-substrate interface begins to crosslink, the adhesion starts to decrease and eventually reaches a constant value again.

---

*Instron Corp., Canton, MA.

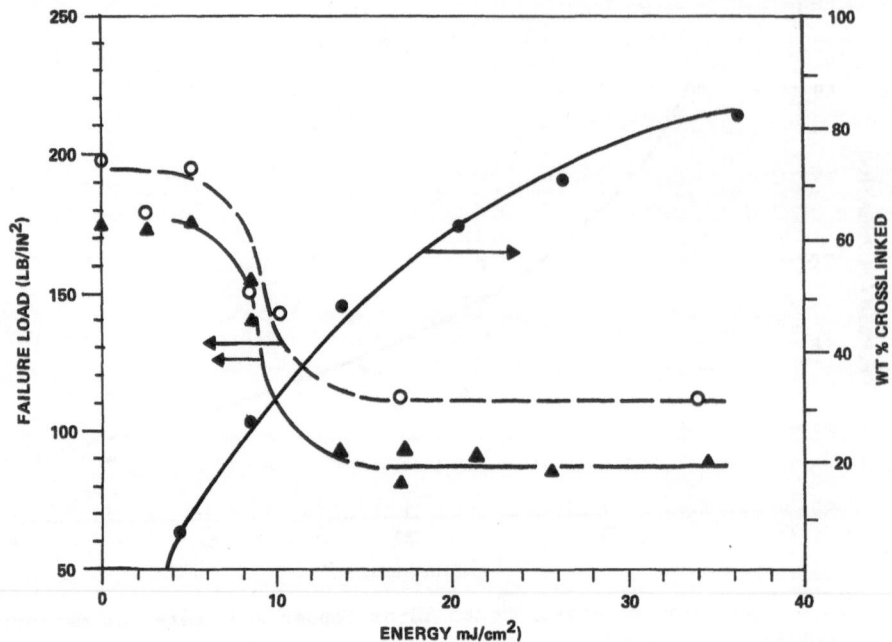

Figure 5. Adhesion of Riston® to a copper (▲), and an organic (○) substrate measured by the load at failure as a function of energy. Also shown is weight % crosslinked at the resist-substrate interface (●).

CONCLUSIONS

We have determined the weight fraction of the monomer that has reacted in different thicknesses of spin-coated Riston® photoresist films by a dual-solvent extraction procedure. This, in conjunction with a simple subtractive procedure, allowed us to determine a crosslink gradient as a function of depth in a 2.2-mil-thick resist film. We have also carried out adhesion measurements using two different techniques. The adhesion of the resist film to both a copper and an organic substrate has been demonstrated to be inversely correlated to the crosslinking of the resist at the resist-substrate interface. The calculated crosslink gradient suggests that a minimum extent of crosslinking at the resist-substrate interface is required before the adhesion of the resist to the substrate will be affected.

ACKNOWLEDGEMENTS

This work was carried out, as part of the Clarkson University-IBM cooperative exchange program, with the advanced manufacturing engineering group at Endicott. We appreciate the encouragement and support of J. Hoffarth, J. Welsh, J. Brauer and W. Alpaugh. Support of R. Day for the adhesion measurements was invaluable.

REFERENCES

1. R. D. Small, J. A. Ors and B. S. H. Royce, in "Polymers in Electronics," T. Davidson, editor, ACS Symposium Series No. 242 pp. 325-344, Washington, DC (1984).
2. P. Dhamelincourt, F. Wallart, M. Leclercq, A. T. N'Cuyen and D. O. Landon, Anal. Chem., 51, 414 (1979).
3. J. H. O'Donnell and P. W. O'Sullivan, Polymer Bulletin, 5, 103 (1981).

4. M. B. Moran and G. C. Martin, Polymer Preprints, 24 (2), 141 (1983).
5. R. L. Geary, Master's Thesis, Dept. of Chem. Eng., Clarkson University, Potsdam, NY, 1985.
6. J. L. Anderson, in "Characterization of Metals and Polymer Surfaces," L. H. Lee, editor, Vol. 2, p. 409, Academic Press, New York, 1977.
7. V. Srinivasan and S. V. Babu, Photograph. Sci. Eng., 28 (5), 175 (1984).
8. K. L. Mittal, Solid State Tech., 89-95, 100 (May 1979).
9. K. L. Mittal and R. O. Lussow, in "Adhesion and Adsorption of Polymers," L. H. Lee, editor, pp. 503-520, Plenum Press, New York, 1980.
10. C. A. Deckert, in "Adhesion Measurement of Thin Films, Thick Films and Bulk Coatings," K. L. Mittal, editor, pp. 307-340, American Society for Testing and Materials, Philadelphia, 1980.
11. J. L. Parker, Jr. and R. B. Ranes, in "Proc. of the American Electroplaters Society 2nd Design and Finishing Printed Wiring and Hybrid Circuits Symposium" held in San Francisco, CA, January, 1980.

PART III. ADHESION ASPECTS OF THIN FILMS AND METAL-POLYMER INTERFACES

RADIATION ENHANCED ADHESION OF THIN FILMS

J.E.E. Baglin

IBM Thomas J. Watson Research Center
Yorktown Heights, New York 10598

It has been shown that dramatic thin-film adhesion enhancement
can be produced by irradiating the film/substrate interface with ion
beams, electron beams or photons. The origin of this interface bonding
has been the subject of much experimental study, notably in systems remarkable for their lack of bulk chemical affinity or interaction. In this
paper, many of these studies are reviewed, with special attention directed
to understanding the mechanisms of adhesion enhancement, the effects
of interface contaminants, and the scope for future optimization of
interface chemistry and bonding.

INTRODUCTION

The integrity, durability and stability of a tremendous diversity of manufactured
products depends on the nature and quality of adhesion of thin deposited coatings.
This is true for optical coatings, protective layers and most particularly for conducting
metal films deposited on ceramic or polymer or semiconductor substrates in order to
serve in forming contacts and interconnects, or in a variety of configurations for device
chip packaging.

The factors which govern the ability of two materials to form an adhering interface are complex, and have not been universally specified, beyond such general beliefs
as the principle that chemically reacting materials should bond well (often but not
necessarily true); and that metals will often bond to other metals in the absence of the
ubiquitous native oxide or contaminant layer (except that in some cases an oxide or
contaminant layer itself makes a successful "glue"). It would seem that the criterion
of "minimum interface energy" would be safe, were it not for possible factors of
interface morphology, whereby an irregular (perhaps interlocked) interface of weak
intrinsic bonding could display good adhesion in practice. Ultimately, most tests of
adhesion performance depend directly or indirectly on the propensity of the interface

under test to nucleate and propagate cracks, either along the interface or within the adjoining solid - a process depending greatly on the structural condition, grain size, voids, stress and elastic moduli, of both film and substrate[1].

In light of the diversity of these considerations, any one of which might govern the adhesion of a particular film/substrate couple, it seems all the more remarkable when a process is discovered which appears to be capable of significantly improving adhesion at most interfaces on which it has been tried. This seems to be true, indeed, for the process of interface irradiation. There now exists a developing body of evidence showing adhesion enhancement produced by directing beams of energetic ions, electrons or possibly photons through the region of a pre-formed interface; also, in a different process, low energy ion bombardment of complex substrates prior to film deposition has been shown to produce greatly enhanced adhesion in certain cases.

The basic mechanisms responsible for this adhesion enhancement remain a fertile field of study, both experimental and theoretical. However, some progress has been made in recent years, and some of the results will be outlined in this review. It should be stressed at this point that the complete generality of the ion beam enhancement effect is not yet established, and the collected data may well embody a variety of unrelated processes which can independently influence thin film adhesion.

TECHNIQUE

In most applications of beam-enhanced adhesion, the thin film (typically 100Å to a few hundred Å thick) is deposited on its substrate by some standard technique, and then a beam of energetic ions, electrons or photons is introduced, penetrating into the substrate, usually well beyond the interface of interest. Such a situation is illustrated in Fig. 1. After the exposure, adhesion has usually been improved, particularly in cases where there is initially little or no bonding displayed, such as that of a Cu film vapor-deposited directly on glass. Often, a mild heat treatment after irradiation has been found to increase adhesion still further. The most remarkable systems to respond to this treatment are those in which no chemical affinity or solubility exists between film and substrate materials, and such systems are the focus of this review.

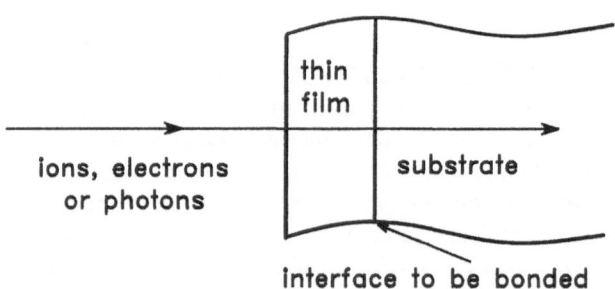

Fig. 1   Radiation enhancement of adhesion (schematic).

OVERVIEW

Systems Bonded

A summary of experimental reports on a great variety of film-substrate systems is shown in Table 1. In 1969, Collins et al.[2] reported increasing adhesion of Al(500Å) on soda glass substrates by a factor of 50 to 100, using modest doses of 120 keV Ar ions. They also noted that the ion dose required for adhesion could change by a factor of 20 depending on whether the substrates were washed in organic solvents or in water prior to Al deposition. In 1972, Gukelberger and Kleinfelder[3] patented the process in which a beam of Ne ions at 100 keV would produce adhesion of Cu films on SiO_2. Since that time, especially in recent years, there has been a lot of research activity in this field, beginning with broad-ranging surveys to examine the scope of the applicability of the ion-enhanced adhesion process, moving to studies of systematic behavior which might suggest the basic nature of the process, and very recently including efforts to microcharacterize the state of bonding at such enhanced interfaces.

For much of the work, adhesion enhancement has been observed by using the Scotch Tape test. A sample "passes" the test when the tape will detach from the film in preference to stripping the film from the substrate. A similar threshold test has also seen much use, in which a Q-tip or cotton bud is rubbed under pressure across the film; if the film will not detach from the substrate, it "passes". More information is available from quantitative tests which attempt to measure the adhesion strength, either by measuring the force required to peel off a strip of the film[49,50], or by measuring the stylus load required to produce a scratch that detaches the film[51].

In examining the data of Table 1, we are presented with a great variety of film-substrate combinations, and a great variety of radiations (ions, photons, electrons). In looking at the ion species, one should not be preoccupied with their chemical nature, since as far as we understand, the ions themselves become buried too deep to participate chemically in the interface bonding. Rather, we should be considering the nature of their energy deposition as they pass through the interface in question ---- how does their *passage* improve adhesion?

General Applicability

Does irradiation enhance adhesion in *all* systems? Early results may have led to this attractive assumption. However, as can be seen from Table I (refs. 24 - 33), no adhesion enhancement could be found for several situations cited. Furthermore, despite the overwhelming good news of systems in which Scotch Tape or Q-tip threshold tests were satisfied, it should be remembered that both those thresholds still represent very weak adhesion (peel strength a few gm/mm). Neither test represents assurance of a technologically strong (20 to 200 gm/mm) bond, and only in a few instances to date is there quantitative confirmation that irradiation has produced such *strong* adhesion where previously there was little or none. With these caveats, it is fair to conclude that some radiation enhancement does seem to be possible for most interfaces between materials normally regarded as non-reacting and immiscible. There is evidently great variability in the proportional increase of adhesion achieved and in the resulting adhesion strength possible. For example, Baglin et al[11] measured high peel strengths and large enhancement for Ne ion bombardment of Cu on Al_2O_3 (0.2 gm/mm initially → 20 gm/mm), while similar treatment of Cu on SiO_2 (fused quartz) was much less effective (1.2 gm/mm → 6 gm/mm). A further indication that even enhanced adhesion is subject to system chemistry may be found in studies such as those of refs. 6, 8 and 9,

Table I. Radiation enhanced adhesion -- experimental summary. Unless otherwise noted, beam dose shown represents approximate threshold dose D_{th} required for the sample to satisfy the type of adhesion test stated.

Film	Substrate	Beam	Threshold Dose or Range of Doses (cm^{-2})	Adhesion Test	Reference	Comment
Al	soda glass -organic wash -water wash	Ar 120keV	~6E13 ~1E15	Scratch, Pull	Collins et al.[2]	Measured adhesion vs. dose Increase x(50-100) max.
Cu	SiO₂	Ne 100keV	?	Scotch Tape	Gukelberger and Kleinfelder[3]	
TiB₂	stainless steel	Kr 300keV	<1E15	Pull	Padmanabhan and Sorensen[4]	Measured adhesion vs. dose
Ag	soda glass	Ar 70keV O₂ 70keV	to 2E16	Scratch	Laugier[5]	Adhesion increasing with dose
Al	soda glass	Ar 50keV O₂ 50keV	to 2.5E16	Scratch		Adhesion increasing with dose
Au	Teflon	He 1MeV H 2MeV F 5MeV	4E13 ≤5E14 ~3E12	Scotch Tape	Griffith, Qiu[6] and Tombrello	
	SiO₂	Cl 20MeV F 5MeV	5E14 >5E15			
	Ni-Zn ferrite	Cl 20MeV F 5MeV	3E13 2E15			
	soda-lime glass	Cl 20MeV	>1E15			
	Al₂O₃ sapphire	Cl 20MeV	1E16			Slight effect only
Cu	Al₂O₃	Cl 20MeV	3E15			

Film	Substrate	Beam	Threshold Dose/ Range of Doses	Adhesion Test	Reference	Comment
Au	glass	S 12MeV	1E12 to 1E15	Scotch Tape	Jacobson et al.[7]	Cu-Al adhesion strong as deposited if no Al oxide
Cu	Al with native oxide	O 12MeV	1E14 to 1E15	Scratch Scotch Tape Scratch		
Au	Teflon	S 12MeV	1E12	Scotch Tape		
Au	Si	Cl 20MeV	5E14	Scotch Tape	Mendenhall[8] and Tombrello[9]	Propose $D_{th} \sim \left[\left(\dfrac{dE}{dx} \right)_{el} + Q \right]^{-2}$
	SiO$_2$	Kr 107MeV	1.5E13			
	InP	Cl 20MeV	<5E14			
	ferrite	Cl 20MeV	1E15			
	Ta	Cl 20MeV	2.5E13			
Ag	Si	Cl 20MeV	2E15			
	SiO$_2$	Cl 20MeV	<2E14			
	Al$_2$O$_3$	Cl 20MeV	≤5E15			
Pd	Al$_2$O$_3$	Cl 20MeV	<1E15			
Au	Si	Cl 20MeV	1E15	Scotch Tape	Werner et al.[10]	
	SiO$_2$	Cl 20MeV	2E15			
	InP	Cl 20MeV	5E14			
	ferrite	Cl 20MeV	3E13			
	ferrite	F 5MeV	2E15			
	W	Cl 20MeV	2E14			
	CaF$_2$	Cl 20MeV	2E14			
	Al$_2$O$_3$	Cl 20MeV	5E15			
	GaAs	Cl 20MeV	5E14			
	Teflon	He 1.5MeV	1E13			
		H 1MeV	1E14			
Ag	Si	Cl 20MeV	5E15			
		F 10MeV	1E16			
	SiO$_2$	Cl 20MeV	2E15			
Pd	InP	Cl 20MeV	5E14			
	Al$_2$O$_3$	Cl 20MeV	2E15			

Table I. Continued

Film	Substrate	Beam	Threshold Dose/Range of Doses	Adhesion Test	Reference	Comment
Cu	alumina	He 200keV	to 6E16	Peel force measurement	Baglin et al.[11,12]	Adhesion rises at ~5E15 cm⁻²; saturates at ~2E16 cm⁻²; heating 450°C 1 hr increases adhesion x 10 for He, x 2.5 for Ne.
		Ne 280keV	to 6E16			
	glass-ceramic	He 200keV	to 6E16	Peel force measurement		
		Ne 280keV	to 6E16	measurement		
	SiO$_2$ (suprasil)	He 200keV	to 6E16	Peel force measurement		Poor adhesion; heating gives x3
		Ne 280keV	to 6E16			
	Teflon	He 200keV	to 6E16	Peel force measurement		Substrate damage by 5E15 leads to detachment
		Ne 280keV	to 6E16			
Cu	Cr with thin oxide	P 1MeV	<2E15	Scotch Tape	Bottiger et al.[13]	Peel force implies strong bond
		Ne 250keV	5E15			Peel force implies poor bond
		He 200keV	~2E16			Peel force implies weak bond
		e 7keV	5E17			
Pt (800Å) (3000Å)	yttria stabilized zirconia	He 2MeV	2E14	Q-tip	Mitchell et al.[14,15]	Approx. independent of electron energy
		He 2MeV	<1E15			
Pt (800Å) (1860Å)	yttria stabilized zirconia	e 5-30keV	~1E17			
		e 5-30keV	~5E16			
Pt	alumina	He 2MeV	1E14			
	glass	He 2MeV	4E16			
	glass	e 10keV	3E18			
	Si	He 2MeV	1E15			
	Si	e 5-30keV	~6E16			
Au	Si	He 2MeV	>2E15			
	Si	e 5-30keV	5E17			
	glass	He 2MeV	>6E16			
	glass	e 5-30keV	5E17			
	GaAs	He 2MeV	7E17			
	GaAs	e 5-30keV	1E15			
Sn		He 2MeV	4E17			
		e 5-30keV				

Film	Substrate	Beam	Threshold Dose/Range of Doses	Adhesion Test	Reference	Comment
Au	Teflon	γ 0.2-0.3MeV (1MeV mean) e 240eV to 3.5keV	5E6 rads 1E16	Pull	Sofield et al.[16]	20Å interface zone; XPS shows ternary complex
Au Ag Cu Al	Mo with native oxide	Ni 1MeV	$4E14<D_{th}<1E15$ $9E14<D_{th}<4E15$ $9E14<D_{th}<2E15$ $9E14<D_{th}<2E15$	Scotch Tape	Pronko et al.[17]	No extensive interface mixing
Au Pt	Si Si	$h\nu$ 10eV,21eV $h\nu$ 21eV	<1E16 <1E16	Q-tip	Mitchell et al.[18]	
Pt	yttria stabilized zirconia Al_2O_3 alumina	He 2MeV e 5-30keV He 2MeV	~2E14 ~1-2E17 1-2E14	Q-tip	Sood et al.[19]	
Al	SiO_2 on Si	e 20keV	3E16	Q-tip	Sai-Halasz and Gezecki[20]	±80V bias over interface increases adhesion, decreases D_{th}
Au	SiO_2 (vitreous) GaAs n-type p-type (Zr) (Cr)	Cl 5MeV Cl 18MeV	5E12-5E13 2E14 1E14 no adhesion 5E13	Scotch Tape/Scratch Scratch	Tombrello[21] Livi[22]	Two regions. Lower dose associated with microcracks in SiO_2. Effects later shown (Wie et. al[23]) to be entirely governed by oxide/OH at interfaces

Table I. Continued

Film	Substrate	Beam	Threshold Dose/ Range of Doses	Adhesion Test	Reference	Comment
Au	Si	N 6.5MeV N 3.4MeV C 3.3MeV F 3.0MeV P 3.4MeV	9E15 2E15 <1.3E16 <5E15 2E15	Q-tip/ Scotch Tape	Berkowitz et. al[24]	
	SiO$_2$ (vitreous)	N 6.5MeV N 3.4MeV C 3.3MeV F 3.0MeV P 3.4MeV P 6.8MeV	>1.5E16 >1E17 >4E16 >2.5E16 2.5E16 <1.3E16	Scotch Tape		No adhesion found No adhesion found No adhesion found No adhesion found
Al	glass	He 2MeV e 10keV	~1E14 9E16	Q-tip	Sood et al.[25]	
Au	glass	He 2MeV e 10keV	>2E17 8E17			No enhancement found
Pt	glass	He 2MeV e 10keV	>4E16 1E18			No enhancement found
Cu	Al$_2$O$_3$ alumina -clean -water wash -ethanol wash	Ne 280keV	to 2E16	Peel force measurement	Baglin[26]	Interface contaminant effects; see Tables II and III
Au	Al$_2$O$_3$ alumina -clean -water wash -ethanol wash	Ne 280keV	to 2E16	Peel force measurement		Interface contaminant effects; see Tables II and III

Film	Substrate	Beam	Threshold Dose/ Range of Doses	Adhesion Test	Reference	Comment
Al	Si	hν 3.5-6eV	not known	Q-tip	Kellock et al.[27]	Adhesion enhanced
		21eV	not known		Gazecki et al.[28]	Adhesion enhanced
	GaAs	hν 3.5-6eV	not known	Q-tip		Film detached
		21eV	not known			
	glass	hν 3.5eV	not known	Q-tip		Adhesion enhanced
	SiO₂ on Si	hν 6-21eV	not known	Q-tip		Voltage bias often improved effect
Au	SiO₂ (Puropsil A)	He 2MeV	>2E18	Q-tip and Scotch Tape	Battaglin[29]	No enhancement observed
		e 2keV	1E18		Battaglin et al.[30]	Only secondary e's reached interface
	SiO₂ newly fractured surface	e 5keV	>2E18			No enhancement observed
	SiO₂	e 2.5keV	2E18			Adhesion by dissociating hydrocarbon contaminants?
Fe	soda-lime glass	H 100keV	fixed test dose 1E16	Scratch	Battaglin et al.[30,31,32]	Enhancement x4; diminished Fe-C bonding
	SiO₂	H 100keV	1E16		Carbucicchio et al.[33]	Enhancement x3
Al	SiO₂ on Si	H 100keV	1E16			Enhancement x3
	glass	H 100keV	1E16			Enhancement x5
	SiO₂	H 100 keV	1E16			Enhancement x2
Cu	glass	H 100 keV	1E16			Negligible effect
	SiO₂	H 100keV	1E16			Negligible effect
Au	glass	H 100keV	1E16			No enhancement
	SiO₂	H 100keV	1E16			No enhancement

Table I. Continued

Film	Substrate	Beam	Threshold Dose/ Range of Doses	Adhesion Test	Reference	Comment
Au	Ta with native oxide	2.85MeV/amu for C O Si Cl Ni	5E16 1E16 1.5E15 3.3E14 6.4E13	Scotch Tape	Stokstad et al.[34]	find $D_{th} = 10^{17}(dE/dx)^{-3.0}cm^{-2}$
	Si with native oxide	2.5MeV/amu for Si Cl Ni	9.7E15 2.5E15 2.5E14			$D_{th} = 6x10^{18}(dE/dx)^{-4.1}cm^{-2}$
	SiO_2	Ni	2.3E13			
Cu	Mo	O 2MeV Si 3MeV Ni 3MeV Au 5MeV	2E16 5E15 1E15 5E14	Scotch Tape and pin pull	Ingram and Pronko[35]	

where the "threshold" dose (to satisfy the Scotch Tape test) for a single ion species will vary over two orders of magnitude depending on the substrate and film materials. It is unlikely that van der Waals forces alone could enable a bond to satisfy the Scotch Tape test, and it seems reasonable therefore to presume that the stronger forces of chemical or metallic bonding are responsible for enhanced adhesion.

Mechanisms

It is well known that ion irradiation of *reactive* interfaces will produce "ion beam mixing" to form an intermediate layer. The rate of mixing has been shown[36] to be related closely to the thermodynamic driving force for intermediate compound formation, however, and no such forced mixing has been found for non-reactive or immiscible systems. Analyses of the surfaces produced by peeling in well radiation-bonded systems such as Cu on Al_2O_3[11,12,26] have shown transfer of less than a single monolayer of substrate material or film material. Not only does this remove the possibility that induced adhesion must be due to physically mixed or interlocked interface structures; it again supports the idea that the adhering interface consists of atoms of both film and substrate which have been persuaded to intermix to the extent of perhaps a monolayer, enabling stable chemical bonding configurations to occur. Such configurations supposedly represent an energetically favorable *redistribution* of bonding ligands in preference to that of the virgin terminated substrate structure presented to the as-deposited metal film. The resulting lowered interface energy is, of course, equivalent to an enhanced adhesion (in the absence of other factors such as stress). Chemical bonding associated with enhanced adhesion has in fact been verified by Sofield et al.[16] for the case of Au on Teflon after irradiation in a high energy γ-ray flux. Their XPS analyses of irradiated Au(20Å)/Teflon samples display dramatic line shifts, implying the formation of Au-C=O, C-O-C-Au or other similar complexes.

The *thermal stability* of such an interface structure is of critical interest, for practical applications as well as for learning more about the nature of the layer. Baglin et al.[11] reported a substantial improvement in adhesion as a result of heating irradiated samples in an inert atmosphere at 450°C. This succeeded with Cu on Al_2O_3, Cu on glass-ceramic and Cu on fused quartz. However in contrast Ingram and Pronko[35] reported loss of adhesion of Cu on Mo after heating beyond 500°C. Heat treatment will, of course, accelerate relaxation of the system, whether by formation of interface bonds, or segregation of interface contaminants, or dissociation of unstable bonding, or by introducing stress changes associated with annealing or with differential thermal expansion of film and substrate. In such possible complexity, it will clearly be necessary to test each film-substrate combination separately before stability is assumed for practical purposes.

It should perhaps be mentioned here that thermal treatment may soften or anneal the film material. Even in the absence of any interface energy changes, in subsequent adhesion testing, peel strength and scratch test performance can be greatly increased simply because of changed mechanical properties of the film. This effect will have contributed to some extent to the thermal enhancement of ref. 11. However, such mechanical ambiguities are absent from the interface-wetting test (Baglin and Clark[12]), in which a 100Å layer of Cu on Al_2O_3 was irradiated and then heated at 450°C. The film stayed flat and continuous, indicating that the radiation had lowered the interface energy sufficiently to prevent the balling-up that occurred in samples not irradiated before heating.

221

From a variety of directions, therefore, the evidence seems to imply that irradiation leads to a redistribution of chemical (or metallic) bonds among both film and substrate species in one or two monolayers at the interface, forming a low energy interface whose adhesion strength and stability will depend on the chemistry involved. Being confined to the junction plane, and only locally ordered, the composition and structure of this "mixed" layer need not be assumed to correspond to those of any particular phase familiar to us in bulk form.

ELECTRONIC OR COLLISIONAL?

In order to understand better the mechanism of radiation enhanced adhesion, it would help to know whether the bonding is initiated mainly as a result of the ionization ("electronic") or nuclear ("collisional") energy deposition at the interface.

It is clear from the comments above, and from the "saturation" as a function of ion dose shown in refs. 2 and 11, that the radiation functions as a facilitator to help the interface monolayers reach a new bonding configuration. After that stable arrangement has been established, further irradiation does not affect adhesion (unless other effects are involved e.g. deformation or dissociation of the substrate). We are therefore looking for mechanisms capable of producing rearrangement of bonds and/or atoms over a distance no more than 1 or 2 atomic spacings.

Table I is well populated with reports of adhesion improvement by irradiation with electrons and photons. Bonding with photons has been achieved at energies ranging from 4 eV[18,27,28] to 3 MeV[16] and it seems likely that the photoelectron or secondary electron flux produced near the interface is responsible for the bonding process. Battaglin et al.[30] report using 2.5 keV electrons to bond Au(400Å) on SiO_2, while not succeeding with higher energies, implying an energy threshold or window for that system. However, in general the reports of success involve electron energies ranging from 200 eV to 40 keV. Sai-Halasz and Gazecki[20] also tested the effect of electron irradiation in the presence of a potential difference applied through the $Al/SiO_2/Si$ sample. Biases of $\pm$ (50 $-$ 100)V markedly improved the resulting adhesion, the preferred polarity of Al with respect to SiO_2 being negative. They suggest that bonding is produced via the greatly increased net electron current flowing across the interface when the electron beam enters the biased interface, and the authors comment that adhesion might alternatively be enhanced by other direct means of generating electron flux across the interface such as avalanche injection, Fowler-Nordheim tunneling or light assisted injection of carriers. Unfortunately, none of these papers quotes absolute adhesion strengths (all use Scotch Tape or Q-tip tests), so one can not say for sure that the adhesion produced by photons or electrons is as strong as that produced by other beams. However, it is clear that the purely electronic interactions of electrons and photons *can* produce enhanced adhesion. This presumably takes place by disruption of existing bond structures in the substrate surface, some of which promptly re-form in configurations involving neighboring atoms of the thin film, having lower intrinsic energy. Minor atomic displacements would also be expected to result from this re-bonding.

A similar kind of process would be expected from the ionization ("electronic" energy loss) produced by a heavy ion in passing through the interface. For 20 MeV Cl, 107 Mev Kr and even 5 MeV F (refs. 6, 8, 9) nearly all of the specific energy loss at the interface is indeed electronic in nature and, as pointed out by Tombrello et al., these ions must surely achieve interface modification primarily by electronic interactions.

For low energy heavy ions, however, a large part of the specific energy loss at the interface involves nuclear collisions, which may lead directly to recoil displacements in the solid. Will such recoils be notably more effective in producing adhesion enhancement? The work of Baglin et al.[11] addressed this question by comparing the ultimate adhesion strength obtainable for Cu on Al_2O_3 using He or Ne ions at low energies chosen so that, at the interface, both species would have the same *electronic* energy loss ($dE/dx \simeq 50$ eV per (10^{15} atom/cm^2)), while the Ne alone would in addition have appreciable *nuclear* energy loss ($\sim$ 28 eV per (10^{15} at/cm^2)). If the only interaction of interest were electronic, both species would produce the same effect. If nuclear collisions would dominate, only the Ne would be useful. As shown in Fig. 2, Ne produced adhesion 6 times stronger than He could produce (regardless of dose). However, after subsequent heating, when both systems became stronger, the He-treated sample actually exceeded the strength of the Ne. Clearly, the nuclear energy loss has a powerful effect on the result, and can overshadow the electronic effects when it is present. One mechanism to consider might be the following. As He is almost incapable of producing direct atomic displacement at the interface, it succeeds only in disrupting the substrate surface bonding, which (at room temperature) occasionally leads to the formation of Cu-Al and/or Cu-O links with neighboring atoms; heating then provides enough mobility for the interface atoms to form new configurations of stable binary or ternary bonds. Ne combines some of these functions by moving atoms ballistically as well as creating neighboring bond disruption. It has simply less to gain from the subsequent heat treatment.

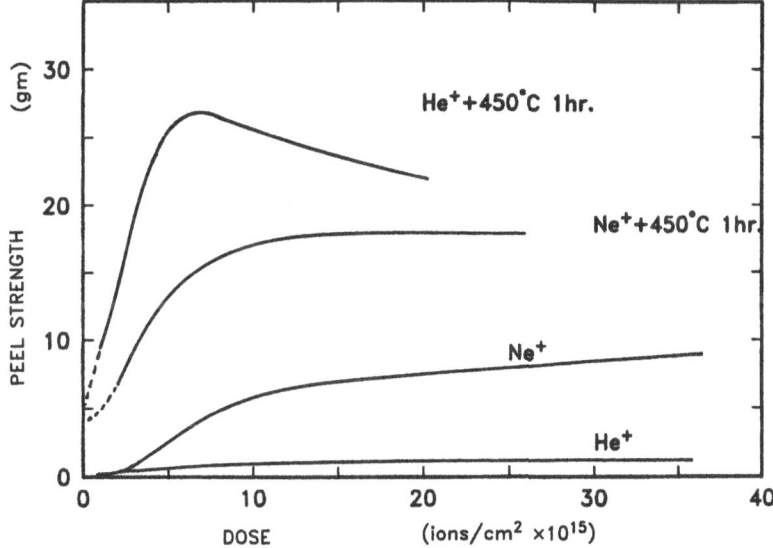

Fig. 2   Peel strength produced for Cu(700Å) on alumina as a function of dose for both He(200 keV) and Ne(250 keV) bombardment. The upper lines show the increased peel strength resulting from heat treatment at 450°C for 1 hour in pure flowing helium.

223

Table II. This table displays correlations between the critical ion dose required to produce "Scotch Tape test" adhesion of Cu on polished Mo, and the form and magnitude of specific energy loss of ions at the interface (from ref. 35).

Ion species	Electronic energy loss (eV ion^{-1} nm^{-1})	Nuclear energy loss (eV ion^{-1} nm^{-1})	Ion dose required for 1 d.p.a. (ions cm^{-2})	Critical dose to pass Scotch Tape test (ions cm^{-2})
2 MeV helium	450	0.8	6.6×10^{17}	none found
2 MeV oxygen	1800	70	7.5×10^{15}	2.0×10^{16}
3 MeV silicon	2700	120	4.4×10^{15}	5.0×10^{15}
3 MeV nickel	2700	650	8.1×10^{14}	1.0×10^{15}
5 MeV gold	2200	3000	1.7×10^{14}	5.0×10^{14}

Ingram and Pronko[35] more directly argue that nuclear recoil effects are dominant in their work on adhesion of Cu on Mo. They used a variety of ion beams (2 MeV He, 2 MeV O, 3 MeV Si, 3 MeV Ni and 5 MeV Au) offering different proportions of electronic and nuclear energy loss at the interface, and compared the threshold adhesion dose D_{th} among them. (See Table II.) As the authors point out, if electronic energy loss alone were significant, D_{th} for the Si, Ni and Au ions could be identical, whereas in fact D_{th} varies by a factor of 10 between them. Correlation of D_{th} with the nuclear energy loss alone seems much more plausible, and the ion dose calculated to produce one displacement per atom (d.p.a.) at the interface for each species seems to scale excellently with D_{th}. At least in this case then, it seems that collisional energy deposition is primarily responsible for producing adhesion.

Seiberling and Headrick [37,38,39] have shown, using transmission ion channeling, that MeV ion bombardment, even at energies where the energy deposition is almost entirely electronic, can indeed cause disordering of Si atoms in one or two atomic layers at a metal/Si interface, at doses which enhance adhesion. This is partly attributed to the creation of new bond structures as a result of the bond disruption produced by the passing ions. The result serves to support the concept of interface reordering (or disordering) even if only electronic excitation is available; very small atomic movements may enable the formation of an interface bonding layer whose chemistry determines the improved adhesion.

THRESHOLD SYSTEMATICS

A model of the initial stages of the adhesion enhancement process was proposed by Tombrello[9,40] in an effort to develop a functional relationship between the threshold dose of ions D_{th} required to make a given system pass the Scotch Tape test, and the specific electronic energy loss $(dE/dx)_{el}$ of the ions in the film material at the interface. The model makes several initial assumptions:
i) Electronic interactions alone create new interface bonds. Then (no. of new bonds formed per ion) x (threshold dose (ions/ cm^2))will be a constant determined by the adhesion test threshold chosen.

ii) For adhesion to proceed, exchange of electrons across the interface is needed. The energy to do this is the energy delivered by the bombarding ion.

iii) Electronic exchange across the interface can be represented by the Richardson-Dushman equation:

$$\text{electron current} \propto T^2 e^{-\phi/kT} \tag{1}$$

where T is the local temperature, raised by the various energy loss processes of the ion in electronic collisions, and ϕ is the difference in work function for the two materials.

This electron current is then proportional to the number of new bonds formed per ion; T is proportional to ($(dE/dx)_{el} + Q$) where Q is the energy supposedly gained back from the interface upon increased (exothermic) binding.

For metal-metal interfaces, $\phi \simeq 0$, and

$$D_{th} \propto ((dE/dx)_{el} + Q)^{-2} \tag{2}$$

For most semiconductors and insulators, $\phi > kT$, and hence

$$D_{th} \propto ((dE/dx)_{el} + Q)^{-2} \cdot \exp(\frac{\phi}{kT}) \tag{3}$$

The experimental Scotch Tape threshold data for a variety of ion species bonding Au(500Å) on to Ta were obtained,[9] with results shown in Table I, and simple best fits were obtained, of the form
$$D_{th} \propto (dE/dx)^{-(1.6\pm0.2)}$$
or $\qquad D_{th} \propto ((dE/dx) + 0.15)^{-2},$
where Q=0.15 MeV/(mg cm^{-2}) corresponds to 70 eV liberated per incident ion as the interface bonds are formed.

One difficulty in applying this model lies in the concept of localized temperature T as it is elevated by each ion. For example, a "thermal spike" picture for a heavy ion penetrating a solid might predict an equivalent transient temperature of $10^3 - 10^4 °K$, lasting for perhaps 10^{-11} sec. and implying large atomic mobility for this period. It may be asking a lot for equation (1) to apply in such a highly non-equilibrium transient condition. A further problem is the absence in the model of any contribution to the bonding process resulting from collisional interactions, although considerable effect is demonstrated in refs. 11 and 35.

A series of careful experiments was reported recently by Stokstad et al.[34], who used a variety of ion species to obtain the best-fit constants for these D_{th} - dE/dx relations for films of Au(500Å) on oxides of Si or Ta. For Au-TaO$_x$, they found $D_{th} \sim (dE/dx)_{Au}^{-(3.0\pm0.2)}$ and for Au-SiO$_2$, $D_{th} \sim (dE/dx)_{Au}^{-(4.1\pm0.3)}$. The exponent for Au-TaO$_x$ differs significantly from the value 1.6 found for Au-Ta by Tombrello, and the disparity between all exponents and the original predicted value of 2.0 raises questions about the generality of the basic relationship. However, one notes that for oxide substrates, the form of eq. 3 is predicted, with substantially greater slope than eq. 2 In this sense the data might be reconciled. Alternatively, Stokstad et al. point out that a similar form with exponent 4.6 might be predicted by considering collective processes analogous to track formation in insulators.

A further problem is raised by the measurements of Sood et al.[19] in which D_{th} was compared for He ions and electrons, bombarding identical samples of Pt on zirconium. They observed D_{th} (30 keV electrons)/ D_{th} (2 MeV He) = 600, in serious disagreement with the value of 5000 predicted from eq. 2, or the larger value implied by eq. 3. It seems at present that different systems lead to a variety of values for the exponent in eqs. 2 or 3. However, the exponent value alone can not be said to imply with any certainty the existence of a uniquely defined mechanism of adhesion enhancement. The empirical fact that a power-law dependence (eq. 1) seems to be observed consistently with ions but not with electrons leaves open the interesting question of what mechanism is responsible for it.

CONTAMINANT LAYERS

Effects of contaminants

We have established that the adhesion promoted by irradiation in non-reacting systems is strictly an interface phenomenon, involving at most a few monolayers, and possibly an atomically abrupt bonding plane. Under these conditions, sub-monolayer quantities of adsorbates or contaminant atoms at the interface will completely determine the chemical bonding structure available for adhesion. Sadly, only a very few of the experiments of Table I (e.g. refs. 29-31) have involved preparation of samples under UHV conditions, without which it can be presumed that *almost all the adhesion enhancement work on which our discussion is based involved unspecified substrate surface contaminants capable of dominating the bonding process.* There are, in fact, large discrepancies among experimental results on nominally identical systems e.g. Au on SiO_2, where a large range of D_{th} values is reported, but where Battaglin et al.[30] using well-characterized UHV-prepared samples found *no* bonding enhancement possible with He and H beams, and a very large $D_{th} = 1 \times 10^{18}/cm^2$ for 5 keV electrons. Baglin[26] has also reported problems with quantitative reproducibility of peel strengths from experiment to experiment, using non-UHV deposition conditions.

The power of substrate surface contaminants to enhance or defeat all efforts at bonding is indicated in reference 26. There, substrates of alumina were prepared, by means of sputter cleaning and exposure to O_2 or water or alcohol, so that their surfaces would be expected to be (i) clean, oxygen-terminated Al_2O_3, or (ii) coated with adsorbed H_2O or (iii) coated with adsorbed hydrocarbon. After prompt deposition of Cu or Au films, adhesion enhancement irradiations and heat treatments were carried out, with the results shown in Table III. For Cu/Al_2O_3, the striking effect is that of hydrocarbon, defeating all bonding procedures. By far the preferred surface was "clean" (i). For Au/Al_2O_3, the adsorbed H_2O was inhibiting and hydrocarbons were not tested. Similarly dramatic effects were reported by Collins et al[2], who found adhesion threshold doses of Ar for Al films on soda glass. If the glass had been washed in water before film deposition, D_{th} was found to be about 10^{15} ions/cm^2; following cleaning with alcohol and vapor degreasing, D_{th} had been lowered in this case to about 6×10^{13} ions/cm^2.

That significant amounts of insulator surface contamination may be expected in any but UHV conditions is highlighted in the review paper of Henrich[41], in discussion of molecular chemisorption on oxide surfaces. Some illustrative examples follow: While O_2 will not adsorb on most insulating oxides, for conducting or semiconducting

Table III. Effects of adsorbed contaminants introduced at the surface of alumina substrates by washing in water or alcohol, prior to deposition of Cu or Au films. The ability of ion irradiation (2×10^{16} Ne/cm^2) and subsequent heat treatment to enhance adhesion depends markedly on such "contaminants" (from ref. 26).

Metal film	Substrate surface preparation	Peel strength (arb. units)			
		not heated		450°C 1hr	
		As deposited	Ne$^+$ 280 keV	As deposited	Ne$^+$ 280 keV
Cu	Normal surface: (Sputter cleaned; O$_2$ exposure)	3.5	6.4	13.5	13.0
	Hydrated surface: (Water washed)	0.9	4.1	5.3	20.2
	Hydrocarbon: (Ethanol washed)	0.0	0.0	2.2	4.5
Au	Normal	5.6	7.5	.	.
	Hydrated	0.0	1.9	.	.
	Hydrocarbon	0.0	2.8	.	.

oxides the interaction is strong, and typically exposure of a vacuum-fractured surface to 1 L (10^{-6} Torr sec) of O$_2$ will lead to complete coverage with added oxygen. H$_2$O is readily chemisorbed (via the O atom) on vacuum-cleaved surfaces of the corundum oxides such as Ti$_2$O$_3$ and V$_2$O$_3$, interaction being complete at 0.5L; in a few cases OH$^-$ is adsorbed. Adsorption of hydrocarbons occurs readily on insulating oxide surfaces, and the adsorption of ethanol on ZnO, for example, promptly leads to dissociation to either ethylene and H$_2$O or acetaldehyde and adsorbed H$_2$.

Oxide or glass substrate preparation for most adhesion processing reported in Table I (and incidentally in most technological applications), has involved final "cleaning" with either water or alcohol. Even when surface cleaning by heat or sputter etching has been used, such surfaces may re-contaminate in seconds in a vacuum of 10^{-6} Torr frequently used in many deposition systems. Carbucicchio et al[32,33] quote their experience with substrates of soda lime glass and fused quartz which were cleaned by HF(10%)- H$_2$O dip followed by 20 minutes in HCl: H$_2$O$_2$:H$_2$O =1:1:5 at 80 °C and subsequent ultrasonic washing in trichlorethylene, acetone and alcohol. After this cleaning, they found surface hydrocarbon contamination amounting to 4.4×10^{15} atom/cm^2 carbon, implying complete coverage.

Similarly, in preparing metal substrates for deposition of bonding films, it will often (depending on the metal reactivity) be a challenge to the cleanliness of a UHV system to produce an interface free of contaminant metal oxide (or other compounds).

In examples such as those of Pronko et al[17], monolayer coverage of native oxide on Mo was in fact measured for polished and carefully cleaned substrates.

In retrospect then, it now seems likely that many of the systems where radiation enhanced adhesion has been observed must have incorporated an initial interface contaminant layer involving oxygen, hydrogen and/or carbon. In these cases, the radiation supposedly served to break down the bonding environment, enabling a reconstruction involving atoms of contaminant species as well as those of both film and substrate. Mixing of species in a joining monolayer would not require ballistic displacement of atoms by the ion beam. In the case of a thicker contaminant e.g. a 50 – 100Å metal oxide, either ion beam mixing involving nuclear recoils or bonding at each surface of the oxide layer would be required. All are, in principle, possible. An example of ion enhanced adhesion achieved in the presence of a thin native Cr oxide, yet not attained for thicker layers, is given by Bottiger et al[13]. Explicit testing of the ability of the ion beam to overcome a native oxide layer was also reported by Jacobson et al[7]. They deposited Cu on Al on which a thin native oxide remained and also on sputter-cleaned Al. Adhesion was enhanced on the oxide sample by MeV-ion bombardment, however, initially good adhesion on the clean sample was not affected by irradiation.

Contaminants: Good or bad?

We have proposed that radiation treatment disrupts interface bonds and promotes the formation of local bonded clusters of interface atoms. These can be thermally stable even if the joining materials have no bulk chemical affinity. For example, at the Cu/Al_2O_3 interface, Cu-Al-O ternary environments akin to $Al_2O_3.CuO$ (spinel) may be produced, bridging the interface. However, the chemistry of some systems will offer no such bonding options. (Au/SiO_2 may be one - - see ref. 30.) Then, the *addition* of a new contaminant species of high chemical activity such as O, C or H might *enable* the formation of a stable interface layer. Such means of creating metal-to-oxide bonds have long been used, for example, in hot joining or deposition processes carried out in oxygen atmospheres[1,42]. Mattox[42] in fact produced successful bonding of Au to clean SiO_2 by depositing the Au in an oxygen rich atmosphere. In such a case, Borom and Pask[43] proposed a model of the interface as a transition region of mixed oxides of the metal and substrate.

Of course, the process can readily go the other way, in which a contaminant (often an oxide) prevents the contact of two otherwise reactive species. In that case, ion treatment might promote the perforation of that layer with bonding links, and subsequent segregation of the contaminant under heat treatment. In both cases, the radiation process would serve to bond the contaminated interface to some degree. Naturally there will also be cases where interface H_2O or C will bond only to the substrate or the film, preventing adhesion. However, it seems at least equally likely that the accidental or *deliberate* introduction of "contaminant" species (sub-monolayer amounts) will enhance the possibilities of creating a stable adhesion layer. In that case, each system will probably possess some optimum interface "stoichiometry" for adhesion. The best bonding would be obtained by tailoring the interface composition in advance before radiation- or thermal-mixing .

INTERFACE CHEMISTRY

Beam-induced changes

Few explicit analyses demonstrating that radiation interaction has altered chemical bonding at interfaces have been published.

Battaglin et al.[29,32,33] examined samples of Fe deposited on carbon-contaminated glass and SiO_2 substrates. Depth-selective Mossbauer spectra contained 6 peaks found for Fe in its pure environment, plus a broad feature attributable to Fe atoms in non-stoichiometric Fe-C compounds. After irradiation with 1×10^{16} H/cm²(100keV), adhesion increased greatly, and the Fe-C peak had been dispersed. It is concluded that this was one case where contaminant bonds were destroyed by the beam and replaced by stronger film-substrate bonds.

Sofield et al.[16] examined γ-ray and electron effects at the Au-Teflon interface, using XPS. Irradiation produced improved adhesion, accompanied by a new spectral feature in the region of the C(1s) lines which was attributed to new C-O or C-F bonds, and at the same time the Au(4f) lines were found to be shifted to lower energy, implying a change in Au bonding. It was concluded that the data suggested the formation of Au-C-O complexes following solid state radiolysis of Teflon.

Interface Tailoring

As mentioned above, it would seem desirable to *pre-condition* a substrate *before* film deposition with appropriate surface activation (i.e. bond-breaking) (e.g. ref. 44) and if necessary appropriate surface atoms added (sub-monolayer) so that upon deposition and subsequent mixing via heat and/or radiation, the interface adhesion layer will be formed with preferred chemical abundances. This might be achieved by depositing new adatom species on a sputter cleaned surface or, in the case of some compound substrates, adjusting the surface abundances by preferential sputtering.

The effect on adhesion due to adatom species introduced at the metal/Al_2O_3 interface has been examined by Pepper[45]. After sputter cleaning in UHV, he exposed surfaces of Cu, Ni and Fe to either oxygen, chlorine, ethylene or vinyl chloride gas for 200 - 1000L. The metal was then drawn over a clean Al_2O_3 surface under load to measure the coefficient of friction μ. Changes of over 60% from the "clean value" $\mu = 0.75$ were found (Table IV) and attributed to the adsorbates (whose presence in sub-monolayer quantities was verified from Auger spectra). Notably, added oxygen produced stronger adhesion in all cases, suggesting the formation of interface layers of complexes of metal oxide and Al_2O_3 in the nature of spinels.

Baglin et al.[26] have tried pre-sputtering sapphire (Al_2O_3) substrates with a 500 eV Ar beam in situ immediately prior to deposition of a Cu film. At 500 eV, Ar sputtering would be expected[46] to deplete the surface oxygen concentration of Al_2O_3. (However, this effect has not yet been explicitly verified.) The result of this processing is a spectacular enhancement of the Cu-sapphire adhesion, several times stronger than any obtained by irradiation treatments, and no longer subject to improvement by implantation. Heat treatment (450°C) produces further improvement, up to a peel

Table IV. Change produced in the coefficient of friction μ of Al_2O_3-metal interfaces following exposure of the metal to various gases. At the clean interface, $\mu = 0.75$. (From ref. 45)

Gas	Metal		
	Cu	Ni	Fe
Oxygen	+0.25	+0.41	+0.45
Chlorine	-0.30	-0.25	-0.40
Ethylene	0.00	0.00	+0.30
Vinyl Chloride	0.00	-0.15	-0.16

strength of over 100 gm/mm. The interface interactions accompanying these processes have been studied in UHV by Schrott, Thompson and Tu[47]. They prepared sputtered substrates of sapphire, coated with a few Å of Cu, and studied the system with XPS both before and after heating. From a new dominant peak which develops near the Cu Auger lines, and from other evidence, it is concluded that the pre-sputtering and subsequent heating lead to the formation of a well-defined ternary bonding environment involving Cu, O and Al. It may be supposed that this structure is the one that creates the links responsible for the very strong adhesion found for thicker films. Its structure and composition have yet to be clearly understood.

SUMMARY

The ability of radiation processing to enhance the adhesion of thin films on non-reacting substrates has been widely celebrated even though so far most systems tested may have involved undocumented interface contaminants. It now seems likely that the radiation serves primarily to disrupt the electronic structure of chemical/metallic bonding of interface atoms, enabling new bonding configurations to develop involving an adhesion monolayer or two where atoms of both film and substrate species are stably bonded. In this situation, the strength of adhesion can very possibly be improved by the presence of sub-monolayer amounts of chemically active "contaminant" species such as O or H or C which can broaden the scope for finding stable interface "compounds". In appropriate systems, therefore, irradiation can turn a "contaminated" interface into an asset. An example might be the creation of a complex mixed metal oxide layer at a metal-metal or metal-ceramic interface. We reason that all systems should benefit from the displacement mixing offered by the passage of heavy ion beams, although such mixing might be a necessity if a thick oxide layer preventing adhesion needed to be dispersed to allow bonding.

It is therefore fair to say that most practical adhesion systems have a good probability of benefitting from irradiation. However, the system chemistry (not forgetting that of any contaminant layers) will ultimately determine whether the treatment can succeed. Since interface chemistry is so important, each case deserves consideration of how best to optimize it in a controlled way. This might be achieved by suitable washing or sputtering treatment of the substrate, or perhaps by controlled exposure to adsorbate species before the film is deposited, or perhaps by selective ion implantation at the interface.

Because convenience of processing is likely to be a major consideration if engineering applications are to evolve, it is especially interesting to pursue the substrate preparation (or activation) techniques. Heavy ion irradiation suffers the practical limitations of requiring high energy accelerators and very thin films. The energetic ion beam also poses a hazard to the underlying substrate in the form of radiation damage (large physical densification[48] for glass and silica) and contamination. Electron beams penetrate further and are easier to generate at low energies. If the conclusions of Battaglin et al.[30] are correct, an energy threshold or window effect may exist, providing further control over the process via electron energy. However, the problems of heat dissipation during high dose electron irradiations of insulating substrates are not to be overlooked. It remains to explore further the ability of photons (UV light, γ-rays)to produce adhesion enhancement.

We conclude with the comment that the mechanisms of radiation-enhanced adhesion (there are clearly several) have been identified so far only by inference from mostly secondary evidence. A new kind of evidence is needed, in which we seek to document the phenomena in well-defined, uncontaminated systems, and develop an understanding of the chemistry, stoichiometry and bonding structure of the "planar" interface phases and compounds on which future tailored interface construction must surely depend.

REFERENCES

1. D.M. Mattox, Thin Solid Films **18** 173 (1973)

2. L.E. Collins, J.G. Perkins and P.T. Stroud, Thin Solid Films **4** 41 (1969)

3. T.F. Gukelberger and W. J. Kleinfelder, U.S. Patent no. 3,682,729, August 8, 1972

4. K.R. Padmanabhan and G. Sorensen, Thin Solid Films **81** 13 (1981)

5. M. Laugier,Thin Solid Films **81** 61 (1981)

6. J.E. Griffith, Y. Qiu and T.A. Tombrello, Nucl. Instr. and Methods **198** 607 (1982)

7. S. Jacobson, B. Jonsson and B. Sundqvist, Thin Solid Films **107** 89 (1983)

8. M.H. Mendenhall, Ph.D. Thesis, CalTech (1983)

9. T.A. Tombrello, Proc. Mat. Res. Soc. **25** 173 (1984).

10. B.T. Werner, T. Vreeland, M.H. Mendenhall, Y. Qiu and T.A. Tombrello, Thin Solid Films **104** 163 (1983).

11. J.E.E. Baglin, G.J. Clark, J. Bottiger, Proc. Mat. Res. Soc. **25** 179 (1984)

12. J.E.E. Baglin and G.J. Clark, Nucl. Instr. and Methods **B7/8** 881 (1985)

13. J. Bottiger, J.E.E. Baglin, V. Brusic, G.J. Clark and D. Anfiteatro, Proc. Mat. Res. Soc. **25** 203 (1984)

14. I.V. Mitchell, J.S. Williams, D.K. Sood, K.T. Short, S. Johnson and R.G. Elliman, ibid, p.189

15. I.V. Mitchell, J.S. Williams, P. Smith and R.G. Elliman, Appl. Phys. Letters **44** 193 (1984)

16. C.J. Sofield, C.J. Woods, C. Wild, J.C. Riviere and L.S. Welch, Proc. Mat. Res. Soc. **25** 197 (1984)

17. P.P. Pronko, A.W. McCormick, D.C. Ingram, A.K. Rai, J.A. Woollam, B.R. Appleton and D.B. Poker, Proc. Mat. Res. Soc. **27** 559 (1984)

18. I.V. Mitchell, G. Nyberg and R.G. Elliman, Appl. Phys. Letters **45** 137 (1984)

19. D.K. Sood, P.D. Bond and S.P.S. Badwal, Proc. Mat. Res. Soc. **27** 565 (1984)

20. G.A. Sai-Halasz and G. Gazeki, Appl. Phys. Letters **45** 1069 (1984)

21. T.A. Tombrello, Materials Science and Engineering **69** 443 (1985)

22. R.P. Livi, Nucl. Instr. and Methods, **B10/11** 545 (1985)

23. C.R. Wie, J.Y. Tang, T.A. Tombrello, R.G. Grant and R.M. Housley, CalTech internal report No. BB-35, September 1985

24. A.E. Berkowitz, R.E. Benenson, R.L. Fleischer, L. Wielunski and W. A. Lanford, Nucl. Instr. and Methods **B7/8** 877 (1985)

25. D.K. Sood, W.M. Skinner and J.S. Williams, Nucl. Instr. and Methods **B7/8** 893 (1985)

26. J.E.E. Baglin, Proc. Mat. Res. Soc. **47** 3 (1985)

27. A.J. Kellock, G.L. Nyberg and J.S. Williams, Vacuum **35** 625 (1985)

28. J. Gazecki, G.A. Sai-Halasz, R.G. Elliman, A. Kellock, G.L. Nyberg and J.S. Williams, Applic. Surf. Sci. **22/23** 1034 (1985)

29. G. Battaglin, private communication (1986)

30. G. Battaglin, M. Carbucicchio, R. Dal Maschio, F. Marchetti, P. Mazzoldi and A. Valenti, XIV Internat. Congress on Glass, New Delhi, India, March 1986

31. G. Battaglin, P. Mazzoldi and R. Dal Maschio, in "Induced Defects in Insulators", p. 235, ed. P. Mazzoldi, Les Editions de Physique, Le Ulis Cedex (France) 1984

32. M. Carbucicchio, A. Valenti, G. Battaglin, P. Mazzoldi and R. Dal Mascio, Radiation Effects (1986) to be published

33. M. Carbucicchio, A. Valenti, G. Battaglin, P. Mazzoldi, R. Dal Maschio, Hyperfine Interactions (1986) to be published

34. R.G. Stokstad, P.M. Jacobs, I. Tserruya, L. Sapir and G. Mamane, Nucl. Instr. and Methods **B** (1986) to be published

35. D.C. Ingram and P.P. Pronko, Nucl. Instr. and Methods **B** (1986) to be published

36. F.M. d'Heurle, J.E.E. Baglin and G.J. Clark, J. Appl. Phys. **57** 1426 (1985)

37. L.E. Seiberling and R.L. Headrick, (this volume) (1986)

38. R.L. Headrick and L.E. Seiberling, Proc. Mat. Res. Soc. **35** 539 (1985)

39. R.L. Headrick and L.E. Seiberling, Appl. Phys. Letters **45** 388 (1984)

40. T.A. Tombrello, Int. J. Mass Spectrom. Ion Phys. **53** 307 (1983)

41. V.E. Henrich, Rep. Prog. Phys. **48** 1481 (1985)

42. D.M. Mattox, J. Appl. Phys. **37** 3613 (1966)

43. M.P. Borom and J.A. Pask, J. Am. Ceram. Soc. **49** 1 (1966)

44. G.M. Sessler, J.E. West, F.W. Ryan and H. Schonhorn, J. Appl. Polymer Science **17** 3199 (1973)

45. S.V. Pepper, J. Appl. Phys. **47** 2579 (1976)

46. E. Taglauer and W. Heiland, Proc. Symp. on Sputtering, eds. P. Varga, G. Betz, F.P. Viehbock, Vienna (1980)

47. A. Schrott, R.D. Thompson and K.N. Tu, Proc. Mat. Res. Soc. **60** (1986)

48. E.P. Eernisse and C.B. Norris, J. Appl. Phys. **45** 5196 (1974)

49. S. Wu, in "Polymer Interface and Adhesion", Marcel Dekker Inc., New York, 1982, p.531

50. K.S. Kim, Report No. UILU-ENG 85-6003, University of Illinois, March 1985

51. "Adhesion measurement of thin films, thick films and bulk coatings", ed. K.L. Mittal, American Society for Testing and Materials, Philadelphia, STP No. 640 (1978)

MODIFICATION OF SILVER/SILICON INTERFACES DURING MeV-ION

BOMBARDMENT: THE ROLE OF AN INTERFACIAL OXIDE LAYER

L.E. Seiberling[*] and R.L. Headrick[**]

[*] Department of Physics, University of Pennsylvania
Philadelphia, PA 19104
[**] Department of Materials Science and Engineering
University of Pennsylvania, Philadelphia, PA 19104

We have studied the production of disordered Si at Ag/Si
interfaces during 5.9 MeV ^{9}Be ion irradiation. The technique
of transmission ion channeling was used, which is sensitive
to less than one monolayer of nonregistered Si. MeV-ion
irradiation of many thin film/substrate interfaces is known
to significantly improve the adhesion of the film to the
substrate. If atomic mixing or formation of new bonds is
responsible for improved adhesion, then disordered Si should
be produced. The results indicate that the ion bombardment
produces recoiling Ag atoms that cause damage in the Si
substrate. An interfacial oxide layer can act as a barrier
to the recoil damage. For a sample with no Ag layer and a
native oxide on Si, nonregistered Si was produced by electronic excitation during ion bombardment.

INTRODUCTION

It has long been recognized that passing energetic ions through an
interface joining two different materials can significantly alter the
physical and chemical structure of the interfacial region.[1,2] Typically,
ions used to modify solids have been in the energy range of less than
10 keV/amu. More recently, the effects of very energetic ions (roughly
100 keV/amu or greater) on solids have been studied.[3-5] The primary means
by which an ion loses energy when passing through a solid is very different
in these two energy ranges. Figure 1 (taken from Ref. 6) illustrates this
difference. Ions having several keV/amu are in the nuclear stopping regime.
They scatter via the Coulomb force with atomic nuclei making up the solid.
These collisions deflect the ion from its path and impart substantial energy
to the struck atom. MeV/amu ions, which have velocities near $Z_1^{2/3} e^2/ h$,
suffer numerous collisions with atomic electrons. These collisions do not
deflect the ion perceptibly. The electrons are raised to excited states, or
leave the atom altogether and travel distances of typically less than
100 Å.

One of the most important effects of ion beam modification of interfaces is enhanced adhesion of the overlayer to the substrate. Improvements

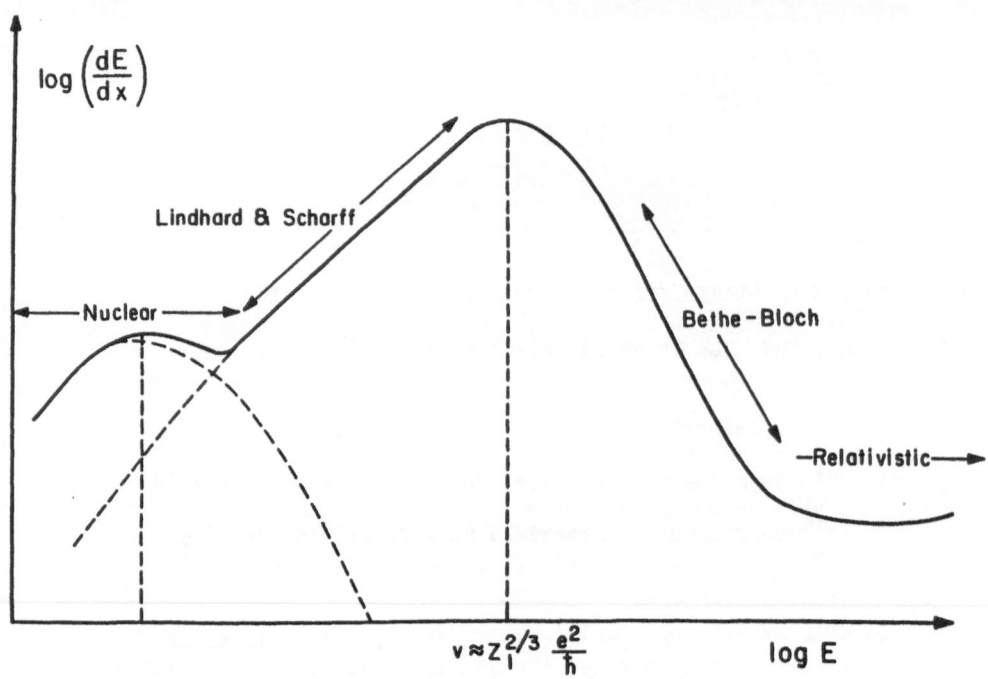

Figure 1. Energy loss of energetic ions in solids. Energy is transferred to the solid by collisions with target atoms (nuclear stopping power) and numerous small energy transfer collisions with target electrons (electronic stopping power). For ions in the MeV/amu regime ($v \approx Z_1^{2/3} e^2/$ h), the electronic stopping power is much larger than the nuclear stopping power. (Taken from reference 6.)

(or degradation) of adhesion caused by keV-ion irradiation is thought to be initiated by ion-beam mixing.[7,8] Nuclear collisions in the vicinity of the interface cause atoms of each material to be scattered across the interface and mixed. It has recently become clear that the extent of the mixing, however, can be dominated by thermodynamic or chemical effects.[2,9,10] The extent to which keV/amu ions can enhance adhesion also appears to depend critically on chemistry. Clark et al.[8] have studied a variety of metals on SiO_2, all of which show poor initial adhesion. The adhesion can be substantially improved by 200–350 keV Xe^+ ions only in those cases for which the metal is able to reduce the oxide and form a silicide phase at the interface. In cases for which this is not true, metal atoms move laterally and coalesce into islands.

MeV-ion enhanced adhesion occurs in an energy range in which the electronic stopping power greatly exceeds the nuclear stopping power. Tombrello et al[11] have shown that for Au films on Ta, the threshold dose for improved adhesion (using the Scotch tape test) scales as the electronic stopping power to the -1.6 power for Cl ions with energies near 20 MeV. Although in some cases, both nuclear and electronic stopping power may affect adhesion,[12] it has been shown unambiguously that electronic effects alone are sufficient to produce improved adhesion in many systems.[13,14] In these experiments, UV photons, or 5 to 30 keV electrons (having energies below the threshold to produce atomic displacements) have been used to irradiate a variety of metal/substrate combinations. Improvements in adhesion were readily observed. The mechanism by which electronic excitation leads to improved adhesion is not well understood, but may also depend critically on chemical effects. Because MeV ions (and 5-30 keV electrons)

236

produce relatively few (no) energetic nuclear recoils, they are able to improve adhesion without the introduction of extensive damage (which may lead to deadhesion in some cases[12]). Thus, adhesion produced by electrons or MeV ions may prove superior in cases where initial adhesion is poor.

For the case of adhesion initiated by electronic excitation, it is of interest to discover whether the adhesion is associated with the formation of a new interfacial phase or involves mixing of atoms across the interface. It has become clear in recent years that MeV ions bombarding the surface of an insulator can produce sputtering and erosion of the surface whereas the same ions passing through a good conductor will not.[3,4] The displaced atoms result from electronic excitation produced by the bombarding ion.[15,16] It has been suggested that this electronic sputtering effect could produce a small amount of mixing at an interface containing an insulating oxide layer, and that the mixing could lead to improved adhesion.[5,17,18] Electronic sputtering yields can be very large (for example, the yield for 6 MeV ^{19}F on H_2O ice is 1400 molecules/ion[16]), however, the energies of the sputtered particles are quite small (less than about 1 eV for 13 MeV Cl on UF_4[16]).

Jacobson et al.[17] have used a scratch test to study the role of an interfacial native oxide during MeV-ion bombardment. Their samples consisted of Cu films on Al. For samples containing a thin oxide, initial adhesion was poor, but improved adhesion was observed after MeV-ion irradiation. When the Al substrates were sputter cleaned in UHV prior to Cu evaporation, the initial adhesion was greatly improved, but no enhancement in adhesion with ion dose could be measured. This seems to imply that the interfacial oxide is a critical component in MeV-ion enhanced adhesion.

A number of investigators have attempted to measure mixing associated with MeV-ion enhanced adhesion.[12,19,20] A Rutherford backscattering experiment with 20 Å depth resolution failed to detect mixing of a Ag film on Si after bombardment with 20 MeV Cl ions.[19] Two attempts have been made to measure metal atoms embedded in a substrate after successfully adhering the metal film using ion irradiation. In the first case, embedded Au atoms were found in the amount of 0.01 of the ion dose.[19] In the second case, no copper was detected in SiO_2 after an adhered copper strip was peeled from an SiO_2 substrate; however, the detection limit in that case was only one monolayer.[12]

We have previously reported the results of a transmission channeling experiment sensitive to monolayer levels of damage produced at a single crystal silicon/metal interface.[20] Disorder at the metal/silicon interface was observed that increased linearly with dose. A dose known to produce substantial adhesion enhancement[11] involved movement of at most a few monolayers of Si atoms. In this paper, we present the results of further transmission channeling experiments designed to test whether the interfacial oxide at the Ag/Si interface is the primary cause of the observed damage. The hope is that an understanding of the relatively simple and well characterized case of Ag on Si will shed light on more complex systems, such as metal/ceramic and metal/polymer interfaces.

EXPERIMENTAL

Device grade 1000 ohm-cm p-type (110) Si wafers were doped with boron at 1000°C in N_2 for 90 min. The wafers were then cut into rectangular pieces and a selective etch[21] was used to produce 4000-4500 Å thick 0.6 cm diameter single-crystal windows in the center of each sample. The uniformity of the thickness of the windows was approximately 10%. One sample was thermally oxidized in dry O_2 at 865°C for 15 min. This resulted in an

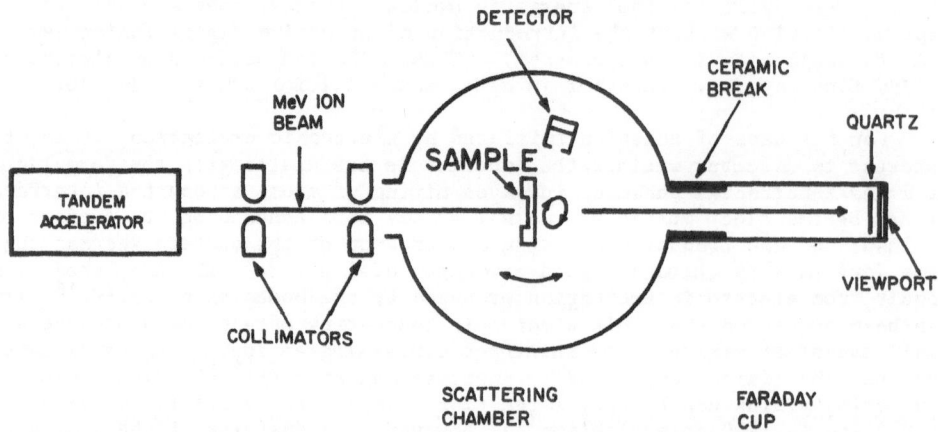

Figure 2. Schematic diagram of the experimental apparatus. The sample is mounted on a goniometer with two axes of rotation and Z translation. The Faraday cup is used for alignment of the sample to a channeling direction and for current integration.

SiO_2 layer approximately 50 Å thick. A silver layer 200 Å thick was deposited over the thermal oxide by evaporation from a Ta boat. A mask was used which covered half of the sample area. The pressure during the evaporation was approximately 1×10^{-6} Torr. A second sample was etched in dilute HF for 1 minute immediately prior to loading in the evaporation chamber. A silver layer was deposited onto half of this window over the native oxide. The four sample areas will be referred to as Ag/thermal oxide/Si, thermal oxide/Si, Ag/native oxide/Si and native oxide/Si.

The samples were mounted on an O-ring sealed two-axis goniometer with the interface of interest on the beam-exit side. Care was taken to load the HF etched sample into the scattering chamber within one hour after the evaporation to avoid oxidation of the silicon due to exposure to air.

Figure 2 is a schematic diagram of the experimental geometry. A 5.9 MeV ^{9}Be ion beam with +2 charge state was produced by the Penn FN tandem Van de Graff accelerator and collimated to less than 0.15 degrees angular divergence before entering the scattering chamber. ^{9}Be ions passing through the thin window were collected in a Faraday cup and used for current integration. The charge state of the ^{9}Be ions increased from +2 to an average of +3.35 after passing through the window. Ions scattered at an angle of 80° were energy analyzed in a silicon surface barrier detector with a solid angle of 1.85 msr. Each sample was aligned in the <110> crystallographic direction to a minimum yield of less than 5%. The task of locating the channeling direction is facilitated by the quartz beam stop and viewport at the end of the Faraday cup. The beam spot forms characteristic "doughnut patterns when the channeling direction is approached,[22] and the axial direction can be located visually to within 0.02 degrees in a few minutes. After alignment of the crystal, the beam was moved to a fresh spot and the sample was irradiated with consecutive doses of 7×10^{15} ions/cm^2. The ^{9}Be ions were used both to create damage in the crystal, and to analyze the damage. After each dose of 7×10^{15} ions/cm^2, an energy spectrum was recorded for analysis and a new dose begun on the same spot.

The use of transmission channeling through thin silicon crystals to investigate nonregistered silicon was first introduced by Feldman and co-

workers.[23] The advantage of this technique is that it allows a substantial
reduction in background and thus greatly enhances sensitivity to nonregistered Si. Consider, for example, a metal layer over a thin thermal oxide on
Si. If the analyzing beam enters through the metal layer, three seperate
scattering processes will add to the background at essentially the same
energy as the signal from nonregistered Si. All three of these sources of
background can be reduced or eliminated if the analyzing beam exits the
crystal through the metal layer.

First is the intrinsic surface peak, a contribution equal to one or
more monolayers of disordered Si, but reduced to a few percent of that
amount in transmission since the channeled beam has no intensity at the
normal lattice sites on the back surface. Second is the contribution from
ions that become dechanneled upon entering the crystal and scatter from
lattice atoms a few monolayers beneath the Si surface. The size of this
nonchanneled background depends on the channeling minimum yield. In the
transmission channeling geometry, an energy difference is introduced between
channeled ions scattering from nonregistered Si and nonchanneled ions
scattering from Si at the interface. This difference in energy is often
sufficient to partially separate the two. The energy shift is possible
because of the anomalously low rate of energy loss of channeled ions. An
ion dechanneled upon entering the crystal will lose more energy in traversing
the crystal than will a channeled ion. Because of the limited energy
resolution of surface-barrier detectors for ^{9}Be ions (approximately 60 keV)
and multiple scattering in the Si crystal and Ag layer, we were unable
to completely separate the signals. However, a substantial reduction in the
background was achieved.

The third reason for a reduction in background using transmission
channeling is a result of multiple scattering in the metal layer. A low
channeling minimum yield requires a very small beam divergence (less than
0.5 degrees). Angular straggling of the ion beam in passing through the
metal layer can significantly increase the minimum yield, and add to the
background in the nonregistered Si peak. In transmission, of course, this
problem is completely eliminated.

For the experiments presented in this paper, transmission channeling
has other advantages related to nuclear recoil damage. One goal of this
investigation was to look for evidence of mixing or rearrangement of bonds
at a silicon/metal interface that resulted from electronic scattering by the
bombarding ion. A small number of nuclear scattering events can cause extensive damage in a crystal and would mask disordered Si produced by
electronic effects. When an ion is channeled, the probability that it will
displace a lattice atom due to a nuclear collision is reduced to zero.[24]
By placing the interface at the beam exit side, over 95% of the ions are
channeled upon reaching the interface. This geometry also eliminates the
possibility that primary nuclear recoils in the (randomly oriented) metal or
oxide layer could directly scatter across the interface. A nuclear contribution to damage of silicon will still arise from the dechanneled component (less than 5%) of the beam; however, this damage will occur throughout the window, not preferentially at the interface. Damage to interfacial
Si produced by nuclear scattering will be caused primarily by collision
cascades that originate in the metal layer and cross the interface.[25]

Figure 3 is a transmission channeling spectrum for the Ag/thermal
oxide/Si sample. Dechanneled ^{9}Be ions that scatter from the Si crystal
are between channels 150 and 300 (represented by the dashed line). Ions
scattered from the front surface (beam entrance side) of the crystal are at
a lower energy than those scattered from the interfacial region because of
the longer path that they must travel through the sample. The peak labeled

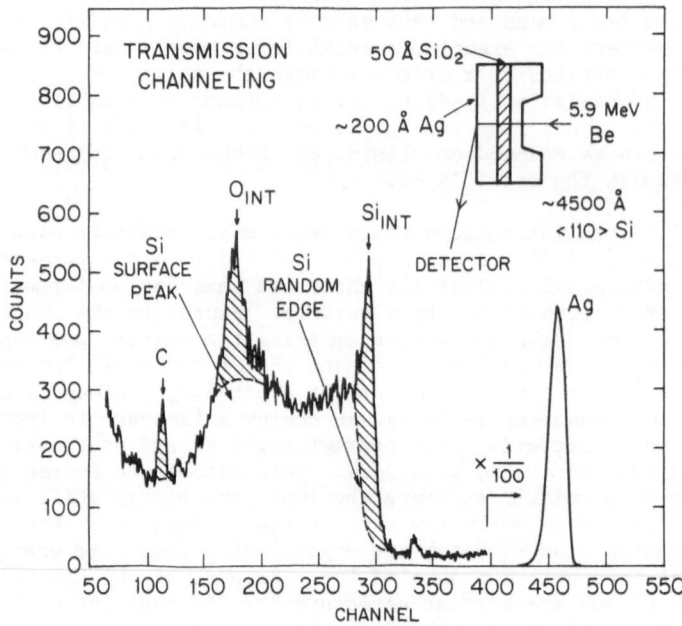

Figure 3. Transmission channeling spectrum of a thermally oxidized 4500 Å Si window with 200 Å of Ag evaporated on the beam exit side.

C is from carbon buildup during the ion irradiation and is on the free surface of the Ag layer. Scattering from oxygen at the interface (in the SiO_2 layer) occurs in the region labeled O_{int}, and coincides with the Si surface peak for this particular geometry and sample thickness. [9]Be ions scattered from nonregistered interfacial Si atoms are in the region labeled Si_{int}. This peak contains Si atoms incorporated in the SiO_2 layer as well as nonregistered Si atoms produced by the ion irradiation. The centroid of the interfacial silicon peak is 53 keV higher in energy than the random edge. This corresponds to a difference in energy loss between channeled and nonchanneled ions traversing the window of 95 keV. Thus, the stopping power of channeled [9]Be ions is approximately 65 percent of the stopping power of the nonchanneled ions. Finally, the Ag peak is shown, reduced by a factor of 100.

Figure 4a is a transmission channeling spectrum for the native oxide/Si sample. In this sample, the interfacial disordered Si (near the beam exit surface) can be partially separated from the nonchanneled background. An expanded view of the interfacial Si peak is shown in Figure 4b, along with a suitably normalized random spectrum. The peak represents 1.2×10^{15} Si at/cm^2, with an uncertainty of less than 0.15 monolayer.

RESULTS

Damage rates have been obtained from the increase of the area of the interfacial Si peak as a function of dose. The signal from bulk silicon introduces a considerable background on the interfacial Si peak. This background was removed by subtracting a suitably normalized random spectrum from

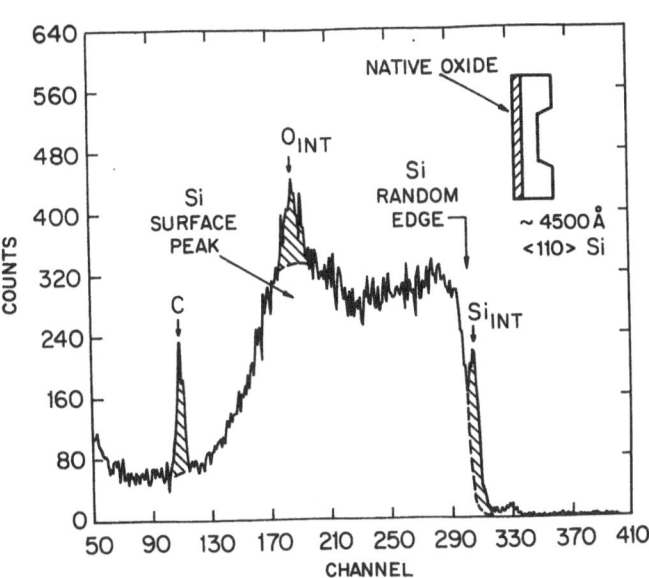

(a)

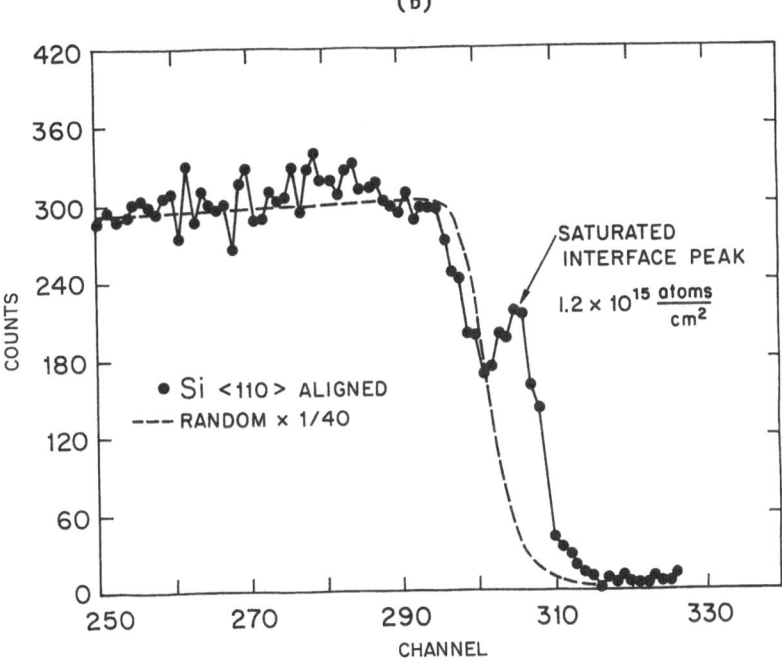

(b)

Figure 4. Transmission channeling spectrum of a 4500 Å (110) Si window that has been etched in HF. (a) The complete spectrum showing the width of the Si signal and the positions of the surface carbon and interface oxygen peaks. (b) An expanded view of the interfacial Si peak region showing the saturated interface peak and the random Si signal.

each channeling spectrum. For the cases shown in figures 3 and 4, the interface peak clearly stands out. This is also true for the thermal oxide/Si sample. However, the interface peaks in the Ag/native oxide/Si spectra are completely obscured by the bulk signal. The background subtraction for this set of data is therefore more uncertain. This uncertainty should not affect the measured damage rate significantly, but may cause some error in the absolute level of disordered interfacial silicon and in the measured saturation value. The damage can be plotted as a function of beam dose as in figures 5 and 6, and the damage rate, initial oxide thickness, and damage saturation (if any) can be determined from the plot.

The theory of radiation damage in solids by nuclear collisions can be used to estimate the nuclear contribution to interfacial damage. The theory of Kinchin and Pease[26] has been used successfully by Iwami et al.[25] to estimate damage production rates in silicon substrates with thin Ag and Au overlayers. They obtained calculated values that were within a factor of two of the experimental damage rates. Their samples consisted of several monolayers of the metal over an atomically clean substrate. We have recently extended these hard-sphere calculations to obtain approximate damage rates for oxidized silicon substrates with thick evaporated metal layers.[18] While these estimates are expected to be rather crude, it may be useful to compare them to the experimentally measured values.

The principal results of this study are presented in Table I. Experimental and calculated damage rates are tabulated for each of the four samples. To obtain the calculated values, the energy to displace a silicon atom, E_d, has been taken to be 12.9 eV.[27] The thermal oxide/Si damage rate shows the best agreement with the calculated value. It is interesting to note that the calculated values are the same for both of the thermal oxide samples, whether or not the Ag layer is present. In other words, secondary recoils produced in the Ag layer are not expected to penetrate the thermal oxide and cause significant damage in the substrate. In spite of this, it is clear that the presence of the Ag layer causes the measured damage rate to increase dramatically (by a factor of 8).

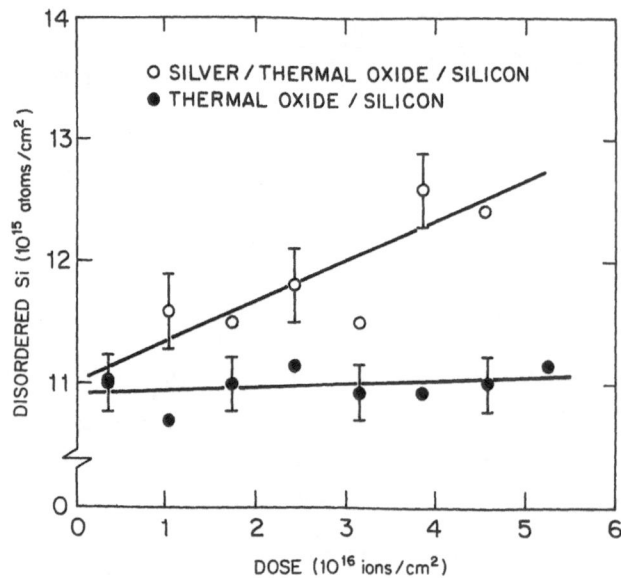

Figure 5. Peak area of the thermal oxide/Si and Ag/thermal oxide/Si interfaces as a function of beam dose.

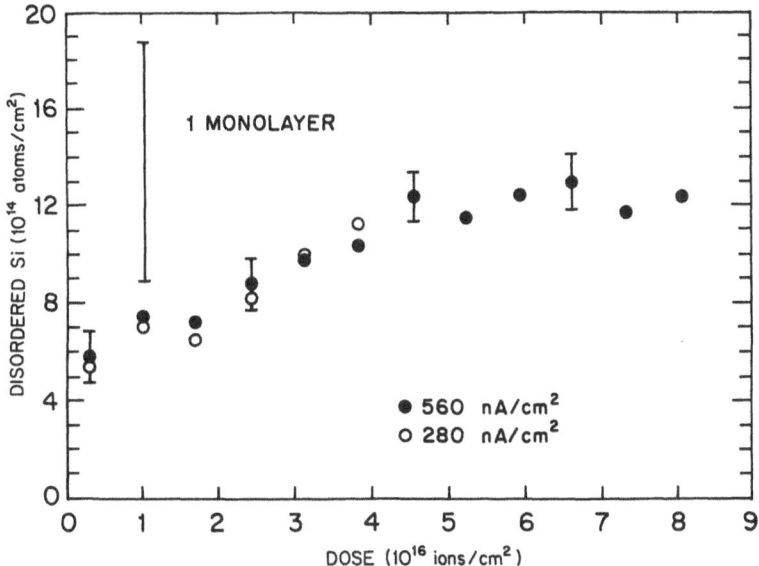

Figure 6. Area of the native oxide/Si interface peak as a function of [9]Be beam dose. The solid points are for a current density of 560 nA/cm[2]. The experiment was repeated at a current density of 280 nA/cm[2] (open circles) to check for thermal effects.

If the observed damage is a result of nuclear recoils, then one would expect to find some Ag atoms implanted into the Si substrate. To check for implanted Ag, the Ag layer was etched off of the Ag/native oxide/Si sample after it had been irradiated with a total dose of 6.3×10^{16} ions/cm[2]. Dilute Aqua Regia ($HCl:HNO_3:H_2O$ 3:1:5) was used to remove the Ag layer, after which the sample was etched in 10 percent HF. Aqua Regia oxidizes several monolayers of silicon, and these oxidized layers are removed by the HF. Therefore, any Ag remaining on this sample after etching must have been originally implanted into the substrate. The sample was re-loaded into the scattering chamber; channeling and random spectra were taken with a beam spot of 0.76 mm by 0.76 mm in the same area of the target that had previously

Table I. Experimental and Calculated Nuclear Damage Rates (R).

SAMPLE	OXIDE THICKNESS (Si at/cm[2])	R_{EXP}	R_{CALC}	$\dfrac{R_{EXP}}{R_{CALC}}$	SURFACE PEAK SATURATION
Thermal oxide/Si	10.9×10^{15}	.004	.007	0.57	*
Ag/Thermal oxide/Si	11.0×10^{15}	.032	.007	4.6	*
Ag/Native oxide/Si	5×10^{14}	.073	.017	4.3	3.6×10^{15}
Native oxide/Si	5×10^{14}	.014	.002	7.0	1.3×10^{15}

* no saturation observed.

243

been irradiated, and also in a fresh spot. These spectra showed 4.8×10^{13} Ag at/cm^2 remaining in the area that had previously been irradiated, and that the Ag remaining in the unirradiated spot was less than 5×10^{11} Ag at/cm^2. A Si interface peak of 1.74×10^{15} at/cm^2 was measured on the previously irradiated spot. This is a factor of two smaller than the saturated interface peak listed in table I, which confirms that some silicon was removed during the Aqua Regia/HF etch. The disordered Si/Ag ratio in the surface layers after the etch is roughly 25 (a correction has been made for the formation of a new native oxide on the surface). An estimate of the amount of Ag recoiling across the Si/Ag interface can be made using Sigmund sputtering theory.[28] We have used the surface binding energy of Ag and assumed that all atoms that would be sputtered from a silver surface (toward the substrate) would be implanted into the silicon substrate. Such a calculation gives 5.0×10^{15} Ag at/cm^2. This can only be consistent with the measured value if most of the Ag were removed from the surface during the etch.

The native oxide/Si sample shows the greatest disparity between calculated and measured values. Damage production in this sample, however, rapidly saturates. Figure 6 is a plot of the disordered interfacial silicon as a function of beam dose for the native oxide/Si sample. The peak area increases linearly up to a dose of approximately 4.5×10^{16} ions/cm^2 and saturates at a value of 1.3×10^{15} Si atoms/cm^2. The damage rate after saturation (zero to within experimental uncertainty) is consistent with the calculated value. These facts have led us to conclude that the initial rapid disorder production in this sample is not caused by nuclear collisions. More will be said about this in the following section.

DISCUSSION

For samples containing an oxide layer, it was expected that electronic sputtering might contribute to substrate damage, since some sputtered particles will travel back toward the Si substrate rather than being ejected from the front surface. Particles ejected during electronic sputtering, however, have energies on the order of 1 eV, and would not be able to displace substrate Si atoms by direct momentum transfer. Thus, any contribution to substrate damage would have to be assisted by chemical effects. From the data in table I, however, it is clear that the primary role of the interfacial oxide is to act as a barrier to substrate damage.

The calculated and measured values of damage for the thermal oxide/Si sample agree rather well. This agreement is probably fortuitous since such a small damage rate is difficult to measure (see figure 5). The damage rates listed in Table I were obtained from a least squares fit to the data; however, in the case of the thermal oxide on Si, the rate is clearly consistent with zero. The measured damage rate could be artificially low since electronic sputtering of the SiO$_2$ layer may not be negligible. The sputter yield, extrapolated from the yield for 20 MeV Cl^{35} is 0.002 Si atoms per incident ^{9}Be ion.[29] In this calculation we have assumed that the sputtering yield scales as the fourth power of the electronic stopping power.[16] If 0.002 is added back into the experimental damage rate, the calculated and measured values agree almost exactly. However, at the pressures prevailing in the scattering chamber during the ion irradiation (greater than 1×10^{-8} Torr), any sputtering would probably have been quenched by surface contaminants.

The measured damage rate for the Ag/thermal oxide/Si sample was 4.6 times larger than the calculated value. Such calculations generally overestimate the amount of energy lost by recoils while traversing an oxide layer. In this case, however, with 50 Å of oxide, it is difficult to see how the Ag layer can directly cause significant damage. For example, the average energy of primary Ag recoils is about 150 eV.[26] The linear range of 150 eV Ag ions in SiO_2, calculated by numerical integration of the Kr-C nuclear stopping power[30], is approximately 17 Å. If we assume that the direct cause of the damage in this case is secondary recoils generated in the oxide layer, there are several mechanisms whereby the Ag overlayer might be able to indirectly cause an enhancement in this effect. First, an excess of electrons in the silicon can cause weakening of the Si-Si bonds. This would enhance the recoil damage produced since the damage rate depends critically on the energy required to displace a substrate atom. Excess electrons could come from secondary electrons scattered from the Ag overlayer. Metals have one or more essentially free electrons per atom, and when bombarded with MeV ions, energetic secondary electrons can be produced with a range of greater than 50 Å. Another mechanism of enhanced damage created indirectly by the Ag layer could be damage in the oxide (which is not measured by channeling) that would result in mobile oxygen atoms that could form new Si-O bonds with substrate Si.

For the Ag/native oxide/Si sample, the measured rate is 4.3 times larger than the calculated rate. Considering that the calculated damage rates for thick (hundreds of angstroms) overlayers are not expected to be very accurate, this difference may not be significant. However, this factor of 4-5 shows up consistently when comparing measured and calculated damage rates for samples with metal overlayers. If the value of $E_d = 12.9$ eV is substituted into the calculations in reference 18, the ratio of measured to calculated damage rates is consistently about a factor of 5, even though both Ag and Au overlayers were used and the damage was created with 14 MeV ^{16}O instead of 5.9 MeV ^9Be. Damage in this sample appears to saturate at approximately 3.6×10^{15} Si at/cm^2, which corresponds to roughly 7 Å of disordered Si. This saturation is consistent with the damage being produced by recoiling metal atoms since they would be expected to have a very short range and would quickly amorphize an extremely thin layer of the substrate near the interface.

The initial fast rate of damage produced in the native oxide/Si sample is apparently not caused by nuclear recoils from the oxide. If it were, then one would expect it to be smaller than the rate for the thermal oxide sample (see Table I). It is also unlikely that the damage is caused by an increase in temperature in the beam spot. The data in Figure 6 show that the same slope is reproduced for two beam currents differing by a factor of two. We have not observed a dependence on pressure between approximately 1×10^{-8} and 5×10^{-7} Torr. We suggest that the increase in interfacial silicon in Figure 6 is caused by ionization induced changes in chemical bonding.

Such ionization effects are not entirely unexpected. There is now a considerable body of literature supporting the idea that ionizing radiation can enhance chemical processes at semiconductor surfaces through a completely athermal mechanism. For example, nonthermal laser enhanced etching of GaAs[31] and enhanced oxidation of GaAs and Si[32,33] have been reported. These investigators suggest that the increased chemical activity is caused by free carriers excited across the semiconductor band gap. A similar mechanism may be responsible for the displaced silicon in our native oxide/Si sample. The native oxide of silicon is not stoichiometric SiO_2, but is probably composed of a monolayer of adsorbed molecular O_2 on a reconstructed silicon surface.[34,35] Electronic stimulation of adsorbed O_2 in the native oxide can cause new SiO_2 tetrahedra to form.[35] This would lead to permanent displacements of Si atoms from their lattice sites.

This picture cannot be confirmed by the present work since residual gasses that contained carbon and oxygen were present in the vacuum chamber (at 10^{-8} Torr). Carbon or oxygen from the vacuum, excited by the passage of the ion beam could react with the surface and form Si-C or Si-O bonds. Carbon was observed on the surface of all samples which increased linearly with dose up to at least several monolayers. The amount of oxygen deposited during the irradiation could not be measured because of excessive background in the region of the oxygen signal. Further experiments performed in a well controlled vacuum would help to clarify this question.

It has been strongly suspected since the pioneering work of Benjamin and Weaver that oxygen can play an important role in adhesion.[36] They found that metals with a large heat of formation of their oxide adhered most strongly to glass slides and that the adhesion of these metals could be improved by the incorporation of oxygen into the evaporated metal films. More recently, in an in-situ UHV measurement, Pepper has shown that a 1 L exposure of atomic oxygen on a Ni sphere produces observable increases in the shear strength of the contact between the Ni and an Al_2O_3 surface.[37] It is interesting to note that O_2 must be in an excited state in order to adsorb onto the noble and near-noble metals Ni, Cu, Ag, Au.[38] It follows that the passage of ionizing radiation through the Ag/native oxide/Si interface could result in the production of atomic oxygen which would improve the adhesion of the metal film to the substrate. This model is also supported by the work of Jacobson on the Cu/Al interface.[17]

CONCLUSIONS

Interfacial damage of silicon substrates by 5.9 MeV ^{9}Be ions has been studied. Comparison of the damage rates for several interfaces has shown that the interfacial damage reported in references 18 and 20 is primarily due to recoiling atoms from the metal overlayer. Assuming that these recoils displace other target atoms by hard-sphere collisions, a rough estimate of the damage rate can be calculated that is consistently too low by about a factor of 5. This may be evidence that the presence of silver on the surface enhances the damage rate by some other mechanism. The interfacial oxide acts as a barrier that inhibits damage to the substrate caused by Ag recoils.

Irradiation of a freshly etched silicon surface with 5.9 MeV ^{9}Be ions causes rapid initial damage that saturates at slightly more than one monolayer of silicon. This damage is probably caused by ionization effects at or near the surface.

ACKNOWLEDGEMENTS

The authors would like to thank J.E.E. Baglin and J.S. Williams for helpful comments and suggestions. This work was supported by the NSF [PHY-8213598 and MRL program DMR-8216718] and the IBM corporation.

REFERENCES

1. B.M. Paine and R.S. Averback, Nucl. Instr. Meth. B7/8, 666 (1985).
2. J.W. Mayer, B.Y. Tsaur, S.S. Lau and L.S. Hung, Nucl. Instr. Meth. 182/183, 1 (1981).
3. W.L. Brown, L.J. Lanzerotti, J.M. Poate and W.M. Augustyniak, Phys. Rev. Lett. 40, 1027 (1978).

4. J.E. Griffith, R.A. Weller, L.E. Seiberling and T.A. Tombrello, Rad. Eff. 51, 223 (1980).

5. J.E. Griffith, Yuanxun Qiu and T.A. Tombrello, Nucl. Instr. Meth. 198, 607 (1982).

6. T.A. Tombrello, submitted to Comments on Nuclear and Particle Physics (1984).

7. L.E. Collins, J.G. Perkins and P.T. Stroud, Thin Solid Films 4, 41 (1969).

8. G.J. Clark, J.E.E. Baglin, F.M. d'Heurle, C.W. White, G. Farlow and J. Narayan, Mat. Res. Soc. Symp. Proc. 27, 55 (1984).

9. S.S. Lau, B.Y. Tsaur, M. von Allmen, J.W. Mayer, B. Stritzker, C.W. White and B. Appleton, Nucl. Instr. Meth. 182/183, 97 (1981).

10. Marc-A. Nicolet, T.C. Banwell and B.M. Paine, Mat. Res. Soc. Symp. Proc. 27, 3 (1984).

11. T.A. Tombrello, Mat. Res. Soc. Symp. Proc. 25, 173 (1984).

12. J.E.E. Baglin, G.J. Clark and J. Bottiger, Mat. Res. Soc. Symp. Proc. 25, 179 (1984).

13. I.V. Mitchell, G. Nyberg and R.G. Elliman, Appl. Phys. Lett. 45, 137 (1984).

14. I.V. Mitchell, J.S. Williams, D.K. Sood, K.T. Short and S. Johnson, Mat. Res. Soc. Symp. Proc. 25, 189 (1984).

15. W.L. Brown, W.M. Augustyniak, E. Brody, B. Cooper, L.J. Lanzerotti, A. Ramirez, R. Evatt and R.E. Johnson, Nucl. Instr. Meth. 170, 321 (1980).

16. L.E. Seiberling, C.K. Meins, B.H. Cooper, J.E. Griffith, M.H. Mendenhall and T.A. Tombrello, Nucl. Instr. Meth. 198, 17 (1982).

17. S. Jacobson, B. Jonsson, and B. Sundqvist, Thin Solid Films 107, 89 (1983).

18. R.L. Headrick and L.E. Seiberling, Mat. Res. Soc. Symp. Proc. 35, 539 (1985).

19. M.H. Mendenhall, Ph.D. Thesis, California Institute of Technology (1983).

20. R.L. Headrick and L.E. Seiberling, Appl. Phys. Lett. 45, 388 (1984).

21. N.W. Cheung, Rev. Sci. Instrum. 51, 1212 (1980).

22. D.D. Armstrong, W.M. Gibson, A. Goland, J.A. Golovchenko, R.A. Levesque, R.L. Meek and H.E. Wegner, Rad. Eff. 12, 143 (1972).

23. L.C. Feldman, P.J. Silverman, J.S. Williams, T.E. Jackman and I. Stensgaard, Phys. Rev. Lett. 41, 1396 (1978).

24. L.C. Feldman, J.W. Mayer and S.T. Picraux, "Materials Analysis by Ion Channeling", Academic Press, New York (1982).

25. H. Iwami, R.M. Tromp, E.J. Van Loenen and F.W. Saris, Physica 116B, 328 (1983).

26. G.H. Kinchin and R.S. Pease, Rep. Prog. Phys. 18, 1 (1955).

27. J.J. Loferski, and P. Rappaport, Phys. Rev. 111, 432 (1958).

28. P. Sigmund, Phys. Rev. 184, 383 (1969).

29. Yuanxun Qiu, J.E. Griffith, Wen Jin Meng and T.A. Tombrello, Rad. Eff. 70, 231 (1983).

30. W.D. Wilson, L.G. Haggmark and J.P. Biersack, Phys. Rev. B15, 2458 (1977).

31. C.I.H. Ashby, Appl. Phys. Lett. 45, 892 (1984).

32. S.A. Schafer and S.A. Lyon, J. Vac. Sci. Technol. 19, 494 (1981).

33. W.G. Petro, I. Hino, S. Eglash, I. Lindau, C.Y. Su and W.E. Spicer, J. Vac. Sci. Technol. 21, 405 (1982).

34. A. Redondo, W.A. Goddard III, C.A. Swarts, and T.C. McGill, J. Vac. Sci. Technol. 19, 498 (1981).

35. H. Ibach and J.E. Rowe, Phys. Rev. B10, 710 (1974).

36. P. Benjamin and C. Weaver, Proc. R. Soc. London A254, 177 (1960).

37. Stephen V. Pepper, J. Appl. Phys. 50, 8062 (1979).

38. Gary G. Tibbets and James M. Burkstrand, Phys. Rev. B16, 1536 (1977).

ADHESION, BARRIER, AND PASSIVATION CHARACTERISTICS OF PLASMA POLYMERIZED ORGANIC THIN FILMS

D. L. Cho and H. Yasuda

Department of Chemical Engineering and
Graduate Center for Materials Research
University of Missouri-Rolla
Rolla, MO 65401

Plasma polymerization is not a means of polymerizing monomers, and materials formed by plasma polymerization (plasma polymers) are not polymers in the conventional sense. Because of very strong interactions which take place between the substrate surface and the depositing reactive species during the plasma polymerization/deposition process, extremely good adhesion of a thin plasma polymer film to substrate surfaces such as metal, glass, and ceramic, to which it is difficult to achieve good adhesion of (conventional) polymer films, can be obtained. Furthermore, a plasma polymer can be utilized as a primer to create water-insensitive adhesion of conventional polymers to other materials. Because of the unique advantage that other plasma treatments and reactions can be used in the pretreatment of substrate surfaces and/or the post-treatment of deposited plasma polymers in an essentially one step process coupled with their excellent barrier characteristics, plasma polymerization can be utilized to obtain remarkable levels of corrosion protection of metal surfaces. Adhesion, barrier, and passivation characteristics of plasma polymerization are reviewed.

ADHESION CHARACTERISTICS

Plasma polymer films can be made to have good adhesion characteristics on a wide range of substrate materials. Although the exact mechanisms of adhesion of plasma polymer films have not been elucidated, and reliable methods for quantitative evaluation of adhesive strength are not established yet[1], the adhesion characteristics of plasma polymers may be explained in the following way.

1. Most plasmas have a cleaning effect on substrate surfaces, although there are some differences in extent. Especially, plasmas containing oxygen can remove the organic contaminants which commonly exist on surfaces of various materials.[2] Contaminants on the substrate surface significantly decrease the adhesive strength of coated films.[3]

2. Plasma polymers are generally highly cross-linked. High cross-linkage results in high cohesive strength near the adhesive joint, which is ideal for production of strong adhesion.[4] Also, high cross-linkage causes an increase in the dispersion contribution of surface energy, which is favored for better adhesion, through an increase in film density at the interface between a coated film and a substrate[5] and provides a tight network with the substrate surface.

3. Kinetic chain length of plasma polymers in the gas phase is extremely short because of thermodynamic limitation.[6] Therefore, relatively small molecules are adsorbed on the substrate surface, and high polymers are formed by the reactions on the solid surface. As a consequence, contact of the plasma polymer films to the substrate surface is better than that of conventionally applied polymer films.

4. Depending on the chemical composition of substrate surface, some plasma polymers can form chemical bonds with the surface. Bombardment of highly energetic species in the plasma and UV irradiation from the plasma cause formation of free radicals in a polymeric substrate. Free radicals in the substrate cause grafting of a plasma polymer. Further, bombardment of highly energetic species combined with UV irradiation can cause cross-linking in a thin layer of substrate surface, which provides a strong adhesive joint.[4] Chemical bonding is also possible between organosilicon plasma polymers and metal oxides.[7]

5. A high extent of interdiffusion between a plasma polymer and the substrate surface, which is the most important factor for good adhesion between different materials, can be obtained by controlling the energy level of the plasma. This is explained by the atomic interfacial mixing (AIM) principle proposed by Yasuda et al.[8] AIM is more than interdiffusion over a few monolayers; it involves modification of surface chemistry. AIM is accomplished by injecting atoms or ions into the substrate surface, radically disturbing bonding, perhaps dislodging, or ejecting atoms and forming new compounds or alloys. Thus, AIM significantly alters the surface free energy of the substrate.

6. Electrode materials can be sputtered during polymer deposition by use of a conductively coupled system (magnetron). The electrode material is thus embedded in the deposited polymer film under high energy input. The embedded metal can increase the adhesion of plasma polymer film to a metal substrate, depending on the matching of metals.

Evaluation of Adhesive Strength

One of the problems encountered in the study of thin film adhesion is measurement of adhesion or, more precisely, the difficulty of measurement of adhesion. There is no generally acceptable method of measurement. Among the various test methods proposed to evaluate adhesive strength of thin films,[1,9] the direct-pull test,[10-12] the lap-shear test,[13,14] and the Scotch tape test[15,16] are used most frequently for the evaluation of adhesive strength of plasma polymer films.

Quantitative values of adhesive strength are obtained by the direct-pull test, the lap-shear test, and the scratch test. However, the quantitative evaluation of adhesion of plasma polymers, particularly of ultrathin layers (less than 0.1 μm thick), is extremely difficult because of the many artifacts involved in measuring methods and procedures. In the direct-pull test and the lap-shear test, adhesive strength is determined by measuring the force needed to separate the ultrathin layer from the substrate surface after preparing test specimen. The following factors become important to properly evaluate the adhesive strength of a plasma polymer to a substrate (used as one end of a joint): 1) adhesive strength at the interface of a plasma polymer and a substrate, 2) cohesive strength of a plasma polymer, 3) adhesive strength at the interface of a plasma polymer and an adhesive (glue), 4) cohesive strength of an adhesive used, and 5) adhesive strength of an adhesive to a material used on the other part of a joint.

It is not always easy to judge which of the factors mentioned are responsible for failure, even on examination with an optical or a scanning electron microscope. In many cases the failures of adhesive joints do not occur in a simple, clear-cut manner. Consequently, the measured value of force does not necessarily correspond to the adhesive strength.

Poor adhesion can be easily detected by the simple "adhesive tape test," which is adopted as an ASTM method.[17] If the ASTM procedure is followed, the test provides semi-quantitative results given in the grades 0 to 5 (5 meaning the film cannot be peeled off the substrate with the tape). The Scotch brand adhesive tape is usually used as the adhesive tape in this method. The measurable upper limit of adhesive strength by this method is relatively low compared to the other methods, because the adhesive strength of the tape is relatively low.

An indication of the adhesive characteristics of the plasma polymer which passes the adhesive tape test can be obtained by the following simple test used by Sharma and Yasuda.[18] A plasma polymer is deposited on a glass microscope slide. A cross-shaped scratch is made on the plasma polymer layer with a razor blade according to the procedure described in the ASTM method for the adhesive tape test. Then, the slide-glass is immersed in boiling water and periodically examined to see if the coated layer starts to peel off. If there is no peeling after eight hours of boiling, adhesion is generally considered to be excellent. Since most metal surfaces show better adhesion than the glass slide, this simple test provides a reasonably reliable guide to what kind of adhesion characteristics could be expected if the same plasma polymer were deposited onto some other substrates. It should be noted that the adhesive characteristics judged by this method depend on water sensitivity of adhesive joint, and that dry adhesive strength and wet adhesive strength do not always correspond.[19] Wet strength is often more important than dry strength in practical applications.

Dependence of Adhesion on System Conditions and Surface Characteristics of Substrate

Adhesion characteristics of plasma polymers is closely related to the mechanism by which they are deposited. Therefore, for a given monomer, adhesion is highly dependent on the conditions of plasma polymerization, including the design factor of the reactor, the location of the substrate in the reactor, the frequency of the electric power source, and other factors. Therefore, some system conditions which are optimum for obtaining good adhesion of a plasma polymer cannot be necessarily used to obtain good adhesion of the same plasma polymer in other reaction systems. The following summary is based on reported experimental data.

Adhesive strength of plasma polymers generally increases with energy level[19-21] (W/FM value,[22] where W is the discharge power, F is the monomer flow rate, and M is the molecular weight of the monomer). Clear indication of this trend was observed by Sharma and Yasuda.[21] As shown in Table I, best adhesion to both glass and platinum was obtained at the lowest monomer flow rate, thus at the highest energy level (W/FM value). And adhesion to platinum was invariably poor. In a separate study, Sharma and Yasuda found that the film deposition should be conducted at an energy level (W/FM value) beyond 10^{10} J/kg in order to achieve good polymer-platinum adhesion.[8] The better adhesion characteristics of plasma polymers at higher energy levels may be a consequence of higher cross-linkage of a plasma polymer, greater interdiffusion between plasma polymer and substrate (explained by the AIM principle[8]), higher extent of grafting and cross-linking in the substrate in the case of organic substrate, and the embedded electrode material in the case of metal substrate.

It is more difficult to obtain good adhesion with a relatively thick layer than with a very thin layer. According to Dynes and Kaelble's study,[23] lap-shear strength of plasma polymers decreases as film thickness increases, and levels off at a film thickness of 1,600A, at which point lap-shear strength is approximately 50% of the highest value (Figure 1). However, it should be noted that the film thickness should be greater than the minimum thickness which can cover the entire surface area of a substrate, which will depend on the system employed. Ross[24] reported that approximately 155A is required to obtain complete coverage of a metal surface.

Very clean, micro-rough, and highly energetic conditions of the substrate surface are required to obtain good adhesion of plasma polymers, especially to nonpolymeric substrates. Cleaning of a substrate may be accomplished by treatment with appropriate organic or inorganic solvents. Sometimes, however, cleaning with solvents does not provide a surface which is clean enough for good adhesion, and solvent residue can be left after the cleaning procedure. Cleaning by treating surfaces with nonpolymerizing gas plasmas, such as Ar or O_2 plasma, is known to be one of the most effective methods. If this method is used for cleaning of a substrate, there is the additional benefit that film deposition by plasma polymerization can be directly carried out without exposing the substrate to contaminated environments. The surface of a substrate can be roughened

Table I. Adhesion of Glow-Discharge-Polymerized Tetramethyldisiloxane.*

Polymerization conditions		Adhesion to	
Flow rate(s) (cm³min⁻¹)	Power (W)	Glass	Platinum
0.71	83	5B	2B-3B
1.24	78	5B	2B
2.90	84	2B-4B	2B
4.76	90	1B-2B	0B-1B
1.15(TMDSO) + 0.65(O_2)	78	2B-3B	--
1.20(TMDSO) + 1.20(O_2)	74	2B-3B	--
1.20(TMDSO) + 2.34(O_2)	80	2B-3B	--

*Adhesion is characterized by the wet adhesion test described in ref. 18 which rates adhesion on 0-5B ratings. From Sharma and Yasuda (21).

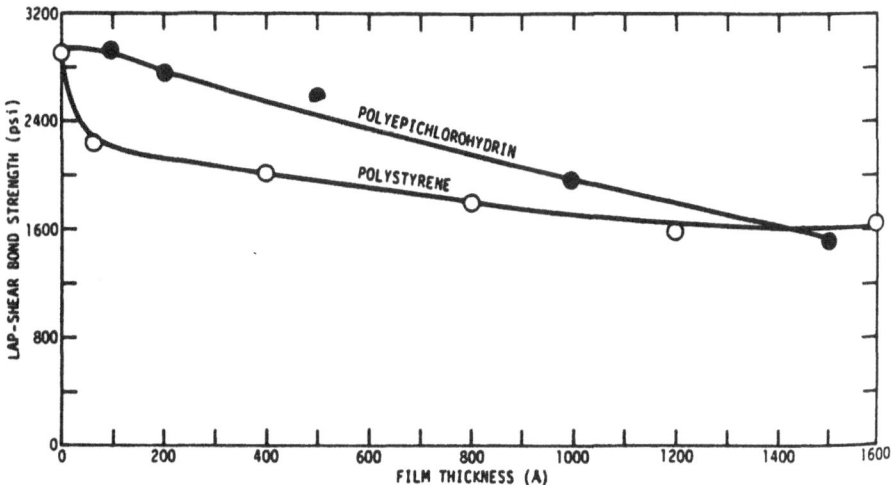

Figure 1. Plot of lap-shear bond strength versus film thickness of PPE and PPS on FPL etched Al 2024-T3, bonded with HT≈424 epoxy-phenolic adhesive. From Dynes and Kaelble (23).

by chemical etching, plasma etching, or mechanical polishing. Although micro-roughness of the substrate surface improves adhesion of plasma polymers, it should be noted that adhesion can be decreased by increased roughness beyond a certain level. Smith et al[13] used several methods to provide roughness on various metal substrate surfaces, and found decreasing adhesion with increased roughness. The level of roughness for optimum adhesion was not found.

Adhesion of plasma polymers depends on substrate temperature and/or system temperature. Because mechanisms of growth/polymerization of plasma polymers depend on substrate temperature and/or system temperature, different chemical compositions and structures of plasma polymers are obtained at different temperatures. Wrobel et al[26] observed very strong adhesion of plasma polymers of organosilicones at a substrate temperature of 800°C. This was explained by high cross-linkage formation and a decrease in the relative organic content of the film at high substrate temperature.

Finally, adhesion of plasma polymers can be improved by mixing a properly selected carrier gas with a monomer gas. For instance, formation of polar groups, decrease of internal stress, and enhancement of AIM can be attained by mixing N_2 gas, Ar gas, and O_2 gas, respectively, which increase adhesion of plasma polymers.[7,8]

Improvement of Adhesion by Plasma Polymers

The excellent adhesion of plasma polymers to a wide range of substrates, together with their high cohesive strength, can be utilized to improve the adhesion characteristics of other materials. For example, if a plasma polymer adheres tenaciously to a metal surface (e.g., platinum), the surface characteristics of metal are changed to those of the plasma polymer. Thus an adhesive which did not adhere well to the metal surface will adhere well to the plasma polymer surfaces, making it possible to glue two pieces of metal together. By using this principle, many heterogeneous joints (e.g., platinum-Teflon, stainless steel-polyethylene) can be made by the use of a plasma polymer primer applied to both surfaces. Results of experimental priming and bonding by Inagaki and Yasuda[14] are shown in Tables II and III. Remarkable increases were observed in lap-shear strength of all combinations of substrates (3 to 10 times) by plasma polymer priming. Similar work has been carried out to improve the adhesive bonding of polymer films and Kevlar laminates, with considerable increase in adhesive strength.[27,28]

Matsuda and Yasuda[29] studied the use of plasma polymers as adhesion promoters when applying alkyd paint to titanium plates. Adhesive strength of the paint was improved when plasma polymers were used as a primer,

Table II. Lap-Shear Strength of Adhesive Bonding Between Various Plasma Polymer Coated Substrates.*

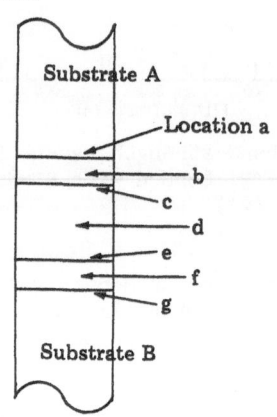

Construction of adhesive joint		Lap-shear strength (kg/cm²)			
Substrate A	Substrate B	Uncoated	CH_4 (2000)**	$CH_2=CH_2$ (2000)	$CH\equiv CH$ (2060)
PE[†]	PE	5.89(b)[††]	24.2(b)	24.4(b)	20.7(b)
PE	AL	16.6(b)	47.6(f)	49.5(f)	48.2(b)
PE	ST[†]	28.5(b)	45.1(f)	47.8(f)	46.5(b)
T[†]	T	2.47(b)	17.9(b)	17.8(b)	17.8(b)
T	AL	15.8(b)	36.8(f)	40.4(f)	41.6(b)
T	ST	3.80(b)	38.0(f)	39.8(f)	40.3(b)

*From Inagaki and Yasuda (14). Operational conditions: flow rate, 2.0 cm³(STP)/min; current of af power, 250 mA.
**Film thickness, in A.
[†]PE: polyethylene; T: poly(tetrafluoroethylene); AL: aluminum; ST: stainless steel.
[††]Location of failure, the symbol defined as follows illustrates the location of failure:
a: interface between substrate A and plasma polymer
b: cohesive failure of plasma polymer layer on A side.
c: interface between plasma polymer on A side and adhesive.
d: cohesive failure of adhesive.
e: interface between plasma polymer on B side and adhesive.
f: cohesive failure of plasma polymer layer on B side.
g: interface between substrate B and plasma polymer layer.

Table III. Deterioration of Adhesive Bonding Between Various Plasma Polymer Coated Substrates in Water at 70°C.

Construction of adhesive joint Substrate A	Substrate B	Uncoated	CH$_4$ 2000			CH$_2$=CH$_2$ 2000			CHCH 2000	Used gas Film thickness, A
			1.0	2.0	10	1.0	2.0	10	2.0	Flow rate, [cm^3(STP)/min]
			120	90	90	120	90	90	90	af power, W
			(400)	(250)	(250)	(400)	(250)	(250)	(250)	af current, mA
PE†	PE†	20(b)**	169.5(b)	178(b)	47.5(b)	79.8(b)	91(b)	19(b)	23(b)	
PE	AL†	20(b)	80.5(f)	178(b)	38(f)	40(b)	53(f)	19(b)	23(b)	
PE	ST†	20(b)	169.5(f)	141(b)	19(f)	79.8(f)	65(f)	19(f)	23(b)	
T†	T	20(b)	143.5(f)	178(b)	47.5(b)	48(b)	91(b)	19(b)	23(b)	
T	AL	20(b)	103.7(f)	127(b)	28.5(f)	40(b)	32(f)	19(b)	23(b)	
T	ST	20(b)	129(f)	103(b)	19(f)	79.8(f)	91(f)	19(f)	23(b)	

*From Inagaki and Yasuda (14).
**Location of failure, the definition is the same as shown in Table II.
†PE: polyethylene; T: poly(tetrafluoroethylene); AL: aluminum; ST: stainless steel.

compared with the value measured on control surfaces. Adhesive strength depended on film thickness and system conditions. This indicates the high potential of plasma polymer films as adhesion promoters, particularly for paint-substrate combinations which do not presently have appropriate primers.

Effect of Water on Adhesion

The most serious problem in many thin film applications is the failure of adhesion due to the action of water molecules, which always exist in the surrounding media as liquid or vapor, especially when nonpolymeric substrates are used. Adhesion of polymer coatings to nonpolymeric substrates is achieved, in most cases, by the secondary bonds or relatively weak forces, such as hydrogen bonding and van der Waals force. Consequently, water molecules, which are one of the strongest hydrogen bonding agents, can easily break such bonds between a substrate surface and a polymer coating. The effect of water on adhesion can be demonstrated by comparing the adhesive bond strength measured in a dry condition with that measured in a wet condition.

Inagaki and Yasuda[14] applied plasma polymer films to various substrates and evaluated the adhesive strength by lap-shear tests and by immersion in water at 70°C, and compared the measured values. As shown in Figures 2 and 3, there was no correlation between the values measured by lap-shear test (in a dry condition) and the values measured by immersion time (in a wet condition). The best adhesion under wet conditions was obtained when the most hydrophobic plasma polymer (plasma polymer of methane) was applied to the most hydrophobic substrate (Teflon or polyethylene). Adhesion was independent of the monomer used. The worst adhesion under dry conditions was obtained when Teflon was used as a substrate. This indicates that interaction with water significantly influences adhesion of thin films.

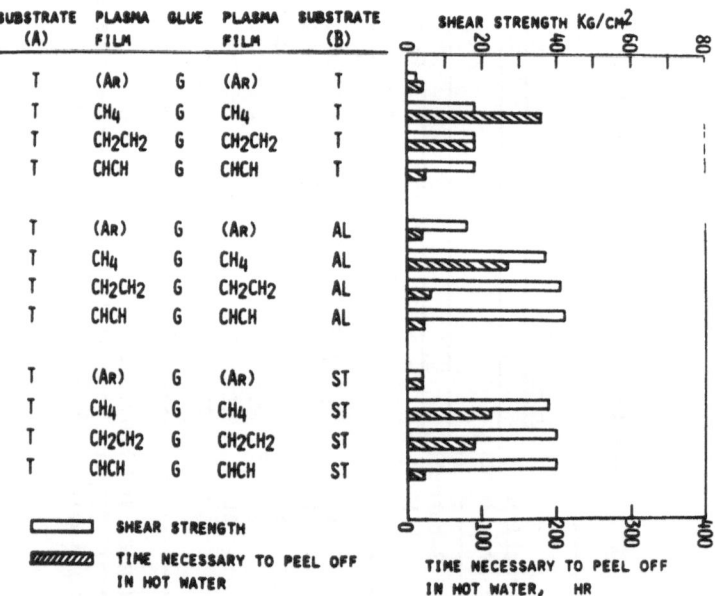

Figure 2. Lap-shear strength and deterioration in hot water as a function of material and gas used for polymerization; T, polytetrafluoroethylene; AL, aluminum; ST, stainless steel; (AR), argon etching; CH_4, plasma polymer from methane; CH_2CH_2, plasma polymer from ethylene; CHCH, plasma polymer from acetylene. From Inagaki and Yasuda (14).

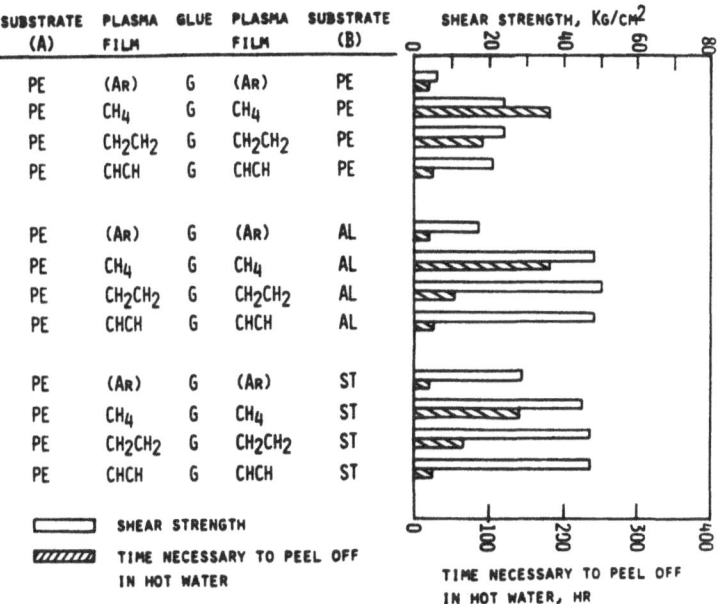

Figure 3. Lap-shear strength and deterioration in hot water as a function of material and gas used for polymerization; PE, polyethylene; AL, aluminum; ST, stainless steel; (AR), argon etching; CH_4, plasma polymer from methane; CH_2CH_2, plasma polymer from ethylene; CHCH, plasma polymer from acetylene. From Inagaki and Yasuda (14).

Thin film coating by plasma polymerization has unique advantages when water resistant or water insensitive adhesion is desired. Because polymers deposit in a vacuum, low or intermediate molecular weight reactive species adsorb on the substrate surface. Therefore, the typical surface energetic response of the substrate material to wetting during the coating process may be minimized, and the most intimate interface contact may be obtained. In other words, wetting of a coating solution or of a polymer melt, and consequent void formation at the interface, can be virtually eliminated. And the surface energetics of a substrate can be readily altered in the plasma polymerization process by modifying AIM[8,19] through energy control and proper monomer selection.

Several investigators have studied the preparation of water-insensitive plasma polymer coatings[8] and their application to improve adhesion of Parylene,[18,30,31] which has excellent mechanical, chemical, electrical, and thermal properties, but has poor adhesion characteristics. Good water-insensitive adhesion of plasma polymers to a metal substrate (Pt) was obtained when the energy level exceeded a critical value, which is different depending on the monomer (10^{10} J/kg for methane and 2×10^8 J/kg for tetrafluoroethylene). The adhesion of Parylene to various substrates was improved by the use of water-insensitive plasma polymer films as a primer.

Several plasma polymer films were boiled in a 0.9% NaCl solution and then subjected to a pull test. The restuls are shown in Figure 4. Adhesion of Parylene on various substrates coated with plasma polymers of methane is listed in Table IV. Among the plasma polymers, the plasma polymer of methane was the best, probably due to its hydrophobic characteristics and low molecular weight. If molecular weight is low, it is easy to increase the energy level (W/FM value). In this respect, the role of plasma polymers, particularly that of methane prepared at high energy

257

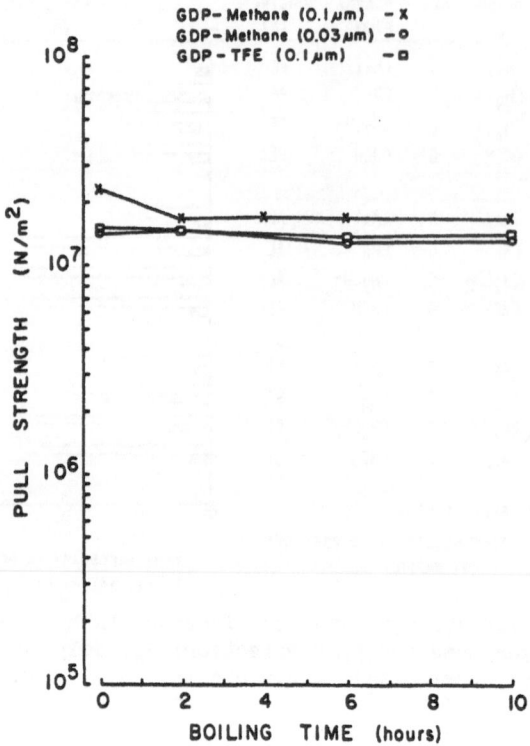

Figure 4. Comparative pull strengths for various glow discharge polymers boiled in 0.9% NaCl solution prior to pull test. From Yasuda (19).

level, is unique in that it provides high water insensitivity (highly durable) when coated to a substrate or used as a primer on various kinds of substrate materials.

BARRIER CHARACTERISTICS

Unlike most conventional polymers, which are formed by molecular polymerization of corresponding monomers, plasma polymers have very short segments (no polymeric segments) between cross-links and can be characterized as being between organic polymers (with high long-range molecular mobility) and inorganic materials (with no long-range mobility). Therefore, they can be formed in an extremely tight network. The degree of tightness seems to be far greater than that resulting from the ordinary cross-linking of conventional polymers. The unusually tight network is characterized by absence of rotational mobility of the polymeric molecules at the surface, as found in the plasma polymer of methane film,[32] and by the presence of a sieve effect of the polymer network. As the permeant molecules increase in size, they reach a critical point beyond which they are too large to be accommodated among the molecules of the free volume.[33] Because permeation through a plasma polymer occurs differently from solution-diffusion principle, and is controlled by diffusivity of permeants,[34] such an unusually tight network would be ideally suited for reducing the permeability of various environmental materials.

Table IV. Effect of Methane Glow Discharge Treatment of Substrate Surface
on Adhesion of Poly(p-xylylene); A, Dry Test; B, Wet Test.*

Substrates	Adhesion Test Results	
	Test A	Test B
Untreated poly(tetrafluoroethylene) (PTFE)	0A	5B
Methane glow discharge treated PTFE	5A	5B
Untreated polypropylene (PP)	2A	5B
Methane glow discharge treated PP	4A	5B
Untreated polyethylene (PE)	0A	5B
Methane glow discharge treated PE	5A	5B
Untreated poly(methyl methacrylate) (PMMA)	1A	3B
Methane glow discharge treated PMMA	4A	
Untreated poly(ethylene terephthalate) (PET)	0A	2B-3B
Methane glow discharge treated PET	5A	4B-5B
Untreated Nylon-6	0A	1B-2B
Methane glow discharge treated Nylon-6	5A	5B
Untreated glass	0A	0B
Methane glow discharge treated glass	5A	2B

*From Sharma and Yasuda (18).

As shown in Table V plasma polymers of methane are significantly less
permeable to water vapor than are conventional polymers. Note that permeability to water vapor of conventional polymers passes through the minimum as the cohesive energy density increases (the polymers are arranged in
order of increasing cohesive energy density). Permeability of plasma
polymers of acetylene and of acrylonitrile to some gases (CO_2, O_2, and H_2)
was measured by Yasuda and Hirotsu.[35] Very low permeability constants
($10^{-13} \sim 10^{-11}$ cm^3(STP)$\cdot$cm/cm$^2\cdot$sec$\cdot$cmHg for plasma polymer of acrylonitrile and $10^{-14} \sim 10^{-12}$ cm^3(STP)cm/cm$^2\cdot$sec$\cdot$cmHg for plasma polymer of acetylene) were observed (see Table V for the comparison with those through
conventional polymers). Permeability depended on system conditions of
plasma polymerization, especially the monomer flow rate.

The good barrier characteristics of plasma polymers, together with
good adhesion and passivation characteristics, provide plasma polymer
films high potential as candidates for protective coatings. Protection of
low carbon steel by various plasma polymers from 10^{-4} M LiOH solution and
4% NaCl solution was investigated by Schreiber et al.[36,37] They compared
the protection ability of plasma polymers with that of polytetrafluoroethylene film, which is one of the most highly corrosion resistant materials
conventionally used. A plasma polymer of thiophene (3 min treated) with-
stood 3 days of exposure to a 10^{-4} M LiOH solution at 275°C without any
deterioration, and a sample of low carbon steel coated with 2 μm thick
plasma polymer of hexamethyldisiloxane was totally free of corrosion after
immersion for 28 days in a 4% NaCl solution. Polytetrafluoroethylene
coated low carbon steel showed serious deterioration on exposure to 10^{-4} M
LiOH solution at 275°C. Samples, on which the plasma polymer of hexamethyldisiloxane thinner than 2 μm was coated, were corroded to some extent.

Table V. Oxygen Permeability and Water Vapor Permeability of Polymers.*

Polymer	$P_{O_2} \times 10^{10}$**	$P_{H_2O} \times 10^{10}$**
Poly(dimethylsiloxane)	605	43,000
Natural rubber	23.3	2,290
Polyethylene, d = 0.914	2.88	90
Poly(trifluorochloroethylene)	0.025	0.29
Poly(vinylidene chloride)	0.0053	0.50
Poly(acrylonitrile)	0.0002	300
Plasma polymer of CH_4		0.0076
Sputtered Cu layer		0.0026

*From Yasuda and Stannett, Polymer Handbook 2nd Edition.
**Data at near 25°C; P in $cm^3(STP) \cdot cm/cm^2 \cdot sec \cdot cmHg$.

Plasma polymers as moisture barriers for alkalihalide optics[38,39] and paper[40] have been studied by several investigators, and some plasma polymers were found to be effective in this utilization.

Another indication of the high barrier characteristics of plasma polymers can be found in so called "diamond-like" amorphous carbon films. According to Angus et al,[41] very significant amounts of hydrogen (up to 50%), much of which was not chemically bound to the carbon, was detected in "diamond-like" film prepared by rf discharge of methane. This was considered to be caused by the unusually low diffusion coefficient of hydrogen, which was found to be on the order 10^{-18} cm^2/sec, in this film.

Although their tight network can significantly reduce the permeability of plasma polymers to corrosive species from various environments, as described above, a more pronounced role for plasma polymer films as a barrier in protective coatings may come from their selective permeability, because the thickness of a plasma polymer film in practical applications usually ranges from 100 to 5000A. Since the transport rate of the permeant is a function of permeability and film thickness, barrier protection cannot be expected to play any role in protection of a substrate when such an ultrathin film is applied.

The degree of ionization of acid dye permeants was found to have a marked effect on the permeability of a plasma polymer film of methane. Yasuda et al[33] applied ultrathin films (70 to 200A) of a plasma polymer of methane to a Nylon-6 film and measured the permeability of acid dyes and a nonionic dispersion dye. Permeability was measured in both directions. As shown in Table VI, a remarkable degree of vectored permeation was observed for most combinations of film and dye. The trend is most pronounced with an acid dye. The variation in permeation cannot be explained by the size of the permeating molecules, because from the uncoated side Brilliant Scarlet 3R, which has larger molecular size (molecular weight: 604), permeates faster than Celliton Fast Yellow 5R, which has smaller moleculer size (molecular weight: 316), while from the coated side the permeation of the former is completely blocked and the latter can pass. This may be explained by the differences between the degree of ionization of the acid dyes in the aqueous phase and that in the polymer phase. In the aqueous phase, acid dye molecules are highly ionized and the plasma polymer generally acquires oxygen after the polymer is removed from the reactor. These films have the characteristics of a partially ionized

Table VI. Apparent Permeability, P_a, of Various Dyes Through Plasma Polymer Coated Nylon-6 (Film Thickness 50 μm).*

Dye			Film	P_a x10^{12}(mm²/sec)		
Name	M	M$^{2/3}$	No.	W/FM$_{-10}$ x 10 (J/kg)	from coated side	from uncoated side
Solar Orange R	364	50.98	Nylon-6, untreated	--	174.09	174.09
			Nylon-6/PPM-9	0.076	0.34	5.08
Acid Blue Black 10B	616	72.40	Nylon-6, untreated	--	10.13	10.13
			Nylon-6/PPM-9	0.076	0	1.71
Brilliant Scarlet 3R	604	71.45	Nylon-6, untreated	--	37.40	37.40
			Nylon-6/PPM-7	0.796	0	23.33
			Nylon-6/PPM-8	0.191	0	5.36
			Nylon-6/PPM-10	0.074	0	8.31
Celliton Fast Yellow 5R	316	46.39	Nylon-6, untreated	--	1.82	1.82
			Nylon-6/PPM-9	0.076	0.82	1.65

*From Yasuda et al (33).

polymer because they incorporate oxygen-containing functional groups such as carbonyl and carboxyl. Consequently, Coulombic repulsion or Donnan exclusion would play a role when ionic solutes approach the barrier. These effects are absent on uncoated side because acid dye molecules do not ionize in the nonionic polymer matrix.

The achievement of selective permeation of a plasma polymer by blocking the permeation of ionic species provides the plasma polymer an unique advantage over nonionic conventional polymers in protective coating applications because the corrosion rate increases as the concentration of ionic species increases at the substrate surface.

PASSIVATION CHARACTERISTICS

It should be reemphasized that plasma polymerization is an ultrathin film technology. Therefore, the unique advantage of plasma polymerization is in ultrathin film applications (e.g., thickness 100 to 5000A, particularly the 100 to 500A region), where application of polymer films by conventional methods is very difficult and is often economically unfavorable. Conversely, the use of plasma polymers for applications in which a thicker layer (greater than a micrometer) of polymer is required is not considered as their most advantageous application. Because of the very small thickness of ultrathin coating of plasma polymer films, the barrier protection obtainable is generally limited, and relatively small penetrants such as SO_2, NO_2, and H_2O cannot be completely blocked. Therefore, in applications of plasma polymer films as protective coatings, passivation of a substrate surface is more important than barrier protection.

Passivation of the substrate surface is believed to occur during the plasma polymer coating process, possibly because of the tenacious adhesion between a plasma polymer film and the substrate surface and/or the change in chemical structure by interfacial atomic mixing. That is, if tenacious adhesion of a plasma polymer film is obtained, the surface characteristics of a substrate change to that of a plasma polymer, which is highly inert to various environments; and by interfacial atomic mixing, the surface chemistry of a substrate can change to that of new compounds or alloys. This passivation effect can be seen clearly in the case of water insensitive adhesion, described earlier, and in the following examples.

Freitag et al[42] applied plasma polymer films with a thickness less than 0.1 μm on magnetic discs (of Ni and Fe), and observed that such thin plasma polymer films protected against corrosion by NO_2 and SO_2 in air while all conventional polymers applied at this level of thickness had shown no effect, as shown in Figure 5. Since permeation of NO_2 and SO_2 through such ultrathin films can occur within a fraction of a second, protection by barrier cannot be expected. A clearer indication of the passivating effect of plasma polymers can be observed in the work of van Lier.[43] Plasma polymers were coated on a copper surface, and the surface was exposed to a KOH electrolyte. Resistance to electrochemical corrosion was determined by measuring KOH creepage. As shown in Figure 6, a remarkable decrease of KOH creepage was obtained by plasma polymer coating of methane film, and the best resistance was observed at the smallest film thickness (100A). Before the plasma polymer film coating, H_2 plasma was used to reduce copper oxides to copper, and the reduction of copper oxides was considered to reduce the KOH creepage derived from reduction reactions. Unless there is passivation of a substrate surface by plasma polymerization, such results cannot be obtained.

Passivation of the substrate surface can also be accomplished by treatment with nonpolymerizing plasmas. Freon plasma, for example, causes the substrate surface to become more hydrophobic. Freon plasma also removes hydroxide groups, which exist on the surface of substrates such as

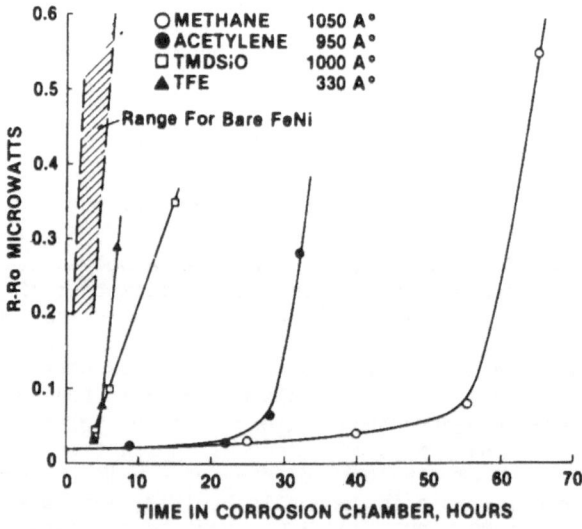

Figure 5. Corrosion of FeNi by NO_2 and SO_2 observed by light reflectance as a function of exposure time. From Freitag et al (42).

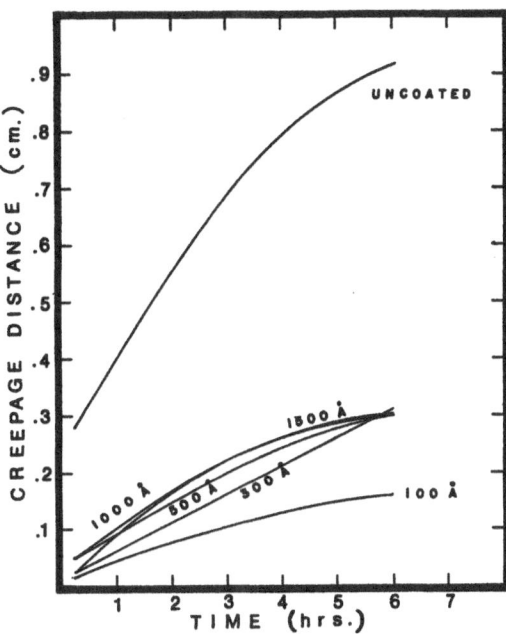

Figure 6. Creepage of plasma polymer of methane coated copper strip versus time, W/FM of methane. Plasma; 1.0 x 10¹⁰ J/kg. From van Lier et al
(43).

alkali halide and causes corrosion when contacted by water vapor.[38] It
should be noted that passivation characteristics, like other characteristics, can be controlled by the system conditions of plasma polymerization.

CONCLUSIONS

Plasma polymers can be made to have good adhesion characteristics to
a wide range of substrate materials due to following characteristics: 1)
cleaning effect of plasma, 2) high cross-linkage, 3) adsorption of relatively small molecules onto a substrate surface, 4) possible chemical
bonding of plasma polymers with a substrate, 5) high interdiffusion by
AIM, and 6) embedding of sputtered electrode material.

Evaluation of adhesive strength should be carried out using proper
test methods, depending on the actual environmental media which surround
samples. Adhesion in dry conditions does not correspond to adhesion in wet
conditions.

Adhesion of plasma polymers is highly dependent on the condition of
the system in which they are deposited and on the nature of the substrate.
Better adhesion can be obtained: 1) at high energy levels; 2) with small
film thickness; 3) with clean and micro-rough surfaces; 4) in dry conditions when the surface energy of a substrate is high and in wet conditions
when surface energy of a substrate is low; and 5) by mixing a proper carrier gas.

Adhesive bonding of materials where adhesives adhere poorly can be improved by applying plasma polymer films on the surface. In addition, plasma polymers can be used as enhancers of adhesion.

Water in surrounding media causes adhesion failure of thin film coatings. Therefore, highly water insensitive adhesion is necessary to obtain highly durable film coating, and this can be accomplished by plasma polymer film coating.

Plasma polymers also have good barrier and passivation characteristics. These characteristics, together with good adhesion, provides a high potential for protective films, particularly when a thicker coating cannot be employed.

REFERENCES

1. K. L. Mittal, editor,   "Adhesion Measurement of Thin Films, Thick Films and Bulk Coatings", ASTM, Philadelphia, 1978.
2. M. A. Baker, Thin Solid Films, 69, 359 (1980).
3. L. C. Jackson, Adhesives Age, 34 (Sept. 1978).
4. H. Schonhorn and R. H. Hansen, J. Appl. Polym. Sci., 11, 1461 (1967).
5. A. M. Wrobel and M. Kryszewski, J. Macromol. Sci., Chem., 12, 1041 (1978).
6. H. Yasuda and C. R. Wang, J. Polym. Sci., Chem., 23, 87 (1985).
7. A. K. Hays, Thin Solid Films, 84, 401 (1981).
8. H. Yasuda, A. K. Sharma, E. B. Hale, and W. J. James, J. Adhesion, 13, 269 (1982).
9. D. S. Campbell, in "Handbook of Thin Film Technology", L. I. Maissel and R. Glang, editors, McGraw-Hill, New York (1970).
10. R. Jacobsson and B. Kruse, Thin Solid Films, 15, 71 (1973).
11. R. Jacobsson, Thin Solid Films, 34, 191 (1976).
12. K. Bhasin, D. B. Jones, S. Sinharoy, and W. J. James, Thin Solid Films, 45, 195 (1977).
13. T. Smith, D. H. Kaelble, and C. L. Hamermesh, Surface Sci., 76, 203 (1978).
14. N. Inagaki and H. Yasuda, J. Appl. Polym. Sci., 26, 3333 (1981).
15. A. K. Sharma, F. Millich, and E. W. Hellmuth, J. Appl. Phys., 49, 5055 (1978).
16. A. K. Sharma and H. Yasuda, J. Vac. Sci. Technol., 21, 994 (1982).
17. ANSI/ASTM D-3354-76.
18. A. K. Sharma and H. Yasuda, J. Adhesion, 13, 201 (1982).
19. H. Yasuda, in "Adhesion Aspects of Polymeric Coatings", K. L. Mittal, editor, Plenum Publishing Corp. (1983).
20. A. F. Diaz and R. Hernandez, J. Polym. Sci., Chem., 22, 1123 (1984).
21. A. K. Sharma and H. Yasuda, Thin Solid Films, 110, 171 (1983).
22. H. Yasuda and T. Hirotsu, J. Polym. Sci., Chem., 16, 743 (1978).
23. P. J. Dynes and D. H. Kaelble, J. Macromol. Sci., Chem., 10, 535 (1976).
24. D. L. Ross, RCA Reviews, 39, 136 (1978).
25. L. W. Crane and C. L. Hamermesh, in "Adhesion Measurement of Thin Films, Thick Films and Bulk Coatings", K. L. Mittal, editor, Amer. Soc. for Test. Matls., Philadelphia, PA, 1978.
26. A. M. Wrobel, J. E. Klemberg, M. R. Wertheimer, and H. P. Schreiber, J. Macromol. Sci., Chem., 15, 197 (1981).
27. A. Moshonov and Y. Avny, J. Appl. Polym. Sci., 25, 771 (1980).
28. M. R. Wertheimer and H. P. Schreiber, J. Appl. Polym. Sci., 26, 2087 (1981).
29. Y. Matsuda and H. Yasuda, Thin Solid Films, 118, 211 (1984).
30. R. K. Sadhir, W. J. James, H. Yasuda, A. K. Sharma, M. F. Nichols, and A. W. Hahn, Biomaterials, 2, 239 (1981).

31. M. F. Nichols, A. W. Hahn, W. J. James, A. K. Sharma, and H. Yasuda, Biomaterials, 2, 161 (1981).
32. H. Yasuda, A. K. Sharma, and T. Yasuda, J. Polym. Sci., Polym. Phys. Ed., 19, 1285 (1981).
33. T. Yasuda, T. Okuno, and H. Yasuda, J. Membrane Sci., 23, 93 (1985).
34. H. Yasuda, J. Membrane Sci., 18, 273 (1984).
35. H. Yasuda and T. Hirotsu, J. Appl. Polym. Sci., 21, 3167 (1977).
36. H. P. Schreiber, M. R. Wertheimer, and A. M. Wrobel, Thin Solid Films, 72, 487 (1980).
37. H. P. Schreiber, Y. B. Tewari, and M. R. Wertheimer, Ind. Eng. Chem. Prod. Res. Dev., 17, 27 (1978).
38. F. G. Yamagishi, D. D. Granger, A. E. Schmitz, and L. J. Miller, Thin Solid Films, 84, 427 (1981).
39. J. R. Hollahan, T. Wydeven, and C. C. Johnson, Appl. Optics, 13, 1844 (1974).
40. R. Liepins and J. Kearney, J. Appl. Polym. Sci., 15, 1307 (1971).
41. J. C. Angus, J. E. Stultz, P. J. Shiller, J. R. MacDonald, M. J. Mirtich, and S. Domitz, Thin Solid Films, 118, 311 (1984).
42. W. O. Freitag, A. K. Sharma, and H. Yasuda, ACS Org. Coat. Appl. Polym. Sci. Proc., 47, 449 (1982).
43. J. A. van Lier, D. L. Cho, and H. Yasuda, Paper presented at IUPAC Symposium on Plasma Chemistry, Eindhoven, Netherlands, July 1985.

NUCLEAR SCATTERING PROFILES OF POLYIMIDE-METAL INTERFACES

C. Chauvin, E. Sacher, and A. Yelon

Groupe des Couches Minces
Département de génie physique
École Polytechnique
C.P. 6079, Succursale "A"
Montréal, Québec
Canada H3C 3A7

R. Groleau, and S. Gujrathi

Groupe des Couches Minces
Laboratoire de Physique Nucléaire
Université de Montréal
C.P. 6128, Succursale "A"
Montréal, Québec
Canada H3C 3J7

Nuclear scattering profiles of polyimide-metal interfaces were obtained using the Elastic Recoil Detection and
Rutherford Backscattering techniques. Metallized polyimide
targets were prepared by evaporation of Al, Cu or Au onto
dry and hydrolyzed polyimide film. The nuclear scattering
profiles show polymer-metal mixing at the interface. The
structure, width and composition of the interphase thus
formed depends on the type of metal used. The relation
between the extent of mixing and the adhesion of the
metallization, as determined by a standard tape test, is
discussed.

INTRODUCTION

Adhesion at polyimide-metal interfaces plays an important role in
determining the long term properties of multi-layered microelectronic
devices. Humid environments are known to have a deleterious effect on
interfacial adhesion. In the case of polyimide (PI) deposition onto
metals, silane ester promotors may be used to promote adhesion. In the
case of metal deposition onto polymer, however, no adhesion promotors
exist. The interfacial reactions between metal and polymer are thus
specific to the materials used[1,2], and their effect on bond strength and
bond durability differs for each metal[3].

Over the past few years, various surface analytical tools have been used to study the structure of polymer surfaces and polymer-metal interfaces[3]. However, the use of ion-beam techniques for the characterization of organic materials and binary systems has been rather limited. In this study, the Elastic Recoil Detection (ERD) and Rutherford Backscattering (RBS) techniques were used to analyse the structure of PI-metal interfaces. These ion beam techniques were chosen because they permit simultaneous profiling of light and heavy elements both in the metal layer and in the polymer substrate. First, the effects of radiation damage on the polymer substrate were evaluated and experimental conditions were derived to obtain reliable results. Then, the effect of various temperature and humidity treatments on the penetration of Al, Cu and Au into the PI film were investigated. Finally, tape tests were performed on the metal-coated PI in order to correlate interface structure and adhesion properties.

The nuclear scattering results indicate that the effects of heat treatment and humidity exposure on the interfacial structure differ for each metal. In contrast, the tape tests results show that the effect of a particular treatment on the adherence of metallization is independent of the type of metal used.

2.   EXPERIMENTAL SET-UP AND PROCEDURE

2.1   Sample Preparation

Polyimide-metal interfaces were prepared by evaporation of Al, Cu or Au onto Du Pont Kapton type H polyimide film. Before metallization, these films were subjected to one of the following pre-treatments: drying at 120°C for 24 hrs (to remove any residual water and convert polyamic acid to polyimide), or exposure at 100% RH at RT for 10 days (to hydrolyse the surface, converting part of the polyimide to polyamic acid). Both dried and hydrolyzed surfaces were cleaned with spectral grade isopropyl alcohol prior to metallization, a standard technique for removing surface contamination without reaction. Evaporation was carried out using resistively heated tungsten boats or spirals containing high purity metals. In each case, the product of the metal density and its thickness was about 100 $\mu g/cm^2$: this gave thicknesses of 275 nm for Al, 150 nm for Cu and 50 nm for Au. Metallized samples were variously stored in dry or humid atmospheres, or heat treated at 150°C for 5 days and then stored in dry or humid atmospheres until used. These preparation conditions are summarized in Table I. They were chosen because they reflect possible fabrication processes conditions or potential user's conditions of multilayered microelectronic devices. In subsequent discussions, the treatments will be designated T, H or D, as in Table I.

Table I.   Metallized Polyimide Sample Preparation Conditions

Pre-treatment

T: 120°C / 24 hours
H: RT   / 100% RH   / 10 days

Post-treatment

D: RT   / dessicator / stored until used
H: RT   / 100% RH   / stored until used
T: 150°C / 5 days   / stored in D or H until used

D - Dry, H - Humid, T - Temperature

2.2 Adhesion Measurements

The standard Scotch tape test was used to evaluate the effect of the various pre- and post-treatments on the adherence of the metal films to the PI substrates. A strip of tape was applied to the metal, left for a few seconds and then removed by hand in a 90° peel. The removal of metal by the tape demonstrates poor adhesion at the PI-metal interface. This method is obviously only qualitative and gives no indication of the relative magnitudes of the adhesive forces at the interface. The tape test, however, permitted us to ascertain whether a particular treatment enhanced or decreased adhesion, for a given PI-metal couple.

2.3 Principles of the Elastic Recoil Detection Technique

The Elastic Recoil Detection (ERD) technique is used to measure the concentration of various elements vs depth. It is based on the ejection of the light elements recoiled from a target by an energetic heavy ion beam. The energy of the incident ion is such that a collision between an impinging particle and a target nucleus can be described by the Rutherford scattering of two point charges. For a collision at the surface of the target, the energy E_s of a recoil atom of mass M_2 depends on the energy E_0 of the incident ion of mass M_1 according to the expression:

$$\frac{E_s}{E_0} = 4 \frac{M_1 M_2}{(M_1 + M_2)^2} \cos^2 \Psi = k, \tag{1}$$

where Ψ is the angle at which M_2 is scattered with respect to M_1, in the laboratory frame of reference. For a given geometry, the energy ratio given by Equation (1) is a constant, K, called the kinematic factor.

As the incident ion penetrates the target, it gradually loses energy through collisions with the electrons in the solid. Similarly, the recoil atom loses energy as it emerges from the sample (Fig. 1). The rate of energy loss of these particles in the target ($\delta E/\delta X$) is a function of the energy of the particle, and depends on the target composition. For a collision at depth X from the surface of the target (Fig. 1), the detected energy, E_r, of the recoil atom is given by:

$$E_r = K(E_0 - \Delta E_0) - \Delta E_r, \tag{2}$$

where K is the kinematic factor defined in Equation (1) and ΔE_0 and ΔE_r correspond, respectively, to the amount of energy lost by M_1 and M_2 in the target material. The detected energy E_r, therefore, depends on the position of the recoil atom in the solid. Further, the number of recoil atoms of mass M_2, detected as a function of energy, can be related to the concentration of atoms of that mass in the target, as a function of depth. The energy spectra of recoiled elements, measured by the ERD technique, can thus be used to calculate concentration profiles.

In the present study, the energy spectra obtained for metallized PI targets were used to analyse the structure of PI-metal interfaces, and to evaluate the effect of various temperature and humidity treatments on the penetration of metal in the PI film. The interface width and composition and the concentration of carbon and oxygen in the metal film were evaluated by a conventional deconvolution technique, similar to the method generally used for the analysis of RBS spectra[4]. The rates of energy loss for various ions in the target material were calculated from the electronic stopping power values tabulated by Northcliff and Schilling[5].

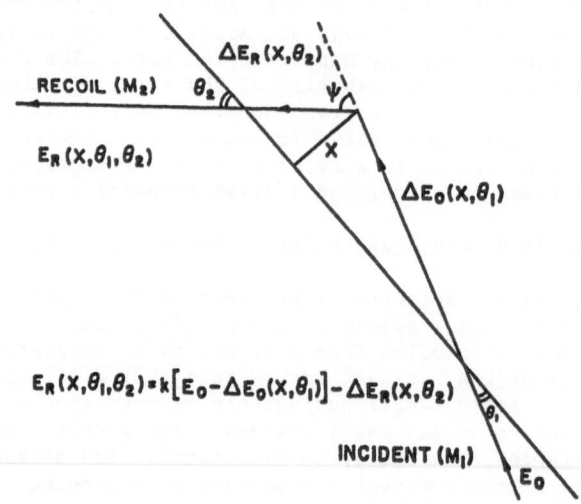

RECOIL (M₂)

$\Delta E_R(X,\theta_2)$

$E_R(X,\theta_1,\theta_2)$

$\Delta E_0(X,\theta_1)$

$E_R(X,\theta_1,\theta_2) = k\left[E_0 - \Delta E_0(X,\theta_1)\right] - \Delta E_R(X,\theta_2)$

INCIDENT (M₁)

E_0

Figure 1. Schematic diagram of the geometry and energies involved in ERD. ΔE_0 is the energy loss of the incident ion, ΔE_R that of the recoil nucleus, and k is the kinematic factor defined in Equation (1).

2.4   Experimental set-up

In the present study, the masses of the recoil atoms were identified by the time-of-flight (TOF) method. The experimental set-up is shown in Fig. 2, and has been described previously[4]. The recoil atoms scattered at $\Psi = 30°$ by a 30 MeV ^{35}Cl beam first pass through a thin carbon foil (10 μg/cm²). The secondary electrons emitted by the foil are collected by a micro-channel plate (MCP) detector, which generates a first signal for the time measurement. The recoil atoms then drift to a low resistivity surface barrier detector (SBD), which gives the second timing signal and the energy of the ions. The mass, M, of the transiting ion is given by:

$$M = \frac{2Et^2}{\ell^2} \ ,$$ (3)

where E is the energy of the particle, t is the time of flight and ℓ is the distance between the carbon foil and the SBD. The calculation of the ionic mass from the measured energy and time signals was carried out using the equation

$$M = C(E_M + E_0) \ (T_M + T_0)^2 \ ,$$ (4)

where C is a constant related to the flight path, E_M is the detected energy, E_0 is a detector offset. T_M is the measured flight time and T_0 represents the various electronic delays and distortions. The parameters E_0 and T_0 are functions of the energy and the mass of the detected ion. These functions can be determined experimentally and used to correct E_0 and T_0 for each event[7]. However, with our present software, these two parameters had to be kept constant for all ions at all energies. Because of this limitation, the mass resolution for a given ion decreases as the range of analysed energies increases. In all of the results presented below, the parameters E_0 and T_0 were chosen to optimize the mass resolution near the M = 12 region.

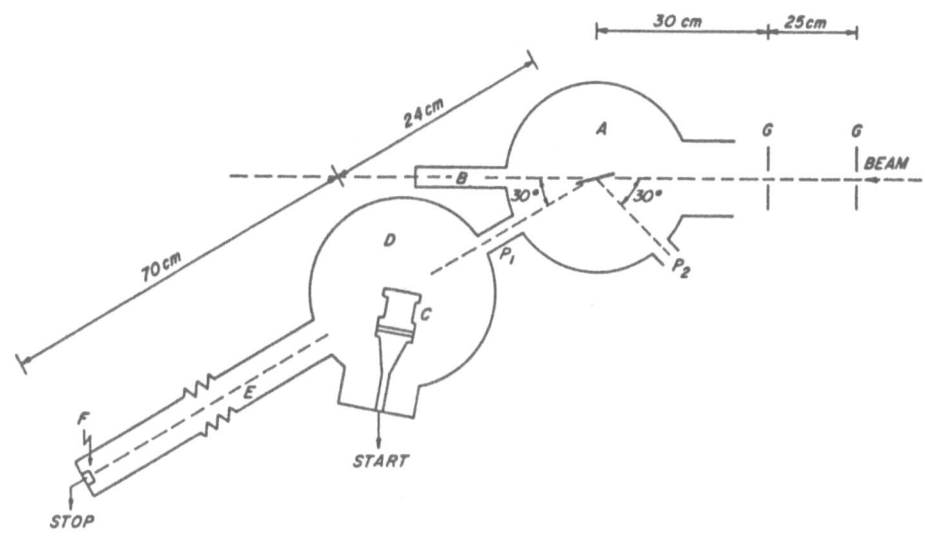

Figure 2. Schematic view of the detection chamber for mass determination of recoil particles by time-of-flight. A: scattering chamber. B: Faraday cup. C: start detector. D: cryopump. E: drift space. F: beam stop and energy detector. G: collimators. Port P_2 can be used for RBS measurements.

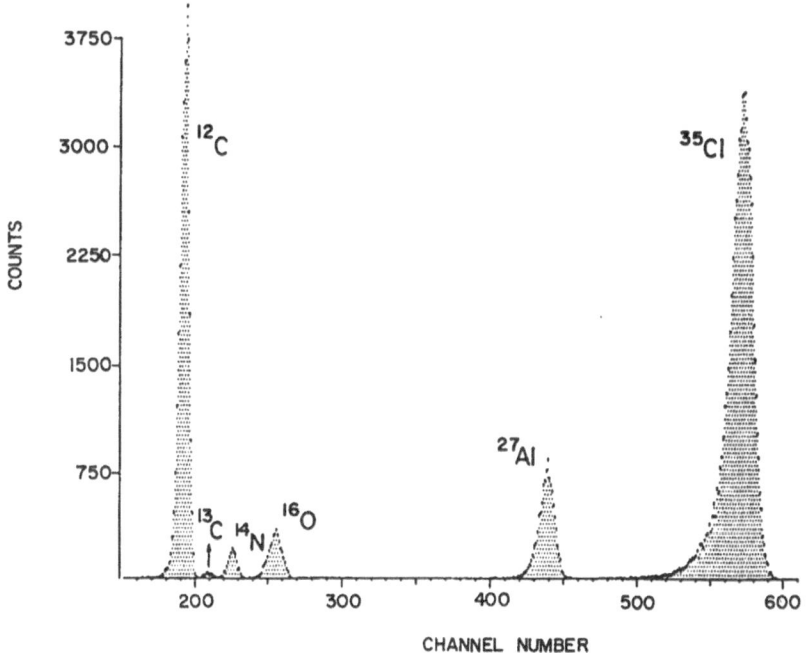

Figure 3. Mass spectrum of an aluminum-coated polyimide target. The ^{35}Cl peak comes from the scattering of incident ions on aluminum.

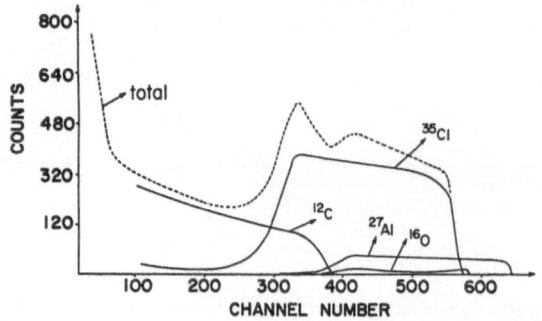

Figure 4. Energy spectrum (dotted line) of an aluminum-coated polyimide target. The separated energy spectra (solid lines) were obtained from the total spectrum by gating on the appropriate mass peaks (Fig. 3).

A typical mass spectrum obtained for a metallized PI target is shown in Fig. 3. The resolution in the ^{12}C region is sufficient to resolve one mass unit, as shown by the separation of ^{12}C, ^{13}C, ^{14}N and ^{16}O peaks. The decrease in mass resolution for ^{27}Al and ^{35}Cl is due to our particular choice of E_o and T_o parameters. However, the mass resolution was always sufficient to separate high mass peaks. For all PI-metal samples, therefore, specific mass windows could be assigned to each mass in the spectrum.

The energy spectra corresponding to the various mass peaks shown in Fig. 3 are presented in Fig. 4. The dotted line represents the total energy spectrum measured under bombardment. The solid lines represent the separated energy spectra obtained with the proper mass windows. Both the backscattered ^{35}Cl and recoil ^{27}Al spectra are representative of the aluminum distribution in the metal layer and at the interface. The high energy ends of these spectra correspond to the free metal surface, as the ^{35}Cl is not backscattered by the light elements for the geometry shown in Fig. 2. The beginning of the polymer substrate corresponds to the high energy end of the ^{12}C spectrum. The ^{16}O spectrum shown in Fig. 4 represents the oxygen detected in the aluminum. The number of counts detected for the ^{14}N and ^{16}O located in the polymer were too low to be shown on the scale of Fig. 4. Most of our discussion and conclusions on the interface structure, width and composition will center around a detailed analysis of the ^{12}C spectrum (for the polymer substrate) and the backscattered ^{35}Cl spectrum (for the metal layer).

3.   EXPERIMENTAL RESULTS

3.1   Nuclear Scattering Results

a)   Polyimide stability

The stability of PI under the ^{35}Cl beam was studied systemati-

cally before profiling the metallized samples. Polyimide targets were
bombarded with various low beam currents (< 20 nA particle), and with a
maximum dose of 5 x 10¹³ cm⁻². The energy spectra obtained for different
integrated doses, at the same beam current, were compared. The carbon
spectra showed no dependence on the beam current, instantaneous or
integrated, incident on the target. For doses larger than 10¹³ cm⁻², a
loss of H and O content was detected to a depth of 500 nm, and a decrease
in the initial film thickness was observed. Large doses always decreased
the H and O content relative to C, even at low beam currents. This effect
has been observed by several authors[8,9], and could be due to radiation-
induced gas release from within the polymer itself[10]. For small doses,
however, the various energy spectra showed no evidence of radiation-
induced effects. Compositional analysis yielded constant concentration as
a function of depth for all elements. In our experiments, all measurements were taken at low beam current (1 to 2 nA particle), with doses
lower than 10¹³ particles per cm⁻². Under these conditions, the energy
spectra obtained for different spots on the same target consistently
showed the same structures.

To minimize possible beam effects, the metallized PI targets were
thus bombarded until a dose of 10¹³ particles per cm⁻² was reached, or
until statistically significant results (50 counts per channel) in the
¹²C spectrum were obtained. Consequently, the yields in number of counts

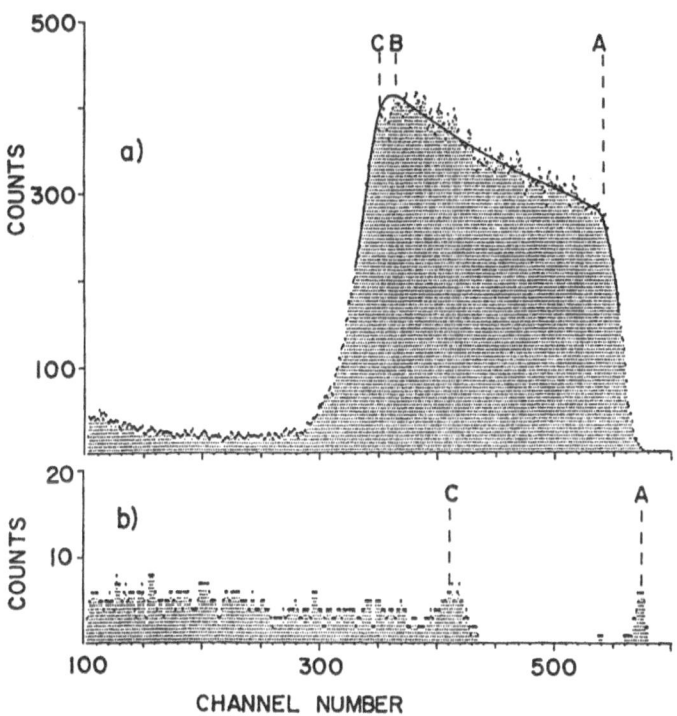

Figure 5. Backscattered ³⁵Cl (a) and recoiled ¹⁶O (b) energy spectra
obtained for a target of aluminum on dry polyimide. The solid line in
(a) represents the ³⁵Cl energy spectrum expected for a free-standing
aluminum film. The letters denote (A) the free metal surface, (B) the
beginning of the mixed region at the interface, (C) the end of the metal
layer.

for the ^{14}N and [16]O located in the polymer substrate were most often too low to have any statistical significance. However, the total concentration of carbon or oxygen in the metal film could always be evaluated with a ~ 10% precision.

b) Aluminum

    Effect of pre-treatment. The effect of PI surface pre-treatment on the PI-Al interface is shown in Figs. 5 through 7. The [35]Cl and [16]O spectra obtained for targets of aluminum on dry and hydrolyzed PI are shown in Fig. 5 and 6, respectively. The corresponding ^{12}C spectra are shown in Fig. 7. A solid line is drawn on the [35]Cl spectra to show the difference between the measured energy spectra and the energy spectrum expected for a free-standing aluminum film.

    For a target of aluminum on dry PI (pre-treatment T), the [35]Cl spectrum (Fig. 5a) shows a sharp decrease in the number of counts at the interface, with little tailing of the metal into the PI. A comparison between the measured energy spectrum and the spectrum expected for a free-standing metal film (solid line) shows a depletion in the number of counts near the interface (between B and C). This structure was consistently seen for all samples, and is, therefore, representative of PI-Al mixing at the interface. The relative concentration of aluminum in this mixed region is ~ 90%, and the depth to which mixing occurs corresponds to about 20 nm of aluminum. The [16]O spectrum obtained for this target (Fig. 5b) shows a small quantity of aluminum oxide at the

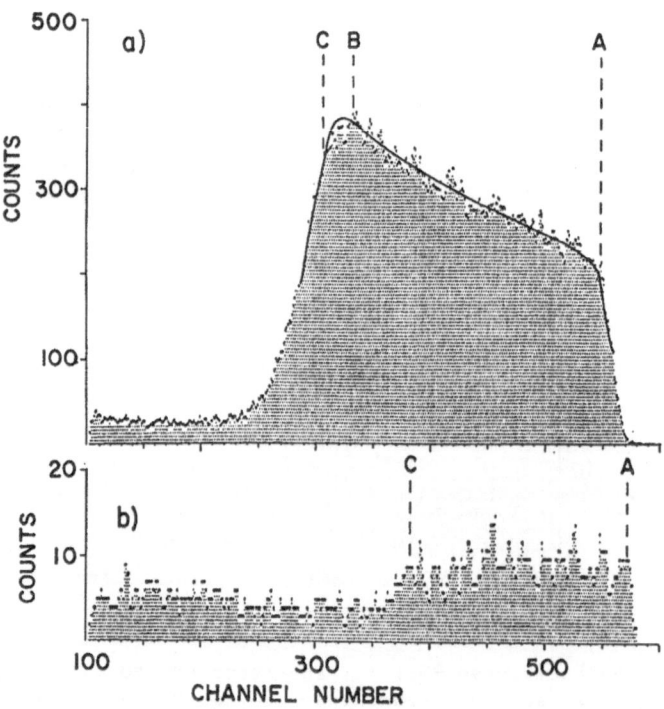

Figure 6. Backscattered [35]Cl (a) and recoiled [16]O (b) energy spectra obtained for a target of aluminum on hydrolyzed polyimide. The features identified by letters correspond to those of Fig. 5.

surface (A), no oxygen inside the metal film, and oxygen enhancement at the interface (C). The increase in oxygen content at the interface was detected for all samples of aluminum on dry PI, and is, therefore, a real effect. This structure, however, was not observed for any of the various PI-Cu and PI-Au targets.

For a target of aluminum on hydrolyzed PI, the ^{35}Cl spectrum (Fig. 6a) shows a broader interface and an increase in metal tailing. The depletion in the number of counts near the interface (between B and C) was again consistently seen for all similar targets. Pre-treatment H increases the width of this mixed region to about 30 nm, and decreases its aluminum content to ~ 80%. The distribution of aluminum between B and C is not uniform, and was seen to vary from point to point on the

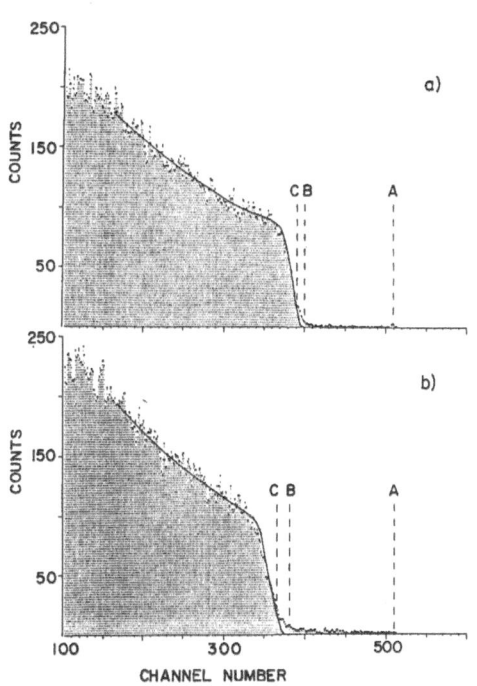

Figure 7. Recoiled ^{12}C energy spectra for targets of aluminum on (a) dry and (b) hydrolyzed polyimide. The solid lines represent the energy spectra expected for a perfectly sharp interface. The features identified by letters correspond to those of Fig. 5.

sample. This could be due to the formation of metal clusters in the swollen polymer layer[11] or to small variations in the amount of absorbed water. The apparent increase in metal thickness observed for this target is related to the presence of oxygen in the metal film. As can be seen in Fig. 6b, the entire metal layer was oxidized during the evaporation process. The oxygen-to-aluminum concentration ratio in the metal layer is ~ 1:5.

The carbon spectra obtained for the same targets are presented in Fig. 7. Here, the solid curve represents the spectrum expected for a perfectly sharp interface. For the dry PI surface (Fig. 7a), the decrease in the number of counts at the PI-Al interface is sharp and uniform. The carbon located in the mixed region near the interface is shown between B and C, and corresponds to the aluminum signal depletion seen between the same labels on the corresponding ^{35}Cl spectrum (Fig. 5a). For the hydrolyzed PI surface (Fig. 7b), the PI-Al interface is larger and the carbon content in the mixed region has increased. An increase of the carbon content in the metal layer is also detected. Close examination of the interfacial region of Figure 7b shows that the spectrum drops in steps. We attribute this to a nonuniform carbon distribution in this region. This is in agreement with the observations made on the ^{35}Cl spectrum.

Effect of post-treatments D and H. The aging of PI-Al samples in 100% RH at RT increases the tailing of the metal into the polymer substrate and decreases the metal content in the mixed region. This treatment, however, does not change the width of the mixed region at the interface. Similar effects were observed for the PI-Cu targets, and the energy spectra obtained for this polymer-metal couple will be presented in the following sub-section.

Effect of post-treatment T. The backscattered ^{35}Cl and recoil ^{12}C spectra obtained for a heat treated sample of aluminum on hydrolyzed PI are presented in Fig. 8a and 8b, respectively. A comparison of the ^{35}Cl spectra obtained for the non-treated sample (Fig. 6a) and the post-treated T sample (Fig. 8a) shows that heating at 150°C for 5 days does not increase the tailing of the metal into the polymer substrate. The heat treatment, however, changes the width and composition of the mixed region at the interface (between B and C). The width of this mixed region decreases from about 30 nm (Fig. 6a) to about 15 nm (Fig. 8a), and the relative atomic concentration of aluminum decreases from ~ 80% to ~ 70%.

The recoil ^{12}C spectrum (Fig. 8b) shows a sharper interface than that observed for the non-treated sample (Fig. 7b). The carbon signal corresponding to the mixed region at the interface (between B and C) is more localized. This is in agreement with the structure observed on the ^{35}Cl spectrum. The amount of carbon lost from the polymer substrate to this mixed region can be evaluated from the number of counts between B and C, or from the difference between the number of counts in the bulk PI (E) and the flat structure near the maximum (D). Both methods yield a relative atomic concentration of carbon in the mixed region of ~ 20%.

The heat treated samples show a uniform distribution of aluminum and carbon in the mixed region at the interface. This was not observed in the case of non-treated samples. The heat treatment does not change the carbon content in the metal layer, and has no effect on the oxygen spectrum. This post-treatment, however, sharpens the interface and localizes the region characteristic of PI-Al mixing. These effects were seen for all heat treated samples, irrespective of the PI surface pre-treatment.

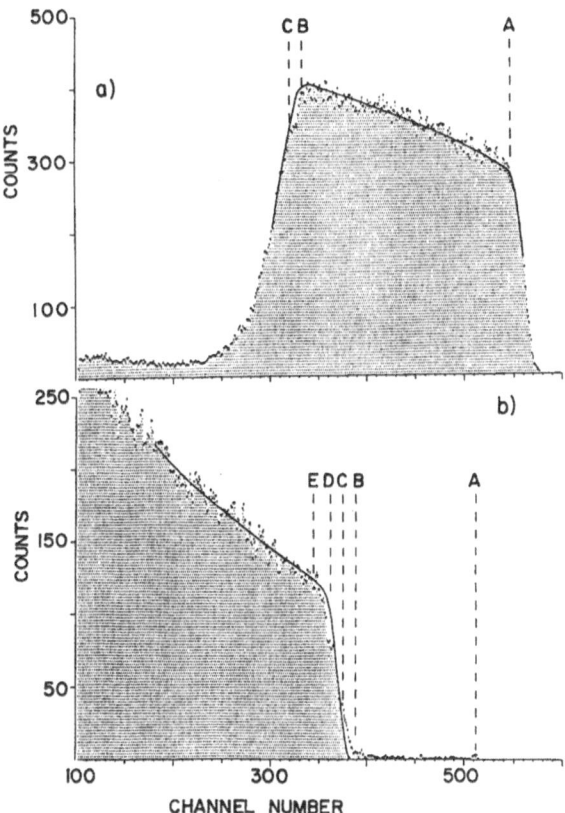

Figure 8. Backscattered ^{35}Cl (a) and recoiled ^{12}C (b) energy spectra
obtained for a heat-treated sample of aluminum on hydrolyzed polyimide.
The features identified by A, B and C correspond to those of Fig. 5.
Other letters denote (D) the polymer substrate close to the interface and
(E) the beginning of bulk polyimide.

c) <u>Copper</u>

   <u>Effect of pre-treatment.</u> The effect of pre-treatment T and H on the
PI-Cu interface is similar to what was observed for the PI-Al targets.
The evaporation of copper on a hydrolyzed PI surface increases the width
of the mixed region from about 15 nm to about 25 nm, and decreases the
relative atomic concentration of copper from ~ 90% to ~ 85%. Pre-
treatment H also substantially increases the tailing of the metal.

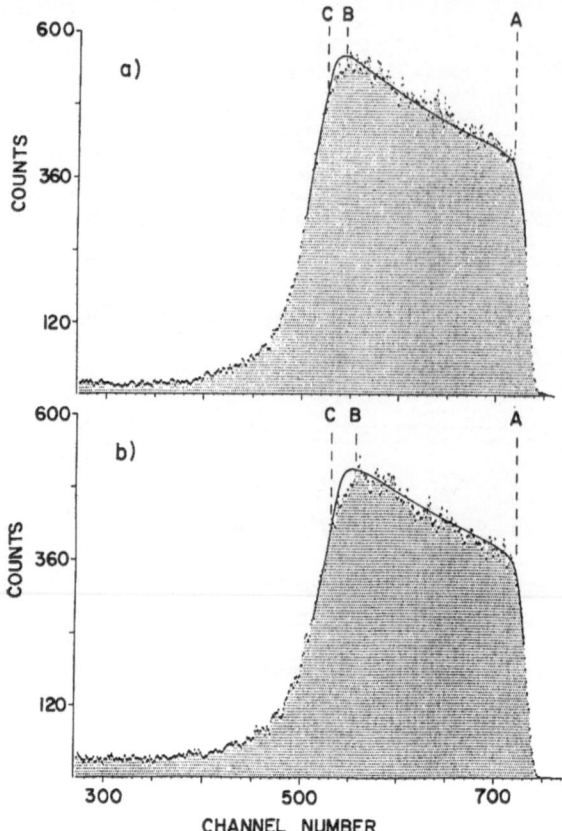

Figure 9. Backscattered ^{35}Cl spectra for targets of copper on dry polyimide stored in (a) dry and (b) humid atmospheres. The letters denote (A) the free metal surface, (B) the beginning of the mixed region at the interface and (C) the end of the copper layer.

Effect of post-treatments D and H. The ^{35}Cl and ^{12}C spectra obtained for PI-Cu samples stored in dry and humid atmospheres are presented in Fig. 9 and 10, respectively. Both ^{35}Cl spectra show a depletion in the number of counts (between B and C) characteristic of PI-Cu mixing at the interface. The humidity exposure has no significant effect on the width of the mixed region ($\sim$ 15 nm) but decreases the relative atomic concentration of copper from $\sim$ 90% (Fig. 9a) to $\sim$ 80% (Fig. 9b). Post-treatment H also increases the tailing of the metal into the polymer substrate (channels 500 and below).

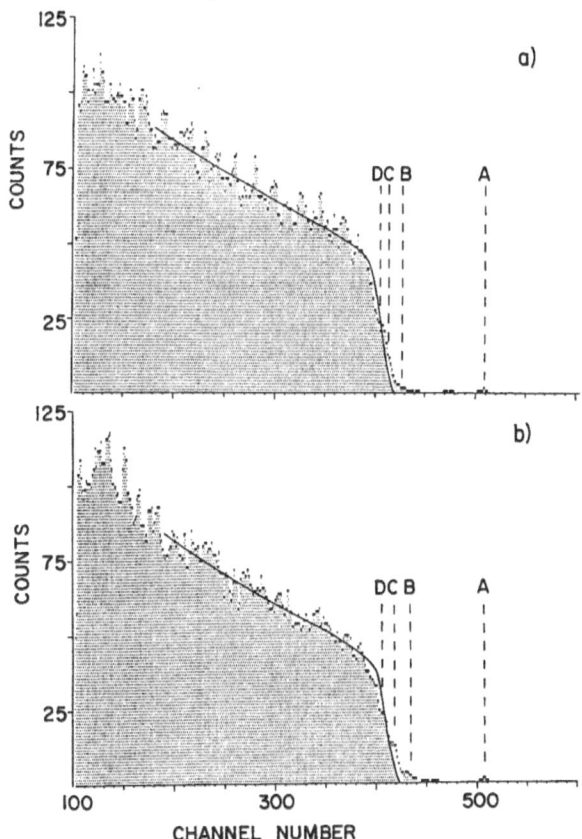

Figure 10. Recoiled ^{12}C spectra for targets of copper on dry polyimide stored in (a) dry and (b) humid atmospheres. The features identified by A, B and C correspond to those of Fig. 9. The letter D denotes the middle of the polyimide-copper interphase.

A comparison of the corresponding ^{12}C spectra (Fig. 10) shows that exposure to water vapor increases the carbon content in the mixed region at the interface (between B and C). Note that in both spectra, the carbon detected in the metal layer near the interface (to the right of D) corresponds to a loss of carbon from the PI film (to the left of D). The ^{16}O spectra obtained for these targets (not shown) were identical. A small oxide layer was seen at the surface, and no oxygen was detected in the copper layer or at the PI-Cu interface.

Effect of post-treatment T. The effect of post-treatment T on the

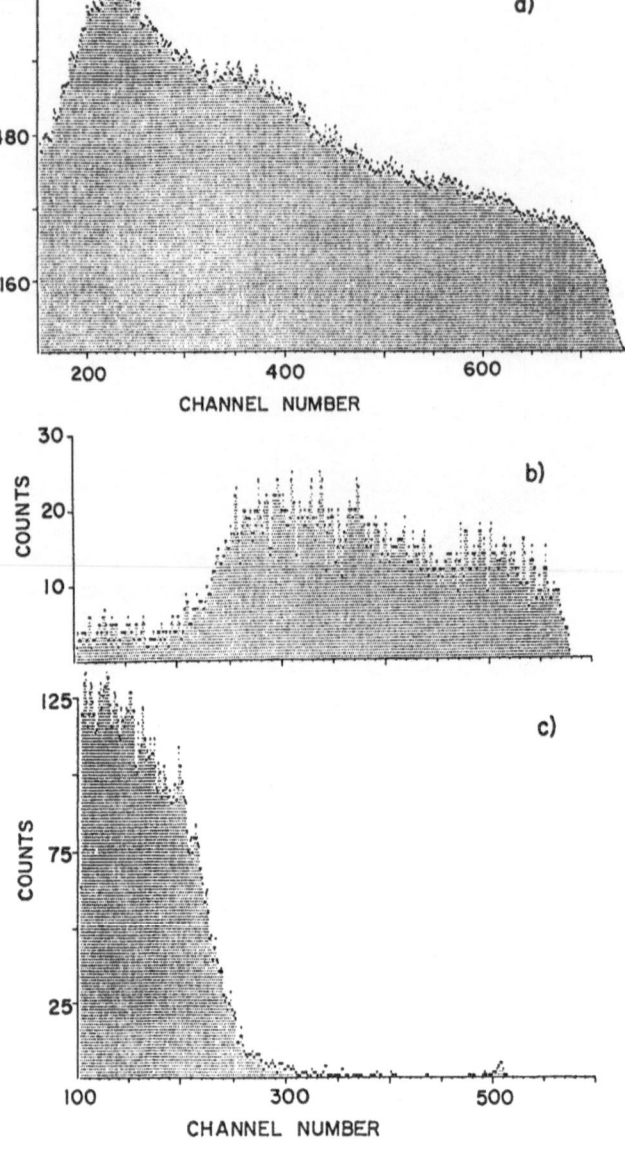

Figure 11. Backscattered ^{35}Cl (a), recoiled ^{16}O (b) and ^{12}C (c) spectra obtained for a heat-treated sample of copper on dry polyimide. The slight bulge in the ^{35}Cl spectrum, mirrored in the ^{16}O spectrum, is occasionally seen in all spectra and may indicate clustering[11].

PI-Cu interface is presented in Fig. 11. The ^{35}Cl and ^{16}O spectra (Fig. 11a and b) show that heating in air for 5 days completely oxidizes the copper layer. This effect was observed for all heat-treated Cu-PI samples. The oxygen concentration in the metal layer is equivalent for all targets and is close to stoichiometric CuO (we shall refer to this as ~CuO).

As can be seen in the ^{12}C spectrum (Fig. 11c), this treatment significantly alters the PI-Cu interface. The interpenetration of PI and ~CuO oxide extends over several tens of nanometers. The tailing of the

¹²C spectrum into the ~CuO film scales with the tailing of the ¹⁴N
spectrum and is, therefore, representative of real ~CuO-PI mixing.

The surfaces of the heat-treated samples were inhomogeneous, and
showed large regions of CuO agglomerates. For copper on dry PI, the oxide
clusters retained a metallic sheen. For copper on hydrolyzed samples, a
nonuniform, powdery oxide was formed, which could easily be removed by
scratching. The spectra in Fig. 11 are typical of the Cu-PI interface in
the oxide cluster regions.

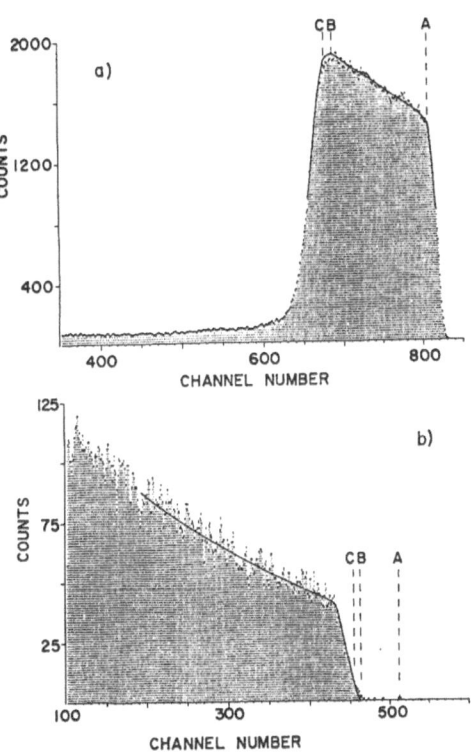

Figure 12. Backscattered ³⁵Cl (a) and recoiled ¹²C (b) spectra obtained
for a sample of gold on dry polyimide. The features identified by
letters correspond to those of Fig. 5.

d) Gold

        The various pre- and post-treatments have very little effect on the
PI-Au interface. Typical ^{35}Cl and ^{12}C spectra obtained for the PI-Au
targets are presented in Fig. 12. As was seen for all samples of
metallized PI, the ^{35}Cl spectrum shows a region characteristic of
polymer-metal mixing near the interface (between B and C). The width of
this mixed region is ~ 5 nm and the relative atomic concentration of gold
is ~ 95%. Both the ^{35}Cl and ^{12}C spectra show a sharp interface. No
oxygen was detected at the metal surface, in the metal film or at the
PI-Au interface.

        Hydrolysis of the PI surface before metallization or exposure of the
metallized samples to water vapor increases the width of the mixed region
at the interface. This increase, however, is much smaller than the one
observed for PI-Al or PI-Cu samples subjected to pre-treatment H or
post-treatment H. Further, the penetration of water vapor has no effect
on the gold content of the mixed region or on the tailing of the metal
into the polyimide.

        No significant changes were observed in the ^{35}Cl and ^{12}C spectra
obtained for heat-treated samples. Prolonged heating (10 days) at 150°C
increased the width of the interphase and the metal tailing. While this
was a small effect, it is opposite to the effect observed for heat
treated PI-Al targets.

        We believe that these results indicate little or no interphase
formation, ~5 nm being the resolution limit of our experiments in the
interfacial region. Thus, broader interphases, as found for the other
metals, must indicate intermixing.

3.2   Tape Test Results

        The tape test results show that the effect of a particular pre- or
post-treatment on the adhesion of metallization does not depend on the
type of metal evaporated. The comparison between pre-treatments T and H
shows that the presence of water at the PI surface is detrimental to
adhesion. The exposure of metallized samples to water vapor is also
deleterious to adhesion. Post-treatment T, however, always increases the
adherence of metallization, irrespective of pre-treatment. Heat-treated
samples are also less susceptible to aging in a humid environment.
Further, the adhesion of metal to a hydrolyzed PI surface may be
increased by prolonged storage in a dry atmosphere.

4.   DISCUSSION

        The nuclear scattering results clearly indicate that the evaporation
of metal onto a PI substrate induces polymer-metal mixing at the
interface except in the case of gold. The energy profiles obtained for
the various PI-metal targets show that the structure and composition of
this interphase is specific to the metal evaporated. The ERD and RBS
results also show that the effect of pre- and post-treatments on the
extent of mixing and depth of interpenetration depend on the type of
metal used.

        The composition and width of the region of mixing can easily be
quantified by the conventional deconvolution technique discussed in
section 2.3. Because of the grazing incidence geometry used in our
experiments, topography effects related to metal film uniformity should
be reflected at both the free surface and the metal-PI interface. A lack

of a broadened metal surface in the ^{35}Cl spectrum indicated good film uniformity. Further, it is to be noted that the gold data demonstrate that the effects of pre-treatments T and H on PI surface roughness do not influence interface structure. The variations in composition and width observed in the interphase can thus be attributed to the effects of the various pre- and post-treatments on the extent of interfacial mixing.

The tailing of the metal into the polymer substrate, however, cannot be readily interpreted in terms of metal diffusion into the PI layer. The low energy tail seen in the various ^{35}Cl spectra may have several origins: multiple scattering occurring in the metal layer[2,12,13], variations in surface and interfacial topography[14,15], and metal diffusion. Beam effects may also account for variations in tailing. For a given metal and metallization thickness, however, the contribution of multiple scattering to the low energy tail is constant. The metallization thicknesses being identical, variations in metal tailing cannot be related to variations in multiple scattering. Further, the energy spectra obtained for different spots on the same target or for different targets prepared under the same conditions consistently show the same low energy tail. The observed variations in metal tailing are, therefore, not artifacts, and can be used to compare the effect of the various pre- and post-treatments on the mutual interpenetration of polymer and metal at the interface.

The exposure of metallized samples to water vapor has the same effect on PI-Al and PI-Cu interfaces. Aging at 100% RH, RT does not change the width of the mixed region at the interface but decreases the metal content in the interphase and substantially increases the low energy tail of the ^{35}Cl spectra. Part of this increase is probably due to the increase of polymer-metal mixing in the interphase. The penetration of water to the interface and into the polymer substrate may also induce random variations in topography at the interface and, consequently, increase the low energy tail. However, for the reasons cited above, the increase in tailing upon aging at 100% RH, RT is not an artifact and is representative of mutual interpenetration at the polymer-metal interface.

The evaporation of a metal layer onto a hydrolyzed PI surface also gives the same interfacial structure for PI-Al and PI-Cu targets. The presence of water at the polymer surface prior to metallization not only increases interfacial mixing and metal tailing, but also increases the width of mixing. This may be attributed to the fact that metal atoms penetrate more readily into polyamic acid than into PI. This would explain why pre-treated H samples show a larger increase in mixing and tailing than post-treated H samples. Thus, the introduction of water at the interface prior to or subsequent to metallization has the same effect on the interphase, the only difference being the extent of polymer-metal interpenetration.

The effect of heat treatment on the interfacial structure is different for the three types of metal studied. For PI-Al targets, heating at 150°C for five days decreases the width and the metal content of the mixed region, and reduces metal tailing. For PI-Cu targets, the heat treatment oxidizes the copper layer and induces ~CuO-PI mixing over several tens of nanometers. For PI-Au targets, prolonged heating causes small increases of both the mixing at the interface and the metal tailing into the polymer. While the ERD technique gives no indication of the chemical state of the metal atoms in the mixed region, these effects may be interpreted in terms of the nature of the chemical interactions occurring at the polymer-metal interface.

It is well known[2] that the condensation energy of evaporated particles is high enough for a reaction to occur between the deposited metal and the polymer layer. Several XPS[14,17,18] studies have shown that the first monolayers deposited react with the surface to give oxidized or partly oxidized metal atoms as well as polymeric reaction products . Some authors[17 18] have suggested that metal atoms react with the carbonyl oxygen atoms in the polymer, forming either metal-organic complexes or metal oxides at the interface. An increase in oxygen content was detected at the interface of PI-Al targets, but not at the interface of PI-Cu and PI-Au targets. This points to specific reactivity of aluminum with oxygen atoms in the PI layer. The ERD results show that the heat treatment of PI-Al targets does not activate the thermal diffusion of reacted aluminum atoms but, rather, induces a contraction of the interphase. Such a contraction might be due to a simple reduction in interfacial roughness through the elimination of polymer swelling. The PI-Au targets showed very little mixing at the interface and were insensitive to the various pre- and post-treatments. These results indicate that gold atoms probably do not react with the polymer and that the metallization process induces mechanical embedding of gold atoms only at the surface of the polymer. In this case, heat treatment seems to activate a limited thermal diffusion of gold atoms from the mixed region to the polymer substrate.

The tape test results indicate that pre-treatment H and post-treatment H induce the weakest interfacial bonds, for the three metals studied. The increase in interfacial mixing observed for PI-Al and PI-Cu targets exposed to water vapor, therefore, does not lead to an increase in adhesion properties. The exact locus of failure, however, is not known. The rupture could occur between the metal layer and the mixed region, in this mixed region, or in the swollen polymer layer. The penetration of water has very little effect the PI-Au interfacial structure but substantially alters its adhesion properties. This suggests that, in this case, the rupture may not occur in the interphase.

The tape test results also indicate that the heat treatment of metallized samples systematically increases interfacial adhesion. However, the ERD results show that the effect of heat treatment on the low energy tail, as well as the width and the composition of the interphase, differs for each metal. This indicates that the extent of interfacial mixing and the depth of interpenetration can not be directly related to the adhesion properties of the various polymer-metal interfaces. Further, an increase in adhesion upon heat treatment was observed for samples that were not exposed to water vapor. The increase in interfacial bond strength, therefore, may not be only attributed to the removal of water in the interphase. Clearly, we must look at specific interfacial interactions to explain adhesion properties.

5.   CONCLUSIONS

Both ERD and RBS techniques were used to analyse the structure of polyimide-metal interfaces. Tape tests were performed on metallized samples to evaluate the effect of various treatments on the adhesion of metallization to the polyimide substrate. Our study has shown that Al, Cu or Au metallization onto dry or hydrolyzed PI induces the formation of a mixed region at the interface. The relation between the structure of this interphase and interfacial bond strength is, however, presently not clear.

The tape test results show that the effect of a particular treatment on the adhesion of metallization does not depend on the type of metal used. Heat treatment always increases interfacial bond strength, while

284

exposure to humidity substantially lowers the adhesion of Al, Cu or Au to polyimide. In contrast, the ERD results show that the effects of heat treatment and exposure to humidity on the composition and width of the mixed region at the interface differ for each metal. This suggests that chemical interactions play a determining role in the formation of the mixed region at the interface. However, the actual interfacial reactions and the extent to which they occur cannot be characterized by these ion beam techniques. To further evaluate the effect of temperature and humidity treatment, the PI-metal interfaces should be studied with surface techniques sensitive to chemical environments.

ACKNOWLEDGEMENTS

The authors wish to thank the technical staff of the Laboratoire de Physique Nucléaire of the Université de Montréal for their assistance. This work was supported by the Natural Science and Engineering Research Council of Canada and by the Fonds FCAR of Quebec.

REFERENCES

1. E. Sacher, P. A. Engel and R. G. Bayer, J. Appl. Polym. Sci., 24, 1503 (1979) and references therein.
2. J. Currie, P. Depelsenaire, R. Groleau and E. Sacher, J. Colloid Interface Sci., 97, 410 (1984).
3. W.J. van Ooij, in "Physicochemical Aspects of Polymer Surfaces", K.L. Mittal, editor, p. 1035, Plenum Press, New York, 1983.
4. J. F. Ziegler, R. F. Lever and J. K. Hirvonen, in "Ion Beam Surface Layer Analysis", O. Meyer, G. Linker and F. Käppeler, Editors, Vol. 1, p. 163, Plenum Press, New York, 1976.
5. L. C. Northcliffe and R. F. Schilling, Nucl. Data Tables, Sect. A, 7, 233 (1970).
6. R. Groleau, S. C. Gujrathi and J. P. Martin, Nucl. Instrum. Meth., 218, 11 (1983).
7. H. Breuer, N. R. Yoder, A. C. Mignerey, V. E. Viola, K. Kwitkowski and K. L. Wolf, Nucl. Instrum. Meth., 204, 419 (1983).
8. T. Hioki, S. Noda, M. Sugiura, M. Kakeno, K. Yamada and J. Kawamoto, Appl. Phys. Lett., 43, 30 (1983).
9. T. VenKatesan, R. C. Dynes, B. Wilkens, A. E. White, J. M. Gibson and R. Hamm, Nucl. Instrum. Meth., B1, 599 (1984).
10. V. Shrinet, U. K. Chaturvedi, S. K. Agrawal and A. K. Nigam, Nucl. Instrum. Meth., B1, 617 (1984).
11. R.M. Tromp, F. Legoues and P.S. Ho, J. Vac. Sci. Technol., A3, 782 (1985).
12. P. Belery, T. Delbar and G. Grégoire, Nucl. Instrum. Meth., 179, 1 (1981).
13. A. Weber, H. Mommsen, W. Sarter and A. Wheeler, Nucl. Instrum. Meth., 198, 527 (1982).
14. R. D. Edge and U. Bill, Nucl. Instrum. Meth., 168, 157 (1980).
15. A. R. Knudson, Nucl. Instrum. Meth., 168, 163 (1980).
16. H. J. Leary, Jr., and D. S. Campbell, Surf. Interface Anal., 1, 75 (1979).
17. J. M. Burkstrand, J. Appl. Phys., 52, 4795 (1981) and references therein.
18. N. J. Chou and C. H. Tang, J. Vac. Sci. Technol., A2, 751 (1984).

A THERMODYNAMIC MODEL FOR PREDICTING FORMATION OF CHEMICAL BONDS BETWEEN METALS AND CURED POLYIMIDES DURING METALLIZATION

N. J. Chou and C. H. Tang*

IBM T. J. Watson Research Center
Yorktown Heights, NY 10598

Adhesion between metals and polymers has been a subject of practical concern in the microelectronic packaging technology. Formation of chemical bonds promises the type of adhesion favored by the new processing techniques being developed in the industry. In an earlier paper, a thermodynamic model was proposed where an effective partial molar energy was defined for the pendent oxygen in the cured polyimide (PI) and its limiting values determined experimentally by in-situ XPS monitoring of interfacial reaction during metallization. Used to predict whether a metal will form chemical bonds with the oxygen in PI, the model was found to be valid for several metals. In this paper a quasichemical approach is used to examine the validity of the assumptions on which the model is based. It is shown that until a detailed quantum mechanical treatment is made, some of these assumptions should be regarded as a mixture of truth and expediency.

INTRODUCTION

In two previous investigations[1,2] we have used in-situ X-ray photoelectron spectroscopy to monitor the interfacial reaction during metallization of cured polyimide and demonstrated that changes in binding energy for the 1s electrons ejected from carbon in the polyimide substrate and the 2p electrons from the thin metal deposit could be used, in conjunction with the expected attenuation or electron escape length effect, to deduce if interfacial oxidation took place during metallization. A thermodynamic model was proposed to predict the possibility of

---

* Present address: Gould Research Center, Rolling Meadows, Il 60008

chemical bond formation by virtue of interfacial oxidation. In the absence of pertinent thermodynamic data, it was assumed that in the case of interfacial oxidation the
reaction kinetics was not a limiting factor and that the free energy for transferring
oxygen from polyimide substrate to its standard state could be determined indirectly
by XPS measurement on a number of selected metal-PI systems (Ag/PI, Cu/PI,
Ni/PI and Cr/PI). Based on a "go or no go" proposition, the free energy of transfer
or the effective partial molar free energy of oxygen was estimated to have values
between ~ 109 and 141 kcal/g-atom. The latter value was then used to evaluate
whether a given metal will form chemical bonds with the oxygen in cured PI during
deposition. Al, Mg and Ti, among other metals, were predicted as those which would
oxidize interfacially [2]. Since the publication of our investigations, several papers
have appeared in the literature, which have dealt with metal-polymer and metal-PI
interactions[3,4,5]. The results of these studies as well as some of our recent observations not only reaffirm our experimental approach and reported findings, but appear
to uphold the predictions of our model. In view of the new experimental evidence
we will use a quasichemical approach to examine the validity of the proposed model.

PROPOSED MODEL

The model postulates that the possibility of a metal forming chemical bonds with
the PI substrate can be evaluated by assessing the thermodynamic driving force for
interfacial oxidation of the form

$$(M) + [O]_{polyimide} = [MO]_{interface,} \qquad (1)$$

which states that the metal vapor reacts with the oxygen in PI to form an oxide at the
interface. Here, ( ) and [ ] denote gaseous and solid state, respectively, and subscripts refer to the polyimide substrate and interface, respectively. To utilize tabulated thermochemical data this non-standard state reaction can be decomposed into
three steps, namely,
(1) condensation of metal vapor: $(M) = [M]$,

(2) transfer of oxygen from polyimide substrate to its standard state:

$$[O]_{polyimide} = \frac{1}{2}(O_2), \text{ and}$$

(3) standard state oxidation of the metal: $[M] + \frac{1}{2}(O_2) = [MO]$

The driving force for interfacial oxidation can then be determined by the sum of the
free energy change of the individual steps:

$$\Delta G(1) = \Delta G_1 + \Delta \overline{G}_2 + \Delta G_3 \qquad (1a)$$

where $\Delta \overline{G}_2$ is the partial molar free energy of oxygen in PI. For the interfacial
oxidation to occur, $\Delta G(1)$ must be negative, which means the free energy of vapor
phase oxidation, $(\Delta G_{vap} = \Delta G_1 + \Delta G_3)$, should be greater than $\Delta \overline{G}_2$ in absolute
magnitude. Of the three terms on the right of Equation (1a), $\Delta \overline{G}_2$ is not available
from the tabulated data [6]. It would be quite simple to obtain the pertinent data if one

could perform the conventional vapor pressure measurement for the oxygen in PI. Unfortunately, the technique is impractical because PI decomposes at elevated temperature with the release of CO and CO_2 [7]. However, the limiting value of $\Delta \overline{G}_2$ can be estimated by a formalism of statistical thermodynamics if we confine our analysis to the pendent oxygen in PI.

Let us use as a reference system according to liquid lattice theory [8] a cured PI consisting of N idealized molecular units:

Its Helmholtz free energy is given by [9]

$$G = \sum_{i \neq j} N_{ij} V_{ij} + G_{el} + G_{vib} + G_{conf}, \tag{2}$$

where V_{ij}'s are pairwise interaction energy (or dissociation energy in quantum-mechanical terms) between atoms i and j, N_{ij} denotes number of similar pairs, and subscripts indicate electronic, vibrational and configurational contributions. A Maxwellian demon is now dispatched to remove, one after another, n pendent (carbonyl) oxygen atoms from the reference system. We then have a PI system with n oxygen vacancies and 2n unbonded valence electrons.

The free energy change between the faulted and the reference system will be

$$\Delta G_t = - n V_{c=o} + \Delta G_{dist} + \Delta G_{el} + \Delta G_{vib} - T\Delta S \tag{3}$$

Here, $-n V_{c=o}$ accounts for the interaction energy associated with the missing C=O pairs; ΔG_{dist} arises from lattice distortion due to vacancy formation; ΔG_{el} is due to the electronic defects; ΔG_{vib} is associated with missing oxygen atoms; and ΔS is the entropy change. It should be noted that an implicit assumption has been made in the preceding discussion, namely, that the intermolecular interactions or the interactions between atoms which are not in the same molecular unit are negligible as compared to intra-unit interactions.

To bring the atomic oxygen to its standard state one more step is required:

$$n(O) = \frac{n}{2}(O_2) \tag{4}$$

Associated with it is the negative free energy of dissociation of molecular oxygen, $- \Delta G_d$:

$$\Delta G_4 = -\frac{n}{2}\Delta G_d \tag{5}$$

Assume that removal of pendent oxygen does not give rise to appreciable distortion. Taking note of the cancelling effect of ΔG_{el} and ΔG_{vib} we now apply the quasichemical approach to the problem at hand and neglect the contributions of $\Delta G's$ in Equation (3). Adding ΔG_4 to ΔG_t yields then the free energy change for transferring n pendent oxygen atoms from PI to their standard state

$$\Delta G_2 = -nV_{c=o} - T\Delta S_c - \frac{n}{2}\Delta G_d$$

where ΔS_c denotes the configurational entropy. Differentiating ΔG_2 with respect to n we obtain the effective partial molar energy for oxygen:

$$\Delta \overline{G}_2 = - V_{c=o} - T\frac{d\Delta S_c}{dn} - \frac{1}{2}\Delta G_d$$

Note that the term 'effective' is used because our analysis treats ether oxygen in PI as if it were a different atomic species. According to the quasichemical treatment, therefore, the free energy change in transferring pendent oxygen from PI to its standard state is simply the sum of the energy required to break the carbonyl bonds and the recombination (association) energy of atomic oxygen plus a entropy term associated with the vacancy formation. Similar treatment was made by Swalin[10] for vacancy formation in face-centered cubic single crystals of Ge and Si. Note, however, that in his case vacancies are formed by moving atoms from the interior to the surface of the crystal so that only half of the number of bonds are considered broken.

Since there are four equivalent or indistinguishable carbonyl oxygen sites in a molecular unit it can be readily shown that $d\Delta S_c/dn = k \ln 3 = 2.2$ e.u. The interaction energy $V_{c=o}$ can be estimated either theoretically by quantum-mechanical calculations or empirically from dissociation energy of C=O bonds in similar molecular structures. The literature value for $V_{c=o}^{11,12}$ varies from 142 to 175 kcal/g-atom. The free energy of dissociation for molecular oxygen is [6]

$$\Delta G_d = 59 - 0.002 \times (T - 298) + 10.5 \times 10^{-3}T$$

which gives $\sim$ 62 kcal/mole or 31 kcal/g-atom at room temperature. Substituting these values into the equation for $\Delta \overline{G}_2$ one obtains a value of $\sim$ 110 to 143 kcal/g-atom. The remarkable agreement between the calculated and the experimentally derived[2] values of $\Delta \overline{G}_2$ is of course fortuitous. No one can expect to achieve a better than 10% agreement in thermodynamic calculations. Nevertheless, the agreement prompted us to take a closer look at the model.

XPS studies of PI surfaces[13,14] have shown that C 1s spectra of fully cured PI surfaces exhibit three XPS-resolvable peaks. Assignment of these peaks was worked out both empirically[13] and theoretically [4] (Fig. 1).

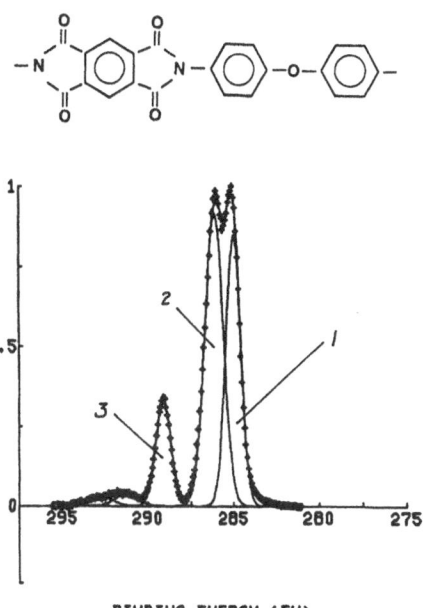

BINDING ENERGY (EV)

Figure 1. C 1s spectrum of PMDA-ODA polyimide and assigment of its resolved peaks: the upper graph shows an idealized molecular unit of PMDA-ODA polyimide and the lower its C 1s spectrum with peaks attributable to (1) aromatic C in ODA rings, (2) aromatic C in PMDA rings and (3) carbonyl C (after P. L. Buchwalter and A. I. Baise, ref. 13)

During metallization of PI, the intensity of these peaks attenuates exponentially with increasing thickness of the metal overlayer due to electron escape length effect. Since these peaks are only few eV's apart they are expected to decrease proportionately if no interfacial reaction takes place. If, howerver, the metal reacts with the pendent oxygen in the PI to form metal-oxygen-carbon complexes or a separate metal oxide phase[1] the carbonyl component in C 1s spectra will exhibit greater change than expected because the change in charge distribution will be accompanied by binding energy change for the relevant C 1s electrons. Indeed our in-situ XPS monitoring of metallization in ultrahigh vacuum environment has shown that when Cr and Ni were deposited on PI the intensity of the carbonyl peak decreased at a higher than expected rate, while the binding energy of the metal 2p electrons exhibited changes consistent with the oxidation of the first few monolayers of metal deposit (Fig. 2).

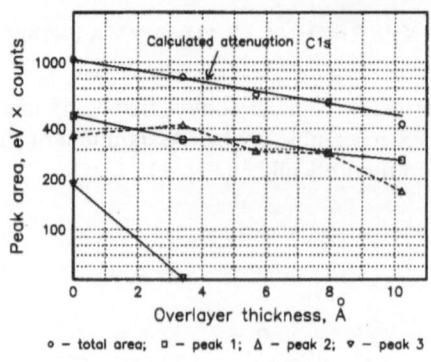

o — total area; □ — peak 1; Δ — peak 2; ∇ — peak 3

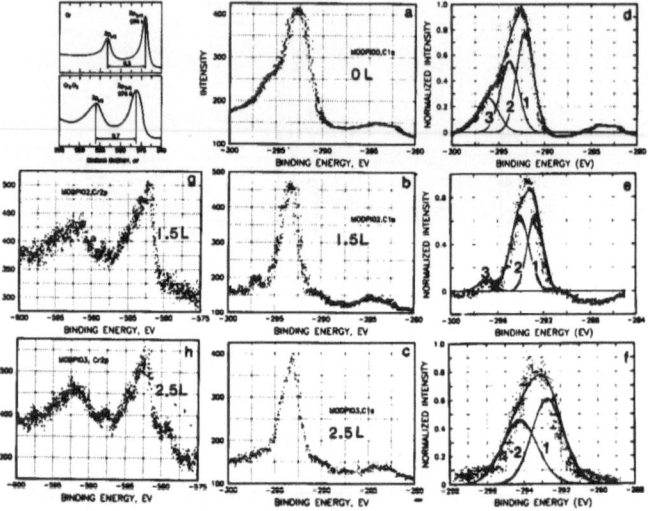

Figure 2. Experimental evidence of interfacial oxidation during Cr metallization of cured polyimide. 2(a) shows the spectral change for C 1s and Cr 2p electrons during the first few monolayers of metallization. The plot in 2(b) compares the observed attenuation of the total and resolved C 1s peak intensities with the calculated values. Note that peak designations in 2(b) correspond to those in Fig. 1.

Cu, on the other hand, showed a completely different behavior. The carbonyl component in the C 1s spectra exhibited proportionate change as other components, and the binding energy of Cu 2p electrons remained characteristic of metallic state during metallization (Fig. 3).

292

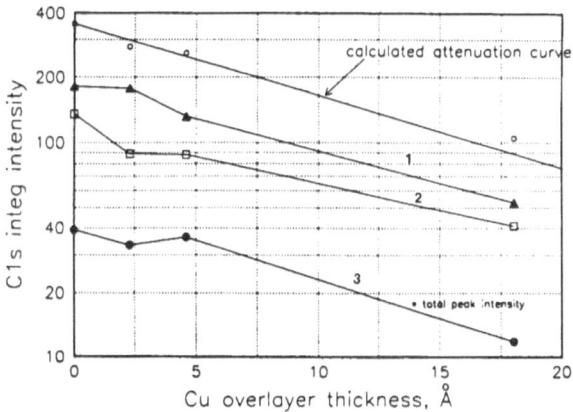

Figure 3. Attenuation of C 1s spectral intensity during Cu metallization of PI: a comparison between the observed and the calculated values. Note that the assignment for resolved peaks 1, 2 and 3 in this plot is the same as in Fig. 2(b)

The interpretation that Cr interacts with pendent oxygen in PI while Cu does not was reaffirmed in three recently published investigations [3] where Cr and Cu were deposited on model compounds similar in structure to the constituents of PMDA-ODA (pyromellitic dianhydride-oxydianiline) polyimide. Observation with the PMDA and ODA model compounds indicated that Cr interacted strongly with PI via carbonyl and less likely with ether groups, whereas Cu only exhibited weak interactions with π orbitals of the carbon rings suggestive of possible charge transfer. Another in-situ XPS study of Al on PI [5] reported that Al reacted with PI via imide group up to 3.5 monolayers (ml) of deposit. The binding energy change of the Al 2p electrons was believed to be consistent with the formation of Al-O-C complex. The study also reported the observation of new components in C 1s and N 1s spectra attributable to the formation of Al-N and Al-C complexes indicative of secondary reactions[2]. Although the experiment was conducted at 300° C, relevant thermodynamic properties can be extracted for comparison via temperature dependence of free energies. It should be noted in passing that the term complex used here is rather tentative since the XPS data do not provide a clear picture of charge transfer.

We have recently investigated the Ti/PI system and found that Ti reacted with PI in a similar way at room temperature. As C 1s spectra in Fig. 4 show, in addition to broadening and reduction in intensity of the carbonyl peak a new lower binding energy component which may be assigned as that of 'carbide' peak[15] was observed after 3 ml Ti coverage. However, the binding energy and Auger parameter of Ti 2p electrons from the first few layers of deposit were found to be 458 and 1294 eV, respectively, which appear to be more consistent with those of TiO_2 than those of lower oxidation states[15] . A detailed study of titanium reaction with PI was recently reported by Ohuchi and Freilich[16] , who observed strong interaction via carbonyl groups at low Ti coverages, followed by the formation of Ti-C bonds as the Ti deposit was increased.

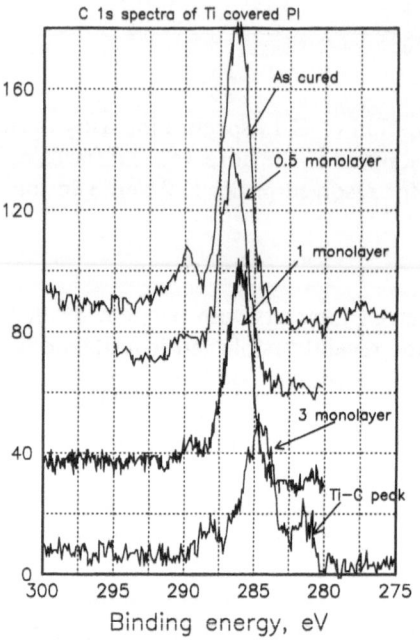

Figure 4. Variation of C 1s spectra for PI with different Ti coverages. Note the broadening and disproportionate reduction of carbonyl peak, follwed by the emergence of low binding energy component (indicated by an arrow), which may be assigned to that of carbide.

According to the proposed model, chemical bonds will be formed between the metal and the pendent oxygen in PI if $| \Delta G_{vap} | > 141 kcal/g - atom(metal)$. Fig. 5 summarizes the results of the aforementioned investigations. As can be seen, the criterion indeed holds true for metals which exhibit evidence of bond formation (Al, Ti, Cr and Ni).

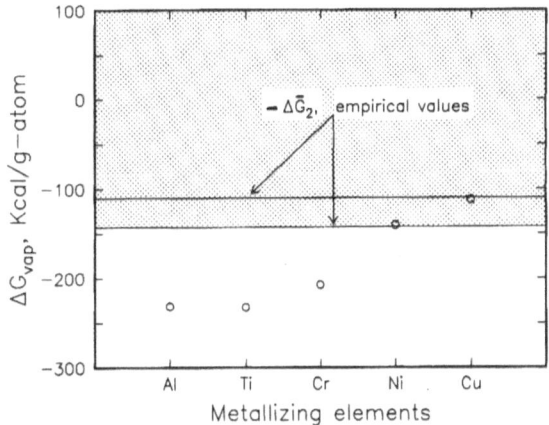

Figure 5. Free energy of vapor phase oxidation for various metallizing elements. Two horizontal lines represent the negative values of effective partial molar free energy of pendent oxygen in PI. As predicted by the proposed model all metals (Al, Ti, Cr, Ni) lying below the lower line exhibit evidence forming bonds with the pendent oxygen in PI.

DISCUSSION

Kinetically, oxidation of metal by pendent oxygen in PI can be envisioned as preceded by electron transfer from the metal atom to the oxygen. As pointed out previously[2], it may result in the formation of metal-oxygen-carbon complex following the scission of N-C bonds, or the precipitation of oxide at the interface after the breakage of carbonyl bonds. In either event the resultant systems are still unstable and secondary reactions may occur at bond breaking sites with outcomes depending on the ambient and the nature of the metal involved. It is not surprising then to observe emergence in XPS spectra of C 1s and N 1s components attributable to metal carbides and nitrides as in the case of metallization with Al and Ti. However, XPS data alone cannot determine unequivocally the end product of the interfacial reaction. For this reason the proposed model has to resort to the concept of faulted (defect-containing) systems.

In terms of transition state theory, the molecular unit which accepts electron or electrons from the metal donor may be regarded as an activated complex whose decomposition leads to two reaction paths shown schematically as follows.

Reported in the literature are numerous quantum mechanical studies on charge transfer and interactions between adatoms and simple metals[17], or between CO molecules and transition metals[18]. No similar treatment has been attempted for metals on polymer surfaces. Qualitatively, we may envision that the charge transfer can take place in different stages of metallization: (1) as metal atoms approach the PI surface electrons may tunnel into unoccupied orbitals of pendent oxygen following resonance broadening of appropriate energy levels[17], or (2) electron transfer may occur after islands of two dimensional metal lattice have formed on the PI surface. When two-dimensional islands are formed, the valence electron levels of individual metal atoms merge into a quasi-continuous conduction band[19] and the tunneling possibility will be enhanced since it is now determined by the position and breadth of the conduction band.

In the proposed model an assumption was made that the barrier height for interfacial oxidation was practically equal for metals studied[2]. This can be true only if the empty orbitals of the pendent oxygen lie within the breadth of the conduction band of the metals in question. In the absence of MO data the assumption should be regarded as that of expediency. The "go or no-go" experimental approach to determination of $\Delta \overline{G}_2$ is not correct because it presupposes the tunneling possibility. A "go only" approach would be more acceptable since the evidence of interfacial reaction establishes a posteriori the charge transfer. A close examination of the reaction paths (a) and (b) indicates that under kinetically favorable conditions the intermediate step (a) will lead to (b). If a similar quasichemical treatment is used to determine the enthalpy difference between systems (a) and (b) one obtains

$$\Delta H_{b-a} = -V_{C-O} + 2V_{N-C} - V_{N=C} + V_{\{MO\}} - V_{M^+O^-}$$

where V_{C-O}, V_{N-C} and $V_{N=C}$ are dissociation energy of C-O, N-C and N=C bonds, $V_{M^+O^-}$ is the interaction potential of the donor-acceptor pair in the complex, and

$V_{\{MO\}}$ the cohesive energy of the metal oxide. If published values of bond energies[12] are substituted into the equation, then

$$\Delta H_{b-a} = 80.8 + V_{[MO]} - V_{M^+O^-}.$$

Without a detailed charge transfer state calculation[20] the value of ΔH_{b-a} can not be ascertained. In view of the large values of $-V_{[MO]}$ (≥ 100 kcal/g-atom) and small values of ΔH for complex formation (≤ 20kcal/g $-$ atom)[20] ΔH_{b-a} is of the order of a few kcal/g-atom, and step (a) is, as a rule, energetically more favorable. The use of the criterion $| \Delta G_{vap} | > 141$ kcal/g-atom (metal) (Fig.5) to predict chemical bond formation is therefore plausible, but one must bear in mind that it does not include other bond forming possibilities.

CONCLUDING REMARKS

A quasichemical approach was used to re-examine the thermodynamic model proposed to predict chemical bond formation between metals and PI during metallization. While new experimental evidence appear to uphold the validity of the model it is shown that the thermodynamic criterion does not constitute the only necessary condition for bond formation. Until a detailed quantum mechanical treatment is available some of the assumptions on which the model is based should be regarded as a mixture of truth and expediency.

ACKNOWLEDGMENT

The authors benefited from discussions with many of their colleagues: Drs. Chin-an Chang, A. B. Fowler, R. Ginsburg, N. D. Lang and B. D. Silverman.

REFERENCES

1. N. J. Chou, D. W. Dong, J. Kim and A. C. Liu, Extended Abstracts, 83-1, Abstr. No. 247, Electrochem. Soc. Meeting, May 8 - 13, 1983, San Francisco, California; see also, J. Electrochem. Soc., 131, 2337 (1984)
2. N. J. Chou and C. H. Tang, J. Vac. Sci. Techn. A2, 751 (1984)
3. P. N. Sanda, J. W. Bartha, B.D. Silverman, P. S. Ho and A. R. Rossi, Mat. Res. Soc. Sym. Proc. 40, 283 (1985);
   J. L. Jordan, P.N. Sanda, J.F. Morar, C.A. Kovac, F.J. Himpsel, and R.A. Pollak, NSLS 1985 Annual Report, Brookhaven National Laboratory, BNL 51947, 98 (1985); and
   J. L. Jordan, P.N. Sanda, J.F. Morar, C.A. Kovac, F.J. Himpsel, and R.A. Pollak, J. Vac. Sci. Technol. A4 1046 (1986)
4. B. D. Silverman, P. N. Sanda, P. S. Ho and A. R. Rossi, J. Polymer Sci., Polym. Chem. Ed. 23, 2857 (1985)
5. J. W. Bartha, P. O. Hahn, F. LeGoues and P. S. Ho, paper presented at American Vacuum Society 31st National Symposium, Reno, Nevada, Dec 4 - 7, 1984
6. O. Kubaschewski and E. LL. Evans, "Metallurgical Thermochemistry", Wiley and Sons, Inc. New York, 1956

7. R. Ginsburg and J. R Susko, private communication, 1983

8. P. J. Flory, "Statistical Mechanics of Chain Molecules", Wiley Interscience, New York, 1969

9. A. I. Kitaigorodsky, "Mixed Crystals", Springer-Verlag, Heidelberg, 1984

10. R. A. Swalin, J. Phys. Chem. Solids, 18, 290, (1961)

11. L. Pauling, "The Nature of Chemical Bonds", Cornell Univ. Press, Ithaca, New York, 1960

12. L. Fieser and M. Fieser, "Organic Chemistry" Reinhold Publishing Corporation, New York, 1960

13. P. L. Buchwalter, A. I. Baise, in "Polyimides: Synthesis, Characterization and Applications", K. L. Mittal, editor, Vol. 1, p.537, Plenum Press, New York, 1984

14. H. J. Leary, Jr. and D. S. Campbell, ACS Sym. Ser. 162, 147 (1981)

15. C. D. Wagner, W. M. Riggs, L. E. Davis, J. F. Moulder and G. E. Mullenberg, "Handbook of X-ray Photoelectron Spectroscopy", Perkin-Elmer Corporation, Eden Prairie, Minnesota, 1979

16. F. S. Ohuchi and S. C. Freilich, J. Vac. Sci. Technol. A4, 1039 (1986)

17. N. D. Lang and A. R. Williams, Phys. Rev. Letters, 37, 212 (1976) and references cited therein

18. See, e.g., S. Bagus and K. Hermann, Solid State Commun., 20, 5(1976)

19. A. B. Fowler, private communication, 1984

20. R. S. Mulliken and W. B. Person, "Molecular Complexes", Wiley- Interscience, New York, 1969

THE MECHANISM OF PLASMA ETCHING OF POLYMERS AND ITS RELEVANCE TO ADHESION

John F. Evans, John G. Newman and James H. Gibson

Department of Chemistry
University of Minnesota
Minneapolis, MN 55455

Experiments combining the use of a quartz crystal microbalance and X-ray photoelectron spectroscopy for the study of
oxygen and water plasma interactions with polymer surfaces
have been carried out on a variety of thin film composites. A
two-step mechanism involving first the oxidation of polymer
moieties, followed by the desorption of these oxidized surface
functionalities, is consistent with the observations made.
The oxidation reaction is confined to the outermost region of
the polymer surface (top 10-20 A of material). The rate of
removal of surface material is found to be considerably higher
for water plasmas under comparable plasma conditions, although
the steady state surface concentration of oxygen containing
functionalities is lower for the water plasmas. The involve-
ment of the plasma sheath in the desorption step is essential
in that little etching is observed unless the polymer is
immersed in the discharge.

INTRODUCTION

Surface treatment and etching of polymers by means of exposure to
ablative plasma discharges are perhaps the most widely used examples of
nonequilibrium plasma processing. Whereas modification of the outermost
regions of a polymer surface is sought in adhesion applications, the
removal of a significant thickness of material is required in the use of
plasma etching in microelectronic device fabrication. In many instances,
the chemical and physical phenomena which contribute to each type of
surface modification are similar for a given material. In fact, they may
be the same phenomena allowed to proceed to different extents, depending
on the desired result.

We describe here the results of a study which indirectly addresses
the mechanism of plasma etching of organic polymer films through the
combined use of a device capable of measuring the rate of removal of
material from the surface of a polymer film exposed to the plasma discharge (a quartz crystal microbalance, QCM), and X-ray photoelectron

spectroscopy (XPS), which allows for chemical analysis of the polymer surface following exposure to the plasma. Both experiments (rate measurement and chemical surface analysis) are conducted on the same sample and without exposure to atmosphere between plasma treatment and surface analysis through the use of a reactor appended to an ultrahigh vacuum (UHV) XPS spectrometer. As such, the XPS spectra provide a measure of the steady state composition of the surface of the polymer film during etching.

Early experiments on spin-coated polystyrene were undertaken to evaluate whether or not the distribution of functionalities on the plasma treated polymer could be affected by the choice of discharge gas (water vapor vs oxygen).[1] The expectation was that a more highly oxidized surface (containing relatively higher coverages of carboxylate groups) might result from oxygen plasma exposure compared to water plasma exposure (where hydroxyl group coverages might be expected to be more predominant). Small differences in the relative concentrations of the various surface functional groups incorporated as a result of exposure were found. However, in comparing water to oxygen plasmas under similar conditions, the more striking observation was the difference in the surface oxygen content following plasma exposure. We observed that polystyrene films which were plasma treated by water discharges to steady state surface composition showed a two to threefold lower surface concentration of oxygen than was found for similar samples exposed to oxygen plasmas (same power, pressure and reactor). Our conclusions were that the same species were likely to be responsible for the surface modification in each case, and that these were lower in number density in the water discharge. We anticipated slower etching of the polystyrene to occur in the water discharge.

Once the QCM rate measurements were commenced,[2] we were initially quite suprised to find that the opposite was true: the water plasma etched the polystyrene films ca 40x more rapidly than the oxygen plasma under the same experimental conditions! Clearly, a more detailed examination of the mechanism of etching in this and related systems was in order. During the course of these investigations, we chose to extend our studies to polymers which contained different oxygen functionalities, both in the backbone of the polymer chain, as well as in the side chains. The effects of aromaticity in the polymer structure, in addition to the inclusion of heteroatom-containing pendant groups, were also examined. We report here the overview of these results. More detailed discussions of these studies will be published elsewhere.[3,4]

EXPERIMENTAL

All experiments were carried out in a reactor appended to either a Perkin Elmer/Physical Electronics Model 548 or Model 5300 electron spectrometer. This reactor (see Fig. 1) was constructed of standard UHV components.[2,5,6] Pumping was achieved through the use of a combination of liquid nitrogen-trapped mechanical and turbomolecular pumps. A base pressure of 10^{-8} Torr could be obtained. Pressure during the discharges was measured using a capacitance manometer. Ultrahigh purity oxygen (99.95%) and distilled, deionized water (subjected to three freeze-thaw cycles before use) were used as the plasma feed gases. The plasmas were ignited and sustained through the use of a Tegal (Novato, CA) Model 100 radio frequency power supply operating at 13.56 MHz. A Tegal L-C matching network was employed to impedance match the output impedance of this power source to the input impedance of the discharge. In all cases, external electrodes were used. The QCM and attendant electronics have been described elsewhere.[6]

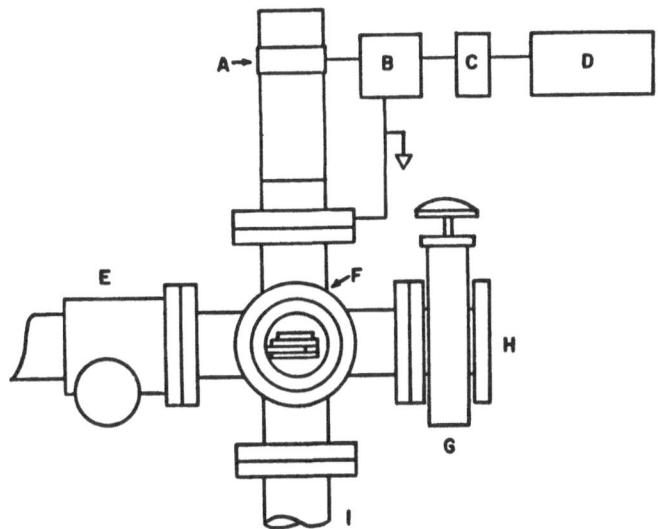

Figure 1. Schematic of the plasma reactor: A - glass to metal transition tube with external copper band electrode; B - impedance matching network; C - directional rf power meter; D - rf power supply; E - linear transport device; F - six way cross; G - gate valve to isolate plasma reactor from UHV analysis chamber; H - flange to UHV chamber; I - to turbo-molecular pump. Gas inlet, pressure transducer and connections to low pressure pump omitted for clarity.

The polymers studied are shown in Table I along with the structure of the repeat unit. Polystyrene (ca 230,000 MW) was obtained from Monomer Polymer Labs (Borden, Inc., Philadelphia, PA). The source of poly(2,6-dimethyl-p-phenylene oxide), poly(vinylmethylketone) and poly(2-vinylpyridine) was Aldrich Chemical Co. (Milwaukee, WI). All other polymers were obtained from Polysciences, Inc. (Warrington, PA). These polymers were spin-coated onto QCM crystals (5 MHz, Sloan Technology, Inc., Santa Barbara, CA) from the appropriate solvent. The silver electrodes on the as-received QCM crystals were replaced with chromium electrodes (deposited by evaporation) onto the QCM faces prior to the spin-coating of the polymer. This procedure was found to be necessary, because the chromium was impervious to plasma oxidation (i.e. chromium passivates), whereas the silver was not.

For XPS analysis a Mg K$_\alpha$ source was used in all cases. High resolution spectra were recorded at 18 or 25 eV pass energy. For angle resolved XPS experiments (Model 5300) a single crystal silicon substrate was used in place of the QCM.

RESULTS AND DISCUSSION

As was mentioned above, the rate of etching of PS was found to be markedly affected by the choice of discharge gas. Figure 2 shows the dependence of etch rate on power for this polymer immersed in water vs oxygen plasmas. Clearly, water is the better etching plasma, even though the surface concentration of oxygen is lower by a factor of 2-3. XPS analysis of the plasma treated surfaces (etched to steady state composition) yields high resolution C(1s) spectra, such as that shown in Fig. 3. Here we see that there are a number of different higher binding energy contributions to this band which are introduced as a result of the plasma

Table I. Repeat Unit Structures and Abbreviations for Polymers Studied

Polyethylene glycol terephthalate
PEGT

Polycarbonate PC

Poly-2,6-dimethyl-p-phenylene oxide
PDPO

Polystyrene PS

Polyethylene oxide PEO

Polymethylmethacrylate   PMMA

Polyvinylmethyl ketone   PVMK

Poly-4-vinylpyridine P-4VP

Poly-2-vinylpyridine P-2VP

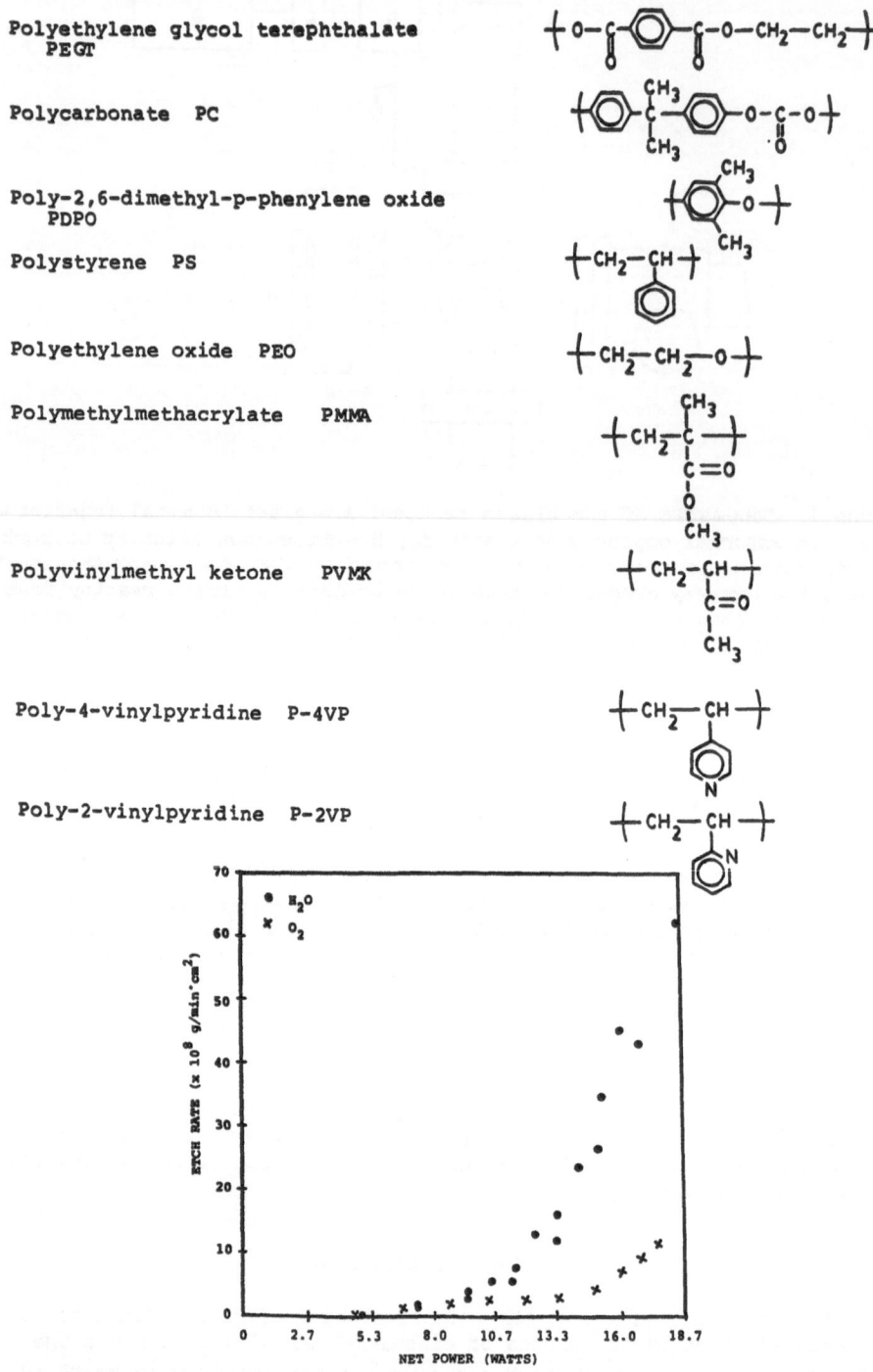

Figure 2. PS etch rate as a function of power. Pressure in all cases was
300 mTorr.

treatment. The vertical lines mark the major components (oxygen containing functionalities). The assignments made below arise from examination
of the C(1s) high resolution XPS spectra of the standard materials listed

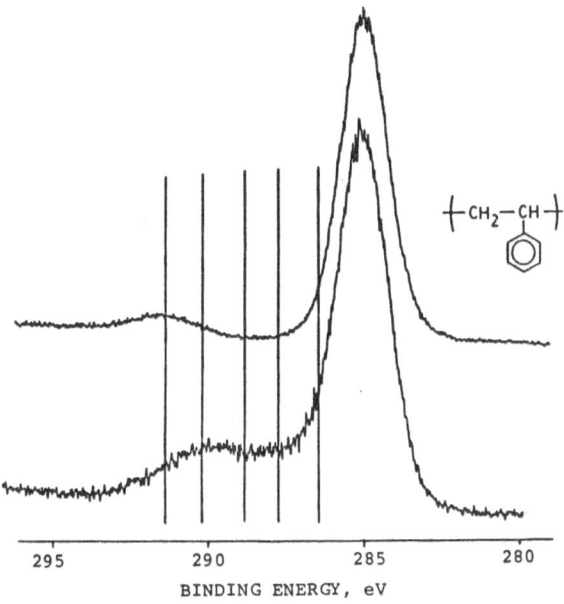

Figure 3. Effect of 5 min, 300 mTorr water plasma on the C(1s) region of the XPS spectrum of PS. Power delivered to the discharge was 13.3 W. Top spectrum is before plasma exposure. Both spectra recorded at 25 eV pass energy.

in Table I, and were found to agree within 0.1 eV of those previously made by Clark, et al.[7,8] From left to right these are: the π-π* shake-up satellite due to the aromatic phenyl pendant group (291.6 eV), carbonate species (290.4 eV), carboxylate species (289.0 eV), carbonyl or O-C-O species (287.9 eV), and ether or hydroxyl species (286.6 eV). From the high resolution spectra taken at various take-off angles (angle of the spectrometer entrance slit relative to the plane of the sample), we see that the depth of incorporation of these various groups is restricted to the top 10-20 A of the polymer film.[9] Typical results are shown in Fig. 4. Curve resolution of these data (e.g. see Fig. 4, ref. 9) show that the π-π* shakeup is still observable at low take-off angle (highest surface sensitivity) and that the distribution of the four types of oxygen containing functionalities is equally populated by each type of species.[9]

The experiments described above were carried out with the PS samples immersed in the glow of the discharge. Experiments were also done under circumstances in which the discharge was confined by a metal mesh (stainless steel grid of 85% transmission) placed at a position ca 4 cm above the sample.[2] Under these conditions, the photons and long-lived reactive species from the discharge will still impinge on the PS surface during the exposure to the discharge, but the ions accelerated across the plasma sheath[10] will undergo many collisions before reaching the surface. As such, effects attributable to low energy (10-30 eV) ion bombardment when the sample is directly immersed in the plasma are extensively diminished when the grid is in place. Under these circumstances, the etch rates observed with the PS coated QCM are extremely low, almost immeasurably low, over the range of conditions shown in Fig. 2. Whereas the etch rate was found to be very low, the XPS spectra acquired after the plasma exposure with the grid in place showed the highest concentration of surface oxygen; the same type of inverse relationship between etch rate and oxygen surface coverage was found in the comparison of oxygen and water plasmas described above.

303

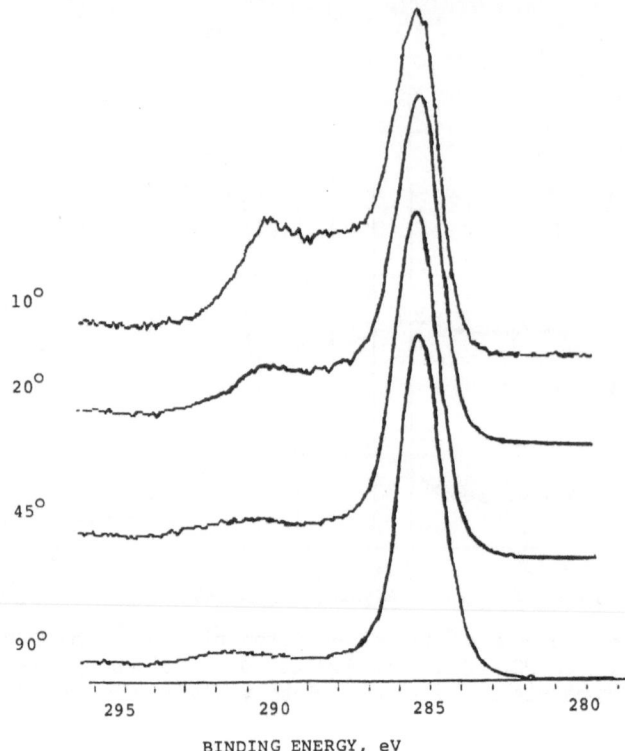

10°

20°

45°

90°

295   290   285   280

BINDING ENERGY, eV

Figure 4. High resolution XPS of the C(1s) region of PS after a water plasma treatment at 14.1 W for 3 min. The spectra are shown for various take-off angles. Note that the lowest angle is the most surface sensitive. Pressure was 300 mTorr during the discharge. Pass energy was set to 18 eV.

A mechanism which is consistent with with these results involves at least two steps, the first of which corresponds to the oxidation of the surface by species other than the ions accelerated across the sheath (thermal chemistry between the surface and the long-lived plasma products), followed by a desorption step which is strongly coupled to the presence of the plasma sheath immediately above the surface. It is likely that some of these desorption pathways depend on nonthermal processes, such as those which can be activated by momentum transfer between the ions accelerated across the sheath and the previously oxidized surface sites.[3] It is very likely that the primary species from the plasma in both types of plasma is atomic oxygen, its concentration being higher in the water discharge than the oxygen discharge under comparable conditions.[11] Furthermore, there may be thermal channels which lead to desorption, as well as those induced by momentum transfer. For instance, atomic hydrogen generated in the water discharge not only contributes to higher atomic oxygen concentrations by virtue of poisoning recombination sites on the walls of the discharge vessel, but also may be involved in the production of water by reaction with the oxidized surface functionalities. We have observed that hydrogen discharges tend to etch oxidized surface organics at higher rates than found for reduced organics.[7,8] For example, there is a 30-fold higher rate for hydrogen plasma etching of PEO vs PS under the same power and pressure conditions[5]. Finally, we find that etching in the cases of water and oxygen plasma interacting with the polymers listed in Table I is very anisotropic, with the maximum rate of etching proceeding normal to the sheath. The etch rate and ion flux reaching the surface of the polymer films are strongly correlated. These results underscore the importance of ion bombardment in inducing desorption from the oxidized surface.

We turn now to an alternative manner of testing the hypothesis of the two step mechanism, with emphasis on testing whether or not oxidation is a prerequisite to etching. That is, concurrent oxidation and desorption arising from ion bombardment might also explain the results thus far discussed. From this point of view we might consider that the precursor to the oxidizing species (water or oxygen) could be adsorbed onto, or sorbed into, the surface, such that the local concentration of these species is quite high. In this model, then, the ions accelerated across the sheath serve to simultaneously excite or rupture bonds in both the oxidant and the polymer surface, such that volatile products (e.g. CO and CO_2) are formed.

Which of these mechanisms is a better description of the etching process may be ascertained by examining the role of polymer structure on the etch rate. Water was chosen as the plasma gas to be used in comparing the etch rate of the polymers listed in Table I. It should be noted that varying degrees of "oxidation" are found in these structures, both in

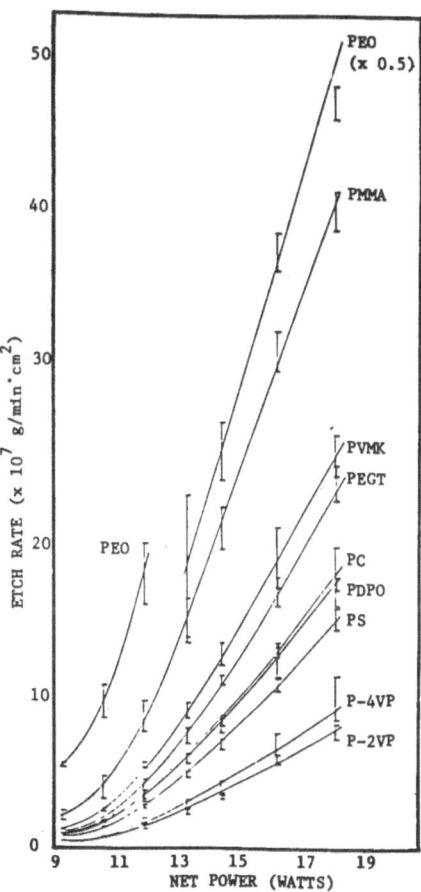

Figure 5. Etch rate as a function of power for the etching of various polymers in a 300 mTorr water plasma. Weight % oxygen of polymers which contain oxygen are: Aliphatics - PEO (36.4), PMMA (32.0) and PVMK (22.9); Aromatics - PEGT (33.3), PDPO (13.3) and PC (18.9).

terms of the backbone and the side chains. Figure 5 summarizes the results of etch rate studied as a function of the power employed to sustain the plasmas.[4,5] It is evident that the polymers which already have oxygen bound as part of their structure are the most rapidly etched at a given power. Within a class of materials (aromatic or nonaromatic) the greater the amount of oxygen present in the repeat unit, the faster the polymer is etched, as indicated in the caption of Fig. 5. In general the aromatic materials are etched more slowly, presumably due to the sluggish kinetics of the ring opening process(es).

In general, we note that those polymers in this set which contain aromatic structure show the presence of carbonate functionality after exposure to the plasma. This can be readily seen in comparing the C(1s) high resolution XPS spectra of plasma treated PS (Fig. 3) and PC (Fig. 6) to the corresponding spectrum of PEO (Fig. 7). For the aliphatic polymer (PEO), there is little evidence for higher oxidation states such as carbonyl and carboxylate, and none for the carbonate species. It appears that the carbonate is generated as a product of the slow action of the plasma in the ring opening reactions required to degrade the aromatic structures into volatile products.

CONCLUSIONS

The experiments described here are consistent with a mechanism for oxygen and water plasma etching of polymer films which is comprised of two successive steps. The first of these involves the oxidation of the surface by long-lived species generated in the plasma (presumably by electron impact processes). This is a necessary, but insufficient, process for the efficient etching of the polymer surface. These thermal processes are followed by a nonthermal momentum transfer-activated desorption step (physical), which may or may not be accompanied by parallel thermal (chemical) channels. These conclusions are in general agreement with those drawn by

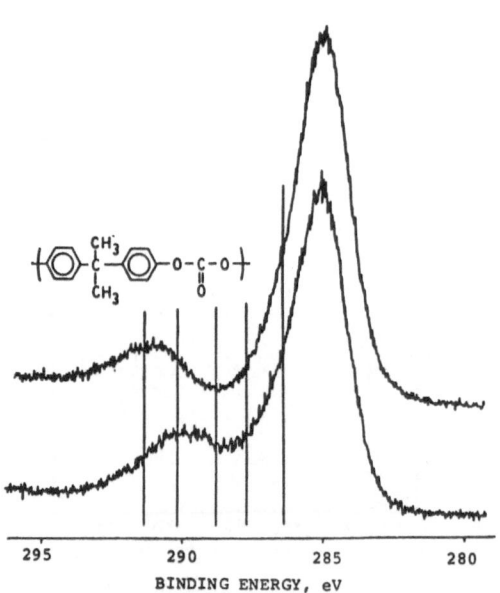

Figure 6. Effects of 5 min, 300 mTorr water plasma on the C(1s) region of the XPS spectrum of PC. Power delivered to the discharge was 13.3 W. Top spectrum is before plasma exposure. Pass energy was set at 25 eV.

Egitto, et al. in studies of polyimide etching in oxygen containing plasmas.[12]

The consequences of these findings for plasma processing of polymers are that if etching is desired, then immersion into the discharge where ion bombardment may be utilized to induce desorption is crucial to obtaining efficient removal of the oxidized polymer. However, in a surface treatment process, where it is desired to increase wettability or adhesion strength at the polymer surface, then treatment in the afterglow will result in maximal surface coverages of the polar surface functionalities which are sought.

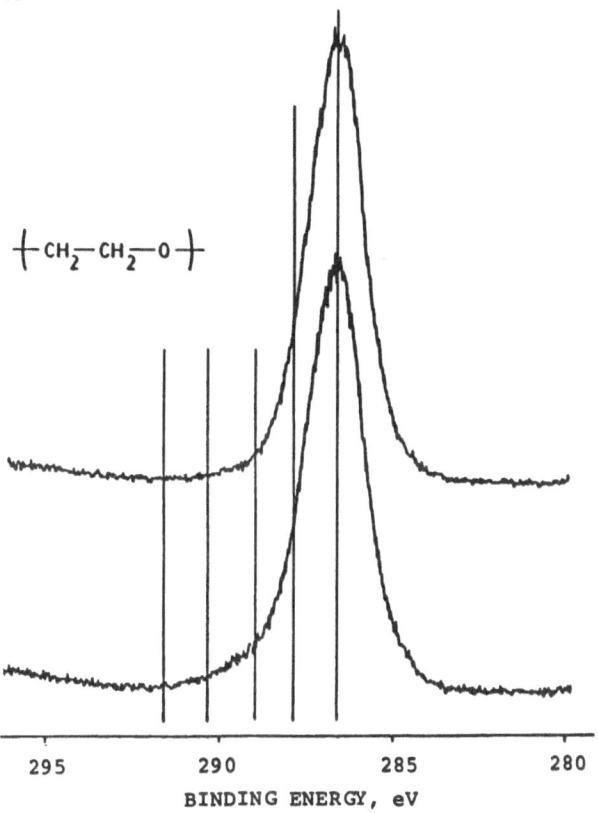

BINDING ENERGY, eV

Figure 7. Effect of 5 min, 300 mTorr water plasma on the C(1s) region of the XPS spectrum of PEO. Power delivered to the discharge was 13.3 W. Top spectrum is before plasma exposure. Pass energy was set at 25 eV.

ACKNOWLEDGEMENTS

Support of this work by the Department of Energy through the Corrosion Center at the University of Minnesota is gratefully acknowledged. The authors also wish to thank J.S. Hammond, J.F. Moulder and T.W. Rusch of the Physical Electronics Division of Perkin-Elmer Corporation for their assistance in the acquisition and fitting of the angle resolved XPS data.

REFERENCES

1. J.E. Fairman, MS Thesis, University of Minnesota, 1983.
2. J.G. Newman, MS Thesis, University of Minnesota, 1984.
3. J.G. Newman, J.E. Fairman and J.F. Evans, to be submitted to J. Vac. Sci. Technol. A.
4. J.H. Gibson and J.F. Evans to be submitted to J. Vac. Sci. Technol. A.
5. J.H. Gibson, MS Thesis, University of Minnesota, 1985.

6.   D.M. Ullevig and J.F. Evans, Anal. Chem., 52, 1467 (1980).
7.   D.T. Clark, B.J. Comarty and A. Dilks, J. Polymer Sci., Polymer Chem. Ed., 16, 3173 (1978).
8.   D.T. Clark and A. Harrison, J. Polymer Sci., Polymer Chem. Ed., 19, 1945 (1981).
9.   J.F. Evans, J.H. Gibson, J.F. Moulder, J.S. Hammond and H. Goretzki, Fres. Z. Anal. Chem., 319, 841 (1984).
10. B. Chapman, "Glow Discharge Processes", Wiley, New York, 1980.
11. R.L. Brown, J. Phys. Chem., 71, 2492 (1967).
12. F.D. Egitto, F. Emmi, R.S. Horwath and V. Vukanovic, J. Vac. Sci. Technol., B(3), 893 (1985).

THERMAL DESORPTION STUDY OF PHYSICAL FORCES AT THE PTFE SURFACE

D.R. Wheeler and S.V. Pepper

National Aeronautics and Space Administration
Lewis Research Center
Cleveland, Ohio 44135

Thermal desorption spectroscopy (TDS) of the polytetrafluoroethylene (PTFE) surface was successfully
employed to study the possible role of physical forces
in the enhancement of metal-PTFE adhesion by radiation.
The thermal desorption spectra were analyzed without
assumptions to yield the activation energy for desorption over a range of xenon coverage from less than 0.1
monolayer to more than 100 monolayers. For multilayer
coverage, the desorption is zero-order with an activation energy equal to the sublimation energy of xenon.
For submonolayer coverages, the order for desorption
from the unirradiated PTFE surface is 0.73 and the
activation energy for desorption is between 3.32 and
3.36 kcal/mol; less than the xenon sublimation energy.
The effect of irradiation is to increase the activation
energy for desorption to as high as 4 kcal/mol at low
coverage.

INTRODUCTION

The adhesion between polytetrafluoroethylene (PTFE) and metals is
of practical importance. Good dielectric and thermal properties make
PTFE useful in electronic applications where it is desirable to increase
the adhesion of metal films to the PTFE. In many mechanical applications
a strong bond is required between a metal substrate and a PTFE film, and
in tribological applications, it is necessary that the transfer film of
PTFE formed during sliding adhere well to the metal counterface. In all
these cases, it would be desirable to increase the normally low adhesion
between metal and PTFE.

It has been found that metal films adhere better to irradiated PTFE
than to virgin PTFE, whether the radiation is ions,[1] electrons[2] or
x-rays.[3] Understanding this improvement could help in understanding
the bond between metals and unirradiated PTFE. The improved adhesion on
irradiated PTFE can be due to one or more of three types of effect;[4]
topographic changes (e.g., mechanical interlocking), chemical interactions, or physical (dispersion) forces. Mechanical forces are unlikely
in the present case, because no transfer of the PTFE to the metal is

observed when the bond fails.[1] Furthermore, the enhanced adhesion is
observed upon irradiation with either x-rays or ions which produce substantially different topographic changes in the surface of the PTFE.[3]
There is some x-ray photoelectron spectroscopic (XPS) evidence for chemical interaction between nickel films and irradiated PTFE.[3] However
the data could not be interpreted unambiguously, and since the PTFE is
damaged by the x-rays used for analysis, the technique is open to question. Physical forces are always present between materials. While much
weaker per atom than chemical forces, they act over the entire contact
area and can thus be an appreciable part of the total bond between macroscopic surfaces. The question is whether the physical forces are
increased by irradiation and whether they can account for the increased
adhesion. To date, there is no evidence on either of these points.

Inert gas adsorption is commonly used as a probe of physical interactions on surfaces. One of the usual techniques is thermal desorption
spectroscopy (TDS).[5] To the best of our knowledge, it has not been used
on polymer surfaces. The purpose of the present study was twofold;
first, to develop the TDS technique for xenon desorption from planar
polymer surfaces and second to use the technique to determine the change
in physical forces at the PTFE surface upon irradiation with electrons.

EXPERIMENT

Thermal desorption experiments were performed in an ultrahigh vacuum
chamber fitted as shown schematically in Fig. 1. The chamber was pumped
with a 150 L/s turbopump and a titanium sublimation pump. A nude ion
gauge, out of the line of sight of the specimen, was used to monitor the
background pressure in the chamber and also to measure the pressure rise
during the TDS experiments. The pressure in the chamber, before beginning a TDS experiment, was less than 8×10^{-11} torr. A cryostat regulated
the sample temperature. Xenon was directed onto the specimen through a
microcapillary array from a calibrated ballast volume. The chamber also
incorporated an electron gun for irradiation of the sample and apparatus
(not shown in Fig. 1.) for x-ray photoelectron spectroscopy (XPS) of the
specimen surface.

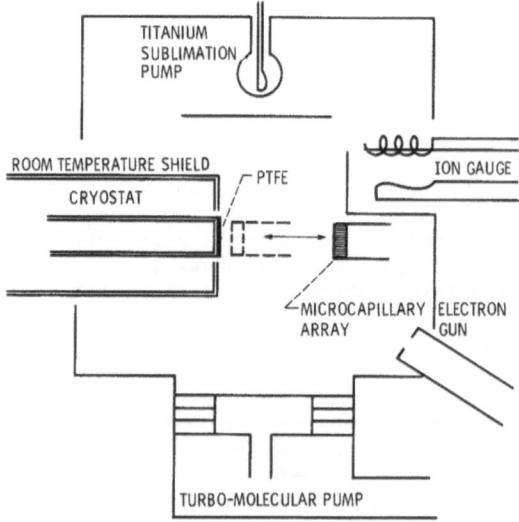

Figure 1. Schematic representation of the thermal desorption spectroscopy
apparatus.

A TDS experiment consisted of exposing the specimen at 30 K to a given dose of xenon. The temperature of the specimen was then increased linearly to 120 K at a rate of 0.1 K/s, and both the specimen temperature and the pressure in the chamber were recorded. Under the rapid pumping speed conditions of this experiment, the pressure rise is proportional to the desorption rate of xenon from the PTFE surface.[6] The resulting curve of pressure versus temperature is the thermal desorption spectrum. The pressure, temperature pairs were digitized and stored in a micro-computer. Some details of the specimen preparation, the dosing procedure and the cryostat are presented here.

Specimen Preparation

The PTFE surfaces used for TDS were prepared by spinning a commer-cial dispersion of PTFE onto copper caps which could be screwed onto the cryostat used to regulate the sample temperature. Before use, the coated caps were annealed in vacuum to between 410 and 415 °C for 15 min to drive off the carrier and sinter the PTFE film. Scanning electron micro-scopy of the specimens showed the filamentary structure typical of spec-imens prepared in this way.[7] They undoubtedly had a surface area greater than the geometric area.

Irradiated PTFE samples were prepared in the same way. They were then exposed to a 1 kV electron beam. The beam was rastered over the surface of the specimen. The effective current density was 2.2 $\mu A/cm^2$, and the sample was irradiated for 180 min. Irradiation was performed in the vacuum system in which the TDS experiments were performed, and the sample was not exposed to air between irradiation and TDS analysis.

Samples prepared for TDS were examined with XPS before and after irradiation. Because the x-ray irradiation used to obtain the XPS spec-tra causes damage to the PTFE,[8] samples that were used for TDS experi-ments were not analyzed by XPS until after the TDS experiments were complete. The XPS spectrum of the unirradiated PTFE specimens was char-acteristic of clean, undamaged PTFE. No impurities could be detected, and there was no trace of copper or oxygen lines from the copper sub-strate. The change upon irradiation was the same as had been observed previously.[8]

Doser

The apparatus used to dose the specimen with xenon consisted of a microcapillary array connected through a bakable valve to an ion pumped ballast volume. Pressure in the ballast volume was monitored with an ion gauge. The amount of xenon to which the specimen was exposed was calculated from the drop in pressure in the known ballast volume during dosing. From the area of the specimen exposed to xenon and the diameter of the xenon atom, the number of monolayers which would result from a particular amount of xenon on a geometrically smooth surface was calcu-lated. In the data to follow, that number of monolayers is referred to as the dose.

The actual coverage of xenon on the PTFE after dosing depended on the true surface area of the specimen and the sticking coefficient of the xenon as well as on the dose. It has already been noted that the true surface area was greater than the apparent area. Furthermore, dur-ing dosing, there was a slight pressure rise in the main chamber which indicated that the sticking coefficient was less than one. However, the pressure rise and the time it lasted were both more than an order of magnitude less than the pressure increase and time of a TDS. Thus, the sticking coefficient must be greater than 0.99. In any case, the actual xenon coverage was proportional to the area under a TDS curve. In all the data below, the coverage of xenon is calculated from the area under the TDS spectrum. The unit is torr-sec. As a point of reference, the coverage produced by a xenon dose of one monolayer was 6.1×10^{-7} torr-sec. That coverage is indicated as 1 mL in the data below.

In principle, the thickness of the xenon layer on the PTFE could be measured by the attenuation of the XPS lines from the PTFE or by angle resolved XPS. Both methods were tried, but it was found that the radiation used for XPS was sufficient to cause progressive damage to the PTFE during the analysis. This produced variable xenon adsorption. Furthermore, the radiation damaged PTFE was not stable during the rather long times required for XPS analysis. Finally, the inelastic mean free path of low energy electrons in xenon, which is required for the analysis, is not well known. As a result, the true monolayer coverage of xenon could not be found precisely. It was found that a dose of 1 mL produced a coverage of between 0.2 and 0.5 mL as determined by XPS.

Cryostat

The sample was mounted on a continuous flow liquid He refrigerator. The refrigerator incorporated a heater for temperature control and a thermocouple for temperature measurement. Because the thermocouple was separated from the sample surface, the actual sample temperature could differ by several tenths of a degree from the indicated temperature. Furthermore, the discrepancy varied throughout a day of operation. The temperature of the peak in the TDS from a standard dose of xenon was used to correct for this temperature drift. Thus, temperatures reported here are consistent to within 0.2 K, but the absolute temperature error could be as large as 0.5 K.

It was found that readsorption of the xenon from the specimen onto other cold areas of the cryostat produced serious artifacts in the TDS. This was a particular problem at the low ramp rate of 0.1 K/sec used here. Readsorption was controlled by adding a room temperature shield to the cryostat as shown in Fig. 1. Clearance between the shield and the cold specimen is minimal, and the amount of desorbed xenon reaching other parts of the cryostat is not detectable in the spectrum. Furthermore, to assure that the cryostat and specimen are cleared of xenon, the specimen temperature is raised to 220 K between runs.

RESULTS

Thermal desorption spectra were obtained for a wide range of xenon doses on virgin and irradiated PTFE. We will consider the general features of the spectra first and then turn to an analysis of the thermodynamic realtionship between the pressure, temperature and xenon coverage represented by each point on a spectrum.

General Features

The TDS curves fell naturally into two families; those for multilayer initial coverage and those for submonolayer initial coverage. Each family will be considered separately.

312

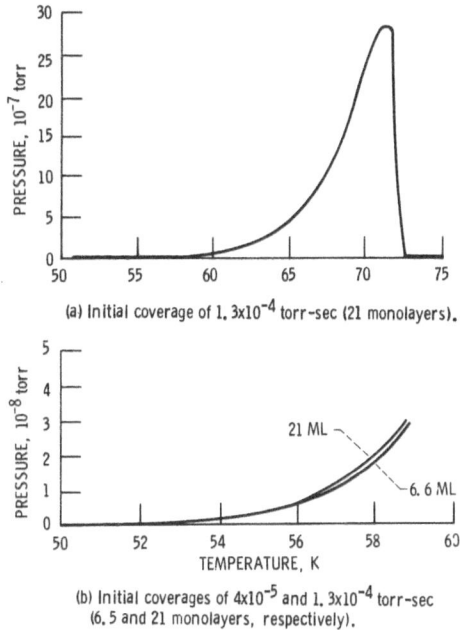

(a) Initial coverage of 1.3×10^{-4} torr-sec (21 monolayers).

(b) Initial coverages of 4×10^{-5} and 1.3×10^{-4} torr-sec (6.5 and 21 monolayers, respectively).

Figure 2. Thermal desorption spectra of xenon on PTFE.

Multilayer coverage. The general shape of the TDS for initial coverage greater than 1.3×10^{-4} torr-sec (about 21 mL) is illustrated by the spectrum of Fig. 2(a). There was a characteristic rapid decrease in pressure at high temperature. Figure 2(b) shows the low temperature (high coverage) region of two such spectra. The pressures and therefore the desorption rates in this temperature range were the same for a wide range of coverage. The variation in temperature of the TDS peak maximum is shown in Fig. 3. In the region of multilayer coverage, the peak temperature increased with the coverage. All of these observations are consistent with zero-order desorption kinetics.[9] In addition, Fig. 3 shows that the TDS peak temperature was independent of substrate irradiation, in the multilayer coverage region.

Low coverage. Spectra obtained at submonolayer coverages are shown in Figs. 4(a) and (b). They did not exhibit shapes characteristic of any simple desorption model. In particular, they did not appear to be produced by first-order desorption. This is confirmed by the variation of the temperature of the peak maximum shown in Fig. 3. At submonolayer coverage, the peak temperature decreased with increasing coverage. In the case of simple, first-order desorption, the temperature of the peak would remain constant. The varying peak temperature could be attributed either to a distribution of binding energy sites on the surface or to fractional order desorption produced by lateral interactions between xenon atoms.[9]

Comparison of the high temperature sides of the two peaks in Fig. 4(a) suggests that irradiation produced some increase in the number of high energy binding sites. The effect of radiation on the substrate is even more apparent in Fig. 4(b). The TDS peak on the irradiated

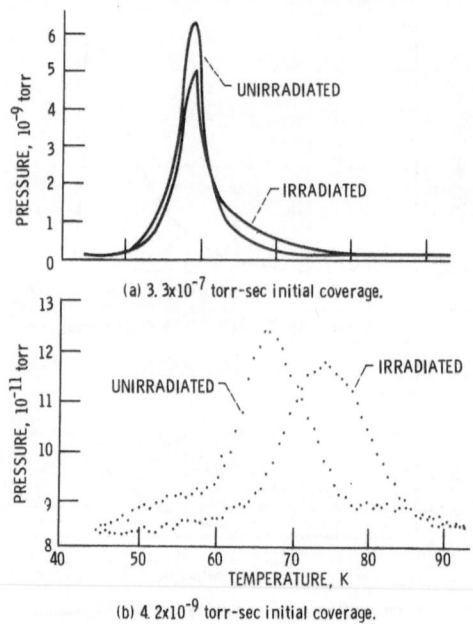

(a) 3.3×10^{-7} torr-sec initial coverage.

(b) 4.2×10^{-9} torr-sec initial coverage.

Figure 3. Dependence of TDS peak maximum on initial coverage.

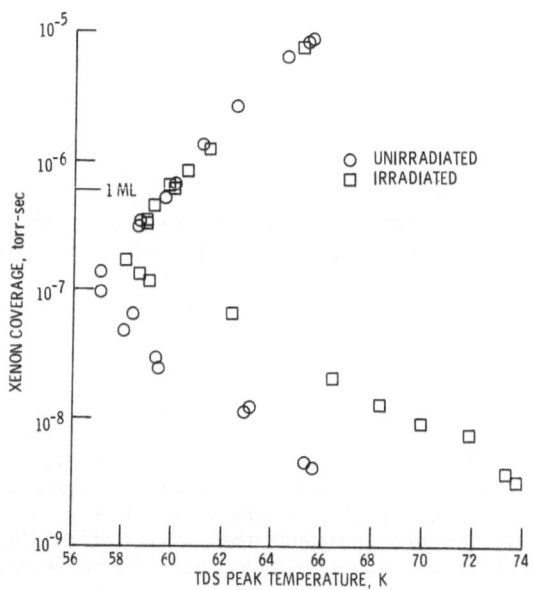

Figure 4. Thermal desorption spectra for irradiated and unirradiated PTFE substrates.

specimen was almost 10 K higher than the peak on the unirradiated specimen. The effect of radiation on the substrate is also evident in the submonolayer coverage region of Fig. 3. Below 2×10^{-7} torr-sec, the maximum pressure always occurred at higher temperatures for the irradiated PTFE than for the unirradiated PTFE.

Thermodynamic Analysis

Because we wish to determine the strength of the xenon-PTFE inter-action, we are particularly interested in the low coverage data. Had the desorption been first-order, a simple Redhead analysis would yield the desorption energy.[6] However, in the present case we resort to analysis based explicitly on the Arrhenius equation:[5]

$$r = \nu C^n \exp (-E/RT), \tag{1}$$

where r is the desorption rate, ν is the "frequency factor," C the coverage, n the order of desorption, E the activation energy for desorption, R the gas constant and T the temperature. Most analyses of TDS assume that ν is independent of coverage, and further, that it has a value of 10^{13} s^{-1}. We do not make these assumptions but proceed as follows.

The pressure at each point in a TDS is proportional to the desorp-tion rate of xenon from the surface. This rate depends on both the tem-perature and the coverage at that temperature. The temperature can be determined directly. The coverage is proportional to the integral under the TDS curve from the temperature of interest to the highest tempera-ture. On each TDS, temperature, pressure and coverage values were obtained at half degree intervals up to the peak of the curve. The pressure and coverage at one temperature were measured on TDS curves for a variety of initial doses. These were combined to produce a plot of pressure versus coverage at that temperature.

A typical plot for virgin PTFE is shown in Fig. 5. The logarithmic coverage scale allows presentation of a wide range of coverage but has no other significance. At very low coverage, the desorption rate must, of course, approach zero. As the coverage increases, so does the desorp-tion rate, but the increase is not linear as it would be for first-order desorption. At a coverage of 9×10^{-6} torr-sec (15 mL), the desorption rate drops and then remains constant for higher coverage. Coverage independent desorption rate is the signature of zero-order desorption kinetics.

Curves of pressure versus coverage were constructed every half-degree for temperatures between 50 and 60 K. Below 50 K, the pressure rise was not large enough to be measured reliably, while there were too

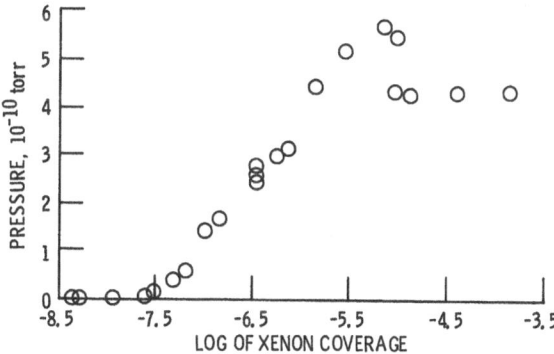

Figure 5. Xenon pressure as a function of coverage at 52 K, unirradiated PTFE surface.

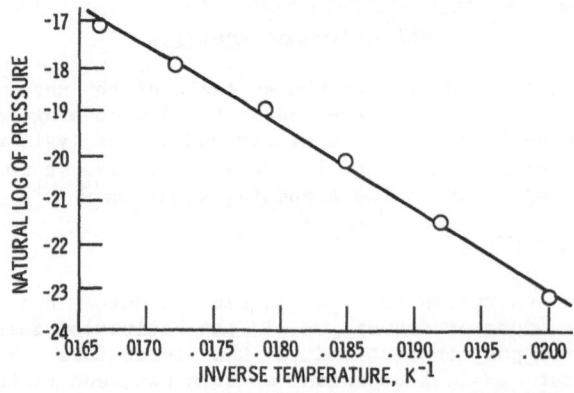

Figure 6. Arrhenius plot: Natural log of the pressure vs inverse temperature for xenon coverage of 1×10^{-5} torr-sec on unirradiated PTFE.

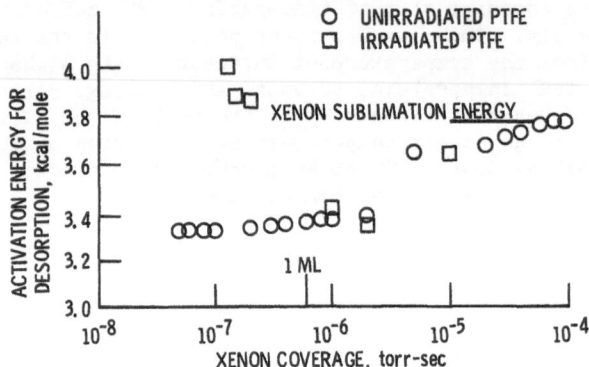

Figure 7. Activation energy for desorption as a function of xenon coverage
in the temperature range of 50 K to 60 K.

few data points above 60 K. Reading the pressure at a particular coverage from each curve in this series yields the pressure as a function of
temperature for that coverage. An Arrhenius plot to the natural log of
the pressure versus the inverse of the temperature can then be made.
Figure 6 is a typical Arrhenius plot. The linearity of these plots is
confirmation that Eq. (1) holds with a single energy in the range of
temperature and coverage analyzed. The slope of the best line through
the data is $-E/R$.

Arrhenius plots were constructed for a range of coverages, and the
desorption energies determined. The result is shown in Fig. 7. For
reference, the sublimation of xenon is also shown.[10] As can be seen,
the desorption energy at high coverage was in reasonable agreement with
the sublimation energy. As the coverage decreased, the desorption energy
decreased until it reached a value of 3.34 ± 0.03 kcal/mol below 1 monolayer coverage.

The same analysis was performed on the TDS from the irradiated PTFE
surface. Because the surface was not stable, fewer TDS curves could be
acquired. However, some desorption energies could be extracted in the
coverage range from 10^{-7} to 10^{-5} torr-sec. These are also shown in
Fig. 7. It can be seen that the energies were the same as those for

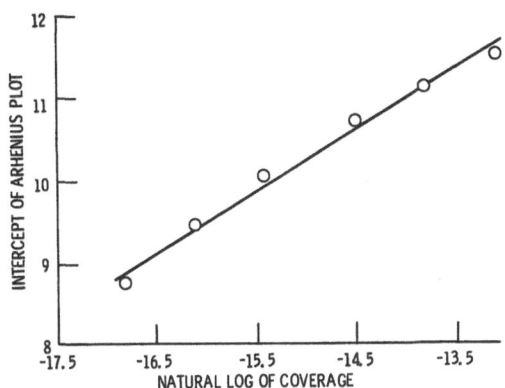

Figure 8. Intercept of Arrhenius plots vs natural log of the xenon
coverage, unirradiated PTFE

virgin PTFE above 1 monolayer. At low coverage, however, the desorption
energy from irradiated PTFE was larger than that from virgin PTFE and
exceeded even the sublimation energy of xenon.

A plot of the intercepts of the Arrhenius plots verses the natural-
log of the coverage is shown in Fig. 8. According to Eq. (1), the slope
of this line is the order n of desorption which is 0.73 in this case.
The linearity of the plot confirms that the order and the preexponential,
ν, are constants. There were insufficient data for a similar analysis
on the irradiated surface.

DISCUSSION

To the best of our knowledge, this is the first report of the
application of thermal desorption spectroscopy to a polymer surface.
Therefore, the first part of the discussion will address the experimental technique, itself. Next, conclusions will be drawn on the nature of
the adsorption of xenon on the virgin PTFE surface. Finally, the effect
of irradiation of PTFE on xenon adsorption and its significance for
adhesion will be discussed.

Experimental Technique

The experiment reported here differs in ramp speed from most
reported flash desorption experiments.[5,6,9] The low ramp speed was
required to assure temperature uniformity in the cryostat used. The
uniformity is demonstrated by the sharpness of the TDS curves. The sharp
drop on the high temperature side of the multilayer peak in Fig. 2(a)
demonstrates the response of which the system is capable. It also shows
that the specimen temperature was uniform.

Because the ramp speed in this experiment was so low, the possibility of readsorption on the sample during the ramp must be considered.[5]
In fact, it was found that the Arrhenius plots displayed a slight depar-
ture from linearity at pressures above 5×10^{-8} torr. This was attributed
to effects of readsorption at high desorption rate, and pressures above
this value were, therefore, excluded from the analysis. The linearity
of the Arrhenius plots is one evidence for the validity of the data. The
fact that the desorption energy at high coverage was nearly equal to the
sublimation energy of xenon is additional evidence.

The nondestructive nature and surface sensitivity of the TDS give it unique capabilities for the analysis of surfaces such as that of PTFE. It was found that the TDS spectrum was sensitive to radiation-produced changes in the surface that could not be detected by XPS. Indeed, the unstable nature of the irradiated surface was only evident from the variability of the xenon desorption spectra.

Effect of Irradiation

For multilayer coverage, the TDS curves are characterized by peak-temperatures that increase with coverage, identical low temperature behavior for all coverages and sharp high temperature edges. These are all features of zero-order desorption. This is confirmed by plots of pressure versus coverage which show that the pressure (and hence the desorption rate) were independent of coverage above 15 monolayers. Zero-order desorption kinetics can be produced by a variety of processes,[11] the most obvious being sublimation from bulk xenon. In the present case, the agreement between the desorption energy at high coverage and the sublimation energy of xenon confirms the obvious interpretation.

For submonolayer coverage, the energy of desorption was less than the sublimation energy of xenon. Clearly, the xenon-PTFE interaction energy was less than the xenon-xenon interaction energy. The low energy of interaction with the substrate and the fractional order of the desorption make it likely that lateral interactions were affecting the desorption energy.[9,11] It seems likely that xenon islands were present on the PTFE surface. In that case, the desorption energy can only be an upper limit to the single-atom, xenon-PTFE interaction energy.

The plot of pressure versus coverage in Fig. 5 clearly shows an abrupt change in desorption rate in the intermediate range of coverage between one and several monolayers. Without further evidence, it is not possible to interpret this transition in detail, but it is certainly related to the change from desorption influenced by the substrate to sublimation of bulk xenon.

Adsorption/Desorption on Virgin PTFE

At coverage less than 1 monolayer, the desorption energy of xenon from irradiated PTFE was 3.85 to 4.0 kcal/mol, whereas it was only 3.32 to 3.36 kcal/mol from virgin PTFE in the same coverage range. Furthermore, the TDS peaks were 10 K higher for the irradiated substrate. It seems clear that the interaction between xenon and irradiated PTFE was stronger than between xenon and virgin PTFE. It is unlikely that chemical bonding was involved between xenon and PTFE, so the increased interaction is attributed to increased dispersion forces.

What changes can irradiation produce in PTFE that would enhance these dispersion forces? The changes in the C(1s) XPS spectrum of irradiated PTFE were previously interpreted as evidence for cross-linking in the surface region of the polymer.[8] The density of the surface region could be increased by crosslinking. The increased density would, in turn, lead to increased dispersion forces.[12] However, the instability of the irradiated PTFE surface when probed by TDS suggests that the change is not entirely structural. Irradiation is known to produce trapped, long-lived radicals in PTFE.[13] It seems probable that the optical polarizability of these radicals would be

318

different from that of PTFE. The presence of radicals then could well affect the dielectric properties and hence the dispersion forces at the PTFE surface.

Whatever the cause of the increased dispersion forces, they can certainly contribute to the adhesion of thin metallic films to PTFE. Since there also seem to be chemical differences in the effect of irradiation on the adhesion of different metals.[2] any explanation of the effect of radiation on adhesion must take into account both chemical and physical forces.

CONCLUSION

Thermal desorption spectroscopy of xenon from irradiated and virgin PTFE has been successfully performed. Analysis of the TDS curves and the desorption kinetics in the 50 to 60 K range showed the following

1. Desorption from multilayers of Xe on either surface proceeded by zero-order kinetics with an activation energy equal to the Xe sublimation energy.

2. At a coverage below 3 to 7 true monolayers the desorption behavior changed abruptly from the multilayer case and was different on irradiated and unirradiated PTFE substrates.

3. For submonolayer coverages on irradiated PTFE the activation energy was 3.85 to 4.0 kcal/mole, which was the highest observed.

4. For submonolayer coverages on unirradiated PTFE the desorption energy was 3.34 ± 0.03 kcal/mole (less than the sublimation energy of Xe), and the desorption was fractional order, indicating significant lateral interactions.

It can be concluded that the Xe-Xe interaction was stronger than the Xe-PTFE interaction on unirradiated PTFE. Irradiating PTFE increased the dispersion forces between Xe and PTFE.

REFERENCES

1. J.E.E. Baglin, G.J. Clark, and J. Bottiger, in "Thin Films and Interfaces II," J.E.E. Baglin, et al., eds., pp. 179-188, North Holland, New York, 1984.

2. Y. Yamada, D.R. Wheeler, and D.H. Buckley, NASA TP-2360 (1984).

3. D.R. Wheeler and S.V. Pepper, J. Vac. Sci. Technol., 20, 442-443 (1982).

4. K.L. Mittal, Polymer Eng. Sci., 17, 467-473, (1977).

5. D.A. King, Surf. Sci., 47, 384-402 (1975).

6. P.A. Redhead, Vacuum, 12, 203-211 (1962).

7. D.C. Bassett and R. Davitt, Polymer, 15, 721-728 (1974).

8. D.R. Wheeler and S.V. Pepper, J. Vac. Sci. Technol., 20, 226-232 (1982).

9. R.G. Jones and D.L. Perry, Surf. Sci., 82, 540–548 (1979).

10. C.W. Leming and G.L. Pollack, Phys. Rev., B,2, 3323–3330 (1970).

11. R. Opilla and R. Gomer, Surf. Sci., 112, 1–22 (1981).

12. K. Hara and H. Schonhorn, J. Adhesion, 2, 100–105 (1970).

13. R.E. Florin and J. Wall, J. Res. Nat. Bur. Stand. Sect A, 65A, 375–387 (1961).

PART IV.   MONOLAYERS AND LANGMUIR-BLODGETT FILMS: RELEVANCE TO
MICROELECTRONICS

PROPERTIES AND APPLICATIONS OF WELL-TAILORED ORGANIZED ASSEMBLIES

G. J. Kovacs and P. S. Vincett

Xerox Research Centre of Canada
2660 Speakman Drive
Mississauga, Ontario
Canada L5K 2L1

In the past several years, the enormous possibilities of the Langmuir-Blodgett (LB) technique for producing ordered supermolecular architectures, with extensive scientific and technological applications, have begun to be appreciated and in some cases demonstrated. In this paper, we review some key developments in the fabrication and utilization of LB films in the last decade or so; we also discuss in some detail various possible applications, particularly in electrical areas such as molecular electronic devices, electroluminescent light sources, and even high temperature excitonic superconductors. Recent LB work at Xerox Research Centre of Canada is described; this includes structural investigations of novel phthalocyanine monolayers and their application to the study of Surface Enhanced Raman Scattering, as well as fabrication of reaction center protein monolayers used to study charge transfer processes in photosynthesis.

1. INTRODUCTION

A Langmuir film[1] may be defined as an ordered insoluble monomolecular layer at a liquid-gas interface, generally introduced by a spreading solvent. The interface is usually kept relatively large when the solution is spread so that solvent evaporation results in formation of a two-dimensional 'gaseous' film . On compression with barriers, this film forms into a quasi-liquid or quasi-solid, accompanied by a dramatic increase in surface pressure. The material must generally be amphipathic, i.e., it must have both a hydrophobic and a hydrophilic end, which orient the molecules into a precise array on compression. A Langmuir-Blodgett (LB) film is one or more Langmuir film(s) transferred to a solid substrate[2,3]. To effect transfer, the substrate is dipped, generally vertically, through the water surface. The film pressure must be maintained constant by a compensating motion of the barriers during deposition. With appropriate materials, one monolayer is transferred on each excursion through the water surface; thus a film of remarkable perfection can be built up monolayer by monolayer. The structure of the transferred film is generally such that adjacent layers have, alternately, their hydrophilic head groups juxtaposed and then their hydrophobic tails juxtaposed.

In the past few years a surge of interest has developed in the properties and uses of LB films[4-8]. A rapidly increasing number of laboratories has become involved in the study of these ultra-thin molecularly ordered layers, with applications ranging from electroluminescent devices[9] to electron beam resists[10]. In this paper, we will attempt to explain the reasons for the present excitement about the possibilities for both scientific and technological applications of LB films: Section 2 reviews the key advances of the last decade or so which have been most responsible for the present promising situation, and Section 4 describes some possible (mainly electrical) device applications of LB films, both as conventional insulators and as electrically active layers. In addition, Section 3 contains an account of some of the current LB work at Xerox Research Centre of Canada, including studies of electric field effects on electron transfer in photosynthesis and of Surface Enhanced Raman Scattering (SERS) from ordered monolayers. Finally, Section 5 contains an assessment of the present LB situation (as of mid-1985) and of possible future developments.

2. SOME KEY ENABLING ADVANCES FOR APPLICATIONS OF LB FILMS

LB films have at least three distinct advantages over other deposition techniques: (i) they are very thin yet very uniform, and stable well-defined films are possible, (ii) film thickness can be accurately controlled, and (iii) extremely precise supermolecular 'organizates' can be fabricated which are difficult or impossible to obtain by other means, e.g. (Section 4) proposed structures for high temperature excitonic superconductivity[4]. These advantages were the predominant motivating factors for the recent major developments in LB film science which are described below.

After its discovery[2,3], the LB technique remained primarily a laboratory curiosity for almost 30 years before many people began to use the technique seriously. The modern era of LB film science was ushered in by the work of Kuhn's group in the 1960's. They basically used very careful procedures and techniques with regard to such things as water purification and ionic content, and elimination of mechanical vibration etc., to overcome the irreproducibilities which had plagued earlier work. Much of their initial work used LB films in distance-keeping applications. For example, donor and acceptor dye molecules were incorporated into fatty acid layers in order to keep the two types of dye molecules at well defined distances from each other. The distance-dependence of energy transfer from donor to acceptor could then be monitored by measuring the distance-dependence of the fluorescence quenching. The measurements[11] were found to be in excellent agreement with the predictions of the classical Perrin-Förster energy transfer mechanism. Predictions of the theory for energy transfer to a primary and subsequent secondary acceptor were also elegantly demonstrated[12]. Another distance-keeping application involved building a staircase structure on a silver mirror from fatty acid molecules, whereby each successive layer is stepped from the preceding one to produce a wedge structure. The wedge was then covered with a layer of dye molecules which were excited by normally incident monochromatic light. Strong fluorescence was then observed from the dye molecules at anti-node positions of the reflected monochromatic exciting light[11]. While these initial distance-keeping applications represented a major development in LB film science, they were relatively insensitive to minor imperfections such as pinholes in the films.

The next major development, which demonstrated the possibility for electrical applications of LB films, was the production of extremely high quality films of fatty acids where pinholes were apparently entirely absent. The high quality of these films was demonstrated by measurements of tunneling currents through the fatty acid insulating layers in MIM sandwich structures[13,14]. According to theory, for an MIM structure the tunneling current should be exponentially dependent on the thickness of the insulating layer. The preexponential factor and slope on a semi-log plot are related to the metal electrode work functions and to the electron affinity of the insulator. The exponential dependence was indeed observed and the slope and pre-exponential factors were found to be in excellent agreement with the predicted values for various metal electrodes[13,14]. Since one pinhole short-circuit would increase the measured tunneling current by several orders of magnitude, the excellent agreement between theory and experiment is evidence that conducting imperfections in the films must be entirely absent. These experiments demonstrated the high electrical quality possible in LB films; however simple insulating layers of highly aliphatic materials such as fatty acids have a limited range of electrical applications.

Up until the late 1970's work on LB films had been devoted almost exclusively to long chain materials. The much more interesting electrical properties of extended π-electron systems (if present at all in the materials studied) were almost completely diluted by large aliphatic regions. This situation motivated the work of a group in England, led by one of us (PSV), which investigated the possibility of forming LB films from anthracene derivatives. They found that, under appropriate conditions, well ordered LB multilayers could be made from anthracene substituted very lightly with only a C_4H_9 tail and a $(CH_2)_2COOH$ acid group[15]. Space-charge-limited currents and recombination electroluminescence (EL) were among the interesting electrical properties which were obtained from these largely undiluted anthracene layers. For the C4 derivative in an MIM structure between Al and Au electrodes, high fields produced enough double injection to give a very steep rise in the I-V curve, beyond which recombination EL was observed, visible under normal room illumination[16]. The I-V characteristics associated with the EL are very similar to those producing EL in evaporated anthracene films. A remarkable 10^8-fold anisotropy of conductivity was also observed, the in-plane value being many orders of magnitude higher than that of anthracene crystals, as expected from the face-to-face in-plane structure of the π-systems.

This work, extending LB materials to lightly substituted aromatics, also demonstrated that the exact molecular arrangement of complex structures is predictable from careful consideration of molecular models and of the deposition conditions[17]. Space-filling molecular models would suggest an initial tilt of the aromatic ring by 35^0 about the long axis to effect efficient packing on the water surface. Indeed the upswing in surface pressure on compression was observed at the corresponding area per molecule of 0.45 nm^2. The models also suggest that the area per molecule can be further reduced by a subsequent tilt of the entire molecule by 35^0 about an in film-plane axis, normal to the long axis. The corresponding area of 0.37 nm^2 per molecule is indeed the minimum obtainable before further compression results in collapse. A somewhat lower second tilt angle would be predicted for the conditions under which LB deposition was carried out. The predicted doubly tilted structure is consistent with the observed isotherm data, and the predicted angles are, in fact, corroborated by polarized absorption spectroscopy and X-ray diffraction data on the deposited LB film. Furthermore, the exact arrangement of the aliphatic chains was predictable, and this arrangement gave rise to an epitaxial ordering between successive layers. This work, therefore, showed that one could confidently use molecular models as a tool to help predict complex packing structures, which were entirely different and had entirely different properties from those of the parent material. Moreover, the structures were highly ordered, remained intact when transferred from water to solid supports, and showed little, if any, interlayer diffusion. All of this suggested that complex, predictable supermolecular structures, using many different kinds of molecules, appropriately substituted to control their orientations, may be possible, with LB films. As a result, an unprecedented range of new applications for LB films was proposed[4] at this time.

Concurrent with this, another major development occurred when actually designed supermolecular structures using LB films began to be demonstrated. A structure was proposed[18] and constructed[19] to mimic the primary photosynthetic process. LB layers of donor and acceptor molecules were proposed to be separated by a low but broad barrier layer (region) which would permit light-excited electron transfer but prevent recombination. A much higher but narrower barrier on the opposite side of the donor layer was intended to prevent excited electron transfer from the donor in the reverse direction, but permit tunneling from an adjacent electron source to replenish the donor and prepare the system to absorb another quantum. The system was constructed using LB techniques and the desired function was in principle achieved[19]. An LB monolayer of ubiquinone was deposited as acceptor on an Al substrate with the isoprene chains forming the low broad barrier. Layers of chlorophyll were then deposited as donors. A single stearic acid monolayer was subsequently deposited to act as the high narrow tunneling barrier. A top Hg electrode acted as the external electron source, and photoresponse was in fact achieved with the structure.

An additional major advance in LB techniques in the last decade was the demonstration of the polymerization of LB monolayers and multilayers, without destruction of the layer structure. This has been demonstrated for a number of materials, e.g. ω-tricosenoic acid, $CH_2 = CH-(CH_2)_{20}-COOH$, where polymerization was induced by electron beam irradiation[20]. This development was important from the fundamental viewpoint of intrinsic interest in two dimensional polymers. It should also extend the applications of LB films, as insulating layers for example, since harder, more stable structures than fatty acids, which are quite soft, will often be required. In addition direct applications for such polymerizable layers are possible, such as e-beam lithography[21], non-linear optical devices[22] and even molecular electronic memory[23].

3. RECENT WORK USING LB FILMS AT XEROX RESEARCH CENTRE OF CANADA

3.1 Structural Studies of Phthalocyanine Monolayers

As mentioned in the previous section, very significant advances have been made in recent years in the extension of LB techniques to produce monolayers and multilayers of largely undiluted aromatic molecules[15-17,24]. However it remained difficult to avoid aliphatic chain substitutions that alter drastically the "crystal" structure and properties of the aromatic (including the thermal and hardness properties), by giving rise to significant aliphatic layers between the aromatic LB layers. A very important advance was subsequently made when phthalocyanines (unsubstituted or lightly substituted by small aliphatic groups that would not be expected to greatly change the "crystal" structure) were formed into LB films[25]. Such layers have obvious electrical and photoelectrical interest[4]; phthalocyanine, Pc, is a well known organic semiconductor and a very efficient photoconductor. In addition they were reported[25] to be extremely hard and abrasion resistant; moreover, the thermal stability of phthalocyanine is well known. These latter properties could enable such films to be used (perhaps as insulators) in the first truly practical electronic LB devices. Furthermore, Pc is an excellent candidate for many more-ambitious device schemes[4].

Despite the importance of these results, the films were admittedly[26] not monolayers, and molecular orientation could not be determined. Hence we undertook work[27] on Pc-LB films at XRCC to try to overcome these limitations. Our results provide evidence for the preparation of single Pc monolayers on water, with the area per molecule expected from space-filling models arranged to be consistent with the X-ray spacings observed in powders of the Pc: this molecular arrangement is very similar to that of unsubstituted Pc crystals. These monolayers have been transferred to slides with every indication of ideal LB deposition, including a unity deposition ratio (the ratio of the film area lost from the water surface to the area of the substrate passed through the water surface). LB multilayers can be formed by repetition of the process. Even these true monolayers (estimated to be only ~1.7 nm thick) are clearly visible to the naked eye, and are very hard and abrasion-resistant. Concurrent with our work and independently, another group has also produced good evidence for the production of LB monolayers of a substituted CuPc[28].

In our work we have used metal-free Pc lightly substituted with t-butyl groups, (t-butyl)$_4$H$_2$Pc. The material is readily soluble in toluene which is used for spreading. A typical isotherm is shown in Figure 1, curve (a). The film pressure becomes measurable at ~0.85-0.90 nm^2/molecule and the onset of collapse occurs at about 0.60 nm^2/molecule. The collapse point (the shoulder on the isotherm) is made much clearer (curve (b)) by adding 10 mole % of arachidic acid to the Pc layer. This slightly disturbs the Pc packing, thus molecularly lubricating the otherwise very rigid film. With the molecular lubricant it was also possible to compress down to a single clear-cut second collapse at almost exactly one-half of the 0.60 nm^2 area. The isotherm behaviors of neat and lubricated films above 0.60 nm^2 are exactly what one would expect for a single monolayer, the compressed 0.60 nm^2 film having the planar Pc molecules stacked face-to-face normal to the water surface; a double-monolayer exists for the lubricated film between 0.60 and 0.30 nm^2.

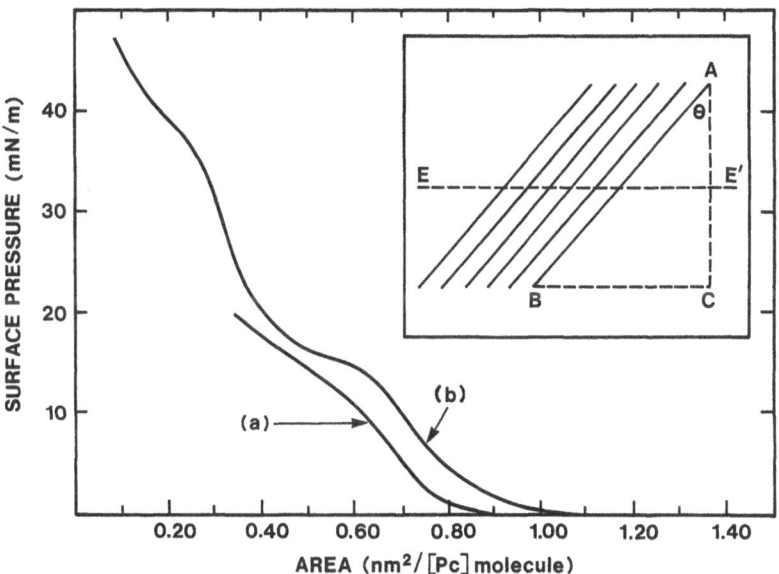

Figure 1. Surface pressure as a function of area per Pc molecule for (a) (t-butyl)$_4$H$_2$Pc, (b) the same compound + 10 mole % arachidic acid. Inset shows stacking of the Pc molecules (the approximately disc-shaped molecules being viewed edge on) and the tilt angle θ. See text for discussion.

X-ray analysis of powder samples of our Pc material shows no long range crystalline structure, only short range order with just three d-spacings of 0.334, 0.547, and 1.715 nm. These data are consistent with the kind of structure usually seen[29,30] in the Pc's, viz. pancake-like stacks of the Pc molecules with a well defined tilt angle, θ. The structure is illustrated in the inset to Figure 1. The presence of only three broad peaks in the X-ray spectrum indicates that while there is order within a stack there is no long range order between neighboring stacks, as there is in unsubstituted crystalline Pc's. The first d-spacing must clearly be the perpendicular distance between adjacent molecules, equal as usual to the well known thickness of benzene rings. If the 0.547 nm spacing is interpreted (analogous to the diffraction data of unsubstituted Pc's) as the distance between adjacent molecules along the stacking direction, then one obtains a tilt angle θ of 52^0, very similar to the equilibrium modification[31] of unsubstituted Pc (45^0). The slightly larger tilt angle may be a result of the somewhat bulky t-butyl groups whose thickness is almost twice that of the π-electron system and which must, therefore, be staggered to effect efficient packing. The tilt angle of 52^0 implies that adjacent molecules are shifted 0.43 nm along the molecular plane. This is precisely the shift required to stagger the t-butyl groups [27]. Finally, the 1.715 nm length can then be interpreted[32] as the average overall in-plane length of this Pc molecule, which is consistent with molecular models.

In the LB films, we propose that at fairly large areas the molecules are oriented in local stacks along the water surface, but with the plane ABC (Figure 1) randomly oriented relative to the water surface, i.e., with random rotations about the axis E-E'. The random rotations occur because the molecules are not amphipathic as is usually the case with LB materials. The strong and directed intermolecular interactions responsible for the stack structure serve to order the molecules on the water surface and no hydrophilic anchoring group is necessary. In this configuration, the maximum area per molecule arises if the plane ABC is perpendicular to the water surface (i.e. Figure 1 with the water surface normal to the page and parallel to E-E'), that is at an area given by the length of a molecule multiplied by the intermolecular spacing measured along the stack. This area is (1.715 x 0.547) or 0.94 nm^2. The minimum area occurs if the plane ABC is in the water surface, so that the Pc planes are normal to the surface (i.e. Figure 1 with the water surface in the plane of the paper). This area is (1.715 x 0.334) or ~0.57 nm^2. Thus, if (prior to final compression) there is a statistical distribution of the angle between ABC and the water surface, then we would expect the onset of significant surface pressure at an area somewhat less than 0.94 nm^2; below this, the pressure would rotate the stacks to a minimum area of 0.57 nm^2, with collapse of the film occurring just below this area. This is just what is observed. Note that 0.57 nm^2 would be expected to be the minimum area for any value of the tilt angle θ, so long as the molecules are normal to the water surface. Thus the isotherm data cannot confirm the actual value of θ in the fully compressed film; however, as we shall see, the optical properties of the transferred films do tend to confirm that the stack arrangement is very similar to that in the microcrystals of the powder. Since the Pc molecules have no hydrophilic anchoring to the water subphase, rotation of the stacks provides a very facile mechanism for lowering the molecular area (hence the very low surface pressures). This mechanism of rotation of the entire stack is far easier than rotation of individual molecules within the stacks about an axis parallel to the water surface and perpendicular to E-E' (to form untilted stacks). This latter mechanism would meet with very large resistance from the bulky, staggered t-butyl groups and probably also from strong π-π interactions between neighboring molecules (virtually all known crystals of Pc's form into tilted stacks).

The combination of unity deposition ratio and ~0.60 nm^2 area per molecule (the isotherm area at the deposition pressure used) was checked by dissolving off the LB film from a known area of the slide and comparing the solution absorption strength with that of standard solutions. This procedure

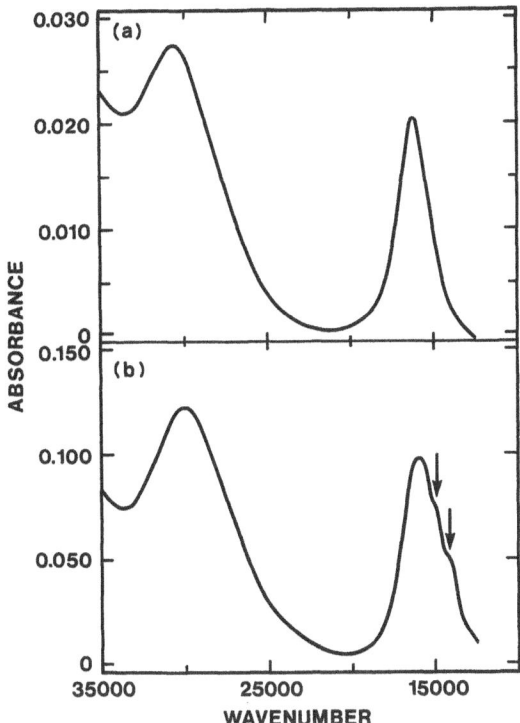

Figure 2. Absorption spectra of (t-butyl)$_4$H$_2$Pc: (a) two LB monolayers (one on each side of slide), (b) film evaporated onto room temperature substrate. Toluene-solution absorption peak positions are indicated by arrows.

(which also, of course, confirms that no chemical changes have occurred to the Pc during the LB procedures) gave an area per molecule on the slide of 0.61 ± 0.03 nm^2, in excellent agreement with the area expected. The absorption spectrum of the LB films (one monolayer on each side of a slide) is shown in Figure 2(a). This spectrum is sharper than those of typical thin evaporated films of the same Pc derivative (Figure 2(b)). When the evaporated film was annealed, its absorption spectrum became essentially identical to that of the LB films. This again suggests that the LB structure is a very stable well-ordered one, and (since quite small changes in the structure of Pc crystals give large changes in absorption spectra) that the LB film structure is similar to that of the microcrystalline material. The two shoulders in the spectrum of the unannealed evaporated film (Figure 2(b)), which correspond in position to the main solution absorption peaks (indicated by arrows) are probably due to "free" molecules in highly disordered positions, which are absent in the LB film.

3.2 Surface Enhanced Raman Scattering Using LB Monolayers

We have used these well defined LB monolayers of Pc to make the first observation of Surface Enhanced Raman Scattering (SERS) from a true LB monolayer. In recent years SERS has been the subject of intense theoretical and experimental investigation[34]. Several models have been proposed[35-39] to

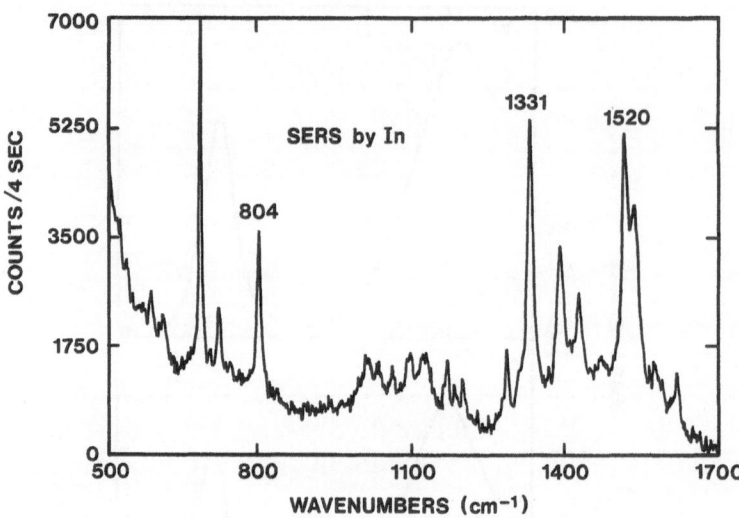

Figure 3. SERS spectrum of a (t-butyl)$_4$H$_2$Pc monolayer on an In island film of mass thickness 15 nm.

explain the interactions that give rise to the enormous Raman intensities associated with SERS (enhancement factor ~10^5-10^6). However the considerable potential of SERS as a surface analysis technique for providing information on adsorbed molecules is partially limited by the absence of a unifying explanation for a wide spectrum of experimental observations. We are, therefore, attempting to use the well defined architectures possible with the LB technique to distinguish between the various proposed mechanisms for SERS. Our initial observations of SERS from LB-Pc monolayers indicate that the electromagnetic mechanism[40] is responsible for the effect.

We have used metal island films of In and Ag evaporated onto glass slides to obtain the enhancement[41]. A series of films of these metals was prepared with mass-thicknesses ranging from 5 nm to 100 nm; the island size is known to monotonically increase with mass thickness. LB films were deposited directly onto the metallized substrates. For both metals, the maximum SERS signal was obtained from 15 nm films. Figure 3 shows the SERS spectrum of one (t-butyl)$_4$H$_2$Pc monolayer on a 15 nm In film. The SERS spectrum from a 15 nm Ag film coated with one monolayer is virtually identical in peak positions and in both overall and relative intensities. The enhancing properties of In are, therefore, very comparable with those of Ag in this case. Assignments of the scattering peaks to molecular vibrations can be found in reference 41.

Since the Pc molecular plane is believed to be perpendicular to the metal surface[27], the surface-monolayer interaction can be classed as physisorption. The normal Raman spectrum of bulk samples of the Pc shows no significant shift in peak positions compared with Figure 3. The lack of frequency shifts in the SERS spectrum is a good indication that there is physisorption and not chemisorption of the monolayer to the underlying metal. That the metal island film morphology can be tuned to produce a maximum enhancement suggests that the electromagnetic mechanism is responsible for the enhancement. Additional strong evidence for the electromagnetic mechanism is provided by our recent experiments[42] using LB spacer layers of arachidic acid to accurately control the distance of the Pc monolayer from the metal surface. Measurements of the enhancement factor as a function of the distance of the Pc molecules from the surface show that SERS is a rather long range effect, consistent with calculations[43] based on the electromagnetic model[40] and contrary to the short-range effects expected from other models (see e.g. reference 39). Additional work is underway to try to unlock the fundamental mechanism of SERS, and a highly sensitive surface analysis technique, based on this work, is being evaluated.

3.3 Field Modulated Charge Transfer in LB Films of Photosynthetic Reaction Centers

After the isolation of the reaction centers (RC's) of photosynthetic bacteria[44], much effort has been invested[45] in elucidating their structural and kinetic properties. The major driving force behind these efforts is the desire for detailed understanding of the primary charge separation process, which proceeds with a remarkable quantum efficiency approaching unity. We have applied electric fields to oriented monolayers of RC's from the photosynthetic bacteria, *Rhodopseudomonas sphaeroides*, in order to perturb the charge transfer reactions[46]. Measurements of the field dependence of these processes are of considerable importance for testing various multi-phonon tunneling theories of electron transfer in photosynthesis[47-49], and for establishing thereby the fundamental nature of the primary photosynthetic mechanism. Application of the external electric field changes the relative energy levels between different sites in the charge transfer chain and, therefore, allows one to study the influence of energy level shifts on the charge transfer processes without introducing the structural changes attendant upon chemical modification.

LB films are a natural vehicle with which to study the effects of large external fields on the charge separation processes in isolated RC's. It has already been established[50] that the RC's in these LB films are oriented in a fashion similar to that encountered in the native membrane except that they are a mixture of two vectorially opposing "up" and "down" states.

Quartz slides with a sputtered ITO (Indium Tin Oxide) transparent conductive layer (~150 nm thick) and a sputtered SiO_2 blocking layer (~500 nm) were used as substrates for LB film deposition. LB monolayers of RC's were deposited as previously described[50]. To allow high electric fields to be applied without cell breakdown, an ~1.5 μm polymer blocking layer was dip-coated[51] on top of the RC monolayer. An evaporated opaque Al layer formed the top electrode of the sandwich. It was possible to apply fields as high as 200 V/μm before sample breakdown occurred. The influence of the electric field on the charge transfer kinetics was studied using the two experiments described below.

In the first type of experiment, the sample was bleached at zero field by a strong light pulse. An electric field was then applied across the sandwich cell and the time-resolved absorption recovery was recorded using an 860 nm probe beam, which passed twice through the RC layer by reflection from the back Al electrode. The loss and subsequent recovery of absorption at 860 nm, the first excited singlet transition of a bacteriochlorophyll dimer within the RC, describes the transfer of an electron from the excited singlet through bacteriopheophytin to the quinone acceptors. This type of measurement gives information about field-modulated <u>recombination</u> processes of already-separated electron-hole pairs.

In the second type of experiment, to be referred to as "quenching of bleaching", the sample bias is applied before and maintained during the bleaching light exposure. The measured quantity was ΔI, the <u>maximum</u> bleaching, recorded immediately after terminating the light pulse and turning off the sample bias. This experiment gives information about the field modulation of the primary charge <u>separation</u> steps. The quenching of bleaching is then defined as the fractional change in the bleaching due to application of the electric field.

Our results for the absorption recovery experiments are given in Figure 4 which shows the logarithm of reflectivity variation versus time for a range of fields applied to the sample. It is obvious that application of the electric field produces a pronounced change in the recombination kinetics in photo-bleached RC's. It is expected that the recombination process should be either hastened or hindered, depending on the RC orientation with respect to the external field. Since the RC's generally have equal populations of up and down

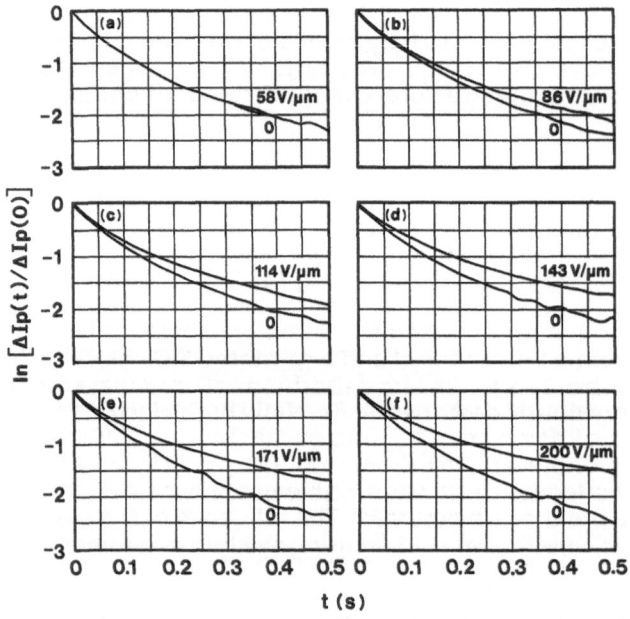

Figure 4. Absorption recovery of RC's as a function of time for various values of the applied electric field. The zero-field reference is given in each case. No detectable shift occurs for fields less than 60 V/μm.

orientations, the time constant(s) observed at zero field should, therefore, split into two (or more). Figure 4 shows that the curves generally shift increasingly upward showing decreasing slope at longer times when compared with the zero field data. This can be understood in the framework of rate constant splitting. At longer times, the time evolution of the decay will be governed by the slow rate constants, leading to a decreased rate of decay as observed.

The variation of the recombination rate constants as a function of the applied field or equivalent energy level shift can, in principle, be obtained by fitting the decay curves. This procedure is generally difficult because of a large number of unmanageable parameters[46], but recent refinements to our experimental techniques[52] have allowed us to confidently extract the rate constants. They vary nearly exponentially with field and change by about an order of magnitude over the range of fields applied (-200 V/μm to +200 V/μm). They show no sign of reaching a maximum as predicted by certain theories of electron transfer[53]. These results should, therefore, provide important information in testing the applicability of various tunneling theories of electron transfer[47-49] to photosynthesis. The rate constant changes themselves are caused by shifts in the energy levels of the quinone acceptors relative to the bacteriochlorophyll dimer donor, on application of the uniform electric field. The shift is given by -d·E, where d is the dipole moment of the electron-hole pair and E is the electric field. If it is assumed that the RC cylindrical axis lies along the electric field direction, a projected separation distance of ±1.5 nm for the pair is reasonable for purposes of estimation. From this separation distance one estimates a substantial change in energy separation of ±0.3 V for the range of fields applied.

Quenching of bleaching is a measure of the field induced change in the yield of the charge separated state. This change is caused by the change in energy level separation between the intermediate bacteriopheophytin level relative to the excited bacteriochlorophyll dimer level and to the quinone acceptor energy level. From relative distance measurements[54], the changes are expected to be somewhat smaller, but comparable, to the changes estimated above between the quinone acceptor energy level and the bacteriochlorophyll dimer ground state.

Our results[46] on the quenching of bleaching show that the electric field decreases the yield (averaged over up and down populations) of the charge transfer state, up to ~13% at the highest fields of 200 V/μm. If one assumes that the yield of one of the two opposing RC populations saturates at unity for large "forward" fields, then the yield of the opposing RC population is decreased by ~26% at the highest field. Nature has apparently found a way to produce a nearly field-independent high quantum efficiency carrier-generating system, the ideal photogenerator. Further experiments are underway to determine the field dependence of the individual rapid charge separation events which contribute to the overall yield. We hope thereby to understand nature's secret design for a highly efficient photogeneration system which is virtually insensitive to external perturbation.

4. SOME POSSIBLE DEVICE APPLICATIONS OF LB FILMS

A large number of electrical and other device applications for LB films have already been experimentally demonstrated. However, many of the more ambitious schemes have yet to be reduced to practice. In general the applications can be classified as those which use conventional insulating-type materials, and those which employ unconventional electrically active films.

4.1 Applications Using Conventional Insulating-Type Materials

Proposed device applications using conventional insulating-type materials include capacitors, hygrometers, MIS devices, ion and gas sensing devices, Josephson devices, electron beam resists, and optical waveguides.

High quality fatty acid films have excellent insulating properties due to their high dielectric strength, small electron affinity and near independence of dielectric constant on frequency. They are, therefore, ideally suited for low voltage, low loss capacitors. A group at the Laboratoire d'Etudes et Recherches Avancées (LERA) in France has described an MIM device for such an application and serious commercial interest has been reported[55]. They claim a capacitance of ~100 nF/cm^2 and a dielectric strength of ~3 x 10^6 V/cm for a single monolayer. The capacitance is sufficiently large that several hundred integrated devices are proposed for a single silicon chip. The same group has investigated similar devices as hygrometers. The capacitance reproducibly undergoes rapid, large, linear variations with moisture content and, therefore, serves as an excellent measure of relative humidity.

Unlike Si, many semiconductors do not have a good native oxide to act as an insulating layer. Fatty acid LB films have been deposited as insulators to produce MIS devices on semiconductors such as InP. In particular LB MISFET's have been made on InP with In source and drain and a Au/Cr alloy gate, and they show excellent I-V characteristics[56]. LB insulators have also been used to make a-Si:H MISFET's with Cr source-drain electrodes and an Al gate[57]. These transistors on a-Si:H have potential use in large area devices, e.g., display controllers. In an effort to achieve improved stability properties, diacetylene-substituted fatty acid monomers have been deposited onto the semiconductors InP and GaP and then polymerized and incorporated into highly stable MIS structures[58].

MIS structures incorporating LB films have also been made into inorganic electroluminescent devices[59]. An LB layer between n-GaP and a semitransparent Au electrode facilitates minority carrier (hole) injection into the n-GaP under forward bias conditions, resulting in greatly enhanced electroluminescence. The ability of the insulating LB layer to support a voltage causes a shift of the relative energy levels of the semiconductor and metal, which assists the minority carrier injection. The thickness of the LB film can be optimized to give a maximum in luminescence efficiency. The optimum thickness for fatty acid layers was ~27.5 nm or 11 monolayers[60], and for an asymmetrically substituted phthalocyanine layer the optimum thickness was estimated to be ~6.0 nm or 7 monolayers[61].

LB insulating layers have also improved the performance of photovoltaic Schottky barriers[62]. The tunneling layer prevents majority carrier dark current, thereby increasing the open circuit voltage without degrading the short circuit current. An optimum efficiency is reached for one fatty acid layer, ~2.5 nm thick, beyond which the short circuit current is greatly decreased and the device performance degraded. A similar effect was achieved with finer thickness-tuning by using 1.2 nm-thick films of the C4 anthracene derivative described earlier. In this case the maximum efficiency was found[8] for two C4 monolayers with a thickness of 2.4 nm.

Applications of insulating LB films to ion and gas sensing devices such as ISFET's (Ion Sensitive Field Effect Transistor) and ChemFET's (Chemical Field Effect Transistor) are somewhat more speculative although encouraging results have been obtained. In an ISFET[63] a reference electrode replaces the gate and an ion-selective (LB) film membrane allows certain ions in solution to penetrate it and interact with the semiconductor surface; this interaction and hence the abundance of the ion species is monitored by the source-to-drain current. Conventional Ba stearate multilayers have been shown to be permeable almost exclusively to one of a range of ions[64] and, hence, may have application in such a device. A ChemFET for gas sensing operates in a similar manner except that the solution, containing ions, is replaced by an atmosphere containing certain trace gases. Clear response of the surface states of certain semiconductors to several gases including NH_3, CO and H_2 has been observed[5].

Polymerized LB monolayers of vinyl stearate have found application in Josephson junction tunnel diodes[65]. Thin films of thickness < 5.0 nm, which are pinhole free and of low conductance, are required. Single monolayers of vinyl stearate, polymerized by Co^{60} γ-radiation, were sandwiched between evaporated Pb-In alloy films. The LB technique allowed the devices to be produced with a 90% yield[66], far better than by any other method.

A very attractive application of conventional insulating LB films is as an electron beam resist for microlithography[67]. Conventional lithography is near its theoretical resolution limit of ~1 µm. Submicron lithography requires a reduction in diffraction effects by the use of X-rays or electrons. However, owing to electron scattering, resist thickness is critical for e-beam irradiation: the thinner the resist, the higher the resolution. LB films can provide the ultra-thin resists necessary to attain high resolution. For this application, ω-tricosenoic acid is very attractive. It is polymerizable by e-beam and a resolution of 50 nm for a 60 nm thick film has been demonstrated[67]. Moreover, fast dipping of ω-tricosenoic acid has recently been demonstrated[68], so that film preparation time is similar to that required for conventional spin coated resins. Although perhaps not as practical as a polymerizable film, Mn stearate films have also been demonstrated as resists whereby the electron beam sublimes the film[69]. In this case a resolution of 10 nm is claimed.

Finally, conventional fatty acid materials have been used as thin film optical waveguides[7,70]. The advantages are an accurately controlled film thickness and a readily tailored refractive index (by CH_2 chain length selection and selected metal ion inclusion in the organic salt). These advantages allow the velocity of light in the waveguide to be precisely controlled[7]. This may be particularly interesting for fine-tuning the characteristics of conventional optical waveguides. However, fatty acids have the disadvantage that they are too fragile for anything other than laboratory use. Attempts to polymerize diacetylene-containing fatty acid monomers to produce more durable films resulted in domain formation in the films, limiting their usefulness[71], but work is continuing. It has been suggested that incorporation of dyes into LB structures of this kind could lead to mode-selection, or even to planar lasers.

4.2 Applications Using Unconventional Electrically Active Films

Potential device applications using unconventional electrically active films include electroluminescent devices, solar energy converters, molecular rectifiers, nonlinear optical devices, molecular electronic memory, biological monitoring devices, gas sensing devices and even high temperature superconductors.

It has long been recognized that thin films of aromatic hydrocarbons hold promise for large area electroluminescent panels[4] since the quantum efficiency of blue recombination electroluminescence from, e.g., anthracene can be greater than 10%. Thin films are needed so that a high driving voltage is not required, which would drastically reduce the power-conversion efficiency. However, thermal evaporation of anthracene produces films with current-reducing traps and often pinholes, and therefore the LB technique holds promise for producing the high quality films necessary for this application. The C4 anthracene derivative films described above have been incorporated into devices and have shown electroluminescence[9]. A C4 anthracene multilayer was deposited onto an aluminized slide and subsequently coated with a semitransparent Au electrode. The Au is believed to be a fairly efficient hole injector and Al a rather inefficient electron injector. In any case, sufficient double injection was achieved to produce electroluminescence clearly visible in room light. Ways to improve such devices have been suggested[4].

Solar energy conversion using unconventional, active films has been achieved by using the light driven electron pumps which mimic the photosynthetic process, referred to earlier. The maximum power conversion efficiency of the chlorophyll - quinone system was improved from ~4 x 10-5 to ~4 x 10-4 with inclusion of the quinone acceptors[19]. Further improvements have been made by using multilayer arrays of chlorophyll and by special treatment of the Al electrode[72]. The possibilities for solar photo-assisted electrolysis with LB films, however, may be much more exciting[4] than those for fully-solid systems, since the factors limiting the efficiency of the latter may be missing and the structure could be very inexpensive.

Specially designed molecules to act as molecular rectifiers have been proposed[73] and the LB technique would be ideally suited for their deposition in the necessary oriented fashion. One such proposed structure consists of a donor π-system and an acceptor π-system separated by a σ-bonded tunneling bridge. Detailed calculations based on a model of the molecular energetics predict very pronounced rectifying properties[73]. Recently, photoconductive p-n diodes have been fabricated using the LB technique[74,75]. Merocyanine and pyrene derivatives have been used as the p-type layers and crystal violet derivatives as the n-type layers. Good evidence of photo-injection and conduction through the junction for both electrons and holes, each with a different decay length, has been obtained[75].

LB films of polydiacetylenes appear promising for applications in nonlinear optical devices[76-79] because of their large third-order nonlinear susceptibility[78]. Entirely-optical signal processing is, in principle, much faster than current-day electronic signal processing and requires materials with large third order nonlinear susceptibility, so that the refractive index can be easily changed using low light power. LB films of diacetylene monomers, $R-C \equiv C-C \equiv C-R'$, can be polymerized with UV or e-beam irradiation to form polydiacetylenes of the form $(= RC-C \equiv C-CR' =)_n$. Films in which R is $CH_3(CH_2)_{15}$ and R' is $(CH_2)_8$-COOH have been polymerized on the water surface and incorporated into optical waveguides ~ 500 nm thick[77]. By measuring the change, as a function of optical intensity, in the coupling angle between an input laser beam and a planar waveguide mode in the film, the intensity dependent index of refraction was determined and thereby the third order nonlinear susceptibility. This susceptibility is large, owing to the highly delocalized π-electron system of the polymer backbone, and has attracted interest for potential all-optical signal processing techniques using this class of material[80].

Application of the polydiacetylene films described above to a molecular electronic memory device has been proposed[81]. The proposed structure contains several polymerized polydiacetylene monolayers each with its delocalized π-electron conduction band (associated with the backbone) and insulating regions due to the long chains and acid groups on either side of the conducting plane. The electronic structure through the multilayer film then consists of alternating insulating and conducting regions. The multilayers are then deposited onto a p-type semiconductor and sandwiched with a top metal electrode. The p-type semiconductor is biased positively and photo-injected electrons from the metal hop from conduction band to conduction band through the polydiacetylene film. Under high fields, there is a well-defined jump time which is not influenced by diffusion. If the exciting light is pulsed with a period equal to the jump time, a time sequence of n light bits is converted into a spatial sequence of n charge bits in the successive conduction bands. If the field is then reduced to zero after all conduction bands have been accessed, the presence or absence of electrons in the n-th layer denotes the 1 or 0 of binary information. The information can be read out by re-applying the field and detecting recombination luminescence as the excess electrons from the conduction bands are emptied into the semiconductor; the spatial sequence of n charge bits is then converted back to a time sequence of n light bits. In principle, the aliphatic chain and acid head groups can be varied to control the well depth and enhance the hold (memory) time before diffusion scrambles the binary information in the conduction bands. It has been estimated that a film of area 1μm x 1 μm and thickness 48 nm can store one 16-bit byte for 134 s. The write and read time for one byte is 25.9 μs.

FET devices have been proposed (similar to the ISFET and ChemFET described earlier) as biological monitoring devices for monitoring immunological response (IMFET) and enzyme-substrate reactions (ENFET)[82,83]. Active biological molecules such as antibodies and enzymes are combined in the insulating region of the device. and the response is again monitored by the source to drain current. The LB technique should be well suited for deposition of the biological molecules, since monolayers and bilayers of lipids are readily made by LB techniques and bilayers form realistic models of cell membranes. In addition, proteins and other cell membrane constituents can now be formed into LB monolayers as our work described in Section 3.3 has shown. BioFET's may also find applications in the field of medicine. For example, an implanted device in a diabetic person might be used to monitor and even restore the level of glucose[6].

Gas-sensing devices based on unconventional electrically active phthalocyanine LB films have recently been demonstrated[84]. LB layers of an asymmetrically substituted CuPc were deposited on Al interdigital electrodes. The conductivity through the film was found to increase greatly and reversibly with exposure to NO_2. The sensitivity of the detection technique was found to be about 1 ppm. The mechanism is thought to be electrophilic attack by the NO_2 and adsorption onto the Pc molecule, which becomes negatively charged. This results in increased hole concentration in the Pc film and, hence, increased conductivity.

The final application of unconventional electrically active LB films which we wish to consider is high temperature excitonic superconductivity. Such an application deserves at least a cursory consideration since the demonstration of a high temperature superconductor would be one of the most important scientific and technological advances which can presently be conceived. As is well known, superconductivity requires an electron-electron attraction to effect pairing. This attraction is via phonons in all known superconductors. However, it is theoretically predicted[85,86] that if this attraction were via excitons then pairing and hence superconductivity could occur at much higher temperatures, even room temperature and above. A structure has been proposed[85,86] in which this exciton-mediated interaction could occur. It consists of a conducting spine[85] or layer[86] 1-2 nm in width or thickness which is surrounded by highly polarizable dyes which are held in a fixed orientation and are separated a specific distance (0.5 - 1.0 nm) from the conductor by an insulating bond. A large component of the transition moment of the dye should be normal to the conductor. An electron moving along the conductor polarizes the surrounding dyes as it goes, creating a polarization wave. A second electron is attracted to follow this polarization wave and is thereby effectively attracted to the first electron; this attraction results in the pairing. The model has never really been tested since the structures have been impossible to fabricate. However, with the promise which the LB technique holds for the design of supermolecular architectures it is now possible to conceive of fabricating the required layered structure, so that the theory can be tested with well-defined structures. For example, a Pc-LB monolayer of the type we have made[27] and described in Section 3.1 could first be deposited onto a suitable substrate. An ultra-thin metallic conducting layer could then be deposited by electron-beam vacuum-evaporation followed by a second Pc-LB monolayer. Pc is, of course, a highly polarizable dye and its orientation in the films, with the molecular plane normal to the film plane, would result in a large component of the transition moment normal to the film plane as required. Moreover the peripheral tertiary butyl groups, or similar groups, could act as the required insulating spacers between the conducting layer and the polarizable π-electron system of the Pc molecule. Alternately, other dyes might be used and the conducting layer might be made from organic materials also deposited by the LB technique[4]. Appropriately-substituted TCNQ molecules might be deposited by the LB technique with a suitable metal ion to form a conductive salt incorporated from the subphase. With TCNQ-TTF, both materials could be appropriately substituted and deposited together if they will form the required segregated stacks on the water surface. Otherwise one could substitute one to allow deposition by LB techniques and then chemically react it with the other to form the thin conductive layer. One might also appropriately substitute both molecules and deposit one and then the other by LB techniques with appropriate molecular orientations being maintained to yield a conducting layer. In principle, the LB technique offers several other possibilities[4] of forming the required structure, which is very difficult, if not impossible, to form with other techniques.

5. CONCLUSIONS

In the past decade or so, huge strides have been made in the development of the LB technique itself and in its applications. In particular the enormous possibilities for electrical and photoelectrical applications have been demonstrated. Designed supermolecular architectures are now a reality and no longer mere speculation. The future should see a continuing number of applications being demonstrated. With the development and use of more robust and stable materials, the first commercial products will begin to appear, maybe within the next decade. These may start with simple insulator materials, and then progress to the more ambitious device structures of the kind described in Section 4.2. The ability to control individual molecules will continue to attract increased attention, including synthetic efforts to produce specially designed molecules for manipulation by LB techniques.

The most important point to be reiterated is that (whether the object is molecular electronic devices, high-temperature superconductors or other systems,) the LB technique now offers us the chance to design remarkably complex supermolecular structures which are unattainable by other means, and (guided by studies of molecular models) gives us a good chance of fabricating them. It seems likely that, as this new ability becomes better-known, it will spur the design of new supermolecular structures for a wide range of physical, chemical and biological functions. We may, in fact, now be more limited by our imaginations than by our molecular-assembly capabilities!

ACKNOWLEDGEMENTS

It is a pleasure to acknowledge our colleagues G. Alegria, R. Aroca, P. L. Dutton, C. Jennings, R. O. Loutfy, Z. D. Popovic and J. H. Sharp, with whom the work described in Section 3 has been a collaborative effort.

REFERENCES

1. I. Langmuir, J. Am. Chem. Soc. 39, 1848 (1917).
2. K. B. Blodgett, J. Am. Chem. Soc. 57, 1007 (1935).
3. K. B. Blodgett and I. Langmuir, Phys. Rev. 51, 964 (1937).
4. P. S. Vincett and G. G. Roberts, Thin Solid Films 68, 135 (1980).

5. G. G. Roberts, P. S. Vincett and W. A. Barlow, Phys. Technol. 12, 69 (1981).
6. M. C. Petty, Endeavour, New Series 7, 65 (1983).
7. C. W. Pitt, Electronics & Power (GB) 29, 226 (1983).
8. G. G. Roberts, Contemp. Phys. 25, 109 (1984).
9. G. G. Roberts, M. McGinnity, W. A. Barlow and P. S. Vincett, Solid State Commun. 32, 683 (1979).
10. I. R. Peterson, IEE Proc. Pt. I, 130, 252 (1983).
11. H. Bücher, K. H. Drexhage, M. Fleck, H. Kuhn, D. Möbius, F. P. Schäfer, J. Sondermann, W. Sperling, P. Tillmann and J. Wiegand, Mol. Cryst. 2, 199 (1967).
12. H. Kuhn and D. Möbius, Angew. Chem. Int. Ed. Engl. 10, 620 (1971).
13. B. Mann and H. Kuhn, J. Appl. Phys. 42, 4398 (1971).
14. E. E. Polymeropoulos, J. Appl. Phys. 48, 2404 (1977).
15. P. S. Vincett, W. A. Barlow, F. T. Boyle, J. A. Finney and G. G. Roberts, Thin Solid Films 60, 265 (1979).
16. G. G. Roberts, T. M. McGinnity, W. A. Barlow and P. S. Vincett, Thin Solid Films 68, 223 (1980).
17. P. S. Vincett and W. A. Barlow, Thin Solid Films 71, 305 (1980).
18. H. Kuhn, J. Photochem. 10, 111 (1979).
19. A. F. Janzen and J. R. Bolton, J. Am. Chem. Soc. 101, 6342 (1979).
20. A. Barraud, C. Rosilio and A. Ruaudel-Teixier, J. Colloid Interface Sci. 62, 509 (1977).
21. A. Barraud, C. Rosilio and A. Ruaudel-Teixier, J. Vac. Sci. Technol. 16, 2003 (1979).
22. G. M. Carter, Y. J. Chen and S. K. Tripathy, Appl. Phys. Lett. 43, 891 (1983).
23. E. G. Wilson, Electron. Lett. 19, 237 (1983).
24. R. Jones, R. H. Tredgold and P. Hodge, Thin Solid Films 99, 53 (1983).
25. S. Baker, M. C. Petty, G. G. Roberts and M. V. Twigg, Thin Solid Films 99, 53 (1983).
26. S. Baker, G. G. Roberts, and M. C. Petty, Proc. Inst. Electr. Eng. 130, 260 (1983).
27. G. J. Kovacs, P. S. Vincett and J. H. Sharp, Can. J. Phys. 63, 346 (1985).
28. J. R. Fryer, R. A. Hann and B. L. Eyres, Nature 313, 382 (1985).
29. M. Ashida, N. Uyeda and E. Suito, J. Crystal Growth 8, 45 (1971).
30. M. Ashida, Bull. Chem. Soc. Japan 39, 2625 (1966).
31. J. H. Sharp and M. Lardon, J. Phys. Chem. 72, 3230 (1968).
32. L. V. Azaroff, Mol. Cryst. Liq. Cryst. 60, 73 (1980).
33. E. Schnabel, H. Noether and H. Kuhn in "Recent Progress in the Chemistry of Natural and Synthetic Colouring Matters and Related Fields ", T. S. Gore et al., Editors, pp. 561-572, Academic, New York, 1962.
34. R. K. Chang and T. E. Furtak, Editors, "Surface Enhanced Raman Scattering" Plenum, New York, 1982.
35. T. E. Furtak and J. Reyes, Surface Sci. 93, 351 (1980).
36. D. P. Dilella, A. Gohin, R. H. Lipson, P. BcBreen and M. J. Moskovits, Chem. Phys. 73, 4282 (1980).
37. H. Chew, D. S. Wang and M. Kerker, Phys. Rev. B 28, 4169 (1983).
38. H. Ueba, S. Ichimura and H. Tamada, Surface Sci. 119, 433 (1982).
39. A. Otto, I. Pockrand, J. Billmann and C. Pettenkofer, in "Surface Enhanced Raman Scattering" R. K. Chang and T. E. Furtak, Editors, pp.147-172, Plenum, New York, 1982.
40. J. Gersten and A. Nitzan, J. Chem. Phys. 73, 3023 (1980).
41. R. Aroca, C. Jennings, G. J. Kovacs, R. O. Loutfy and P.S. Vincett, J. Phys. Chem. 89, 4051 (1985).
42. G. J. Kovacs, R. O. Loutfy, P.S. Vincett, Carol Jennings and Ricardo Aroca, Langmuir, in press (1986).
43. R. Aroca and F. Martin, J. Raman Spectrosc. 16, 156 (1985).
44. D. W. Reed and R. K. Clayton, Biochem. Biophys. Res. Comm. 30, 471 (1968).
45. R. K. Clayton and W. R. Sistrom (editors) "The Photosynthetic Bacteria" Plenum, New York, 1978.
46. Z. D. Popovic, G. J. Kovacs, P. S. Vincett and P. L. Dutton, Chem. Phys. Lett. 116, 405 (1985).

47. M. Redi and J. J. Hopfield, J. Chem. Phys. 72, 6651 (1980).
48. J. Jortner, J. Am. Chem. Soc. 102, 6676 (1980).
49. S. Rackovsky and H. Scher, Biochimica Biophysica Acta 681, 152 (1982).
50. D. M. Tiede, P. Mueller and P. L. Dutton, Biochimica Biophysica Acta 681, 191 (1982).
51. C. C. Yang, J. Y. Josefowicz and L. Alexandru, Thin Solid Films 74, 117 (1980).
52. Z. D. Popovic, G. J. Kovacs, P. S. Vincett, G. Alegria and P. L. Dutton, (1985) to be published
53. R. A. Marcus, Annu. Rev. Phys. Chem. 15, 155 (1964).
54. H.-W. Trissl, Proc. Natl. Acad. Sci. USA 80, 7173 (1983).
55. Electronics, Nov. 23, p.76 (1978).
56. G. G. Roberts, K. P. Pande and W. A. Barlow, Solid State Electron Devices 2, 169 (1978).
57. J. P. Lloyd, M. C. Petty, G. G. Roberts, P. G. LeComber and W. E. Spear, Thin Solid Films 99, 297 (1983).
58. K. K. Kan, M. C. Petty and G. G. Roberts in "Proceedings of the Raleigh Conference on the Physics of MOS Insulators", G. Lucovsky, S. T. Pantelides, F. L. Galeener, Editors, pp. 344-348, Pergamon, New York, 1980.
59. J. Batey, G. G. Roberts and M. C. Petty, Thin Solid Films 99, 283 (1983).
60. J. Batey, M. C. Petty and G. G. Roberts in "Proceedings of the International Conference on 'Insulating Films on Semiconductors'", J. F. Verweij and D. R. Wolters, Editors, pp. 141-144, Elsevier, Amsterdam, 1983.
61. J. Batey, M. C. Petty, G. G. Roberts and D. R. Wight, Electron. Lett. 20, 489 (1984).
62. I. M. Dharmadasa, G. G. Roberts and M. C. Petty, Electron. Lett. 16, 201 (1980).
63. P. W. Cheung, W. H. Ko, D. J. Fung and S. H. Wong in "Theory, Design and Biomedical Applications of Solid State Chemical Senors", P. W. Cheung, D.G. Fleming, M. R. Neuman and W. H. Ko, Editors, pp. 91-118, CRC Press, West Palm Beach, 1978.
64. H. P. Gregor and H. Schonhorn, J. Am Chem. Soc. 79, 1507 (1957); 81, 3911 (1959); 83, 3576 (1961); 85, 3926 (1963).
65. G. L. Larkins, Jr., E. D. Thompson, E. Ortiz, C. W. Burkhart and J. B. Lando, Thin Solid Films 99, 277 (1983).
66. J. B. Lando, (1982), private communication.
67. A. Barraud, Thin Solid Films 99, 317 (1983).
68. I. R. Peterson, G. J. Russell and G. G. Roberts, Thin Solid Films 109, 371 (1983).
69. A. N. Broers and M. Pomerantz, Thin Solid Films 99, 323 (1983).
70. C. W. Pitt and L. M. Walpita, Electron. Lett. 12, 479 (1976).
71. F. Grunfeld and C. W. Pitt, Thin Solid Films 99, 249 (1983).
72. M. F. Lawrence, J. P. Dodelet and L. H. Dao. J. Phys. Chem. 88, 950 (1984).
73. A. Aviram and M. A. Ratner, Chem. Phys. Lett. 29, 277 (1974).
74. M. Saito, M. Sugi, T. Fukui and S. Iizima, Thin Solid Films 100, 117 (1983).
75. M. Saito, M. Sugi and S. Iizima, Jpn. J. App. Phys. 24, 379 (1985).
76. F. Kajzar, J. Messier, J. Zyss and I. Ledoux, Optics Commun. 45, 133 (1983).
77. G. M. Carter, Y. J. Chen and S. K. Tripathy, Appl. Phys. Lett. 43, 891 (1983).
78. G. M. Carter, Y. J. Chen and S. K. Tripathy in "Nonlinear Optical Properties of Organic and Polymeric Materials", American Chemical Society Symposium Series No. 233, D. J. Williams, Editor, pp. 213-228, American Chemical Society, Washington, D. C., 1983.
79. G. M. Carter, Y. J. Chen, J. Georger, Jr., J. Hryniewicz, M. Rooney, M. F. Rubner, L. A. Samuelson, D. J. Sandman, M. Thakur and S. Tripathy, Mol. Cryst. Liq. Cryst. 106, 259 (1984).
80. P. W. Smith, Bell Syst. Techn. J. 6, 1975 (1982).
81. E. G. Wilson, Electron. Lett. 19, 237 (1983).
82. P. W. Cheung, D. J. Fleming, W. H. Ko and M. R. Neuman, Editors, "Theory, Design, and Biomedical Applications of Solid State Chemical Sensors", CRC Press, West Palm Beach, 1978.
83. C. C. Johnson, S. D. Moss and J. A. Janata, U. S. Patent 4,020,836 (1977).

84. S. Baker, G. G. Roberts and M. C. Petty, IEE Proceedings 130, Pt. I, No. 5, 260 (1983).
85. W. A. Little, J. Polymer Sci.: Part C 29, 17 (1970).
86. V. L. Ginzburg, J. Polymer Sci.: Part C 29, 3 (1970).

MICROLITHOGRAPHY WITH MONOLAYERS AND LANGMUIR-BLODGETT FILMS

A. Barraud

CEA-IRDI-DESICP, Département de Physico-Chimie
C.E.N. Saclay, 91191 Gif-Sur-Yvette Cedex
France

This paper is a survey of the main results obtained to date with very thin resists in electron beam microlithography. After an analysis of the influence of resist thickness and accelerating voltage on image resolution, results obtained with different kinds of very thin resists are presented : L.B. films of ω-tricosenoic acid exhibit both high contrast and high sensitivity, and allow a spatial resolution of 500 Å after development and 1000 Å after etching ; adsorbed monolayers show the very limits of electron beam lithography, and new L.B. resists have appeared very recently. Finally the position of these resists in electron beam microlithography is presented together with the associated techniques (3 layer technique, new voltage electron guns) required for electron beam lithography to work below 3000 Å.

INTRODUCTION

High resolution microlithography is a prominent problem in the challenge to submicron integrated circuits. No really universal technique has been found below micron size : optical methods with their well-known advantages are inadequate below one micron, X-rays have source and registration problems, ion beams are slow and expensive and electron beams are also slow and do not have appropriate resists.

Very thin resists, which are virtually of no help for X-ray microlithography, can improve electron beam microlithography substantially. Classical spin-coated resins are limited in thinness to a fraction of a micron by pinhole density and thickness inhomogeneities. Thin solid films such as adsorbed monolayers or Langmuir-Blodgett (L.B.) films do not suffer from the same problems and can be layered in the form of compact films down to 30 Å[1]. Very high resolution can thus be attained with the aid of these films, but other problems do arise. This paper summarizes the results obtained to date with very thin resist films and points out the improvements and the limitations brought by these films.

BLURRING IN ELECTRON BEAM MICROLITHOGRAPHY

Electron scattering spreads the electron energy loss in a pear-shaped volume whose width is roughly equal to the penetration depth[2]. Hence if the electron accelerating voltage is adjusted for the penetration depth to be just equal to the resist thickness, then the spatial resolution is expected to be roughly equal to the resist depth even if the resist exhibits low contrast. For resist films in the micron range, accelerating voltages of 10 to 20 kV fulfill these optimum resolution conditions. For thin resists, the accelerating voltage can be reduced and this improves spatial resolution by decreasing the diameter of the scattering volume. Unfortunately for a 1000 Å thick resist the optimum accelerating voltage lies below 2 kV[3] and no electron beam machine works satisfactorily at such low a voltage. Only short (1 mm) focal length and field effect source electron guns would be able, if they existed, to work satisfactorily at such low voltages.

If the accelerating voltage is increased above the optimum voltage, the beam penetrates into the underlying substrate and backscattered electrons expose the resist significantly. This halo extends to a distance roughly equal to the radius of the scattering volume (fig. 1). This blurring effect is of particular concern in submicron microlithography with classical electron beam machines since the diameter of a 20 kV scattering "pear" is 4 microns. With low contrast resists, this halo cannot be avoided. With high contrast resists, the blurring effect can be cut down or off by a precise choice of the local dose in order to keep the backscattered dose below the exposure threshold of the resist. Most electron beam machines are now equipped with so-called "proximity corrections" which calculate the local dose for best resolution. This points out the major importance of contrast in thin film high resolution

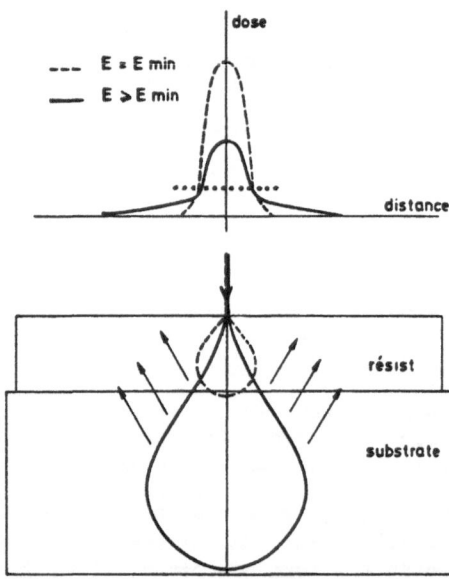

Figure 1. Electron scattering and back-scattering in a thin resist exposed at low (optimum) accelerating voltage (dotted lines) and high accelerating voltage (full lines). Top curves show the relationships between the dose and the distance to the electron beam ; bottom drawings show the scattering volumes.

microlithography : low contrast resists unavoidably require low accelerating voltages, high contrast resists can work satisfactorily at high voltages with proximity corrections.

The requirements of high sensitivity and high contrast, already difficult to fulfill simultaneously in classical spin-coated resists, are even more difficult to meet in thin film resists since only solid state chemical reactions can be performed in these films. Several thin organic films, down to a single monomolecular layer, have been tried to date. The results obtained with them are reviewed hereafter.

ω-TRICOSENOIC ACID

Langmuir-Blodgett mono or multilayers are among the most regular and compact ultrathin films known to date[4]. The compact monomolecular film is made at the surface of water and then transferred onto a substrate by dipping this substrate in or out through the water surface (fig. 2). The film is built up by superimposing successive layers one after another, every layer being some 30 Å thick. The result is a highly organized lamellar solid with, in properly layered films, a very low density of pinholes or defects. The film needs no further treatment and can be used as such.

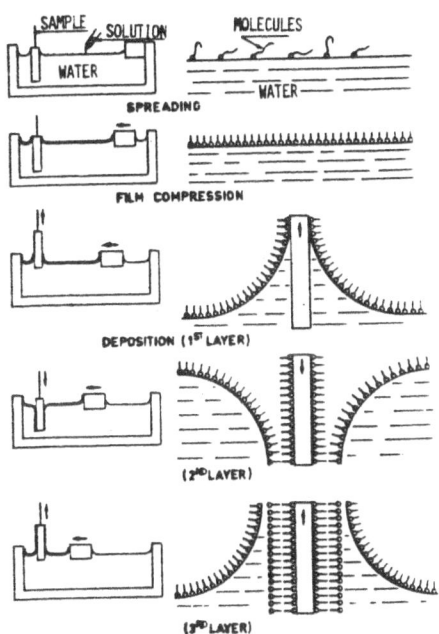

Figure 2. Different steps of the Langmuir sequence (for Langmuir-Blodgett film fabrication) : spreading, compression and transfer.

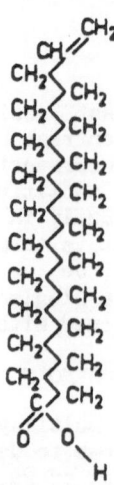

Figure 3. ω-tricosenoic acid.

ω-tricosenoic acid (fig. 3) is a molecule particularly well suited for making L.B. films since it withstands high surface pressure (this makes film fabrication easy) and it allows such a high layering speed (typically 50 times faster than classical fatty acids) that it compares favourably with spin coating[5]. The double bond of omega-tricosenoic acid polymerizes under irradiation by an electron beam[3]. The mechanism of this polymerization is not fully understood yet but the following tentative mechanism, verified by all its consequences, can be put forward. Since propagation is efficient in this solid state polymerization (estimated polymer length : 20 to 50 monomer units), the polymer bond length is geometrically compatible with the monomer lattice parameter (this condition is required for solid state polymerization to take place easily[6]). This gives ω-tricosenoic acid its high sensitivity[7]. But the properties of ω-tricosenoic acid can only be explained if one assumes the polymer

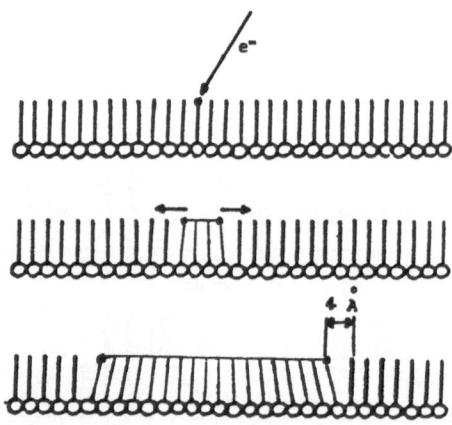

Figure 4. Presumed mechanism of the lattice controlled polymerization of ω-tricosenoic acid. Top : initiation. Middle : propagation. Bottom :stress induced gap which stops propagation.

to be homo or very little dispersed[8]. This may arise from a slight mismatch between the polymer and the monomer lattice which gives rise to a strain increasing at every step of the propagation (fig. 4) until the gap between the chain and the next monomeric unit becomes large enough (4 Å) to stop propagation[9]. This solid state, lattice controlled, stress limited, self induced limitation of the propagation gives rise to polymer units of uniform length. Accordingly this monodispersity gives rise to threshold effects when time limited dissolution is used to develop this resist. This gives ω-tricosenoic acid its high contrast[10].

Typical results with ω-tricosenoic acid are as follows :
- electron beam negative resist.
- high sensitivity (0.5 μC/cm^2 at 5 kV).
- high contrast (2.5 to 3.5 according to development).
- intrinsic spatial resolution (after development) : around 100 Å.
- experimental resolution after development (beam size : 350 Å, thick substrate) : 500 Å (fig.5).
- experimental resolution after plasma etching : 1000 Å (fig. 6).
- plasma etching selectivity factor (no hardening treatment):around 3.
- low pinhole density.
- high adhesion.

Although very attractive, these results need some comment to show the drawbacks and the limitations of this type of material. Firstly when film thickness is reduced below 400 Å in order to try to improve spatial resolution, the film becomes somewhat transparent to active species of the plasma (Cl*). Accordingly, the etching speed is not strictly zero underneath the resist film, although the film remains untouched[11]. High resolution of the resist can only be

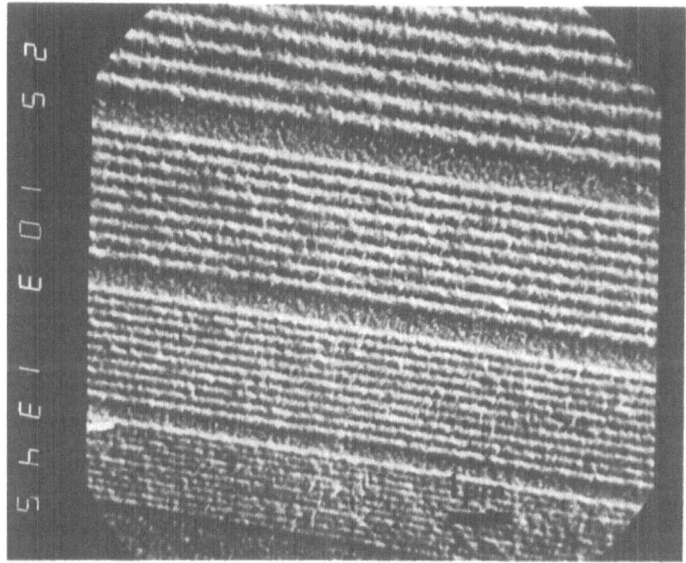

Figure 5. Resolution test on ω-tricosenoic acid after development. Beam size : 350 Å. Film thickness : 1000 Å. Accelerating voltage : 5 kV. Substrate : Aluminum on glass. The last set of lines at bottom left are 500 Å in width, separated by 500 Å gaps.

obtained because ω-tricosenoic acid polymerizes into a linear, non crosslinked polymer : if surrounded by solvent molecules, it dissolves instead of swelling as cross-linked polymers do. Swelling degrades spatial resolution by spreading the polymer network during development. This is confirmed experimentally in L.B. films by a number of reticulable molecules derived from ω-tricosenoic acid : in spite of an acceptable contrast, none of them allows delineation more precise than 2500 Å whatever film thickness. The last comment is a call for further research in the field : all the resists derived from ω-tricosenoic acid are negative resists. Very little data (see below) is available on thin film positive resists. Several molecules are candidates for an L.B. positive resist. A precise characterization of their properties would be of high interest.

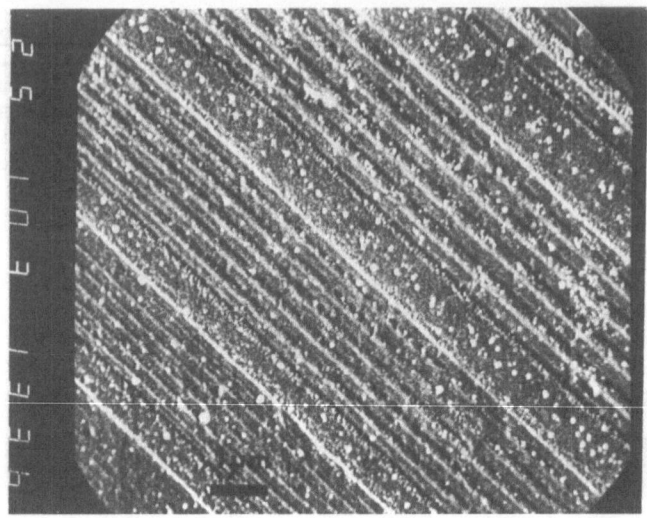

Figure 6. Resolution test on ω-tricosenic acid after plasma etching. Substrate : Al 1200 Å thick on glass. Resist film 800 Å thick. Gas : CCl₄. The last set of lines at bottom left are 1000 Å gaps.

ADSORBED MONOLAYERS

Virtually any surface adsorbs organic molecules from the surrounding atmosphere. The adsorption is generally reversible so that the adsorbed layer is, in most cases, limited to a fraction of a monolayer. Under an electron beam, the adsorbed molecules become highly reactive and can attach to neighbours or to activated substrate groups. This renders adsorption irreversible and molecules accumulate under the beam. Advantage was taken of this effect by Broers[12] to show the ultimate limits of electron beam lithography. Using a very fine beam (less than 10 Å) at a high accelerating voltage (45 kV) on a very thin substrate (no backscattering), he showed that lines 80 Å wide could be obtained. This is the very limit of electron beam lithography since the average range of secondary electrons in the energy range convenient for triggering chemical reactions is around 100 Å.

Table 1. Performances of thin resists

Resist	ω-tricosenoic acid	pollution layer	12-8 diacetylene		octadecyl acrylic acid			C$_{18}$ ω oxiran	
Ref	10,11	12	13	16	14	14	14	15	15
Depos. Method	L.B.	adsorp.	L.B.	single cryst.	L.B.			L.B.	
Film Quality	very good		poor						
Type of resist	Neg	Neg	Neg	Neg	Pos	Pos	Neg	Neg	Pos
Mechan.	Polym.	Activated Adsorption Polym.			film degrad.		U.V. pre polym +polym	Polym.	Degrad.
Develop.	monomer dissol.	no	monomer dissolution		mon. dis.	no	mon. diss.	monomer dissolution	
Dose	1μC/cm²	5mC/cm²	<1μC/cm²	1μC/cm²			1μC/cm²	450 μ/Ccm²	600 μC/cm²
Contrast	3		poor				poor	poor	poor
Resol.(nm)	50	8		5000	50	250	1000	220	150
Plasma Etching Resistance	very good		poor						

OTHER L.B. RESISTS

Virtually any polymerizable molecule can be claimed to be a negative resist and any degradable chain a positive one. However, besides undergoing polymerization or scission, a film must fulfill specific requirements for resist application :

1 - it must be compact and pinhole free.
2 - its contrast must be above 1.5.
3 - its sensitivity must range below 5 $\mu C/cm^2$.
4 - it must withstand plasma action acceptably.

A promising family of L.B. films from the viewpoint of polymerization are films of amphiphilic diacetylenes[13]. Unfortunately most of these films do not fulfill condition 1 and are being progressively abandoned.

A very promising experiment was presented by Lando[14] in 1982. He showed that ∝-octadecyl-acrylic acid worked as a positive resist and that the image was visible by SEM without development. From the pictures published condition 1 seems to be fulfilled. Unfortunately no experiments have been performed concerning requirements 2, 3 and 4.

Very recently a British team[15] showed that ω-epoxy fatty acids in L.B. films worked either as a positive or as a negative resist according to exposure conditions. From the pictures published, condition 1 seems to be fulfilled but contrast seems low, although it has not been measured yet. Requirement 4 has not been checked either.

This number of preliminary data shows that after a few dead years ω-tricosenoic acid is no longer alone and other L.B. molecules are being studied with a glance to resist application. Unfortunately their resist characterization is most of the times very incomplete. Table I summarizes their resist properties when known.

CONCLUSION

The attractive features of L.B. resists are :
1. their extreme thinness.
2. their highly regular thickness and their compactness.
3. their unique chemistry.

Either coupled to a new generation of low voltage electron guns if they exhibit low contrast, or used in conventional electron guns with proximity correction if they exhibit high contrast, they could potentially give a new thrust to electron beam lithography in a resolution range where other techniques are expected to fail, i.e. below 3000 Å. The high layering speed of ω-tricosenoic acid makes it the most attractive of the L.B. molecules presently available. However the advantages of L.B. resists from the viewpoint of resolution have incited several laboratories to undertake research programs on new L.B. resists, especially positive resists, and encouraging preliminary results are being obtained. This is not surprising since L.B. films are undergoing a very fast development in several other fields.

Since silicon circuits are expected to be scaled down later in the vertical direction than in the horizontal, etching problems should arise from the thinness of L.B. resists in spite of their good etching selectivity factor : unless L.B. resists are spectacularly improved, engraving one micron of silicon protected by only 1000 Å of resist seems unrealistic. Therefore the three layer technique (L.B. film, inorganic film, thick polymer) seems the method of choice to provide enough protection to etch thick circuits without losing the resolution advantages of L.B. films.

The need for thin resists is not yet too pressing since largely submicron circuits will enter production only after 1990. However we are clearly witnessing an increasing interest in the field of thin resists as microlithography passes the micron mark.

REFERENCES

1. G.L. Gaines Jr., "Insoluble Monolayers at Liquid-Gas Interfaces" Interscience, John Wiley, N. York (1966).
2. R. Shimizu, T. Ikuta, T.E. Everhart, W.J. Devore, J. Appl. Phys., 46, 1581.
3. A. Barraud, C. Rosilio, A. Ruaudel-Teixier, J. Colloid Interface Sci., 62, 509 (1977).
4. A. Barraud, J. de Chimie Physique 82, 683 (1985).
5. I. Peterson. IEE Proceedings, 130, Pt.I, 252 (1983). Paper presented at the Congress on Polymerizable Films, London, March 13, 1985.
6. H. Morawetz, D. Rubin, J. Polym. Sci., 669 (1962). G. Wegner, Chimia, 28, 475 (1974).
7. A. Barraud, Thin Solid Films, 85, 77 (1981).
8. A. Barraud, Mol. Cryst. Liq. Cryst., 96, 353 (1983).
9. G.M.J. Schmidt, "Reactivity of Photoexcited Organic Molecules" Wiley, Bruxelles, 227 (1967).

10. A. Barraud, C. Rosilio, A. Ruaudel-Teixier, Thin Solid Films, 68, 99 (1980).
11. A. Barraud, C. Rosilio, A. Ruaudel-Teixier, Proceedings of "Microcircuit Engineering 79", I.S.E., Aachen, 127 (1979).
12. A.N. Broers, W.W. Molzen, J.J. Cuomo, N.D. Wittels, Applied Physics Letters, 29, 596 (1976).
13. G. Lieser, B. Tieke, G. Wegner, Thin Solid Films, 68, 77 (1980).
14. G. Fariss, J. Lando, S. Rickert, Thin Solid Films, 99, 305 (1983).
15. B. Boothroyd, P. Delaney, R. Hann, R. Johnstone, A. Ledwith, British Polymer Jour, 17, 360 (1985).
16. R. Mondong, H. Baessler, Chem. Phys. Lett., 78, 371 (1981).

DRY DEVELOPMENT OF MONOLAYER-BASED ELECTRON-BEAM RESISTS

B. M. J. Kellner and G. Czornyj

International Business Machines Corporation
East Fishkill Facility
Hopewell Junction, New York   12533

The application of a dry development process to
Langmuir-Blodgett films used as electron-beam resist
materials is demonstrated. This dry process involves the
use of high vacuum in conjunction with heat. Good
results with regard to resolution and pattern definition
were obtained with vinyl tetracosanoate as the resist
material. The importance of tailoring the physical
properties of the monolayer material to the electron-beam
exposure conditions is emphasized: A deposited film of
vinyl octadecanoate melts under the conditions present in
the exposure chamber, with the result that the pattern lines
coalesce, leading to a poorly resolved pattern, whereas the
corresponding deposited film of vinyl tetracosanoate is
stable under the exposure conditions, and thus gives a
well-resolved pattern.

INTRODUCTION

The potential of utilizing monomolecular resist materials for
high-resolution, electron-beam lithography has been a subject of
considerable interest over the past several years.[1-6] This is mainly due
to the fact that proximity effects do not degrade their resolution
performance as a result of the extreme thinness of these resist layers.
The usual method of developing these resist patterns has been by means of
a conventional wet-chemical process, e.g., selective dissolution in a
solvent.[7] Today, however, for the fabrication of VLSI circuits, there is
a trend toward using "dry" development processes in all areas of resist
work.[8] There are several advantages to such dry processes, the most
obvious of which is the elimination of potentially hazardous materials.
In addition, various defects in developed resist patterns have been traced
to the influence of the multi-step/solvent development procedure. Among
these defects are adhesion failure, delamination, swelling and cracking of
films, and pin-hole formation in the resists. This paper demonstrates the
operability of a dry development process (using heat and vacuum) with one
particular monomolecular resist system. Since the purpose of this work
was simply to evaluate the dry development technique, no attempt was made
to monitor the sensitivity or the resolution of the resist materials per
se or to evaluate the protective properties of the developed resist.

EXPERIMENTAL

Materials

On the basis of earlier work which had demonstrated the[9-11] monolayer/multilayer polymerization of vinyl octadecanoate, two members of the series of vinyl esters of long-chain fatty acids (vinyl octadecanoate and vinyl tetracosanoate) were evaluated as potential electron-beam resists. Both were synthesized from the parent fatty acids and vinyl acetate by ester exchange.[12] The purification procedures and melting points are given in Table I.

Table I. Vinyl Ester Purification and Melting Points

Ester	Purification procedure	Melting Point [^{o}C]
Octadecanoate	1) recrystallization from acetone 2) recrystallization from heptane	36
Tetracosanoate	1) recrystallization from acetone 2) preparative thin layer chromatography using Silica Gel GF plates and a solvent system of 20:80 methylene chloride/hexanes 3) recrystallization from heptane	55-56

Film Deposition

Multilayer films of the esters were built up by means of the well-known Langmuir-Blodgett technique.[13] This method allows one to transfer an oriented monomolecular film (of ca. 3 nm chain length), pre-formed at the air/water interface, onto a suitable solid substrate (such as a silicon wafer) one monolayer at a time. The result is a quasi-two-dimensionally crystalline film which can be built up at will by depositing the desired number of monolayers on top of each other. The automated monolayer apparatus has been described elsewhere.[14] Experimental isotherms at the air/water interface for the materials used in this study are given in Figure 1. Conditions for Langmuir-Blodgett deposition are given in Table II.

Electron-Beam Exposure

All exposures were done using a manufacturing electron-beam exposure tool located at the East Fishkill, New York, Facility of the International Business Machines Corporation. Patterns of repeating lines/spaces of systematically varied dimensions were used as test structures to evaluate the resolution of the developed resists.

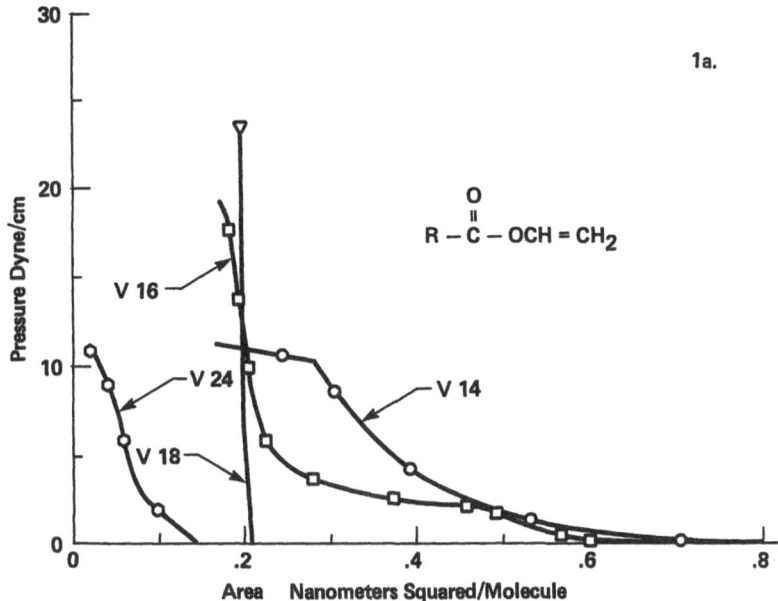

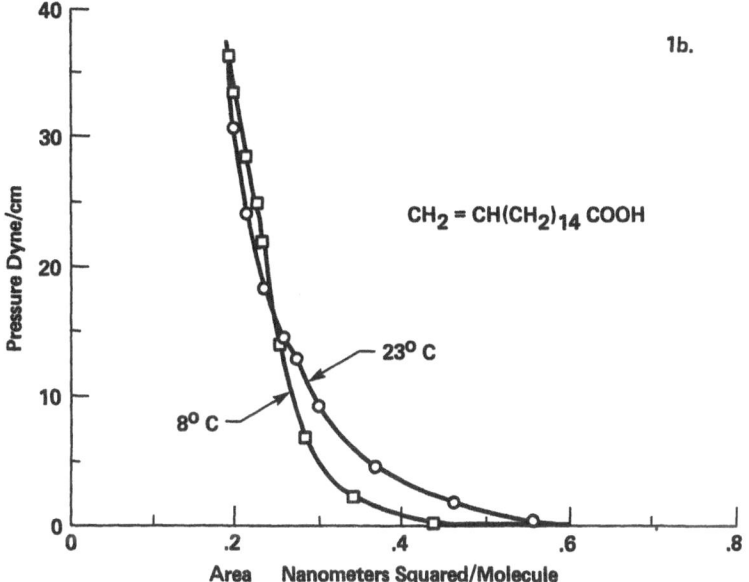

Figure 1. Compression isotherms for the monolayer materials on pH 6
water.
a. Vinyl ester series at 23°C
      V14 = vinyl tetradecanoate
      V16 = vinyl hexadecanoate
      V18 = vinyl octadecanoate
      V24 = vinyl tetracosanoate
    Note that the apparent areas/molecule for the V24 are considerably
      smaller than the cross-sectional area for this molecule. This is
      an experimental artifact caused by the extreme rigidity of the
      film. At 35°C, the isotherm for V24 is effectively identical to
      that for V18 at room temperature.
b. 16-heptadecenoic acid at 8°C and 23°C.

RESULTS AND DISCUSSION

Vinyl Octadecanoate

Initial experiments with films of vinyl octadecanoate which had been exposed at 10-20 $\mu C/cm^2$ exhibited the gross features of the exposed pattern immediately after removal from the electron-beam writing tool (see Figure 2). Even some individual lines were discernible. After a 15-minute development in methanol, however, no additional detail was apparent. The finer lines in the pattern remained unresolved.

Two questions arose from this preliminary experiment: (1) Why, even prior to formal development, was the pattern discernible? (2) Why, after development, was the resolution very poor? The answers to both of these questions lay in the environmental conditions prevailing in the exposure chamber of the electron-beam writing tool. The chamber was under vacuum (10^{-6} Torr), and the ambient temperature was ca. $35^{\circ}C$. It is well known that electron-beam exposure of such resist materials results in cross-linking of the monomers.[1] This in turn raises the sublimation point of the exposed areas vis-à-vis the unexposed areas. The end result is that the material in the unexposed regions sublimes at a faster rate than the material in the exposed regions. The necessary conditions are thus given for the possibility of dry development, even within the exposure tool itself. The appearance of lines written by electron-beam scanning of Langmuir-Blodgett films of manganese octadecanoate was reported earlier,[15] but the interpretation of the phenomenon was not straightforward.

With regard to question (2), it must be kept in mind (see Table I) that the melting point of bulk vinyl octadecanoate is $35-36^{\circ}C$. The

Table II. Conditions for Langmuir-Blodgett Deposition

Monolayer material	Liquid substrate temp. [$^{\circ}C$]	Solid substrate[a]	Dipping film pressure [mN/m]	Dipping speed [cm/min] Layer one	Subsequent layers	Type deposition[13]
Vinyl octadecanoate	22	Al	17	0.3	1.5	Y
Vinyl tetracosanoate	35	Al;C20	7.5	0.3	1-3.75	X
16-heptadecenoic acid	8	Al;C20	27	0.3	1.0	X
			35	0.3	1.0	Y

[a] The following abbreviations are used for solid substrates:
Al = aluminum (250 nm) vacuum-evaporated onto silicon;
C20 = 1 monolayer of cadmium arachidate deposited onto silicon.

In all cases, the liquid substrate was pH 6 water.

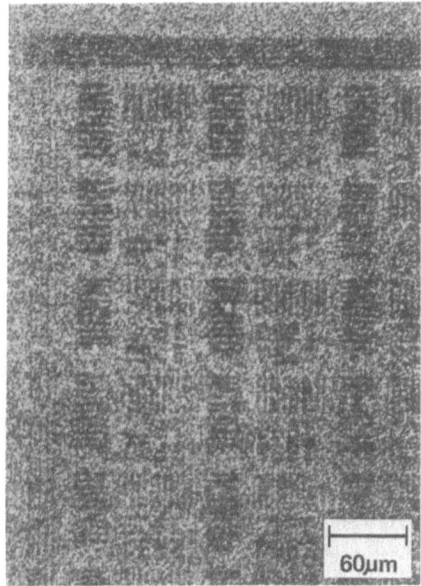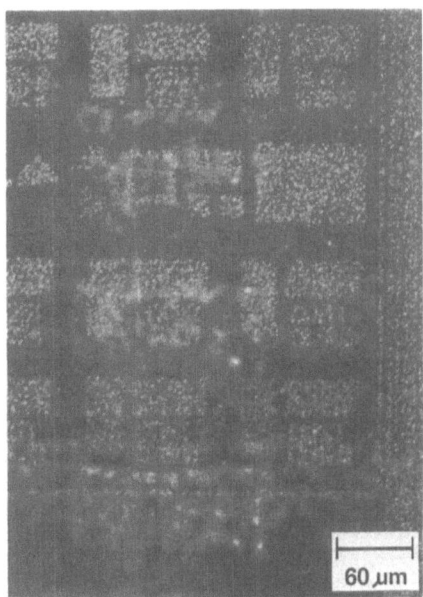

Figure 2. Optical micrographs of an electron-beam exposed film of vinyl octadecanoate before development. The resist [(21 monolayers in (a) and 23 monolayers in (b)] was deposited onto a silicon wafer onto which 250 nm of aluminum was vacuum-evaporated. The exposure dosage was (a) 20 $\mu C/cm^2$ and (b) 10 $\mu C/cm^2$.

practical coincidence of this temperature with the temperature of the exposure chamber in the electron-beam writing tool suggested that the reason for the lack of line resolution in the developed resist was the melting of the resist material prior to or during the electron-beam exposure. The resultant flow of the resist caused coalescence of the pattern lines.[16]

Vinyl Tetracosanoate

By using a longer-chain analog of vinyl octadecanoate, it was possible to test the latter point since the longer-chain compound has a higher melting point. Using vinyl tetracosanoate, a 12-layer film exposed at 1 $\mu C/cm^2$ again resulted in the general features of the exposed pattern being visible to the naked eye without formal development. Development with solvent (methanol for ca. 10 min) resulted in good line resolution, indicating that exposure of this material, coupled with appropriate development, can give good pattern definition and resolution. In order to test the dry development of such a resist material, initially a 12-layer film of vinyl tetracosanoate was exposed at 10 $\mu C/cm^2$ and placed in a standard vacuum oven. The atmosphere in the oven was kept at 40 Torr and 55°C (just below the melting point of the bulk monomeric material). After 23 hours in the vacuum oven, individual fields in the exposed pattern were somewhat resolved (see Figure 3), showing that development was indeed taking place under these conditions, albeit very slowly and incompletely. That the vacuum/heat treatment had not appreciably affected the pattern as such was shown by a methanol development of the sample which had been subjected to this treatment. A very well-resolved pattern was then obtained.

We considered it likely that the incomplete development observed in the vacuum oven was due to the relatively low level of vacuum which could be obtained therein. This hypothesis was tested by attaching a development chamber to a high-vacuum system (shown schematically in Figure 4). The vacuum attainable with this arrangement was ca. 4×10^{-8} Torr. The temperature within the chamber could be adjusted by means of a variable transformer which controlled the current passing through the heating tape wrapped around the chamber. Unfortunately, the arrangement as constructed did not permit the <u>in situ</u> determination of the substrate temperature; only the temperature on the outside of the chamber walls could be monitored. At an outside temperature of ca. 90°C, development of the pattern was almost immediate (within 1 minute) and complete (see Figure 5).

An important point to keep in mind when designing future monolayer resist materials (or, for that matter, any resist material) was emphasized in the case of the two vinyl esters used in this study: The melting point of the resist film must be higher than the operating temperature of the exposure chamber. This fact would appear to be self-evident, but has so far received little notice. When this condition is specifically applied to monolayers, however, it must be emphasized that there is no necessary correlation between the physical state of a monolayer at the air/water interface and the melting point of the material (either in bulk or as a Langmuir-Blodgett film). As an example, at the air/water interface, vinyl octadecanoate forms a solid film at room temperature (Figure 1a), but, due to its relatively low melting point when deposited onto a solid support, it is an unsatisfactory resist material. On the other hand, 16-heptadecenoic acid forms a liquid-expanded film at room temperature at the air/water interface (Figure 1b), thus making it is necessary to carry out Langmuir-Blodgett deposition at a reduced temperature (ca. 8°C); it is a quite satisfactory resist material from the the point of view of resolution.[17]

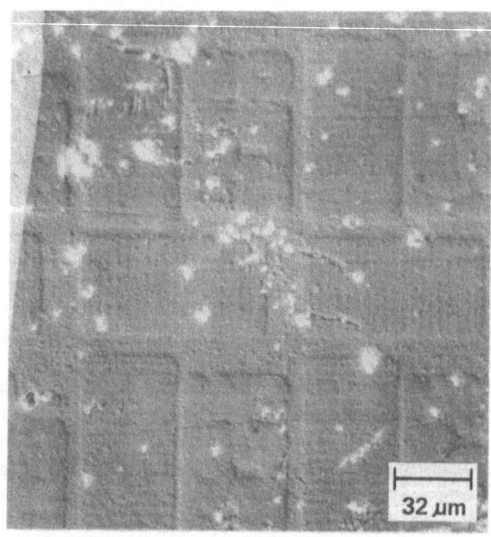

<u>Figure 3</u>. Optical micrograph of an electron-beam exposed film of vinyl tetracosanoate showing incomplete development in a vacuum oven at 55°C and 40 Torr for 23 hr. The resist (12 monolayers) was deposited onto a silicon wafer onto which 250 nm of aluminum had been vacuum-evaporated. The exposure dosage was 10 μC/cm^2.

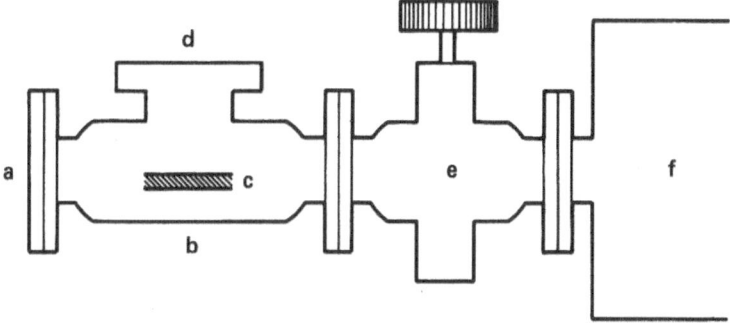

Figure 4. Schematic of the dry development chamber.
a. Sample insertion port.
b. Sample chamber [(wrapped with heating tape (not
   shown)].
c. Sample onto which the exposed monolayer resist had
   been deposited.
d. Observation window.
e. Gate valve.
f. High-vacuum system.

Figure 5. Optical micrograph of an electron-beam exposed film of vinyl tetracosanoate showing successful development in the heated high-vacuum system. The resist (12 monolayers) was deposited onto a silicon wafer onto which 250 nm of aluminum had been vacuum-evaporated. The exposure dosage was 1 μC/cm^2. 250 nm of aluminum was vacuum-evaporated on top of the developed pattern.

CONCLUSIONS

Using an ultrathin electron-beam resist material built up from monomolecular films by the Langmuir-Blodgett technique, a dry development technique was demonstrated. This dry development entails holding the exposed substrate in a vacuum at a temperature somewhat below the melting/flow point of the deposited resist material. Good results with respect to resolution and line definition were obtained using vinyl tetracosanoate as the resist material. Experiments with a shorter-chain analog of this compound (vinyl octadecanoate) emphasized the necessity of care in the selection of the resist material: The melting/flow point of the deposited resist must be higher than the prevailing temperature in the electron-beam exposure tool in order to prevent loss of resolution due to film melting/flow concurrent with electron-beam exposure.

REFERENCES

1. A. Barraud, C. Rosilio, and A. Ruaudel-Teixier, J. Colloid Interface Sci. 62, 509 (1977).
2. A. Barraud, C. Rosilio, and A. Ruaudel-Teixier, Solid State Technology 22, 120 (1979).
3. A. Barraud, C. Rosilio, and A. Ruaudel-Teixier, Thin Solid Films 68, 99 (1980).
4. A. N. Broers, J. M. E. Harper, W. W. Molzen, and M. Pomerantz, I.B.M. Technical Disclosure Bulletin 21, 3035 (1978).
5. G. Fariss, J. Lando, and S. Rickert, J. Mat. Sci. 18, 2603 (1983).
6. I. R. Peterson, IEE Proc. I130, 252 (1983).
7. T. Davidson, Editor, "Polymers in Electronics," A.C.S. Symposium Series No. 242, passim, American Chemical Society, 1984.
8. T. C. Penn, Science 208, 923 (1980)
9. A. Cemel, T. Fort, Jr., and J. B. Lando, J. Polymer Sci. 10, 2061 (1972).
10. S. A. Letts, T. Fort, Jr., and J. B. Lando, J. Colloid Interface Sci. 56, 64 (1976).
11. D. Day and J. B. Lando, J. Polymer Sci. 16, 1431 (1978).
12. D. Swern and E. F. Jordan, Jr., "Organic Syntheses," Collective Volume 4, 977 (1963).
13. G. L. Gaines, Jr., "Insoluble Monolayers at Gas-Liquid Interfaces," Wiley-Interscience, New York, 1966.
14. B. M. J. Kellner, Technical Report TR22.2326, "A Versatile Apparatus for Manipulation of Monomolecular Films at the Air/Water Interface", I.B.M.-East Fishkill, Hopewell Junction, New York, 1980.
15. A. N. Broers and M. Pomerantz, Thin Solid Films 99, 323 (1983).
16. B. M. J. Kellner and G. Czornyj, Colloid Polymer Sci., 263, 413 (1985).
17. B. M. J. Kellner, G. Czornyj, and A. Wu, unpublished observations (1979).

EXPERIMENTS ON LANGMUIR-BLODGETT MAGNETIC MONOLAYERS OF

MANGANESE STEARATE

Melvin Pomerantz

IBM T. J. Watson Research Center
Yorktown Heights, N. Y., 10598

A review is presented of the methods and results of the creation of
magnetic materials of monolayer thickness by the Langmuir-Blodgett
technique. Films of manganese stearate containing as few as a single
layer of magnetic ions have been prepared. Some theoretical considerations on two-dimensional magnetism are mentioned, as well as the relevance to microelectronics.

1. INTRODUCTION

It has been remarked that the motto of IBM should be changed from "THINK" to
"THINK small ". It is now a truism that the improvements in electronic devices are
often related to ever smaller sizes. The most discussed advances are in the fabrication
of electrical devices; this Symposium contains several papers about the possible uses
of Langmuir-Blodgett films as resists to define structures on surfaces. Perhaps less well
appreciated is the the drive for ever smaller magnetic devices. There, too, small size is
a key to greater density of information and switching speed. At stake is a market for
storage devices of about 25 billion dollars annually and growing rapidly. At present the
obstacles to further progress seem to be the extraordinary mechanical requirements of
flying a read/write head very close to the storage disk[1] One can enquire about the requirements on the magnetic media that might arise in the future. Indeed one might ask
whether it will be possible to go the ultimate limit of a magnetic medium that is only
one atom in thickness, and what the physical consequences may be. A seemingly om-
inous effect has been predicted by Bloch[2], and is now also referred to as the Mermin-
Wagner Theorem[3], that for continuous symmetric interactions among the magnetic
ions, in two-dimensional structures there could be no long-ranged magnetic order.

The creation of literally two-dimensional (2-d) magnets in order to test these predictions was finally accomplished in two ways: the adsorption[4] of a monolayer of
gaseous magnetic molecules (O_2) and the deposition of monolayers of organic molecules to which magnetic ions had been chemically bonded[5]. The organic molecule used
was manganese stearate (abbr. MnSt), which was deposited by the Langmuir-Blodgett

technique. Since Langmuir-Blodgett films are one of the themes of this Symposium, this paper will review the methods and results obtained thus far on MnSt. The Langmuir-Blodgett method has distinct advantages if one wishes to prepare a material that is precisely one magnetic atom in thickness. The more conventional approach to making magnetic thin films, evaporation of ferromagnetic metals, is presently the most practical way to make devices. Evaporation is not necessarily the most convenient way to make a monolayer magnet.

The major advantage of the evaporated metal films is that they are ferromagnetic at room temperature, down to at least a thickness of 0.4 nm in the case of Fe[6]. Thinner films may be discontinuous, which is a problem when trying to make monatomic films by the evaporation method. An interesting development is the use of molecular beam epitaxy to grow single crystal films of Fe on GaAs. The thinnest films so far reported[7] are about 2 nm. but they have good crystal structure, albeit with strain due to lattice mismatch, and well-defined magnetic anisotropy. Their ferromagnetic resonance data can be explained by including a surface magnetic mode introduced by Rado. Fe grown in 10^{-6} vacuum on carbon coated polished Si wafers has been reported[8] to be exceptionally smooth. Such films have been found[9] to be ferromagnetic down to thicknesses of about 0.8 nm.

A rather different approach[4] was used by Mc Tague and Nielson to obtain a sample that had monolayer coverage, but a macroscopic amount of sample. Their substrate was grafoil, which is graphite that has been exfoliated, like puff pastry, to expose about 100 m^2/g of interior surfaces of the graphite. This was permeated by known amounts of O_2 gas, such that when it was cooled the O_2 condensed with the desired coverage. This molecule is magnetic, so that they could search for magnetic ordering. They observed neutron diffraction peaks due to antiferromagnetic order below 10 K . Again because of the large quantity of sample available in this method (about 0.1 gram of O_2 per gram of substrate), a range of properties have been measured: magnetic susceptibility[10], x-ray crystal structures[11], specific heat[12], effects of substrates[13], and dilution by other gases[14]. This method is obviously the least practical for devices.

Section 2 will review some of the theoretical predictions of the importance of dimensionality effects in magnetism. In Section 3 I summarize the Langmuir-Blodgett method for the deposition of organic monolayers, the preparation of the magnetic molecule MnSt, and the results of our experiments. In the final section I indicate some future directions for the applications of magnetic Langmuir-Blodgett films.

2. SOME RESULTS OF
THEORY OF THE EFFECTS OF DIMENSIONALITY ON MAGNETISM

The modern theory of magnetic phase transitions hypothesizes that there are only two parameters that determine the nature of the transition[15]. These are the dimensionality of the spatial structure and the symmetry of the interaction between the elementary magnets. The effect of dimensionality was pointed out by Bloch[2] in a calculation of the effect of temperature on the magnetization of a ferromagnet. He assumed highly symmetric Heisenberg coupling between spins, $\vec{S}_i$ and $\vec{S}_j$, such that the energy was $E = J\vec{S}_i \cdot \vec{S}_j$, where J is the exchange constant. He found that at low temperatures the number of thermally excited spin waves increased as $T^{1.5}$ in 3-d, but that in 2-d the number of spin-waves diverged at any finite temperature. Since each spin-wave represents a decrease in the magnetization of the ferromagnet, an infinite number of spin waves meant that the magnet was disordered in 2-d. The prediction of the absence of order was extended to the x-y interactions, and to antiferromagnetism, by Mermin and Wagner[3].

The predicted instability of ordered structures in low dimensions is so well known that the restrictions on the theory are sometimes overlooked. The results apply only to highly symmetrical interactions. They do not apply if there is uniaxial directionality of either anisotropy or exchange. Only in the limit as any uniaxial anisotropy went to zero did Bloch obtain the result that order was absent. The case of uniaxial exchange is the Ising model, interactions of the form $E = JS_z \times S_z$. For this case there is the celebrated analytic solution by Onsager[16] which showed that there is a second order transition to an ordered state in 2-d. Both the Ising interaction and anisotropy introduce an energy gap in the spin wave spectrum, so that a thermal energy comparable to this gap is needed to excite spin waves. This is the physical difference between such models and the Heisenberg and x-y interactions which have continuous symmetry: in the case of continuous symmetry the spins can execute small deviations from each other, which cost an energy which vanishes as the wavelength goes to infinity. It is the high density in low dimensions of the very low energy spin waves that washes out the ordered state. It is also assumed that the interaction is short ranged. The assumption of short-range forces is true for the exchange interaction, but the dipolar interaction is long-ranged, and may cause ordering.

Even in the cases of symmetric interactions there is a question whether the excitation energy can vanish, because the wavevector cannot go to zero since samples are always of finite size. There has been progress in taking finite size effects into account but very high resolution measurements are needed to distinguish short-range from true long-range order.

In addition, there are questions about the effects of randomness in low-dimensional systems. Since randomness tends to oppose an ordered state, one might wonder whether randomness would be sufficient to prevent order in the Ising model, the one case in which it is permitted by theory. This problem is currently being actively studied. Theory distinguishes two cases: random exchange and random fields. Random exchange could arise because of random introduction of non magnetic spins, or of strains. This is likely for 2-d materials since roughness is present. An analysis, due to Harris[17], suggests that in 2-d the only effect of random exchange will be a shift of the transition temperature; the nature of the transition, the critical exponents, will not change. It would be interesting to test this prediction. The case of random fields is more difficult and is still controversial. It can be realised in a dilute antiferromagnet in an external magnetic field[18]. Theory[19] predicts that random fields will destroy the transition of an Ising system in 2-d; the disagreement is whether it will also prevent order in the 3-d Ising model. It is obviously important to establish the 2-d result experimentally.

3. MONOLAYER MAGNETS OF MANGANESE STEARATE

The technique that we[20], and others[21], have used to prepare literally 2-d magnets was developed by Langmuir and Blodgett [22]. Since other talks in this Symposium and a book[23] describe the method, I shall present only some basic ideas and the particular conditions of our experiments. The Langmuir-Blodgett method is based on the tendency of oils and water to separate when mixed. The molecules used in the Langmuir-Blodgett method have one hydrophobic and one hydrophilic end, and are insoluble in the subphase (usually water). On the surface of water the molecules tend to align with the hydrophilic end submerged in the water, and are further oriented when surface pressure is applied. We used the fatty acid $C_{17}H_{35}COOH$, octadecanoic acid, commonly known as stearic acid. Its structure is illustrated in Fig.1a. A dilute layer of such molecules was spread on the surface of water. The Langmuir-Blodgett method consists of compressing the dilute layer with a movable surface piston until a relatively incompressible solid film is formed. This film is transferred from the water surface by inserting and removing a suitable substrate through the surface. When all goes well, the film adheres to the substrate and is pushed onto the substrate by the continual application of the surface pressure. Thus a single monolayer may be deposited, as illustrated in Fig.1a. If the substrate is reinserted another layer attaches, and another upon exiting. The way in which the layers attach to each other and to the substate is that the fatty surfaces attach to each other and the hydrophilic surfaces attach to each other.

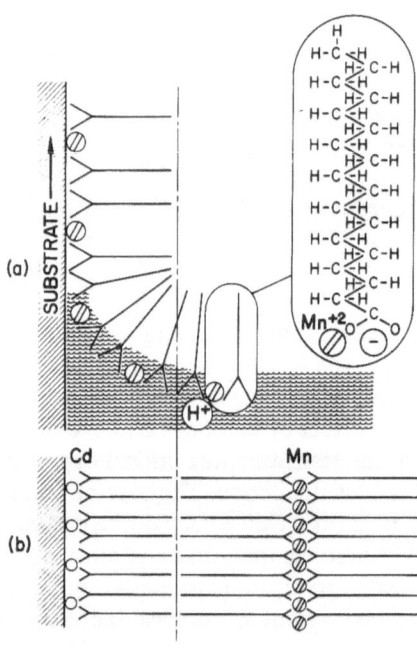

Fig. 1(a) Illustrating the Langmuir-Blodgett method. As the substrate is slowly lifted through the water, a compressed monolayer of long-chain molecules attaches to it. The detail shows the structure of stearic acid with the acidic H removed and a Mn ion attached. (b) The kind of 2-d magnet discussed in this paper was made of a layer Cd stearate and two layers of Mn stearate.

We have made literally 2-d magnets in two ways. The first is shown in Fig. 1a: simply by depositing a single monolayer of MnSt. In the second method, a non-magnetic layer of Cd stearate was deposited and then two layers of MnSt. As illustrated in Fig. 1b, the Mn ions in this structure are in a more symmetrical situation, and also are more dense. Henceforth when I refer to 2-d MnSt I shall be speaking only of the Mn layers formed by the second method. The first type of structure did not show magnetic ordering down to $T = 1.5$ K and so are not of as great interest.

It is important to verify that the layer structure we believe we produce by the Langmuir-Blodgett technique is actually present on the substrate. Such confirmation was obtained[5] by x-ray diffraction. Normally x rays are very penetrating and one would not expect to observe a diffraction effect from a single molecular layer. The Langmuir-Blodgett film has special advantages: its thickness of about 2.5 nm, and a unit cell of two molecules (hence 5.0 nm), results in a Bragg angle of about $1°$ from the grazing angle. This is close to the angle for total internal reflection of x rays and thus the intensity of the interfering beams is relatively large. Fig. 2 illustrates the results we obtained from a 9 layer film on a 4 cm^2 substrate. This required a diffractometer with excellent collimation and monochromaticity, and counting times of about 5 minutes per point. The structure is sufficiently simple that it can be modeled as laminae of the known chemical constituents of the molecule[24]. The resulting calculated diffraction pattern, shown by the solid lines in Fig. 2, is in excellent quantitative agreement with the data.

The in-plane structure of as few as two monolayers of MnSt was observed[20] by Herd and the author using electron diffraction. The structure is similar to that of Pb stearate which has also been measured by electron diffraction[25]. Recently, the in-plane structure of Langmuir-Blodgett films of Pb stearate has been observed[26] by the non-collinear reflection-diffraction of x rays. Some hundreds of layers, and a very heavy metal ion (Pb) in the structure were present in these early measurments, but more general applicability can be expected.

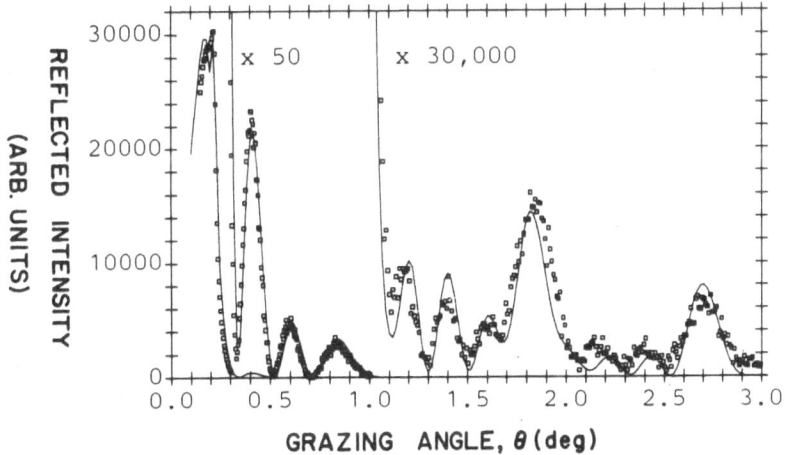

Fig. 2 X-ray diffraction pattern from 9 layers of MnSt$_2$ on a Si substrate. The dots are experimental points, the solid curve is calculated from a model structure described in ref. 24.

To make the monolayers magnetic we bonded magnetic ions to the acidic end of the molecule. Following some results in the literature [27], and our experiments[20], we determined that Mn^{+2} ions would replace the H^+ when the concentration of $MnCl_2$ was 10^{-3} molar and the pH was approximately 7. Under these conditions a monolayer of Mn stearate, $Mn(C_{18}H_{35}OO)_2$, is formed at the surface.

A simple way to study the reaction of the Mn in the water with the stearic acid on the surface was to skim a film on to a Si substrate. The infra-red spectra were recorded on a Beckman spectrograph. As shown in Fig. 3 we could observe the bonding of the Mn to the molecule by the reduction in the COOH carbonyl peak at $6\mu m$ and the growth of an absorption at 6.5 μm, characteristic of bonded metal ions[28]. Using a surface wave technique it was possible to observe ir absorption on as few as one monolayer[29]. The ir absorption showed that Mn was bonded to the stearate molecule, but we had to ensure that there was no other Mn in the samples. This is an unwanted possibility that could occur if precipitates from the bath, or water containing Mn, were trapped in the films. Precautions were taken to avoid this by washing the films with pure water to purge soluble or lodged impurities, and was checked by chemical analyses using both electron microprobe and Rutherford backscattering. The chemical analyses on samples of some tens of layers showed that the Mn concentration corresponded with that expected for MnSt with no excess Mn. Since ir showed that Mn was located at the ionic ends of the molecules, this indicated that Mn was being picked up only in the desired way.

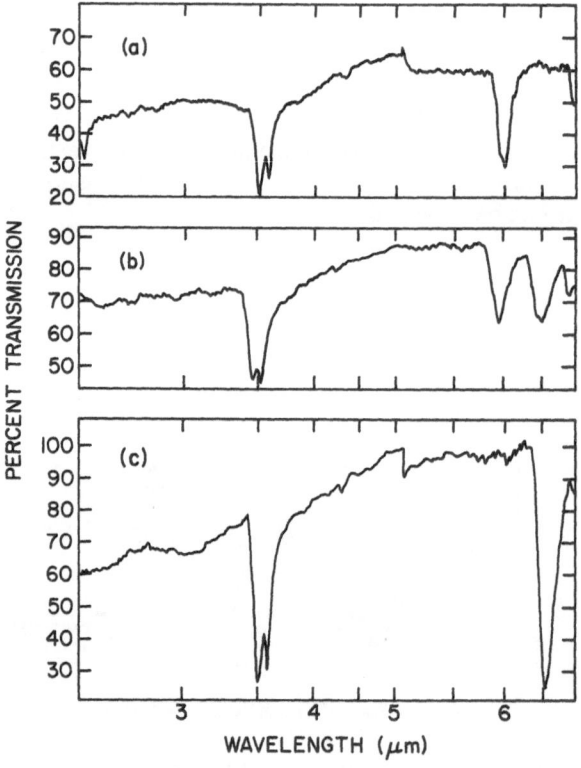

Fig. 3 Infra-red transmission of skims supported by Si wafers. (a) pH = 6.5, no Mn in water. (b) pH = 5.9, 10^{-3} molar Mn^{+2}. (c) pH = 6.8, 10^{-3} molar Mn^{+2}.

Further confirmation that Mn was located in 2-d arrays, and not elsewhere, was found in the electron spin resonance (ESR) in the paramagnetic state. It had been shown[30] that the ESR should possess anisotropy in both the line width, ΔH, and line position, H_0, as a function of θ, the angle of the external magnetic field with respect to the film normal. The line width effect arises because the dipolar broadening depends on spin diffusion, and diffusion depends on dimensionality. In 2-d there are long correlation times which lead to enhancement of the term in the dipole interaction that is at zero frequency, namely the term proportional to $(3 \cos^2\theta - 1)$. In Fig. 4 are measured linewidths vs θ. They fit well to the prediction of a form $\Delta H = A + B(3\cos^2\theta - 1)^2$. The dipolar interaction also contributes an average magnetic field that is anisotropic in 2-d. This results in a shift of H_0 to higher values in the normal direction (to overcome the opposing dipole fields), and to lower fields in the in-plane orientation. This is a small effect but was observed in a multilayer film in the highest field available to us. (See Fig.5.). The significance of these ESR results in the paramagnetic phase is that they demonstrate the anisotropic magnetic characteristics expected for 2-d arrays; there is not a significant amount of Mn in 3-d environments.

ESR was also the method we used[31] to search for spontaneous magnetic ordering. If ordering occurs, it drastically changes the resonance spectrum: the resonance field, the lineshape or the intensity may all be affected. Although ESR has quite high sensitivity, it proved helpful to improve the signal by stacking 50 plates each coated on both sides with a magnetic monolayer. This served to increase the area of the sample, but did not introduce 3-d interactions since the monolayers were separated by macroscopic distances in a partial vacuum. The behavior of the ESR as the temperature was lowered

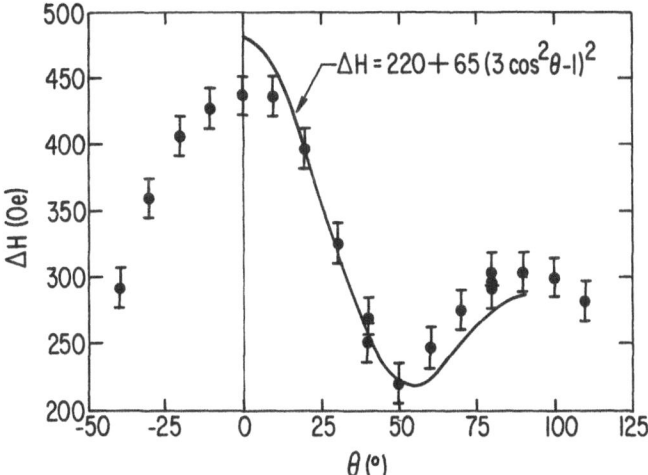

Fig. 4. The measured ESR linewidth, (peak to peak of the derivative) vs. θ, the angle between H_0 and the film normal, of 2-d MnSt in the paramagnetic state. The solid curve is the form $A + B (3 \cos^2\theta - 1)^2$ fitted to the data. $T = 80$ K. $f = 9.3$ GHz.

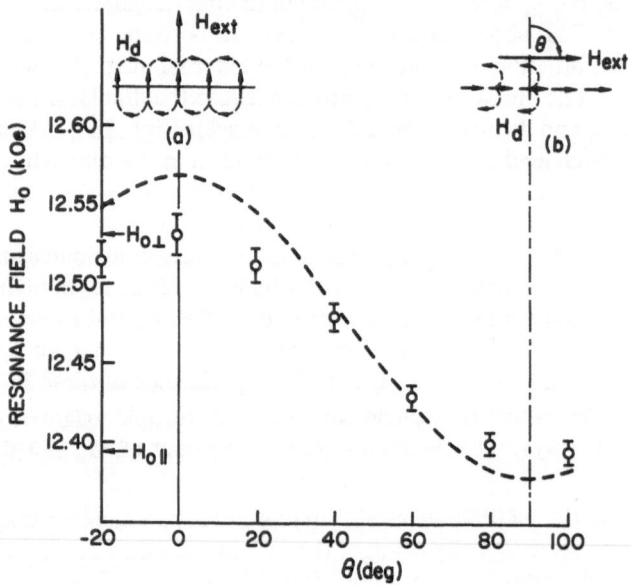

Fig. 5 The ESR field, H_0, vs. θ, of 35 multilayers of MnSt in the paramagnetic state. The dashed curve is a fit to theory of dipolar shifted lines. See ref. 20. T = 80 K. f = 34.8 GHz.

is indicated in Figs. 6 and 7. The line widths was observed to broaden and become independent of orientation. This behavior had been observed in quasi 2-d magnets as they approached a transition to an antiferromagnetic state[30]. The resonance fields for all orientations shifted to lower fields, but the $H_{0\perp}$ showed the greatest shift. Its temperature dependence, shown in Fig. 7, is seen to be very abrupt below about 2 K. This shift of the resonance indicates the rapid development of large internal magnetic fields, which is characteristic of an ordered magnetic state. The anisotropy of H_0 becomes large at low temperatures (Fig. 8), also indicative of an ordered state. I know of no Mn compound that has a remotely similar anisotropy in a paramagnetic phase.

The observations of a large and rapid temperature variation of H_0 and its large anisotropy at low temperatures are the bases for the conclusion that an exactly 2-d magnet of MnSt undergoes magnetic ordering. However the nature of the magnetic state is not certain. We have given[31] arguments that it may be a "weak-ferromagnet". A weak-ferromagnet is dominantly antiferromagnetic, which accounts for the behaviour of the line width, and the susceptibility[20]. However, the opposing magnetic sublattices do not cancel exactly, resulting in a small ferromagnetic moment. This explains why the resonance does not disappear entirely, as happens with an antiferromagnet. To check the plausibility of this explanation we measured the spin resonance of a known[32] weak-ferromagnet, $MnCO_3$, as a function of temperature. We used a powder in order to compare it to the resonances of a powder of MnSt. The derivatives of the resonance absorptions shown in Figs. 9 and 10 are qualitatively similar. They show a line shift to lower fields and broadening which is fairly rapid below

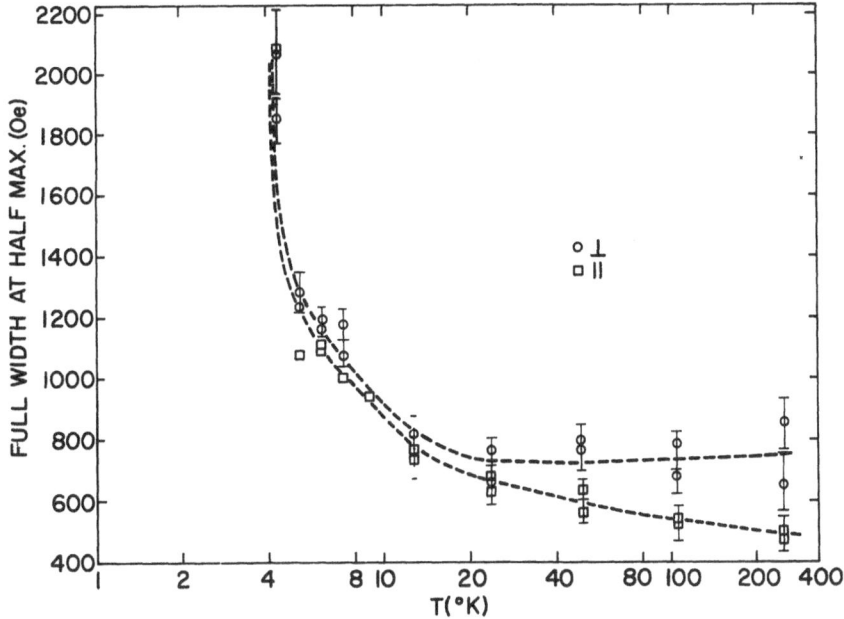

Fig. 6 ESR linewidth (full width at half max. of absorption), of 2-d MnSt vs. temperature. ⊥ and ∥ refer to the direction of the external field with respect to the film plane. f = 9.3 GHz.

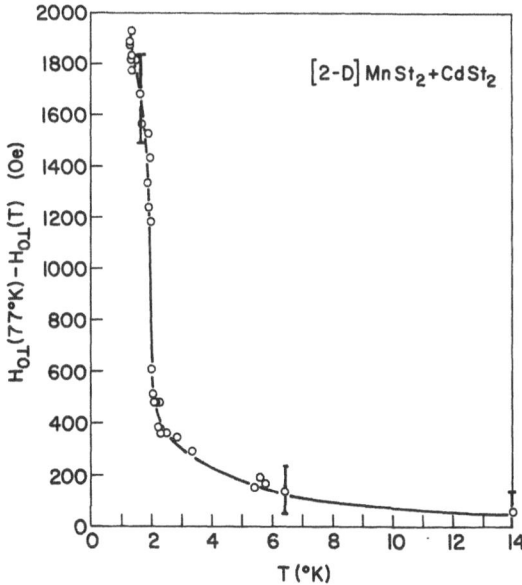

Fig. 7 Temperature dependence of the down-field shift of $H_{0\perp}$, the peak of ESR absorption in perpendicular orientation, for 2-d MnSt. f = 9.3 GHz.

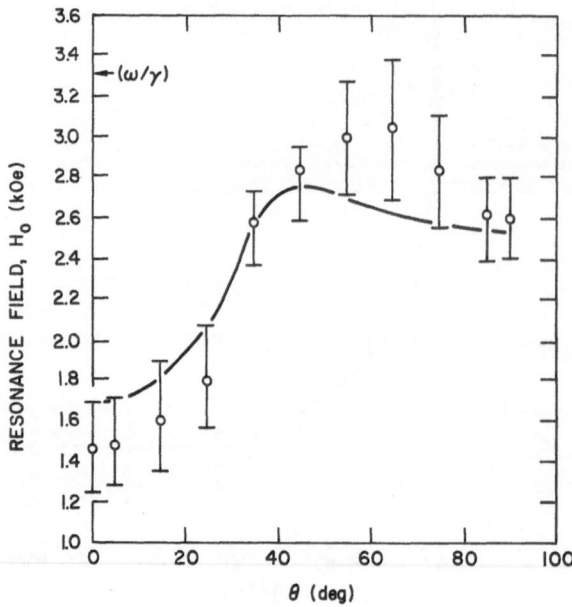

Fig. 8. Angular dependence of resonance fields of 2-d MnSt, at T = 1.4 K . The solid curve is a best fit to the theory of resonance fields of a uniaxial weak-ferromagnet[33] .

some temperature. For Mn carbonate the ordering temperature is known to be 32 K, which agrees well with where strong changes begin. For the MnSt the changes accelerate below about 5 K. Returning to the 2-d films of MnSt, we applied a theory[33] of the resonance fields expected for a uniaxial weak-ferromagnet. The theory contains three unknown parameters which we determined by a fit to the data. The solid curve in Fig. 8 is the best fit we could obtain. Note that it qualitatively reproduces the unusual anisotropy of $H_0(\theta)$. One of the fitting parameters is the temperature of the experiment, if the transition temperature is taken to be known. (We could have assumed the reverse.) If the transition temperatire is taken as 2 K, the computed T = 1.3 K is found to be in agreement with the measured T of 1.4 K. There is some difficulty in reconciling one of the parameters, the exchange field, with an independent estimate derived[34] from the room temperature line widths. We have suggested, alternatively, that the ordered state may be antiferromagnetic, but that there are missing Mn ions which leads to a residual magnetic moment, as proposed by Neel[35]. This idea has some support because antiferromagnetic MnSt has been achieved in a very careful synthesis as a bulk powder[36] . We have not yet modified the deposition of Langmuir-Blodgett films to verify whether films can be made pure antiferromagnets. We also lack a theory of the ESR of a defect antiferromagnet.

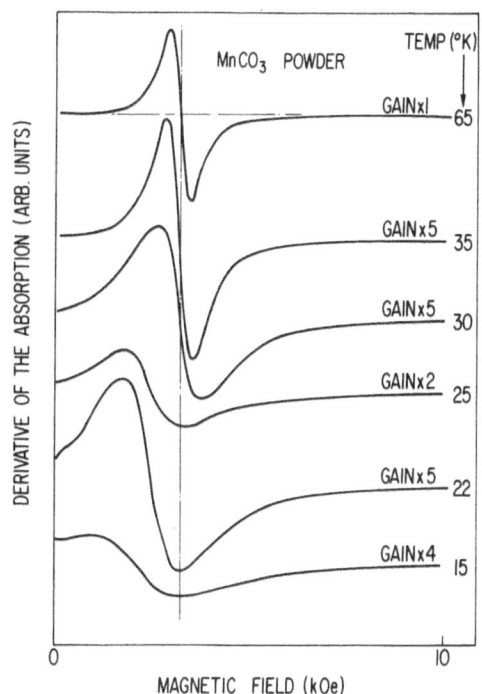

Fig. 9 Electron spin resonances of a powder of $MnCO_3$ at several temperatures above and below the Neel temperature of 32 K.

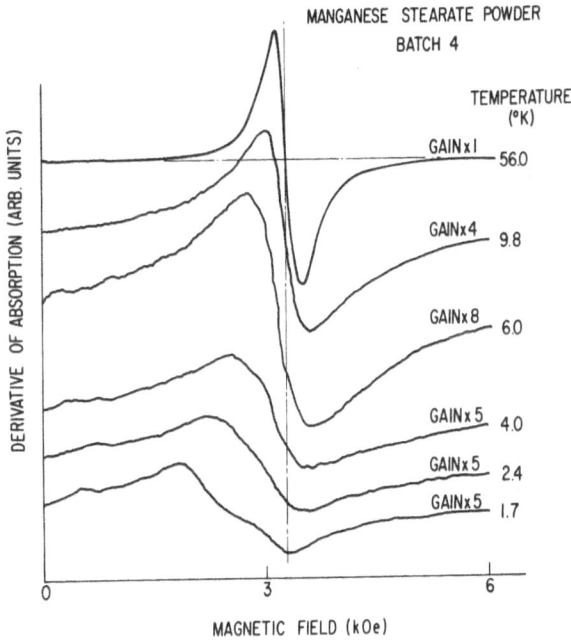

Fig. 10 Electron spin resonances of a powder of Mn stearate (synthesized by A. Aviram), at various temperatures. The behavior is qualitatively like that of $MnCO_3$, but with T_N about 5 K.

4. DISCUSSION

The theme of this Symposium is the relevance to microelectronics of Langmuir-Blodgett films. The work on magnetic films has not yet produced a material that is a candiate for applications. MnSt does seem to undergo magnetic ordering, which was dubious before the experiments were done. These are proof that there is no limit to the thinness that a magnet may have. In fact this is accord with a correct understanding of the theory of two-dimensional magnetism. The impracticality of MnSt arises from the very low temperature at which it orders spontaneously. Our experiments indicated that this may be at about 2 K. Haseda, et al.[21] found an even lower temperature of 0.3 K. However, the Langmuir-Blodgett technique offers a wide range of possible magnetic materials. The most studied case, MnSt described above, should be the best example of a Heisenberg interaction among the transition metal ions, and thus perhaps the worst case for a high ordering temperature. We have examined the other transition metal stearates, and other carboxylates[37], to search for others that might undergo magnetic ordering. The only one that showed clear signs of ordering was ferric stearate[38]. The ordering temperature seemed to be greater than 50 K, much higher than for MnSt. Another class of materials that might have interesting properties are rare-earths, deposited by the Langmuir-Blodgett method. There might be some complications due to the trivalency of these ions. In order to achieve a strong interaction one desires the ions to be close, and thus the minimum amount of hydrocarbon chains. If we are fortunate, the trivalent rare-earths will not bind to three carboxyl chains, but two, as has been found for some trivalent transition metals[38]. Each compound has its peculiarities, so it is difficult to predict which will be interesting and feasible.

5. ACKNOWLEDGMENTS

This work could not have been accomplished without the contributions of many colleagues, including A. Aviram, J. Axe, P. Brosius, Ch. Chen, F. Dacol, F. Ferrieu, G. Grinstein, H. R. Lilienthal, R. Linn, F. Mehran, R. Pelcovits, T. D. Schultz, B. D. Silverman, J. C. Slonczewski, K. W. H. Stevens, and A. R. Taranko. Early work was partially supported by the Defense Advanced Research Projects Agency of the Department of Defense and was monitored by the U. S. Army Research Office under contract No. DAH CO4-75-C-0010.

6. REFERENCES

1. M. H. Kryder and A. B. Bortz, Physics Today, 37, No. 12, 20 (1984).
2. F. Bloch, Z. Phys., 61, 206 (1930).
3. N. D. Mermin and H. Wagner, Phys. Rev. Lett., 17, 1133 (1966).
4. J. P. Mc Tague and M. Nielson, Phys. Rev. Lett., 37, 596 (1976).
5. M. Pomerantz, F. Dacol and A. Segmüller, Phys. Rev. Lett., 40, 246 (1978).
6. G. Bayreuter, J. Mag. and Mag. Mat., 38, 273 (1983).
7. Reviewed by G. A. Prinz, G. T. Rado, and J. J. Krebs, J. App. Phys., 53, 2087 (1982).
8. L. Golub, E. Spiller, R. J. Bartlett, M. P. Hockaday, D. R. Kanis, W. J. Trela, and R. Tachyn, Appl. Optics, 23, 3529 (1984).
9. M. Pomerantz, J. C. Slonczewski, and E. Spiller, J. Mag. and Mag. Mat., 54-57, 781 (1986).
10. S. Gregory, Phys. Rev. Lett., 40, 723 (1978).

11. P. W. Stephens, P. A. Heiny, R. J. Birgenau, P. M. Horn, J. Stoltenberg, O. E. Vilches, ibid., 45 1959 (1980).

12. J. Stoltenberg and O. E. Vilches, Phys. Rev. B,22, 2920 (1980).

13. S. Gregory, Phys. Rev. Lett., 39 1035 (1977).

14. G. N. Lewis, D. D. Awschalom, and S. Gregory, Phys. Rev. B, 29, 3508 (1984).

15. M. E. Fisher in "Essays in Physics", Vol. 4, p,43 (Academic Press New York, 1972) .

16. L. Onsager, Phys. Rev. 65, 117 (1944).

17. A. B. Harris, J. of Phys. C, 7, 1671 (1974).

18. S. Fishman and A. Aharony, J. Phys. C, 12, 1729 (1979).

19. J. F. Fernandez, G. Grinstein, Y. Imry, S. Kirkpatrick, Phys. Rev. Lett., 51, 203 (1983).

20. Reviewed by M. Pomerantz in "Phase Transitions in Surface Films", J. G. Dash and J. Ruvalds, Editors, p.317 (Plenum Press, New York, 1980).

21. T. Haseda, H. Yamakawa, M. Ishizuka, Y. Okuda, T. Kubota, M. Hata, and K. Amaya, Solid State Comm., 24, 599 (1977).

22. K. B. Blodgett, J. Am. Chem. Soc., 57, 1007 (1935).

23. G. L. Gaines, "Insoluble Monolayers at the Liquid-Gas Interfaces", (Interscience, New York, 1966).

24. M. Pomerantz and A. Segmüller, Thin Solid Films, 68, 38 (1980).

25. J. F. Stephens and C. Tuck-Lee, Appl. Chrystallogr., 2, 1 (1969)

26. M. Prakash, P. Dutta, J. B. Ketterson, and B. M. Abraham, Chem. Phys. Lett., 111, 395 (1984).

27. G. A. Wolstenholme and J. H. Schulman, Proc. Farad. Soc., 46, 475 (1950).

28. B. Ellis and H. Pyszora, Nature, 181, 181 (1958).

29. A. Hjortsberg, W. P. Chen, E. Burstein and M. Pomerantz, Optics Comm., 25, 65 (1978).

30. ESR in quasi 2-d is reviewed by P. M. Richards, in "Proceedings of the International School of Physics "Enrico Fermi", Course LIX,", K. A. Müller and A. Rigamonti, Editors. p.539 (North Holland, Amsterdam, 1976).

31. M. Pomerantz, Solid State Comm., 27, 1413 (1978).

32. A. S. Borovik-Romanov, J. Exp. Theor. Phys., 36, 75 (1959).

33. H. Yoshioka and K. Saiki, J. Phys. Soc. Japan, 33, 1566 (1972).

34. F. Ferrieu and M. Pomerantz, Sol. State Comm., 39, 707 (1981).

35. L. Neel, Ann. Phys. Ser. 12, 249 (1949).

36. A. Aviram and M. Pomerantz, Sol. State Comm., 41, 297 (1982).

37. M Pomerantz, A. R. Taranko, and R. J. Begum, Bull. Am. Phys. Soc., 28, 366 (1983).

38. M. Pomerantz, A. Aviram, A. R. Taranko, N. D. Heiman, J. App. Phys. 53, 7960 (1982).

STRUCTURAL MODIFICATIONS OF L-B FILMS:

ORDER-DISORDER TRANSITIONS AND POLYMERIZATION

J. D. Swalen, J. P. Rabe, C. A. Brown, and J. F. Rabolt

IBM Research
Almaden Research Center
650 Harry Road, San Jose, California 95120

ABSTRACT

Langmuir-Blodgett films have become increasingly popular for a number of potential applications because uniform and regularly spaced films of known thicknesses can be constructed with functional groups contained either in the chain at one end or as a side group. In order to determine the molecular structure and orientation, we have used a number of spectroscopic techniques, such as infrared glancing angle reflection and transmission spectroscopy. Changes in the IR spectra with temperature have given evidence for a two step melting process involving the gradual disordering of the hydrocarbon tail followed by the breakup of the head group lattice. With the polymerization of films of either octadecyl fumarate or octadecyl maleate, the vibrational bands associated with the unsaturation disappear and others increase, attributable to the resulting polymer. Initiation by both uv radiation and thermal heating led to similar spectra for both compounds indicating, as expected, that the same product is formed. Our spectral results from some examples will be presented in addition to the details of the experimental method.

INTRODUCTION

Organic thin films are increasingly being used in the electronics and computer industry in the fabrication of microelectronic devices. Some of the applications include organic photoconductors in copiers, photoresists for VLSI circuit fabrication, insulating layers to replace SiO_2, lubricants for magnetic disks, liquid crystal displays, integrated optics, piezoelectric sensors, and methods for surface modification. Now with the continual reduction in size, thinner and thinner continuous films, free from pin holes, are being sought. The Langmuir-Blodgett technique, though first studied in the early 1930's, is now undergoing a renaissance and considerable activity is going on throughout the world[1-5] to develop new films which are useful for scientific and technological applications. The advantages are that these thin molecular films can be made with accurately known thicknesses and that functional groups can be placed at specific positions and with specific orientations. Consequently a number of proposals[3-6] have been made for their application and currently there is a considerable amount of research in these areas.

A major problem with L-B films, however, is their fragile nature, that is, they are basically soap films and as such can be mechanically disturbed by gentle rubbing or abrasion. To understand the limits and possibilities for strengthening these films, we have been approaching this problem in three ways: 1) the measurement of phase transitions as a function of temperature, 2) the polymerization of L-B films, and 3) the deposition of L-B monolayers of polymers. Characterization has been done with infrared spectroscopy,[7-10] both by transmission and by glancing angle reflection.[11]

Structural changes in cadmium arachidate films at elevated temperatures indicated a two step melting process occurs: the chains melt before the head groups.[10] Cycling the temperature led to an almost complete recovery of the L-B structure, until the melting temperature was reached (~ 110° C). Above this temperature the head groups irreversibly disorder and form an amorphous structure. Recent work[12] on other films, such as tetradecyl benzoic acid and 2-nitro-4(N-methyl, N octadecyl) amino benzoic acid,[13] shows that their transitions are at higher temperature. It is our belief that the introduction of aromaticity in the chains influences their crystal packing and their high temperature stability.

An approach to strengthen the films has led to various attempts at synthesizing[14-20] novel compounds, incorporating reactive groups within the fatty acid backbone of the L-B monolayer components. Upon exposure to uv, electron or x-ray radiation, polymerization occurs with the resulting polymer backbone usually lying in the plane of the L-B film, giving enhanced mechanical strength. Polymerized L-B films also can exhibit dimensional stability because the thermal expansion coefficient of most polymers in a direction parallel to the chain axis is quite small. Naegle et. al.[21] reported the polymerization of L-B films of octadecyl fumaric acid (ODF), the trans isomer. Our work includes a reinvestigation of this isomer, as well as the cis isomer, octadecyl maleic acid (ODM),[22] for contrast and comparison. The intention of the present work was to compare the structure and orientation of these compounds in L-B films, before and after polymerization. Infrared spectra of the two isomers were found to be quite different but both produced the same polymerized product, as expected. In addition, a parallel study of these two materials and their mixtures provided an opportunity to study the macroscopic behavior of polymerizing L-B films, both on the water surface and on a solid surface, the results of which are reported here.

FILM PREPARATION

ODF was synthesized by adding purified octadecanol in ether over a three hour period to a 3- to 4-fold molar excess of fumaryl chloride in an ether solution at 0°C. The reaction was continued for one to two hours until thin layer chromatography showed no remaining octadecanol. The product was isolated by hydrolysis in the presence of tetrahydrofuran as a co-solvent and partitioned between water and ether. After evaporating the ether, a tan to brown solid was obtained. Column chromatography with a mixture of methylene chloride, ether, ethanol, and acetic acid was used twice for purification. The preparation of ODM was done in similar manner by adding purified octadecanol to an excess (3 molar) of maleic anhydride in tetrahydrofuran. The mixture was refluxed until thin layer chromatography showed no remaining alcohol and the product was isolated in a manner similar to that for the ODF. Both materials were free of alcohol and their isomeric purities were verified by proton and carbon NMR, IR, thin layer chromatography, and gas chromatography of the methyl esters of the acids. In addition, both had elemental microanalyses in agreement with the empirical formulas.

The L-B films were prepared on a commercially available film balance (Joyce Loebl), the enclosure of which was purged with argon to avoid possible oxidation. The water was doubly distilled and contained 2.5×10^{-4} mol/ℓ $CdCl_2$ and 4×10^{-4} mol/ℓ $KHCO_3$, as well as NH_4OH to maintain the pH around 8. ODF and ODM were both spread from a chloroform solution. Polymerization on the water surface was performed with a uv-lamp (Spectroline R-51), emitting monochromatic radiation at $\lambda=254$ nm and with an intensity of 1.5 mW/cm^2, which was uniform over an area of 16×8 cm^2. Under these conditions, the surface pressure of the saturated fatty acid salt did not change during an hour of uv illumination, indicating no photolysis was occurring. A filter, cutting off at wavelengths shorter than 254 nm, was necessary, however, because without it a reduction in area over time was observed indicating some photolysis was occurring.

Monolayers of the Cd-salts of ODF and ODM were transferred as Y-type films (head-to-head, tail-to-tail) onto substrates of either KRS-5 or 200 nm of aluminum or silver evaporated onto microscope slides at a dipping speed of 3 mm/min and a pressure of 30 mN/m. At the same pressure, the molecular area was smaller for the trans compound (ODF) than for cis (ODM), probably reflecting the structurally larger head group for the cis compound. Both isomers could further be compressed to an area of 0.20 nm^2, the molecular area of an extended hydrocarbon chain. The films could be irradiated *in situ* with uv light or heated by a thin film heater to initiate polymerization.

Uv polymerization on the water surface resulted in an increase in area for ODF and a decrease for ODM, but the isotherms of the resulting polymers were identical within experimental error, as expected. Mixtures of ODF/ODM with 65%, 50% and 35% ODF gave an increase in area with uv radiation. An ODF content as low as 25% was necessary to maintain the area constant. This mixture is, therefore, a candidate for a film which could be polymerized on a solid substrate without cracking.

INFRARED SPECTROSCOPIC RESULTS

FTIR studies were a combination of Grazing Incidence Reflection (GIR),[23] originally suggested by Greenler,[11] and transmission spectroscopy to determine the orientation of molecular segments in the L-B films. In Figure 1, the reflection attachment is schematically shown. The monolayers are supported on an aluminum coated glass slide positioned at a glancing angle with respect to the incident infrared radiation. Two plane mirrors and an off axis paraboloïd mirror return the reflected beam to the original optical path of the spectrometer. The whole optical system was evacuated and all spectra were recorded at a resolution of 2 cm^{-1} with a room temperature DTGS detector. With the summation of 1000 scans, spectra with high signal-to-noise ratios were obtained.

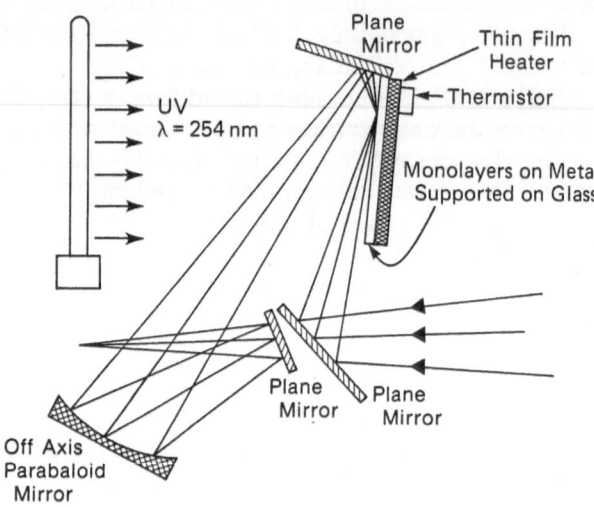

Figure 1. Schematic for the grazing incidence infrared reflection attachment.

The isotropic spectra of Cd-ODF and Cd-ODM were measured on KBr pellets with the salt deposited from solution. Spectra with $E_\perp$, i.e., only those groups with vibrations having a component oriented perpendicularly to the plane of the films, were observed with the L-B films deposited on aluminum surfaces. Transmission through films on KRS-5 gave spectra with $E_\parallel$ to the surface. These measurements provided spectra at two orthogonal directions of the electric field and the pronounced differences show the high degree of molecular orientation, as seen in Figure 2. The ODF chains appear to stand almost upright on the surface with a small tilt of the aliphatic backbone. Spectra of the monomeric Cd-ODM can be seen to be quite different from that of the ODF. A significant degree of tilt exists for the aliphatic chains of Cd-ODF, as seen with the stronger methylene stretch bands observed in reflection. This larger tilt of ODM compared to ODF is consistent with the fact that its molecular area on the water surface at 30 mN/m is approximately 10% larger than for ODF.

The lack of a pronounced splitting of the bending and rocking modes indicated that the subcell packing is not orthorhombic as in Cd-arachidate.[7] The ester carbonyl absorption for ODF is much more intense in transmission than in reflection indicating that the C=O bond is mainly oriented parallel to the surface. While with ODM, the orientation is somewhere intermediate between the two directions. The antisymmetrical C-O-C stretch is more visible in reflection for both ODF and ODM, though more pronounced for ODF. The C-O-C skeleton is predominately in line with the hydrocarbon chain. The absorption of the δ(C-H) in plane bend of the H-C=C-H group[24] is observed in reflection for both ODF and ODM, indicating that the C-H bond is parallel to the surface and the bending motion is in the direction of $E_{\perp}$. Therefore, the double bond is also oriented primarily perpendicular to the substrate surface. The orientation of the carboxyl group is clear since, in the reflection spectrum, the area under the symmetrical stretch peak is much larger than that under the antisymmetrical stretch, indicating that the carboxyl group is symmetrically attached to the substrate surface.

After uv irradiation both monomers gave essentially the same spectra in reflection, as shown at the bottom of Figure 2. Polymerization is depicted schematically for the cis monomer (ODM) in Figure 3, where in-plane chains are formed. A considerable reduction in the intensity of the in-plane bending of the hydrogens attached to the double bond is ascribed to polymerization during which other structural changes also occur. The intensity of the methylene stretch is also further reduced indicating that the aliphatic chains are now even more upright. We conclude this from the CH stretch region which, after 40 min of uv radiation, appears to be very much like that of cadmium arachidate.[7] Interestingly, monolayers polymerized on the water surface and transferred to the substrate afterwards gave similar spectra indicating that it does not matter where the polymerization occurs.

The spectrum of the thermally polymerized film, though not shown here, is very similar in the head group region to that for uv polymerized films; however, absorption of the methylene stretches in reflection, indicating a significant loss of order. In fact, it appears as if melting with subsequent orientational disorder occurs simultaneously with polymerization resulting in a disordered polymer film.

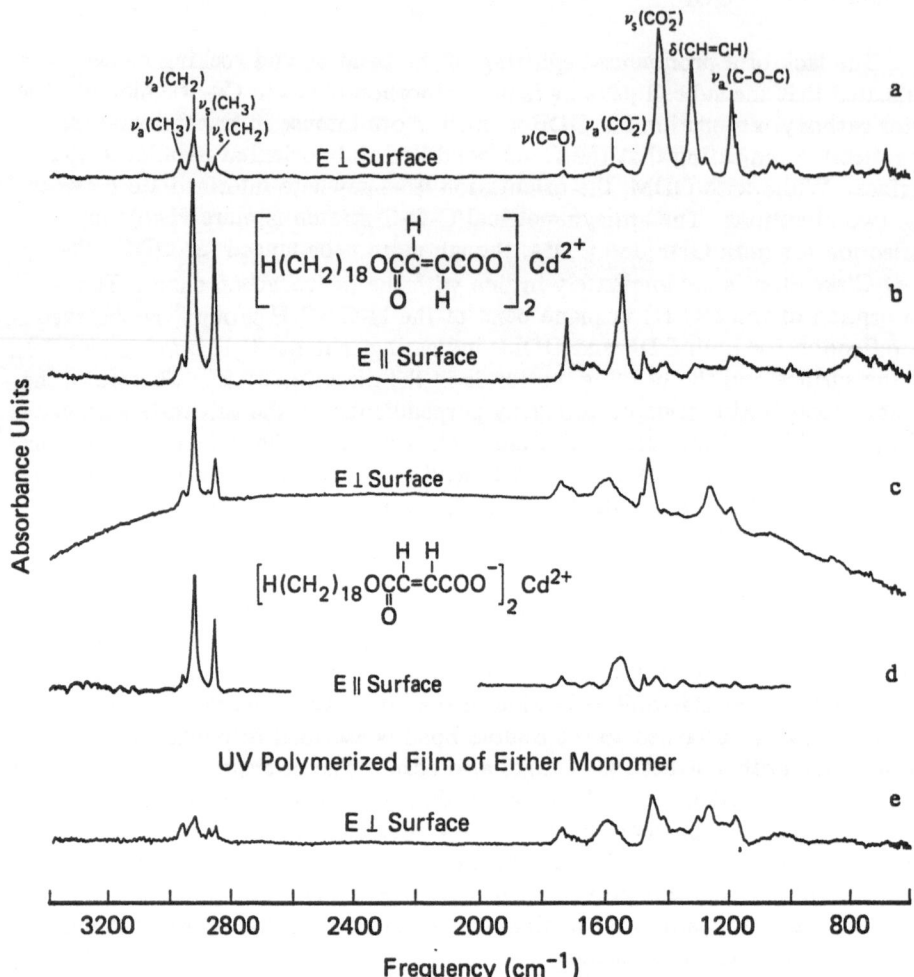

Figure 2. Comparison of the infrared spectra of ODF in reflection and transmission, a) and b) to that for ODM in reflection and transmission, c) and d). Spectrum e) is for the uv polymerized films of either monomer.

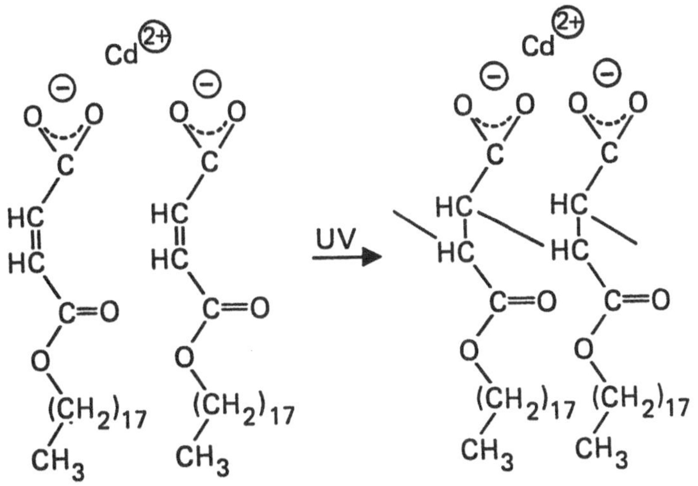

Figure 3. Schematics showing the change in bonding with polymerization of ODM, the cis compound. The polymerization of ODF follows in a similar manner.

CONCLUSIONS

Polymerized mixtures of cis and trans compounds were also formed and gave films which exhibited little dimensional change[25] (see the discussion on isotherms in the Section called FILM PREPARATION). This would indicate that mechanically strong films could be produced without cracking or peeling on a substrate. Work is now underway to make other polymerizerable L-B films and to measure their mechanical strength and dimensional changes. With Brillouin scattering we plan to measure the related phonon modes which are a measure of the elastic constants of the films. Recent experiments have also shown that we can directly transfer a pre-formed polymer monolayer from the water surface and its orientation is definitely different from that of the bulk.[26] This may be an even more attractive way to fabricate usable films with specific properties.

ACKNOWLEDGEMENTS

We wish to thank our colleagues who have contributed to parts of this work: M. Jurich, C. Naselli, S. J. Mumby, and R. J. Twieg.

This work was partially supported by the Army Research Office, Durham, NC under Contract No. DAAG29-83-K-0100.

REFERENCES

1. H. Kuhn, D. Mobius, and H. Bucher, "Spectroscopy of Monolayer Assemblies," *Physical Methods of Chemistry*, A. Weissberger and B. W. Rossiter, Editors, Part IIIB, Chapter VII, Wiley-Interscience, New York, 1972.
2. G. L. Gaines, "Insoluble Monolayers at Liquid-Gas Interfaces," *Interscience*, New York, 1966.
3. P. S. Vincett and G. G. Roberts, Thin Solid Films **68**, 135 (1980)
4. See the Proceedings of the First International Conference on Langmuir-Blodgett Films, Durham, Thin Solid Films **99**, 1 - 329 (1983).
5. The Second International Conference on Langmuir-Blodgett Films, Schenectady, New York, July 1 - 4, 1985.
6. G. G. Roberts, *Sensors and Actuators*, **4**, 131 (1984).
7. J. F. Rabolt, F. C. Burns, N. E. Schlotter, and J. D. Swalen, *J. Chem. Phys.*, **78**, 946 (1983).
8. J. F. Rabolt, F. C. Burns, N. E. Schlotter, and J. D. Swalen, *J. Electron Spectros.*, **30**, 29(1983).
9. J. D. Swalen and J. F. Rabolt, "Order in Thin Films by Fourier Transform Infrared Spectroscopy," *Fourier Tranform Infrared Spectroscopy: Applications to Chemical Systems*, J. R. Ferraro and L. J. Basile, Editors, Academic Press, New York, 1985.
10. C. Naselli, J. F. Rabolt, and J. D. Swalen, *J. Chem. Phys.*, **82**, 2136 (1985).
11. R. G. Greenler, *J. Chem. Phys.*, **44**, 310 (1966).
12. C. Naselli, J. P. Rabe, J. F. Rabolt and J. D. Swalen, *Thin Solid Films*, **134**, 173 (1985) and *J. Chem. Phys.*, **84**, 4096 (1986).
13. This compound was prepared by R. J. Twieg.
14. M. Puterman, T. Fort, Jr. and J. B. Lando, *J. Colloid Interface Sci.*, **47**, 705 (1974).
15. B. Tieke, G. Wegner, D. Naegele and H. Ringsdorf, *Angew Chem.*, **15**, 764 (1976).
16. B. Hupfer, H. Ringsdorf and H. Schupp, *Chem. Phys. Lipids*, **33**, 355 (1983).
17. R. Elbert, T. Folda and H. Ringsdorf, *J. Am. Chem. Soc.*, **106**, 7687 (1984).
18. A. Barraud, C. Rosilio and A. Ruaudel-Teixier, *Thin Solid Films*, **68**, 91 (1980).
19. G. Lieser, B. Tieke and G. Wegner, *Thin Solid Films*, **68**, 77 (1980).
20. B. Tieke, G. Lieser and K. Weiss, *Thin Solid Films*, **99**, 95 (1983).
21. D. Naegele, J. B. Lando, and H. Ringsdorf, *Macromolecules*, **10**, 1339 (1977).
22. J. P. Rabe, J. F. Rabolt, C. A. Brown and J. D. Swalen *Thin Solid Films*, **133**, 153 (1985).
23. J. F. Rabolt, M. Jurich and J. D. Swalen, *Appl. Spectrosc.*, **39**, 269 (1985).
24. N. Sheppard and G. B. B. M. Sutherland, *Proc. Roy. Soc.*, **A196**, 195 (1949).
25. A more detailed discussion will be published elsewhere.
26. S. J. Mumby, J. F. Rabolt and J. D. Swalen, *Macromolecules*, **19**, 1054 and *Thin Solid Films*, **133**, 161 (1985).

PART V. INTERFACIAL ASPECTS IN PRINTING

HYDROCARBON BASED INKS FOR ELECTRONIC PRINTING

J. M. Duff, R. W. Wong and M. D. Croucher[†]

Xerox Research Centre of Canada
2660 Speakman Drive
Mississauga, Ontario L5K 2L1, Canada

Electrostatically assisted liquid inks have firmly
established themselves as a viable developer option for
certain printer applications. In this process a particle,
either a pigment or a latex particle, is dispersed and
electrostatically charged in a dielectric fluid. Under the
influence of an electrostatic field within a development
housing the particles undergo electrophoresis to neutralise
a latent image that resides on the surface of a dielectric
substrate. This results in a hard copy of the latent image
being obtained. In this paper we review the properties
demanded of a liquid toner, methods of preparing such inks,
their colloidal behaviour and their electrical and imaging
characteristics. The trade-offs that need to be made when
optimising such developers is also discussed. Some
conclusions are drawn regarding future materials directions
for this toning technology.

1. INTRODUCTION

The past decade has witnessed a rapid increase in the use of
electronic systems in the office environment. This technological change
has resulted in a demand for improved methods of obtaining hard copy
images of these electronic signals. Numerous printing technologies have
emerged to meet this demand. They include ink jet, laser, thermal
transfer and electrographic printing as well as the more traditional
impact printers[1].

In this paper we focus our attention on electrographic printing
which is illustrated diagrammatically in Fig. 1. In this printing method
a latent electrostatic image is deposited onto dielectric coated paper by
an array of metal stylii which are selectively discharged according to
the electronic input the stylii receive. The ions that form the latent
image are caused by the dielectric breakdown of the air between the
stylii and the paper. The physics of this process is beyond the scope of
this paper but has been discussed by numerous authors[2-5]. Following the
formation of the latent image the dielectric paper is charged to about
300 volts. This latent image is then passed through a development zone
in which liquid developer comes into contact with this latent image.

[†] to whom inquiries should be addressed

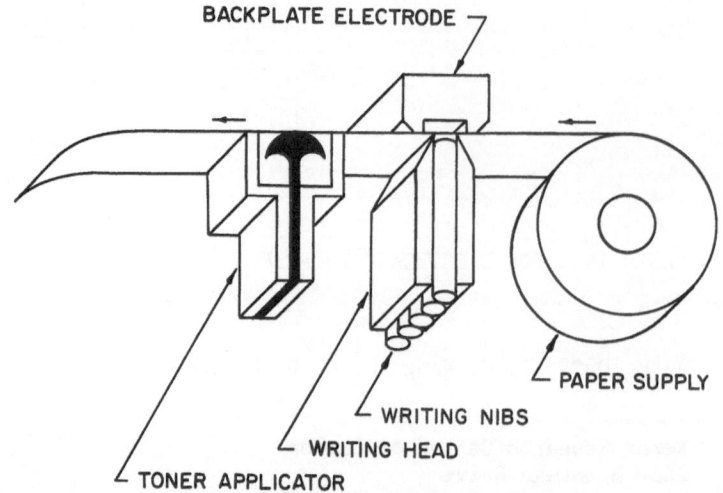

BACKPLATE ELECTRODE

PAPER SUPPLY

WRITING NIBS

WRITING HEAD

TONER APPLICATOR

Figure 1. Schematic diagram of an electrographic printer.

Under the influence of an electric field and (perhaps) the turbulence
created in the development zone, the particles that form the liquid
developer undergo electrophoresis. Since the particles are
electrostatically charged with the opposite polarity to that of the
latent image, the charge on the paper is neutralised by the developer to
give a legible hard copy output. Once out of the development zone the
hydrocarbon fluid on the paper is then allowed to evaporate leaving a dry
image.

     In this paper we discuss the requirements demanded of a liquid
developer. This is followed by a description of the design options
available to the formulator, together with a discussion of the
physicochemical and imaging characteristics to be expected from such
systems.

2. OVERVIEW OF ELECTROGRAPHIC TONER TECHNOLOGY

(i) Functional Requirements of Inks

     Hydrocarbon based liquid immersion developers (LID) are usually
colloidal dispersions of particles that are contained in a dielectric
fluid. There are numerous properties these developer materials should
possess, namely

*   the particles should be colloidally stable and preferably non-
    sedimenting.

*   the particles must be electrostatically charged so as to undergo
    electrophoresis in an electric field.

*   the particles must be in the correct size range. For instance, in
    direct printing onto dielectric paper submicron size particles are
    preferred. For a transfer process from an electroreceptor surface
    onto plain paper the particles should be in the micron size range in
    order to effect efficient transfer of the particles.

- the particle should possess the correct colour characteristics whether it be black or the process colours cyan, magenta and yellow.

All of the above requirements have to be satisfied in order to formulate a LID toner that would be commercially acceptable.

(ii) Design of LID Toners

The requirements stated above place considerable demands upon the materials for such a developer. A diagram of the components of an electrographic ink is shown in Fig. 2 and indicates that they consist of an insulating fluid, a surfactant to stabilise the particles against flocculation, a charge control agent and a particle. Each of these components must meet specific criteria which are detailed below.

Dispersion Medium. This has to be an insulating fluid with a resistivity greater than 10^9 ohm cm so that it does not discharge the latent image. It should also be of low viscosity, i.e., less than 2.5 $mN.s.m^{-1}$ to allow for fast migration of the particles through the fluid. The dispersion medium that is almost universally used for liquid developers are the Isopar hydrocarbons from Exxon, the G grade being extremely popular for these ink systems.

Surfactant. The purpose of the stabilizer is threefold (i) it is to help disperse the dry particles in the dispersion medium, (ii) to stabilise the particles against flocculation, and (iii) to fix the toner particle to the paper after development of the latent image. As we shall discuss later this third function is unnecessary in certain formulations since the toner particle is self-fixing on paper.

In principle, both high and low molecular weight compounds can be used as the steric stabiliser although polymeric materials are usually to be preferred. In order for it to be effective it should also have amphipathic characteristics so that part of the molecule can anchor the surfactant to the particle surface. Block and graft copolymers are obviously the most useful materials.

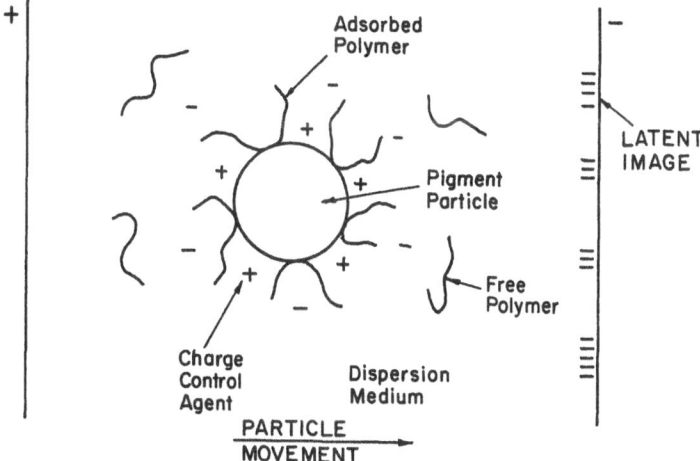

Figure 2. A schematic diagram indicating the components of a LID ink.

Charge Control Agent. This must impart an electrostatic charge to the particles in order for them to undergo electrophoresis in an electric field. Various postulates as to the mechanism of charging exist and these are discussed in another section. The technique of charging the particles that we have used is to add a material to the dispersion medium that is capable of undergoing adsorption and ionisation to impart charge to the particles. The types of materials used as charge control agents are usually of the metal soap variety such as iron naphthenate, zirconium octoate and calcium alkylbenzene sulfonate with many other examples being quoted in the literature[6,7]. It is also believed by numerous workers that the charge control agent should be insoluble in the dispersion medium in order to impart charge. However, this aspect of charging particles has not been well researched or documented.

Particle. The function of the particle is as a colourant and in providing the gain mechanism in the printing system, i.e., large ink particles adhere to the small electrostatic charge during the toning process. Two types of particles have been used in LID toners:

- inorganic or organic pigment particles.
- latex particles.

The majority of commercially available LID inks use pigment particles with carbon black being used almost universally for black liquid developers. The advantages of carbon black are availability, cost and a rich surface chemistry which allows the particles to be readily electrostatically charged in an aliphatic hydrocarbon based liquid. Numerous workers have given considerable thought to controlling the surface chemistry of the pigment by surface modification or encapsulation by a resin in an effort to have a better control over the charging properties of the pigment. It should be noted that the particle concentration in a working ink is approximately one per cent by weight.

In recent years the idea of using latex particles instead of commercially available pigments has invoked a good deal of interest[8-11]. The advantage of such a system is the excellent control over particle size and size distribution, control over the thermo-mechanical properties of the particle and a knowledge of the surface chemistry of the latex. The disadvantage is that the particles have to be coloured in a separate step. This has been discussed in previous publications[8,11].

(iii) Preparation of LID Inks

The traditional method of making LID inks is to take the components discussed previously and attrite these materials until a colloidal dispersion of a specific particle size has been obtained. It is often the case in a grinding operation that the order of mixing the components often affects the final product and care has to be excercised in this process. This will be discussed further in a forthcoming publication[12]. While this method of manufacturing LID toners is relatively straightforward it often leads to batch-to-batch quality control problems. A more elegant technique of making these inks which circumvents this problem is via the nonaqueous dispersion (latex) polymerisation[8]. This method of preparing particles in aliphatic hydrocarbon media was developed by ICI[13] and the process is shown schematically in Fig. 3. The first stage involves the preparation of an amphipathic block or graft copolymer which will become the steric stabilizer for the particle. This amphipathic polymer consists of two moieties, one of which is extremely soluble in the dispersion medium while the other moiety would by itself be insoluble in the dispersion medium. The second stage of this reaction consists of polymerizing the

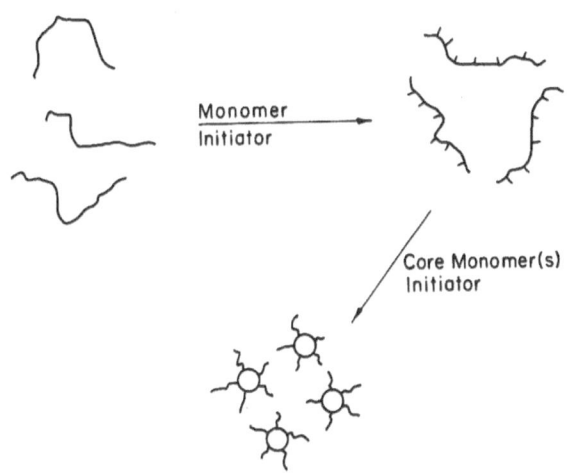

Monomer
Initiator

Core Monomer(s)
Initiator

Figure 3. Schematic representation of the steps involved in the dispersion polymerisation process. The first step involves grafting a monomer onto a homopolymer to form an amphipathic stabiliser while the second step involves the formation of the particles. For example, the homopolymer used could be poly(2-ethylhexyl methacrylate) while the core monomer used could be vinyl acetate although many other material combinations are possible.

core monomer in the presence of this copolymer. The only condition that is placed on the dispersion polymerization is that the monomer or monomers which will form the core particle should be soluble in the dispersion medium, but that its polymer should be insoluble. Therefore, after initiation of the polymerization the monomer grows to a specific chain length after which it phase separates and forms the nucleus of the core particle. The amphipathic copolymer in solution is then adsorbed onto this nucleus which then grows as a discrete particle. After the reaction is complete a milky white latex of particle size between 0.1 - 1.0 µm diameter is obtained as is shown in Fig. 4.

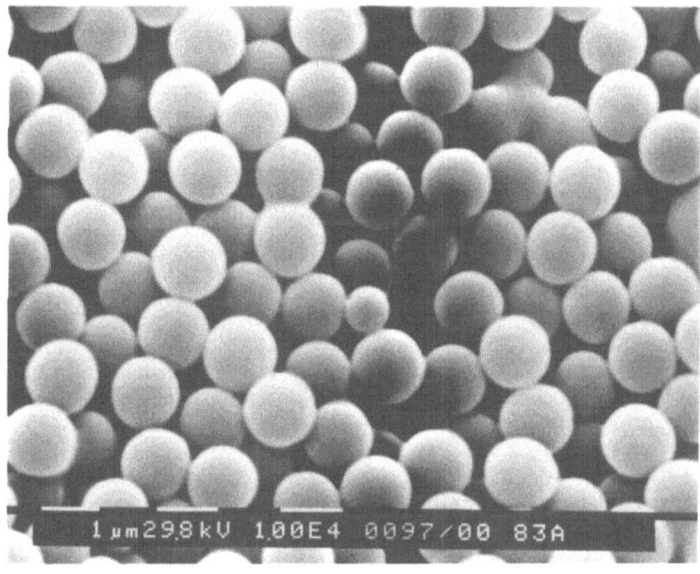

Figure 4. A scanning electron micrograph of latex particles which are used as the basis for LID inks. The white marker indicates 1 µm.

In order to turn this particle into a LID toner, the particle has first to be colored. This can be carried out by either ball-milling colored pigments into the particles or by dyeing the particles. We have used both techniques with varying success[8,11]. The disadvantage of ball-milling in a pigment is that it is an inefficient process and unattached particles are left in the dispersion medium. If they are not removed they can cause problems during development. A much better procedure to colour the particles is to dye them and specific procedures have been worked out to do this[1]. The advantage of dyes is that a broad range of colors can be readily obtained which will not bleed from the particle provided the dye chosen has the correct solubility characteristics in the Isopar and the polymer particle.

The final step in this process is to add a charge control agent to the coloured dispersion of essentially neutral particles in Isopar type solvents so that they undergo electrophoresis in an electric field.

3. PHYSICOCHEMICAL AND IMAGING CHARACTERISTICS

In order to be functionally useful the particles must be collodially stable, electrostatically charged and image in a reprographic device. In this section we will discuss these subjects in order to understand the interplay and trade-offs that need to be made with these developer systems.

(i) Colloid Stability

The colloidal particles constituting the developer should remain as discrete entities over the lifetime of the toner. This is not an insignificant problem since by definition colloidal dispersions are almost always thermodynamically metastable. If naked colloidal particles are dispersed in a fluid medium, they coagulate quickly due to the attractive van der Waals forces (V_A) that exist between the particles. In order to stabilize the dispersion, a repulsive potential (V_R) must be introduced between the particles such that $V_R > V_A$. This can be achieved either by introducing a steric barrier, or by introducing a repulsive electrostatic potential between the particles. It is also possible that both mechanisms are operative. The total particle interaction, V, of two particles in a dispersion is, to a first approximation, given by[14].

$$V = V_A + V_R \qquad (1)$$

and

$$V_R = V_E + V_S \qquad (2)$$

where V_E is the double layer interaction contributed by the charge control agent and V_S is the contribution of the steric stabilizer to the particle interaction. In theory, the charge control agent can generate electrokinetic potentials greater than 50 mV, which translates into a repulsive interaction, V_E, that is larger than the attractive van der Waals interaction V_A. Therefore, theoretically there is no need for a steric stabilizer. *However, since electrostatic charge is used to facilitate transport through the fluid medium, we choose not to rely on this technique to colloidally stabilize the particles but to utilize a steric stabilizer for this purpose.*

A first order approximation for the V_A between two shperical particles is given by the expression[14]

$$V_A \backsim - A^* a / 12H \qquad (a >> H) \qquad (3)$$

where a is the particle radius, H is the surface to surface distance between the particles and A^* is the effective Hamaker constant which includes the medium in which the particles are dispersed. Calculations for A* for inorganic sols, e.g., carbon black[6], show that A* is of the order of 5 - 10 kT while for latex particles A* is of the order of kT. V_A plots for 1 μm carbon black and 1 μm latex particles are shown in Fig. 5a and indicate that it is easier to stabilise latex particles than it is pigment particles. Equation 3 also indicates that V_A is directly proportional to the particle radius, a. In general, this means that the larger the particles the more difficult they are to stabilise as shown for carbon black particles in Fig. 5b. This indicates that the particle size is a parameter that needs to be carefully controlled.

The adsorption of a polymeric surfactant from solution onto the surface of the pigment gives rise to a repulsive steric barrier between the particles. It has been found that steric stabilization forces are of a shorter range than attractive forces[15] and increase rapidly as soon as the steric barriers start to interpenetrate. Numerous workers have contributed to the theory of steric stabilization with the various expressions being able to be written in the simple form[15]

$$V_S = B(\tfrac{1}{2} - \chi)aSkT \qquad\qquad (4)$$

where B is a function of the molecular parameters of the steric stabilizer and the dispersion medium and is always a positive quantity; S is a function that accounts for the distance dependence of the particle interaction, and χ is a parameter which is a measure of the antipathy between the stabilizing polymer and the dispersion medium. Equation 4 predicts that the dispersion will be stable when V_S is positive, which means that $\chi < \tfrac{1}{2}$. When $\chi > \tfrac{1}{2}$, V_S is negative and the system flocculates. The condition $\chi = \tfrac{1}{2}$ corresponds to the theta (θ) point of a polymer solution. Therefore, the θ point for the sterically stabilizing polymer in the dispersion media represents the limit of stability of the nonaqueous dispersion[15]. This analysis is based on the assumption that the polymeric stabiliser does not desorb from the particle surface. For pigment based LID inks where we rely on effective adsorption of the stabilizer, desorption is always a potential problem of which we should be aware. For latex based inks the steric stabiliser is <u>irreversibly attached</u> to the particle which is, therefore, thermodynamically stable

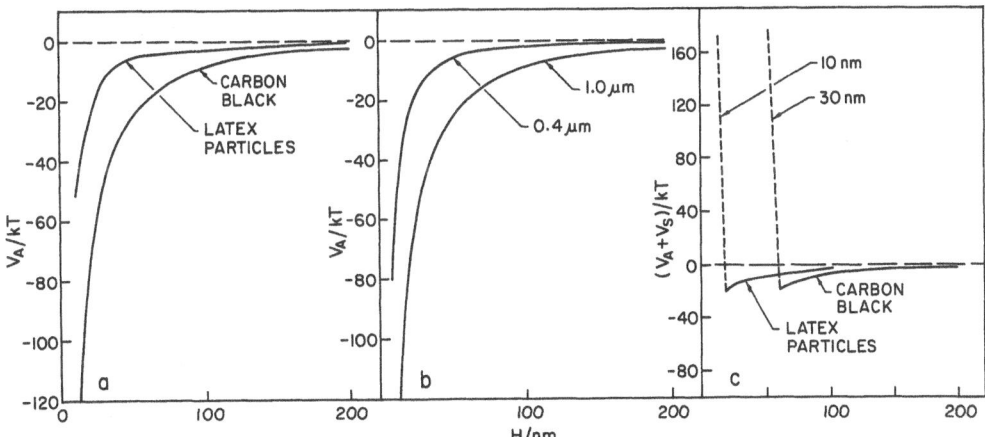

Figure 5. (a) Shows the attractive potential, V_A, between two carbon black particles and two latex particles while (b) highlights the effect of particle size on the attractive potential between two carbon black particles and (c) indicates the effect of steric interactions on the interaction between carbon black particles and latex particles. The A*, B, S and χ values used in eqns. 3 and 4 are given in reference 6.

rather than metastable as in the case of an adsorbed stabilizer. This gives latex particles a considerable advantage in terms of developer shelf-life and gravitational settling instabilities since destabilisation of the ink because of stabiliser desorption cannot occur. As was mentioned earlier the range over which steric interactions occur are extremely short. For the steric barrier to be effective it should effectively screen the attractive forces between the particles. This means that for a latex particle the steric barrier can be much thinner, i.e., of a lower molecular weight, than a stabiliser for a pigment particle as is indicated in Fig. 5c. A good stabiliser needs to be of sufficient thickness to stop the particles flocculating in a secondary minimum which has an energy well greater than a few kT. This is an important practical consideration in designing functionally useful LID toners.

(ii) Particle Sedimentation

In a printing engine the toner particles should ideally be non-sedimenting so that the ink does not need to be agitated when not in use. Since the concentration of particles in the developer is usually less than ∽ 1% by weight we can use the Stokes equation[16] to analyse the settling behaviour, viz.,

$$v_0 = \frac{2a^2(\rho_p - \rho_l)g}{9\eta} \tag{5}$$

Equation (5) shows that for the settling velocity to be small $v_0 \rightarrow 0$ then a should be as small as possible and/or $\rho_p \simeq \rho_l$ where ρ_p is the density of the particle and ρ_l the density of the liquid. Alternatively, the viscosity of the fluid (η) should be large. However, this is not a realistic practical alternative for an electrophoretic process. We have found that extremely stable carbon black based inks[6] do not settle at all, i.e., $v_0 = 0$, while poor dispersions settle extremely fast because of their dependence upon a^2 in Equation (5). The situation where $v_0 = 0$ is the ideal situation for a liquid toner in order that it may have a long shelf-life. We have also found[6] that latex based inks, while undergoing slow settling, also have a long shelf life since the stabiliser cannot desorb as in the case of an adsorbed steric stabiliser. It should also be noted that settling of small particles can be impeded by the convective stirring due to thermal gradients under normal conditions of storage. Because of convection, v_0 need not be zero to eliminate settling, but it must be less than a cricital value. This aspect of particle settling has been discussed in detail in Verwey and Overbeek[14].

(iii) Electrical Characterisation

In order for the toner to image it must be transported under the influence of an electric field. Consequently, the particle must be charged. It is usual in LID toner technology to measure the charge/mass (Q/M) ratio of the ink and to relate this to imaging performance in a printer. Measurements of this ratio have been described before and can be related to the fundamental electrokinetic parameters of the dispersion[6]. Thus

$$\frac{Q}{M} = \frac{3\varepsilon\zeta}{a^2\rho_p} \tag{6}$$

392

where ε is the dielectric constant of the dispersion medium and ζ is the zeta potential of the particles. Equation (6) can alternatively be written as

$$\frac{Q}{M} = \frac{9\eta u}{2a^2 \rho_p} \qquad (7)$$

where u is the electrophoretic mobility of the particles. Thus, Q/M is proportional to u/a^2, which is a function of the electrophoretic mobility of the system. Qualitatively, we can say that for rapid development of the image, Q/M should be large. However, the optical density of the developed image depends upon the number of particles deposited per unit area, and this will obviously be dependent upon the Q/M value. Therefore, there is obviously a trade-off between development speed and optical density. This should be borne in mind in designing a commercial toner. In commercial printing devices the Q/M ratio can vary from 10 - 1000 μC g^{-1} depending upon the application for which it is to be used.

Of perhaps more fundamental importance for these ink technologies is understanding how to effect and control the electrostatic charge on a particle in a dielectric medium. This is a subject which is not well understood at the present time. It is thought that particles in dielectric media can be charged in numerous ways

- by contact electrification
- dissociation of surface groups
- acid-base interactions
- by adsorbtion and dissociation of ionic surfactants.

It is known that when dielectric liquids flow down pipes electric currents are produced. If the liquid is of low electrical conductivity then significant potentials can be produced which can give rise to an incendiary discharge. Thus, electrostatic charge due to contact electrification is to be avoided in such systems. While dissociation of surface groups also gives rise to a charged surface it is unwise to rely on this as a mechanism for controlling the charge.

A subject of considerable interest in recent years is the effect of acid-base interactions. Fowkes has published extensively on this subject[17,18] and only a summary of his ideas will be presented here. The mechanism of acid-base interactions giving rise to a charged particle is shown schematically in Fig. 6a. The first step involves the adsorption of a surfactant (which is depicted as having a basic character) onto the surface of a particle (which is shown as having an acidic surface chemistry) together with a transfer of protons from the acidic sites on the particle to the basic group of the adsorbed surfactant. The second step required for particle charging is desorption of the proton-carrying polymer into the solution leaving a negative charge on the particle. Evidence for this mechanism[17] was provided using ^{14}C tagged surfactant materials to follow the adsorption-desorption process. While this mechanism certainly has an intuitive appeal it is by no means certain that it is correct or is the only mechanism. A more conventional view is shown in Fig. 6b in which a surfactant is adsorbed at the solid-liquid interface which is followed by dissociation of an extremely small fraction of the adsorbed molecules to leave a charged particle. It has been estimated[6,19] that about one molecule in $\sim 10^4$ adsorbed molecules dissociates in this process. Considering the fundamental importance of understanding the mechanism of the electrostatic charging of particles,

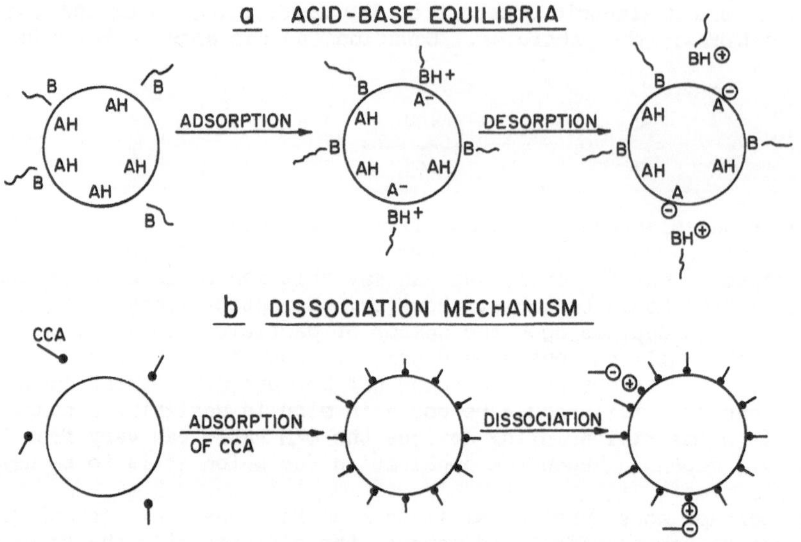

Figure 6. Schematic representation of two proposed mechanisms of generating electrostatic charge on particles in dielectric media.

it is perhaps surprising that so little effort has been directed towards this subject.

Most of the work emanating from this laboratory has involved metal soaps as the charge control agent since we have mainly been interested in imparting a positive charge to particles. They also give rise to relatively large equivalent conductances. It has not been established whether the conduction is due to simple ions or to some multiple ionic species in solution. In the systems we have studied it is always the cation that is in the adsorbed state while the anion is in solution in equilibrium with, or incorporated into, the non-adsorbed species in solution. A measure of the affinity of the cation for the particle surface can be obtained by measuring Q/M (which is indirectly a measure of the amount of charge control agent adsorbed) as a function of the concentration of added charge control agents. It has been found that in many cases Q/M increases almost linearly until at a critical concentration of charge control agent Q/M reaches a constant value as shown in Fig. 7a. The conductivity of the developer also indicates a slow rise in the conductivity of the toner until Q/M becomes constant, i.e., no more adsorption of surfactant can take place, at which point the conductivity rises more steeply as more free ions become available in solution.

From measurements of Q/M it is possible to obtain some rather crude estimates of the electrokinetic parameters of the particles[6,17]. These measurements have shown that zeta potentials for both pigment and latex based inks range from 50-250 mV while electrophoretic mobility values are in the range of $10^{-8} - 10^{-9} m^2 V^{-1} s$. These zeta potential values are extremely large and indicate that particles in dielectric media can, in principle, be charge stabilised. However, as we have mentioned previously we prefer to sterically stabilise particles while allowing the charge to facilitate transport of the particles in the development zone.

(iv) Fixing Properties and Image Density

The adhesion of the LID image to paper must be adequate to avoid rub-off of the image on handling. Three factors are thought to contribute to the fixing characteristics (a) the roughness of the paper surface (b) the properties of the dielectric layer, and (c) the film forming characteristics of the polymeric stabiliser/fixant.

The pigment based inks we have prepared have all been found[6,7] to have adequate fixing characteristics but could be smeared when continuously rubbed. However, the latex based inks were prepared with a glass transition temperature, T_g, such that they would lose their spherical shape and form a film on paper[8,11]. Since the polymers also wetted the paper, excellent adhesion of these toners to dielectric paper and other dielectric substrates has been obtained[11] which is independent of paper roughness contrary to what has been found with pigment based inks[6].

The optical density or absorbance of the toned black image should be in the range 1.1 - 1.4 and this can readily be obtained with both carbon black and dyed based latex developers[6]. The Q/M ratio of the toners has to be carefully controlled since it has been found that if excess ions are in the developer they compete with particles to discharge the latent image in the development zone. A graph of optical density against concentration of charge control agent is shown in Fig. 7b. It can be seen that there is a fairly narrow window of charge control concentration which is optimum for a specific LID ink formulation.

The resolution in LID images can be exceptionally good and 200-line pairs per millimetre has been achieved. This high resolution toning process makes LID an attractive proposition for technologies where resolution is important such as in micrographics or color printing.

4. CONCLUSIONS

Electrographic printing is a process of toning a latent electrostatic image which utilizes the electric field migration of a charged particle through a dielectric fluid dispersion medium. This concept is able to provide for a low cost high resolution image on paper or on transparent film.

This method of toning a latent electrostatic image is conceptually extremely simple and appealing. In practice, it is necessary to prepare a dispersion of particles which have specific electrical and colloidal properties in suspension but possess specific optical and adhesive properties when attached to a substrate. The materials problems involved are relatively complex, with interactions occurring among the numerous constituents of the toner which are not completely understood from a colloid chemistry viewpoint.

There are, at the present time, numerous electrographic printers in the marketplace that make use of liquid toners for developing a latent electrostatic image. These toners have been made on a trial and error basis with little thought given to the physicochemical principles involved in such systems. The purpose of this paper has been to delineate what we consider to be the important issues and problems with regard to liquid developers. Only by taking such a systematic approach will advances in such developer materials be made.

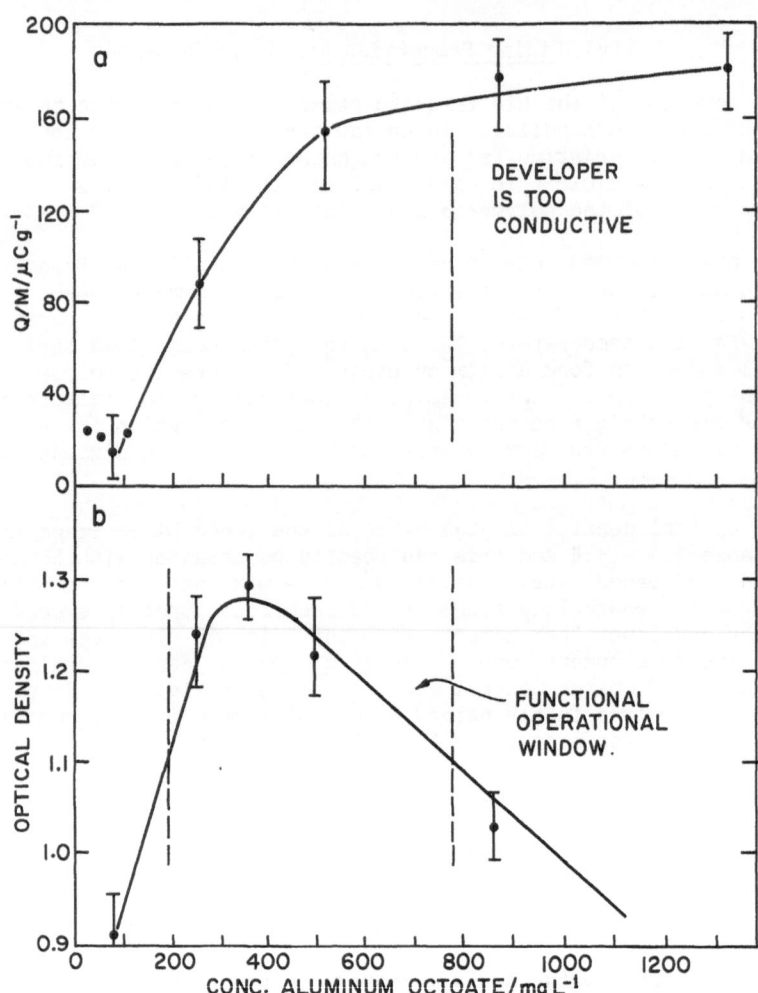

Figure 7. (a) Shows the Q/M ratio of a carbon black based LID toner as a function of the concentration of charge control agent, while (b) shows the optical density of the same toner when imaged.

REFERENCES

1. J. Gaynor, Editor, "Advances in Non-Impact Printing Technologies for Computer and Office Applications", Van Nostrand Reinhold, New York, 1982.
2. J. A. Dahlquist and I. Brodie, J. Appl. Phys., 40, 3020 (1960).
3. V. Novotny in "Colloids and Surfaces in Reprographic Technology", M. L. Hair and M. D. Croucher Editors, ACS Symposium Series No. 200, p. 281 (1982).
4. I. C. Roselman and W. Tait, Paper presented at SPSE Conference, Washington, D.C., (1977).
5. R. B. Crofoot and Y. C. Cheng, J. Appl. Phys., 50, 6583 (1979).
6. M. D. Croucher, S. Drappel, J. Duff, K. Lok and R. Wong, Colloids and Surfaces, 11, 303 (1984).
7. M. Croucher, S. Drappel, J. Duff, G. Hamer, K. Lok and R. Wong, Photogr. Sci. Eng., 28, 119 (1984).
8. M. D. Croucher, K. P. Lok, R. W. Wong, S. Drappel, J. M. Duff, A. Pundsack and M.L. Hair, J. App. Poly. Sci., 29, 593 (1985).
9. G. E. Kosel, U. S. Patent 3,900,412 (1975).

10. K. Tsubuko, T. Kurotori, T. Kimura, T. Kawanishi and Y. Kareko, U. S. Patent 4,081,391 (1978).
11. M. D. Croucher, J. M. Duff, M. L. Hair, K. P. Lok and R. W. Wong, U. S. Patent 4,476, 210 (1984).
12. M. D. Croucher, to be published.
13. K. E. J. Barrett, "Dispersion Polymerisation in Organic Media", Wiley, London (1975).
14. E. J. Verwey and J. Th. G. Overbeek, "Theory of the Stability of Lyophobic Colloids:, Elsevier, Amsterdam (1948).
15. D. H. Napper, "Polymeric Stabilisation of Colloidal Dispersions", Academic Press (1983).
16. P. C. Heimenz, "Principles of Colloid and Surface Chemistry", Marcel Dekker, New York (1977).
17. F. M. Fowkes, F. W. Anderson, R. J. Moore, H. Jinnai and M. A. Mostafa, in "Colloids and Surfaces in Reprographic Technologies", M. L. Hair and M. D. Croucher Editor, ACS Symposium Series, No. 200, p. 307 (1982).
18. F. M. Fowkes, see this volume for review.
19. V. Novotny, Colloids and Surfaces, 2, 373 (1981).

EFFECT OF SURFACE CHEMISTRY ON THE ADHESION OF THERMOPLASTIC

POLYMERS TO PAPER

J. Borch

International Business Machines Corporation
Information Products Division
Boulder, Colorado 80302

The sizing treatment of printing papers to make
them resistant to penetration by liquids, for example,
water or printing inks, also weakens the bond strength
between the polymer and the paper in printing
processes in which solid polymers create the image on
plain, bond-type papers. The adhesion characteristics
of a series of business papers treated with either
rosin sizes or synthetic, cellulose-reactive sizing
agents are described. It is demonstrated that the
decrease in bond strength correlates with a decrease
in paper wettability, as measured by capillary
penetration of organic liquids of suitable surface
tension, or as defined by surface energetics using the
elution-gas-chromatography technique. Examples are
presented, which show that these analytical techniques
are effective in specifying paper performance in
nonimpact printing technologies (for example,
electrophotographic printing and resistive-ribbon,
thermal-transfer imaging).

INTRODUCTION

Many modern printing processes are based on imaging by polymer
spreading and adhesion onto the surface of the paper. In
electrophotographic printing and copying, micrometer-sized, polymeric
toner particles are heat-sintered (fused) onto paper fibers of the
surface of uncoated, plain, bond-type papers[1]. In resistive-ribbon,
thermal-transfer technology (correctable or noncorrectible R2T2
techniques), the image transfer is effected by the transfer of polymeric
ink from a ribbon to paper through the localized joule heating of the
ribbon structure[2]. For correctable R2T2 printing, erasure is
accomplished by the transfer of the image back to the ribbon surface.

We had previously proposed[3,4] that the sizing treatment of
commercially manufactured, bond-type papers with organics to restrict
penetration by liquids, for instance, water or printing inks, also
affected the polymer-paper adhesion in electrophotographic printing.
The present study was carried out to (a) further determine the effect of

paper wettability on polymer-paper adhesion, and (b) identify analytical methods suitable for image-fix characterization, based upon paper wettability in nonimpact printing technologies such as electrophotography and the R2T2 technique.

EXPERIMENTAL

A capillary-penetration method was applied to compare the wettability of several paper types. Surface energy measurements were carried out via the elution-gas-chromatography technique.

Wettability

Paper wettability was assessed via a two-liquid-penetration measurement technique previously described by Aberson[5]. Theoretically, the penetration rate into a paper strip suspended in a fluid (by wicking) can be described by the Lucas-Washburn equation, according to which

$$h^2 = (r\delta\cos\theta t)/(2\eta) \tag{1}$$

where h = Liquid rise (height at time t)
  r = Effective capillary radius
  δ = Liquid surface tension
  θ = Contact angle between the liquid and the solid
  η = Liquid viscosity.

Rewriting Equation (1) for liquids 1 and 2 and deriving the ratio between the individual equations

$$(h_1/h_2)^2 = (\delta_1\cos\theta_1 t_1\eta_2)/(\delta_2\cos\theta_2 t_2\eta_1) \tag{2}$$

eliminates r.

Choosing liquid 2, for which $\cos\theta=1$ (complete wettability) provides the contact angle of liquid 1 as

$$\cos\theta_1 = (h_1/h_2)^2 (\delta_2\eta_1 t_2)/(\delta_1\eta_2 t_1) \tag{3}$$

Using benzyl alcohol (δ_1=39.0 mN/m, η_1=5.8 cp) and hexadecane (δ_2=26.7 mN/m, η_2=3.3 cp), and assuming that $\cos\theta_2$=1 for hexadecane, the $\cos\theta_1$ scale will range from 0.2 to 0.9 for bond-type printing papers.

Benzyl alcohol was chosen for capillary penetration because of its relative low polarity and a surface energy value comparable to those of various printing papers[3]. Experimentation using water mixed with lower alcohols such as methanol, ethanol, and glycol was less successful, especially in the case of rosin-sized papers, in which the liquids reacted with sizing components and swelled the fiber structure (nonconstant r in Equation (1)). The nonpolar hexadecane seems to be of sufficiently low surface energy to ensure instantaneous spreading on the fiber surfaces ($\cos\theta_2$=1), and the vapor pressure is sufficiently low to ensure that penetration is not affected by solvent evaporation during the test procedure.

All papers, except certain high-quality business-paper types, described in the following obeyed the theoretical requirement between penetration time and height given by Equation (1).

Surface Energetics

The gas-adsorption technique described by Dorris and Gray[6] was used for the surface-energy characterization of a range of bond-type papers ised for correctable R2T2 printing[2]. This technique gives the contribution due to the London dispersion force to the surface energy, which is relevant to this printing method because of the nonpolarity of the printing ribbon surface (see "Discussion" in the following).

Materials

Commercially manufactured bond-type papers were analyzed as in our previous study[3,4]. In addition, samples were produced in the laboratory on a centrifugal, dynamic, vertical-sheet former using paper constituents designed to simulate four common bond-type papers of different internal sizings and each of four levels of sizing, as shown in Table I.* The sizing samples contained both an internal filler (10-12% clay or calcium carbonate) and an external surface size (2% starch), as is common practice in the manufacture of bond-type papers for printing applications.

The paper types analyzed for R2T2 printing included a wider range of bond papers, including cotton-type bonds.

Table I. Laboratory-Made Paper Types *

Sample	Internal Sizing	Sizing Level
A1	Soap Rosin + Alum	0.75% Rosin + 1.4% Alum
A2		1.5% " + 1.6% "
A3		3 % " + 1.9% "
A4		6 % " + 3.3% "
B1	Dispersion Rosin + Alum	0.75% " + 1 % "
B2		1.5% " + 1.5% "
B3		3 % " + 1.5% "
B4		6 % " + 1.5% "
C1	Alkyl Ketene Dimer (AKD) + Cationic Starch	2% AKD + 1% Cationic Starch
C2		4% " + 1% " "
C3		8% " + 1% " "
C4		16% " + 1% " "
D1	Alkenyl Succinic Anhydride (ASA)	0.075% ASA
D2		0.15 % "
D3		0.3 % "
D4		0.6 % "

*Prepared by Centre Technique de l'Industrie des Papiers, Cartons et Celluloses, Grenoble, France.

Table II. Examples of Chemically Different Paper Types Used for Electrophotographic Printing and Copying.

Sample	Size Type	Fuse Rating	Cos θ*
1	Rosin	Very good	0.84
2	AKD	Good	0.77
3	AKD	Good	0.64
4	ASA	Fair	0.65
5	AKD	Poor	0.54
6	Rosin	Poor	0.51
7	ASA	Poor	0.35
8	ASA	Very poor	0.23

* As defined by Equation (3) using benzyl alcohol as liquid 1.

RESULTS

Eight commercially made paper types were imaged in electrophotographic copiers and printers (IBM Copier III and 3800 Printing Subsystem). Using criteria for fuse quality (polymer-paper adhesion) as described by Prime[7], the fuse rating was assessed as being very poor to very good (Table II, Column 3). The sizing type was identified by infrared spectroscopy (Table II, Column 2).

Wettability

Using benzyl alcohol as liquid 1 for Equation (3), the cosine of the contact angle θ varied from 0.23 to 0.84 (Table II, Column 4). The cosine value for laboratory-made papers covered a somewhat narrower range (0.23 to 0.68), as seen in Table III, Column 2.

The resistance to water uptake was measured using the Cobb sizing-test procedure (Table III, Column 3), which is often used as a quality assurance test for bond-type printing papers (TAPPI Method T 441 OS-77).

Surface Energetics

Because the procedure to measure surface energies via gas adsorption is quite complex and time consuming, we only measured selected samples of papers used for electrophotographic imaging. A cursory examination of seven paper types produced the values shown in Table IV. Energy levels ranged from 36 to 50 mN/m, with the AKD-sized paper showing the lowest level. The value for unsized filter paper was in good agreement with measurements previously obtained by Dorris and Gray[6].

Table III. Wettability of Laboratory-Made Papers by Benzyl Alcohol
(cosθ according to Equation (3)) and Water (Cobb Sizing Test).

Sample	Cosθ	Cobb Sizing Degree g/m/60 s - wire/felt)
A1	0.51	20/17
A2	0.48	14/14
A3	0.49	13/12
A4	0.47	13/12
B1	0.57	*
B2	0.59	34/22
B3	0.41	21/13
B4	0.38	14/12
C1	0.61	17/14
C2	0.48	14/14
C3	0.36	12/12
C4	0.30	12/11
D1	0.68	**
D2	0.44	29/16
D3	0.30	16/12
D4	0.23	14/12

*   Paper soak-through at 5 seconds.
**   Immediate soak-through.

A correlation between erasability and surface energy of a wider
range of printing papers used for correctable R2T2 printing experiments
is shown in Figure 1. The set tested included high-quality cotton-
containing bond papers for which wettability measurements via capillary
penetration could not be carried out because penetration was erratic and
nonuniform. The time-liquid rise relationship according to Equation (1)
was not fulfilled, and the gas-adsorption measurements appeared to be
less reliable (irreversible adsorption, specifically for high molecular
alkanes). Consequently, only C_7 and C_8 adsorption peaks [2,6] were used
for the surface energy calculations illustrated in Figure 1.

In the case of correctable R2T2 printing, erasability (the ordinate
in Figure 1) was measured by remelting and lifting off the polymer from
the paper[2]. In the case of noncorrectable R2T2 printing, the ink
penetrates further into the paper producing mechanical interlocking,
which minimizes the effect of paper surface energy on adhesion.

Table IV. Examples of Printing Papers Analyzed by Gas-Adsorption
Technique.

Sizing Type	Dispersion Surface Energy (mN/m)
None - Whatman No. 4 filter paper	50.3
Rosin - No. 4 Xerographic (A)	45.5
Rosin - No. 4 Xerographic (B)	42.5
Rosin - No. 1 Xerographic (A)	38.8
Rosin - No. 1 Xerographic (B)	38.7
ASA - No. 4 Register Bond	40.0
AKD - No. 4 Bond	35.8

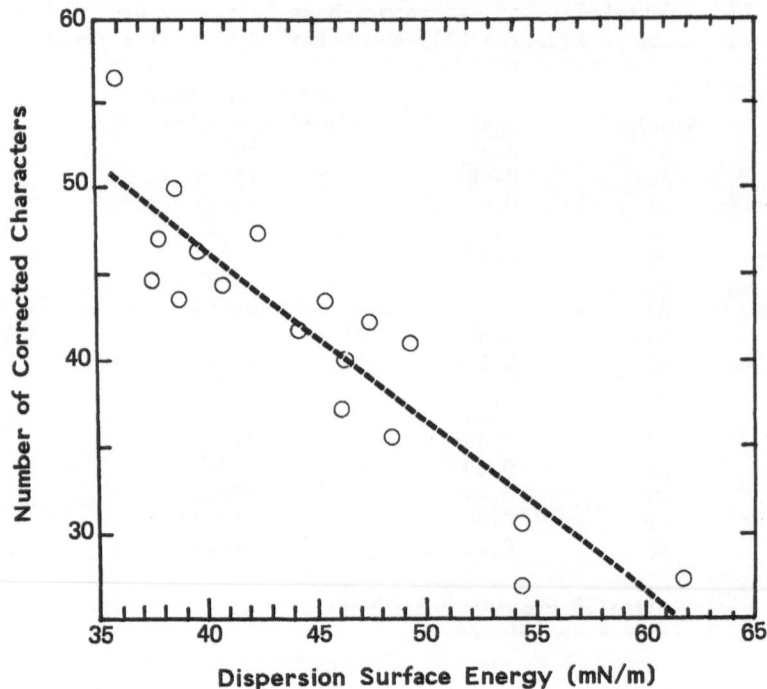

Figure 1. Resistive-ribbon thermal-transfer correction of bond-type papers of varying surface energy.

DISCUSSION

The effect of surface energetics on paper wettability and polymer-paper adhesion was reviewed in our previous study[3]. The data shown in Figure 1 confirm that paper chemistry through surface energetics plays a decisive role in the bond formed between the paper surface and the printing ribbon in R2T2 printing. Because the gas-adsorption technique only defines dispersion energetics, adhesion forces seem to be dominated by the attractive forces of a dispersive nature across the interface. This is substantiated by wettability measurements on the ribbon surface, which show that the polymer surface is of a nearly nonpolar nature. Possibly, correctable R2T2 printing might lend itself particularly well to a thermodynamic interpretation of polymer-paper adhesion via surface energetics because of the method of localized heating that fixes and removes the print. Such an analysis is beyond the scope of the present study but has been attempted by Wu[8] for selected polymer pairs, and by Warburton[9] for adhesion between cellulose fabrics and acrylate polymers.

In electrophotographic printing and copying the heating process is less well-controlled, and viscosity effects often override the effect of paper chemistry when polymer-paper adhesion is assessed[7]. Nevertheless, Table II indicates a correlation between wettability and polymer-paper adhesion when organic liquids such as benzyl alcohol penetrate the paper structure. Because a lowering of $\cos\theta$ corresponds to lower surface energetics of the substrate, the interpretation is similar to that of Figure 1, that is, a low-energy paper surface provides less adhesion.

It is apparent that both rosin and AKD and ASA sizes may provide inferior adhesion (Table II). The experimentally prepared papers (Tables I and III) more fully detail the effects of size nature and level on wettability. Both AKD and ASA sizes lower the wettability towards benzyl alcohol when the water uptake, as indicated by Cobb-sizing testing, decreases (Figure 2). AKD sizing creates a very rapid decrease in wettability at a relatively narrow range of Cobb-sizing degrees; ASA sizing develops low wettability for all the tested papers, except one, which did not develop water resistance either (D1 in Table III). This confirms our previous concern over the use of cellulose-reactive sizes for printing-paper sizing[4] and validates observations by others when paper and board products treated with these chemicals have been coated or otherwise subjected to molten polymers[3].

Current papermaking trends are to generate a more neutral papermaking process for both rosin and cellulose-reactive sizing (pH 6-8), which leads to the substitution of dispersion-rosin-size types for traditional soap or paste rosins. Table III demonstrates that even here wettability towards organics such as benzyl alcohol is affected. Soap-rosin-sized paper remains relatively nonsensitive to size addition levels, whereas the wettability of dispersion-rosin-sized paper decreases much faster with size addition, as measured by Cobb-sizing degree (Figure 3). No effort was made to define dispersion vs. soap-rosin procedures for commercially made paper in this study (samples 1 and 6 in Table II). Our experience shows that both types of rosin-sized papers provide low fuse-fix levels at high sizing degree, for example, when the paper is intended for lithographic printing (preprinted forms bonds of relatively high Cobb-sizing level). Rosin-sized paper may show lower polymer-paper adhesion than synthetic-sized paper (sample 6 in Table II).

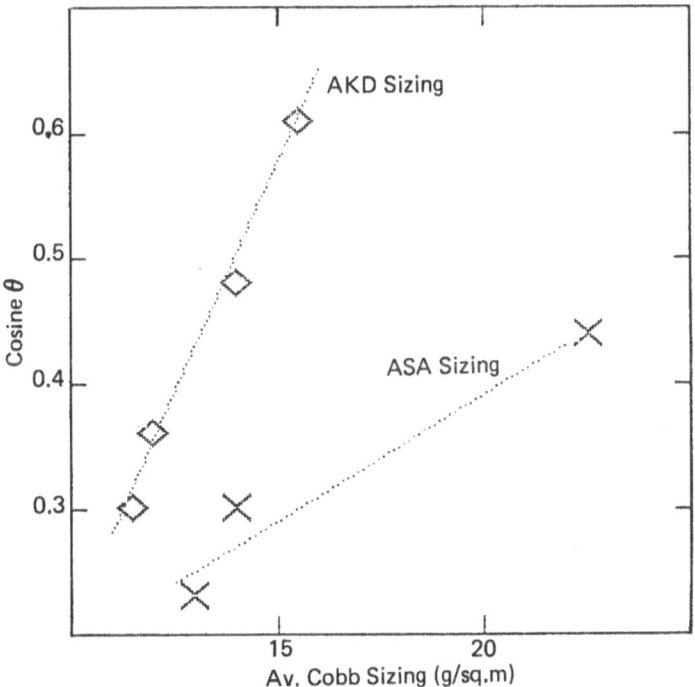

Figure 2. Correlations between wettability by benzyl alcohol and Cobb-sizing degree for AKD- and ASA-sized papers.

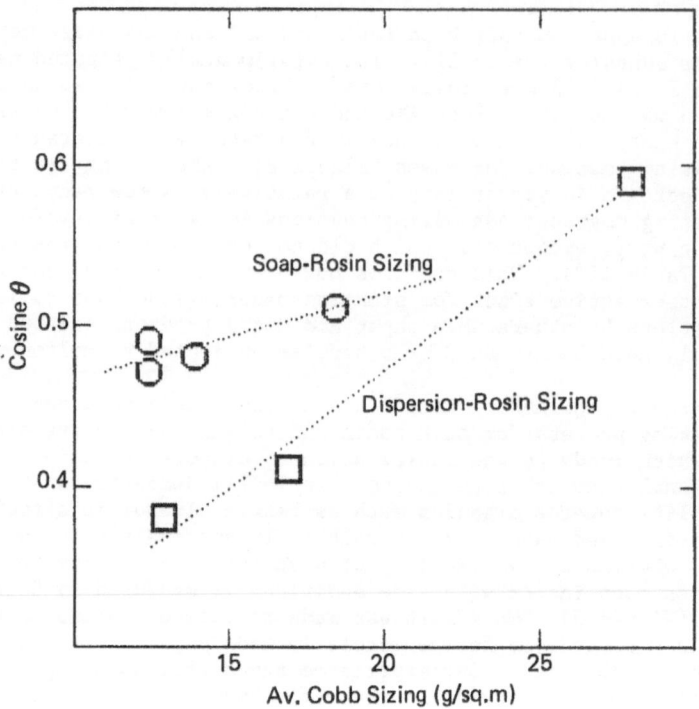

Figure 3. Correlation between wettability by benzyl alcohol and Cobb-sizing degree for soap-rosin-sized and dispersion-rosin-sized papers.

CONCLUSIONS

Our results demonstrate that it is both feasible and desirable to characterize polymer-paper adhesion via wettability measurements using suitable organic liquids or the surface energy measurements technique using gas adsorption. Data obtained using meaningful analytical procedures should encourage papermakers and users to further develop realistic testing procedures to determine the effect of paper characteristics in modern printing technologies.

ACKNOWLEDGMENTS

The author thanks A. S. Campbell for providing R2T2 printing data and for many helpful discussions in the course of this study, D. T. Covington and F. L. Rodgers for carrying out the gas-adsorption analysis, and J. L. Cadilhac for obtaining laboratory-prepared papers and Cobb-sizing data from the Centre Technique de l'Industrie des Papiers, Cartons et Celluloses, Grenoble, France.

REFERENCES

1. J. Borch and R. G. Svendsen, IBM J. Res. Develop., 28, (3) 285 (1984).
2. S. Applegate, J. Bartlett, A. Bohnhoff, A. Campbell, and J. Molloy, IBM J. Res. Develop., 29, (5) 459 (1985).
3. J. Borch, in "Colloids and Surfaces in Reprographic Technology," M. Hair and M. D. Croucher, Editors, ACS Symposium Series No. 200, p. 475, American Chemical Society, Washington, D.C., 1982.
4. J. Borch, TAPPI, 65, (2) 72 (1982).
5. G. M. Aberson, TAPPI, 8, 282 (1970).
6. D. G. Gray, in "Colloids and Surfaces in Reprographic Technology," M. hair and M. D. Croucher, Editors, ACS Symposium Series No. 200, p. 421, American Chemical Society, Washington, D.C., 1982.
7. R. B. Prime, Photogr. Sci. Eng., 27, 19 (1983).
8. S. Wu, J. Adhesion, 5, 39 (1973).
9. C. E. Warburton, Jr., J. Adhesion, 7, 109 (1975)

INK/PAPER INTERACTIONS IN INK JET PRINTING (IJP)

J. F. Oliver

Xerox Research Centre of Canada
2660 Speakman Drive
Mississauga, Ontario
Canada L5K 2L1

IJP is among one of several non-impact printing technologies which has recently established a very definite niche in the competitive and burgeoning computer printer market. The basic principle of IJP is to force an inviscid ink through a small nozzle or array of nozzles onto paper and form an image by means of computer control. Basically there are two modes of generating ink drops, viz., continuous and impulse. In view of the current and projected range of print speeds, drop sizes, print resolution and impact velocities, continuous and impulse IJP demands some rather unique paper properties to fulfill print quality requirements.

To illustrate some of the subtleties of porous printing substrates this article will draw upon results from several dynamic methods developed to study ink/paper interactions. Attention will be focussed on the phenomenology of ink/paper interactions especially in terms of the role of surface chemistry and capillary processes in porous media and their overall influence on image quality.

INTRODUCTION

With the advent of microcomputer technology the need for hardcopy output is increasing very rapidly. To meet this need, a wide variety of printers has emerged ranging from conventional mechanical impact printers to more exotic non-impact printers such as ink jet, capable of output speeds of tens of thousands of characters per second. In contrast to conventional impact printing processes[1], in ink jet printing however, the surface properties of paper play an even more critical role due to the fluidity of typical inks, the relative size of ink drops compared to paper fibers, and the absence of mechanical action.

For economic, practical and aesthetic reasons paper is still foreseen as the substrate which will predominantly be used in electronic printing technologies. The structure of paper, however, is exceedingly complex[2]. Apart from cellulose its chemical composition may include hemicellulose, lignin and various organic extractives, the proportions of which may vary widely depending upon the fiber origin and processing history. Furthermore the addition of flocculants, retention aids and inorganic fillers during papermaking and subsequent size and coating treatments further complicate its chemical composition. Similarly, significant variations in the physical composition of paper arise as a result of the numerous mechanical processes involved during papermaking[3].

With these aspects in mind, the main emphasis of this article will be to review recent studies on the sorption dynamics of ink jet printing from drop impact through to drying and final image development. Many of the unique differences arising in drying and image quality on various grades of paper will be illustrated. These differences will be discussed in terms of the paper structure and its interplay with various surface phenomena.

BASIC CHARACTERISTICS OF INK JET PRINTING

Ink jet printing (IJP) is an electronic non-impact process in which ink is forced through a small nozzle or array of nozzles[4] onto paper forming an image by means of computer control. Since the inception of the first commercial high quality black ink jet printer in 1977 there has been considerable interest in the development of colour IJ printing[5]. Its initial marketplace success, albeit gradual, promises to strengthen with the inevitable increase in innovation resulting in much faster print speeds and improved printhead reliability.

Process

Currently there are basically two modes of propelling IJ drops, viz., continuous and impulse. These are shown schematically in Fig. 1. The main

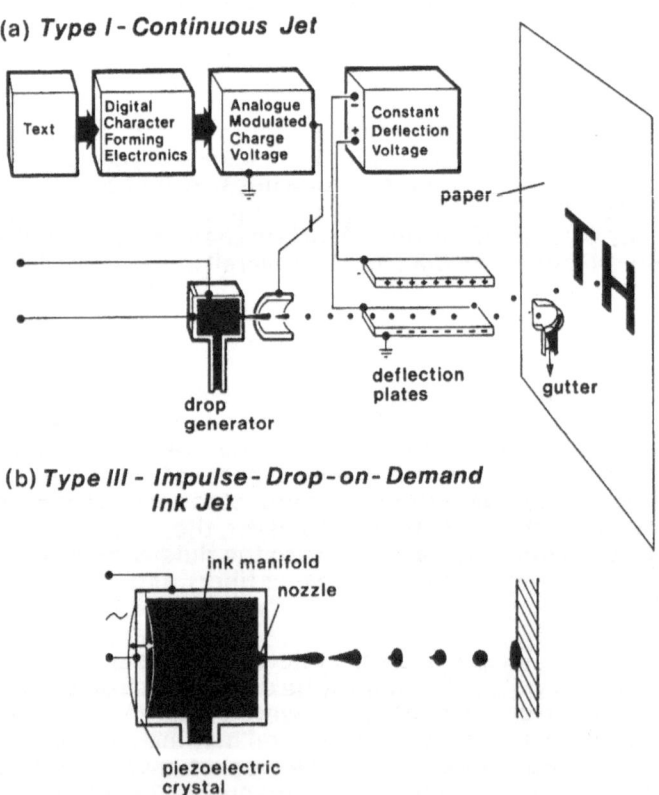

(a) *Type I - Continuous Jet*

(b) *Type III - Impulse - Drop - on - Demand Ink Jet*

Fig. 1.   Basic principle of a) type I, continuous, and b) type III, drop-on-demand (impulse) ink jet printing.

difference between these modes is in the principle of drop generation. In continuous IJP, ink is ejected from a nozzle under modest pressure, excited by a piezoelectric transducer oscillating at high frequency, to form discrete drops which are then charged and subsequently electrostatically steered onto the paper. The propulsive force in impulse IJP is provided by a piezoelectric element or bubble, but in contrast to continuous IJP there is no steering mechanism, drop velocities are much less and only those drops that are needed are ejected.

Inks

IJ inks used in commercial printers are dye-based and employ either water or an organic solvent vehicle system. Many other components are added to meet the multitude of jetting and image requirements[6]. Typically surface tensions range from 40-60 mN/m and viscosities 1 to 10 mPa.s.

Paper Properties

In view of the current and projected range of characteristics of commercial IJ printers[7] (see Table I). IJP requires some rather unique paper properties. The most salient requirements are listed in Table II. As we shall see in the following sections focussing on surface chemical phenomena, IJP presents new challenges for the design of appropriate paper structures.

PHENOMENOLOGICAL ASPECTS OF INITIAL DROP IMPACT

Upon impact, a short time t_q elapses before the inertial forces of an IJ drop dissipate and the drop reaches a mechanically quiescent condition. This period is referred to as the "mechanical regime." Capillary forces then continue to predominate until the drop eventually dries (via inter- and intra-fiber absorption processes). In considering the initial stages of impact it is helpful to separate the phenomena into these two regimes.

Mechanical Regime

For continuous pressurized, electrostatically steered Type I drops[8] with relatively high velocities (i.e., 15-40 m/s), the drop will be initially flattened and have a relatively large surface area. Furthermore, the contact area of the drop at the end of the mechanical spreading regime is determined by the impact efficiency, which is a measure of the transformation of the kinetic energy of the original spherical drop to the final deformational energy of the mechanically distorted drop. In contrast, drop-on-demand Type III drops[9]

Table I. Characteristics of Commercial Ink Jet Printers

Print Speed	1-100 pages/min
Drop Size	30-150 µm
Print Resolution	80-1000 dots/inch
Impact Velocities	3-30 m/sec
Single, Multicolor, Tonal Capability	

Table II. Key Paper Properties in Ink Jet Printing

Property	Goal
Image Resolution	Shallow ink penetration, high and uniform optical density, no feathering (edge raggedness), minimum show-/strike-through.
Drying Time	Fast, no color bleeding or offset.
Color Rendition	Dye hold-out, color saturation, strong dye-paper affinity.
Permanence	Good water-,light-,mechanical-fastness.
Dimensional Stability	Minimal cockle/curl (drop placement).

(with impact velocities $\leqslant$ 6 m/s) will have an approximately spherical profile and, hence, lower contact area. Thus, the local ink concentration/unit pore volume will be higher than for Type I drops of comparable size and presumably result in longer drying times because of the concentration dependence of liquid permeability[10].

Capillary Regime

Once intimate contact between paper and an aqueous liquid is established, surface- and vapor-phase diffusion and fiber sorption are believed to occur immediately and precede capillary flow, which ultimately dictates the rate of penetration. However, there is a delay before all of these processes take effect and liquid penetration commences[11,12]. The overriding factor contributing to this delay is viewed[13,14] as the conversion of the fiber surface to a water-like system or simply the affinity of the cellulosic substrate for water. This so-called wetting delay, t_w, is thus a function of surface hydrophobicity and, not surprisingly, can vary considerably among commercial papers. For example, t_w ranges from as low as 5 ms for chemical pulps to $\geqslant$60 ms for mechanical pulps and may reach several seconds for sized papers. Thus, in the relatively short time intervals between the deposition of individual IJ drops, paper grades with low t_w appear the most desirable.

Single- and Multiple-drop Interactions

As shown in Fig. 2, in considering the initial stages of drop impact in Type I and III printing, several distinctly different situations involving placement of the first and subsequent drops are conceivable. The initial drop impact (Fig. 2(a)) only involves solid (S) and liquid (L) interaction forces, whereas, depending upon drop targeting, the second and subsequent drop impacts (Fig. 2) may involve various combinations of interfacial forces, from impact with a pre-wetted surface to liquid-liquid impact (i.e. coalescence). Since the combined interfacial forces differ in all of these situations, the initial spreading and ultimately drying time, t_d, are bound to vary. In the case

a. Initial Drop Impact

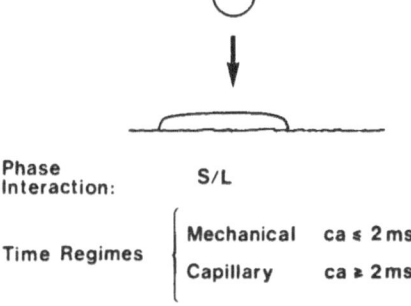

Phase Interaction:	S/L	
Time Regimes	Mechanical	ca ≤ 2 ms
	Capillary	ca ≥ 2 ms

b. Second and Subsequent Drop Impacts

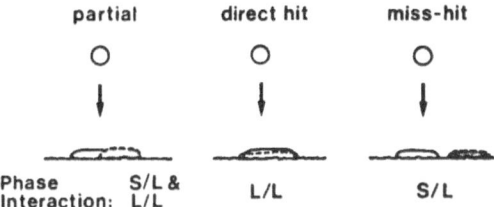

	partial	direct hit	miss-hit
Phase Interaction:	S/L & L/L	L/L	S/L

Fig. 2. Phenomenological aspects of impact of Type I (high velocity) and Type III (low velocity) ink jet drops (Reproduced with permission from Ref. 15. Copyright 1984, TAPPI Press).

of coalescence, the existence of an air film trapped between the drops presents an initial barrier and prevents them from making actual contact. Thus, the rate-determining step is the time for the intervening air film to drain, which may vary considerably as a result of several factors[16]. For the systems described below, the important factors could be the concentration of surfactant materials and mechanical vibrations, both of which have been shown to prolong coalescence.

The significance of some of these factors has been studied by high-speed cinematography[15]. Also, by studying consecutive drops, the effect of drop placement and local moisture preconditioning on the initial impact behaviour can be examined.

Drop Impact and Spreading Development

Figure 3 shows the sequence of events occurring on a highly absorbent paper following impact of drops originating from three consecutive drop pulses triggered within an interval of ~ 1.5s. Table III summarizes the number of drops originating from each pulse, their dimensions, time of impact, and the respective drop quiescence and absorption drying times. In view of the droplet sizes and the relatively large area spreading factor[17] for this system, the second and third pulse experiments involve drop impact on pre-wetted areas of paper. Thus, since the drying time t_d for absorption of the total volume of liquid ejected by the first pulse was only 12 ms, then obviously the drops originating from the second or third pulses will not be impacting a discrete liquid film but rather a fibrous surface in which absorbed ink resides within fiber cell walls or fiber interstices

Corresponding changes in contact angle, θ, drop height h, and base diameter D, immediately following impact, are plotted for each pulse in Fig. 4.

413

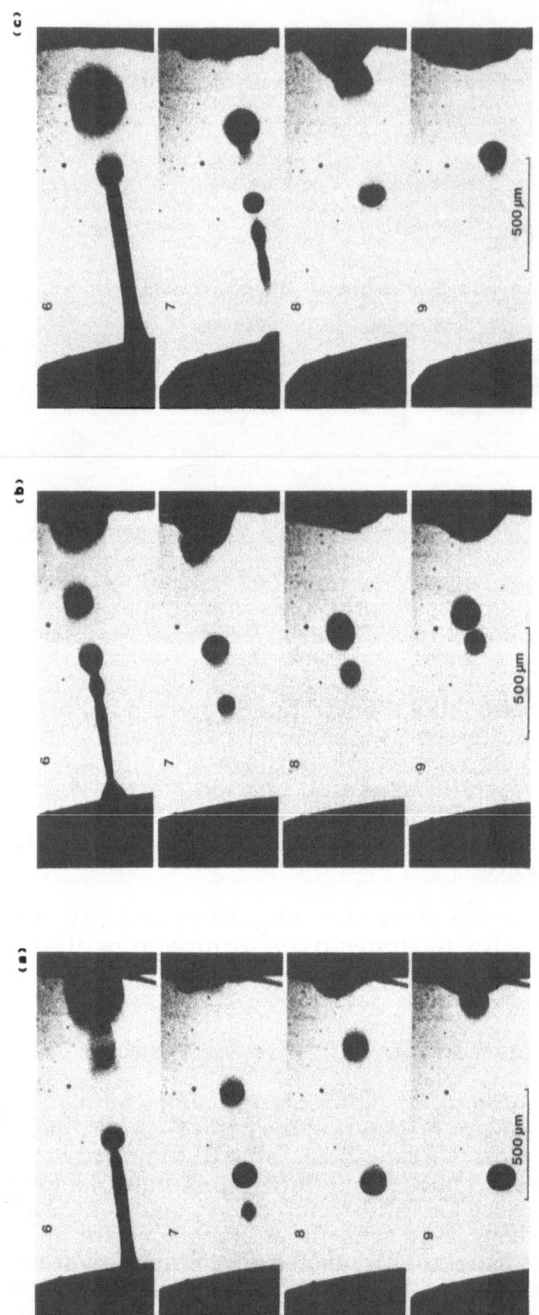

Fig. 3.   Frame-by-frame, high-speed film sequence showing drop impact
of an aqueous IJ ink. A, dry IJ paper; B and C, wet (i.e. pre-wetted by
preceding drop deposition). Frame intervals are ~ 0.25 ms. In each case, the
electronic pulse stimulating liquid ejection produces several drops which
subsequently impact approximately the same area within 0.75 ms.
Reproduced with permission from Ref. 15. Copyright 1984, TAPPI Press).

414

Table III. Drop Dimensions for Consecutive Jet Pulses and Times for Various Stages of Development of an Ink Jet Drop after Impact with Highly Absorbent Ink Jet Paper at a Velocity of ~ 1 m/s (from Ref. 15)

Stage	Dry Paper 1st Pulse		Pre-Wetted Paper 2nd Pulse		Pre-Wetted Paper 3rd Pulse	
	Time, ms	Diam., μm	Time, ms	Diam., μm	Time, ms	Diam., μm
Drop Impact:						
first	0	260	0	260	0	260
second	0.75	124	0.25	124	0.25	124
third	14.5[a]	124	2.0[b]	124	1.5[b]	120
Drop quiescence, t_q	2.5	430 (1.6)[c]	2.5	480 (1.7)[c]	2.25	540 (2.0)[c]
Drying time, t_d	12		36		27	

[a]Drop impacts below jet orifice. [b]Drop impacts paper in a region remote from first and second drops. [c]Splat factor: D_t/D_o is equivalent diameter calculated from the combined volumes of drops 1, 2 and 3.

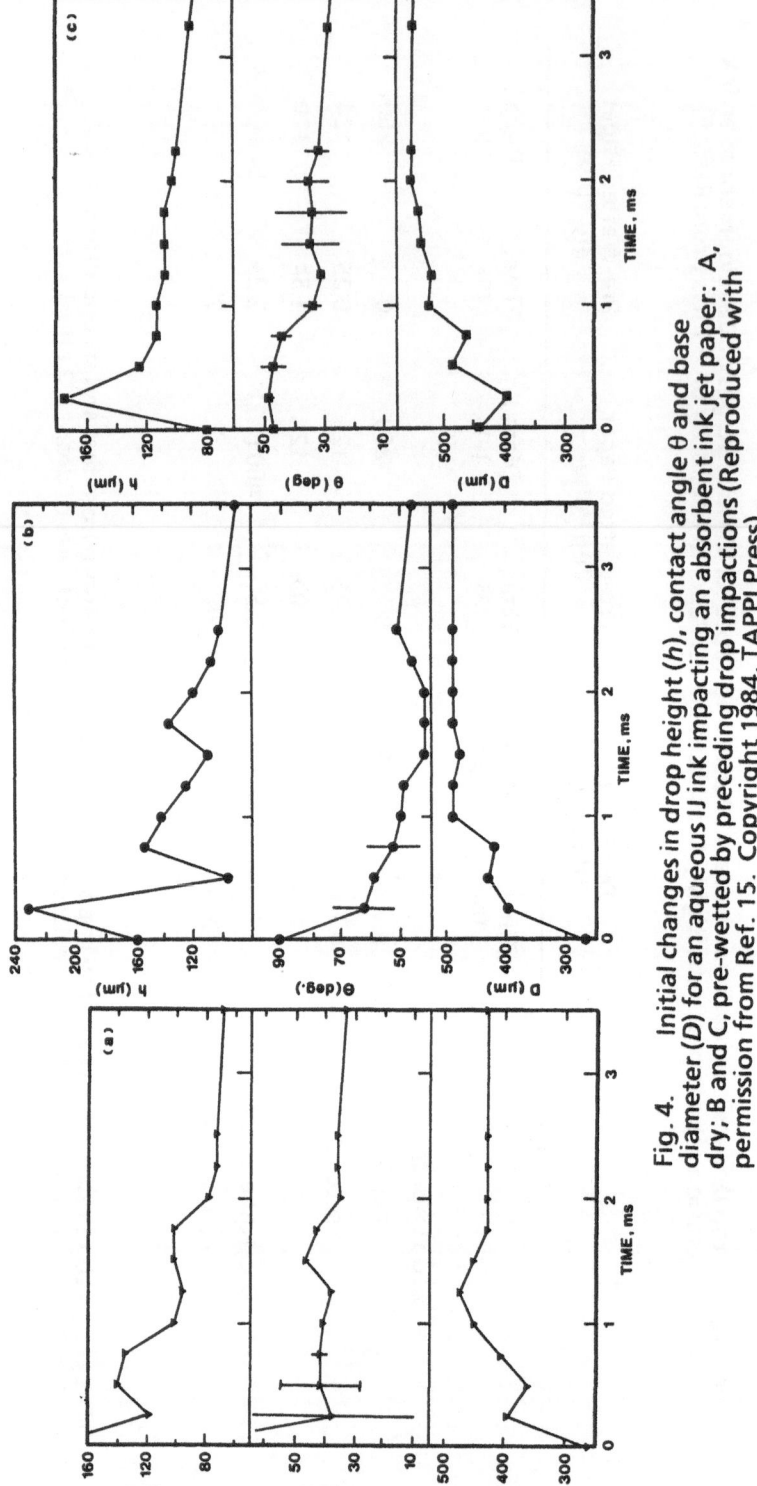

Fig. 4.   Initial changes in drop height (*h*), contact angle θ and base diameter (*D*) for an aqueous IJ ink impacting an absorbent ink jet paper: A, dry; B and C, pre-wetted by preceding drop impactions (Reproduced with permission from Ref. 15.   Copyright 1984, TAPPI Press).

416

On the basis of these graphs and data summarized in Table III various stages of drop development (Fig. 3) can be resolved. These are as follows:

- impact
- formation of cylindrical drop
- drop retraction
- secondary spreading
- formation of quiescent sessile drop
- absorption

As seen in Table III in the latter stage, t_d increased more than twofold on a pre-wetted paper. This difference is attributed to modifications in liquid permeability. The difference in t_d for the second and third pulse drops is probably caused by slight variations in drop volume and the substrate pore structure.

Implications in Type I and III Ink Jet Technologies

Some interesting aspects emerge from high-speed studies which shed further light on modelling and interpreting the ink drying process. Several parameters are likely to be important:

Drop impact. In the case of Type I drops which possess much higher kinetic energies, the impact profile will have cylindrical (Fig. 2) rather than the near-spherical cap form typical of Type III IJ drops. Thus, even with similar drop volumes, the initial ink concentration/unit area of substrate will result in differences in drying times, optical density and image quality. One can anticipate that the most rapid drying situation arises when the ratio of substrate surface area to drop contact area is a maximum.

Liquid permeability. Equally important to the drying process is the question of liquid permeability[10], which primarily depends upon the porous structure, liquid concentration, penetration depth, time, and liquid polarity[18]. As suggested by these experiments, liquid permeability in paper is further complicated by:

a) Its anisotropic structure, i.e., lateral and transverse spreading rates differ appreciably;

b) Swelling (in the presence of a polar liquid) which diminishes the average pore size and, hence, according to Poiseuille's equation for laminar flow, slows down the permeability;

c) Highly tortuous and, in some cases, noninterconnecting capillary network structure, i.e., dead-end pores.

In the context of swelling, if one first considers a rigid porous structure, Cheever[19] has shown that when the capillary supply of liquid is depleted, additional fluid should restore liquid flow to its original steady-state rate. Further complications arise if precipitation or gelation occur within the spreading fluid[20] which is highly probably in the case of IJ inks especially during multiple (color) printing. Also experimental studies on various dry and pre-wetted textile fabrics[21] indicate that in systems where the fibers swell, the viscous drag exerted by the fabric is increased. The latter situation appears consistent with these high-speed studies, viz. t_d increases with pre-wetting (Table III). Similarly one might expect significant changes in t_d if dramatic changes in relative humidity occur.

Drop coalescence. In Type I printing where the drops acquire like charges, there is a possibility that drop coalescence may be prolonged because of relatively slow charge dissipation by the substrate and hence partial repulsion of neighboring drops immediately after impact. Presumably this is offset by the high electrical conductivity of the ink. Charles and Mason[16] have shown, however, that in the presence of an electric field, addition of surfactant can markedly increase coalescence time even in the case of a conductive system.

Initial wetting. Up to t_q, the magnitude of θ for highly absorbent and sized papers[15] is quite similar even though they have widely different roughnesses and t_d values. This behavior suggests that the initial θ value is determined mainly by the drop kinetic energy and is independent of the substrate roughness. Since it is impossible to pinpoint the onset of capillary action or distinguish this from mechanical action up to t_q, it is difficult to assign a meaningful value of θ, which unequivocally is a measure of the surface chemical action. In actuality, θ results from surface wetting phenomena, i.e. variations in topochemistry and roughness, and concurrent absorption processes. Indeed the apparent wetting delay demonstrated by aqueous liquids on paper, which may vary from a few milliseconds to several seconds, depending on the surface chemistry[11-13], results from the interplay of the latter properties. Consequently, θ varies from some finite value to zero between t_q and t_d, respectively. It is, therefore, not the classical thermodynamic value defined by Young and others, and should accordingly be treated with due respect.

As will be discussed in the proceeding sections there are a multiplicity of ensuing events that further compound the utility of θ as a quantitative parameter for predicting drying and the image quality of various paper structures.

WETTING, SORPTION, DRYING AND IMAGE DEVELOPMENT

The next important regime of events concluding with drying is summarized in Fig. 5, for high and low γ IJ inks on various uncoated and coated grades of commercial paper. The contact line spreading contours are derived from ~ 33 fps video[22], rather than high-speed cinematographic recordings[15].

These contours reveal some of the major shortcomings of conventional papers. For offset, rotogravure and internally sized liquid xerographic papers, the initial contours remain near-circular and smooth, whereas for unsized IJ, newsprint and unsized liquid xerographic papers they are ragged. In contrast, for dry xerographic and bond papers which show the most prolonged drying, contact line development is negligible up to t_d, the drying time. This behavior is due to the high degree of sizing and is also evidenced by the progressively lower drop spreading factor[22] for slower drying systems. However, without sizing and/or with a lower γ ink, image quality on plain, bond-type, paper is no better than on the aforementioned uncoated structures.

In contrast, offset and cast-coated papers produce much higher quality images, viz., approximately circular contact lines virtually free of edge raggedness and show-through. Even with a lower γ ink, which dries faster, image quality is more or less preserved. It should be stressed, however, that without the aid of artificial drying, unconventional high surface area pigments such as colloidal silicas[23] are essential in order to attain acceptable drying times and image quality. This is especially true in color IJP where ink loadings may be doubled or even tripled. The coating pore volume is therefore critical and depending upon its receptive capacity will accordingly influence drying, the degree of lateral spreading, color bleeding and hence image quality. In the case of a system with slow pore filling, the surface wettability must be increased, in order to reduce t_d to an acceptable value. However, for uncoated papers, as illustrated in Fig. 5, greater

418

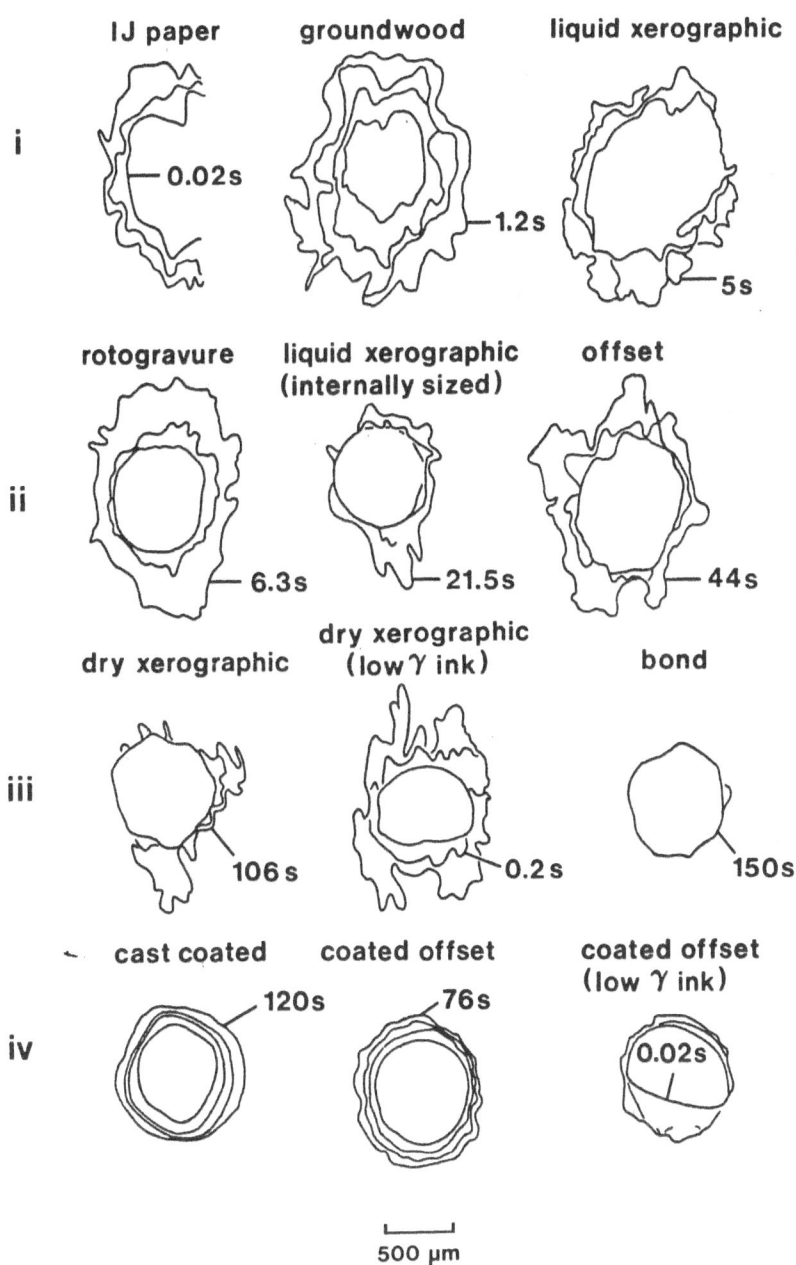

IJ paper   groundwood   liquid xerographic

i — 0.02s   — 1.2s   — 5s

rotogravure   liquid xerographic
(internally sized)   offset

ii — 6.3s   — 21.5s   — 44s

dry xerographic   dry xerographic
(low γ ink)   bond

iii 106 s   0.2 s   150s

cast coated   coated offset   coated offset
(low γ ink)

iv 120s   76s   0.02s

500 µm

Fig. 5.   Contact line development of high surface tension,γ, aqueous ink
jet drops spreading on various uncoated and coated grades of paper.
With the exception of dry xerographic/lowγ and offset coated/lowγ,
contours for uncoated papers (lines (i)-(iii)) and coated papers (line (iv))
respectively, are arranged from left to right in order of increasing t_d. The
outer contours correspond to the final dry image. Contours for the IJ paper
are shown truncated because of much greater spreading (Reproduced with
permission from Ref. 22. Copyright 1983, The British Paper and Board
Industry Federation).

419

wetting/spreading increasingly exploits the inherent structural irregularity of paper (hence feathering) at the expense of image quality. Thus, to achieve the desired image quality and drying time one or other of these mutually opposing ink/paper properties must be compromised!

Sorption Mechanism

Despite the complexity and wide variability among various ink/paper systems, a consistent pattern of events as outlined in Fig. 6 seems applicable. Initially, significant lateral liquid drop spreading is evident for systems with very low t_d values (see Fig. 5). Lateral spreading quickly diminishes as bulk capillary forces take effect. This, has been verified, at least for coated paper, by drop radius data computed according to Cheever's equation[19,22], a modification of the Lucas-Washburn equation to describe radial spreading through open slit capillaries. During this stage the drop literally skims over the substrate surface and penetration is negligible. The corresponding extent of spreading, expressed as the percentage increase in D_o, is found to be negligible for highly sized papers, but appreciable for more absorbent papers. This important initial wetting stage, therefore, plays a critical role in terms of two essential ink-paper properties:

(i) On impact, in association with drop kinetic energy, it governs the initial drop contact area, which in turn determines the ratio of liquid concentration to available pore volume;

(ii) In contrast to (i), greater wetting increasingly exposes the inherent irregularity of the paper structure (i.e. wicking) at the expense of image quality.

The relatively short initial wetting stage is followed by image development (Fig. 6b), during which absorbed liquid undergoes subterranean spreading and/or re-emerges on the surface ahead of the drop. In addition, occasional local capillary channeling along individual fibers is evident on some papers. These localized sites serve to nucleate further irregular development in the image shape for a relatively long period after t_d, i.e. post-drying (Fig. 6d), so that the final image is considerably more ragged (see Fig. 5). This complex behavior is also evidenced by significant variations in optical density around the drop periphery. As noted in previous studies[24] the width of the band of absorbed liquid indicates the relative importance of these phenomena to image development.

In the second and, in a few systems, third stages (Fig. 6b and 6c respectively), liquid spreading terminates and penetration predominates. On the basis of contact area and profile development data[22] none of these systems exhibits penetration rates consistent with the Lucas-Washburn capillary model; rather they show more prolonged spreading and complex behavior. Among the principal factors contributing to this apparently increased porous resistance (or decreased liquid permeability) are the effects of: sharp fibre edges, pore tortuosity, build-up of air pressure particularly in 'dead-end' pores, break-down in liquid supply to spreading front, and fibre swelling which tends to diminish the average pore size. Increased penetration rates shown by several systems during post-drying stem from subterranean absorption processes, e.g., the interconnection of partially filled pores and localized surface fibre wicking. This final stage reflects a lag in the equilibration of bulk and surface capillary forces. In addition, shortly before the drop disappears, many papers become irreversibly swollen (see Fig. 6c). This may result in serious cockle and curl particularly when printing of large solid areas is involved. As an example, among the papers included in Fig. 5, the amount of swelling with the high γ ink (i.e. expressed as the % increase in caliper) ranges from 5-25% for the uncoated papers and 5 to 10% for the coated papers. In situations where IJP of additional information may be required, drop placement (registration) may be seriously affected by this irreversible swelling and, furthermore, result in deteriorating image quality. Finally, excessive curl may also cause runnability problems especially where forced air dryers are used to accelerate drying.

Effects of Paper Structure

A further complication prevalent among many papers is their structural anisotropy, both parallel and perpendicular to the plane of the paper. In contrast to coated papers, which maintain almost circular image development (see Fig. 5), contact line development on uncoated paper is

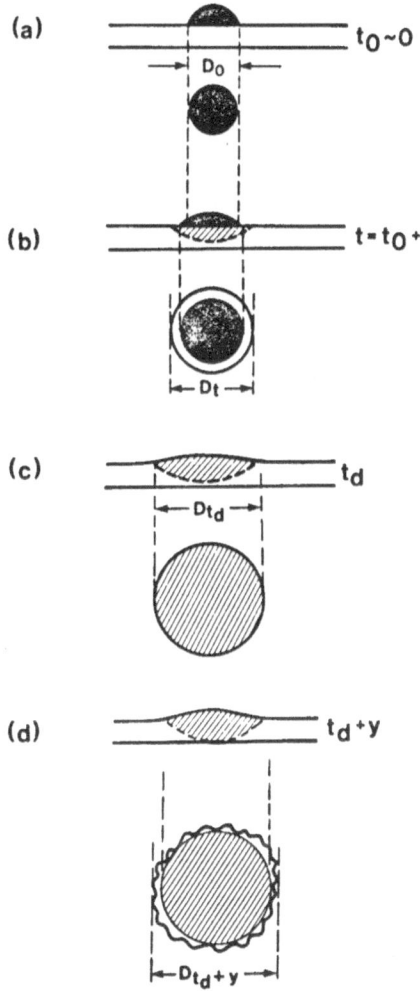

Fig. 6   Major stages of drop spreading, penetration and image development from impact to drying (schematic): (a) lateral drop spreading; (b) lateral and transverse image development; (c) drying, fibre swelling and disappearance of liquid drop; and (d) post-drying - further lateral and transverse image development. Solid areas correspond to the liquid drop and shaded areas to absorbed liquid. For simplicity the contact lines are shown as circular whereas in actuality they are quite anisotropic on most uncoated papers (Reproduced with permission from Ref. 17. Copyright 1983, The British Paper and Board Industry Federation).

invariably anisotropic and manifests to varying degrees the inherent structural directionality of commercial papers.

In many instances the shape of successive drop images is distinctly different, indicating the sensitivity of individual ink jet drops to the structural variability of paper. Evidence for the latter is also revealed by differences between machine (MD) and cross-machine (CD) spreading velocities caused by fibre directionality, the spreading resistance of fibre edges, and fibre surface morphology[24].

Estimates of the transverse spreading velocity (v_t) of the penetrating liquid front during the incipient stages (see Fig. 6) are several orders of magnitude less than for lateral (MD or CD) spreading. For a coated paper (Fig. 7(a)) v_t can be 50-fold greater than v_{MD}, whereas for an uncoated paper (Fig. 7(b)) the difference is only 5-fold. Differences between lateral and transverse spreading rates reflect variations in the effective pore size (r), and as noted by other workers[25,26], follow the trend $r_{MD} < r_{CD} \ll r_t$. Although the rate of penetration increases as pore size decreases; conversely, according to Poiseuille's equation, the resistance to flow is much higher for smaller pores. Thus, the net effect is for lateral spreading to predominate. However, when the capillary resistance is lowered, as for example offset coated paper/low ink (see Fig. 5), or when high surface area hydrophilic coating pigments are used such as silicas[23], v_{CD}/v_{MD} is substantially lower and lateral vs. transverse spreading anisotropy is lessened.

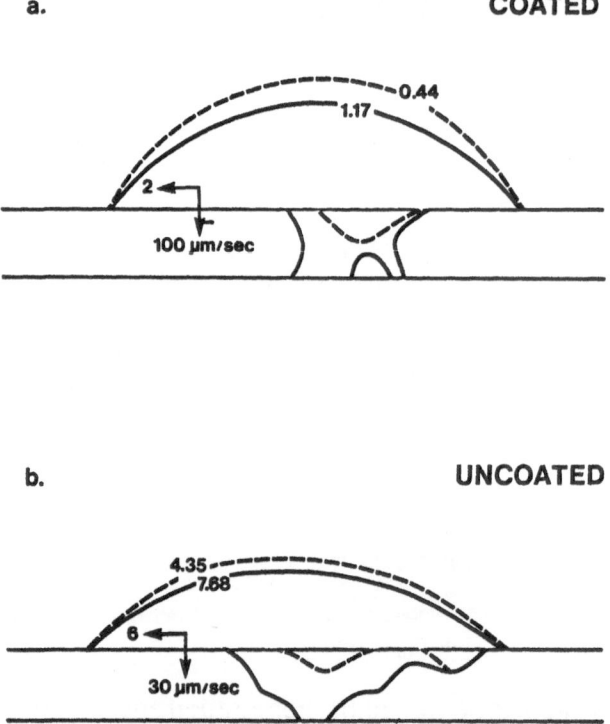

Fig. 7.   Comparison of lateral and transverse spreading development for an ink jet drop on (a) coated and (b) uncoated offset paper. The profiles and respective times (in seconds) correspond to penetration onset and initial show-through of ink, and the velocities to the spreading and penetration rates (Reproduced with permission from Ref. 17. Copyright 1983, The British Paper and Board Industry Federation).

Up to this point our discussion has mainly focussed on ink/paper phenomena that impact the absorption drying process. Attention is now directed towards important factors influencing image quality. In this context it is worth considering a relevant structural picture of paper. According to optical smoothness measurements[27] uncoated papers, depending upon the grade, typically comprise of only 5-15% fibrous solid area. Coated papers are usually somewhat higher, viz. 15-25%. The major area of the surface thus consists of voids, at least during the initial stages of contact with an ink drop. Although underlying strata (typically 7 or 8 fiber layers) may increase the effective solid-liquid contact area somewhat, it is of secondary importance in a non-impact printing process such as IJP since there are essentially no mechanical forces to assist ink transfer.

The net result of this structural composition is the formation of composite solid-liquid and liquid-air interfaces, in which the former depending on surface chemistry, aid wetting, and the latter behave as hydrophobic sites. Systems of this type have been treated by Cassie and Baxter[28] and Dettre and Johnson[29]. It is vividly illustrated in high resolution SEM dynamic studies of an involatile oil spreading on newsprint and bond paper[24]. The liquid which is injected through an ~ 250 μm diameter hole pierced into the paper, is seen to follow a structurally chaotic pattern. Fibers oriented approximately parallel to the direction of spreading tend to enhance the local spreading rate due to their microtopography. In contrast, the advance of the contact line is arrested when it confronts fibers lying across its path until the liquid acquires sufficient spreading pressure to overcome the local spreading resistance, viz. edge effect[30]. On many papers due to these two topographical extremes spreading may be irregular and is manifest in fiber wicking, termed feathering, resulting in poor image acuity.

Similar behavior is also illustrated is Fig. 8 which shows SEM-graphs of a liquid polyphenylether drop deposited on two highly calendered uncoated papers and one coated paper. The degree of contact line irregularity directly relates to local imperfections. However, the coated paper provides a much smoother surface and drop spreading is virtually symmetrical. The shadow-like band surrounding the drop (Fig. 8(c)) is created by the porous nature of the coating and indicates concurrent surface and subterranean (absorption) spreading processes. The width of the absorption band indicates the degree of importance of this behaviour on image development.

The aforementioned spreading behaviour on fibrous surfaces is analogous to that found on surfaces having pre-determined roughness[31]. Figure 9 shows SEM-graphs of a liquid drop on a parallel grooved surface. The drop shape is highly anisotropic because of capillary channeling along grooves and spreading inhibition across groove edges. The marked effect on spreading development for a model porous surface is shown in Fig. 10. The holes tend to impede progress of liquid, rather like "hydrophobic" sites, whereas the solid areas play a directional role in contact line development. As noted earlier this is primarily due to the capillary resistance of voids, which undoubtedly contributes to the wetting delay[11-14] and is substantially lessened with increased wetting.

It is also interesting to note in Fig. 10 how the contact line contour relates to drop size. As is evident from Table I, IJ drops of around 30 μm diameter are conceivable. For drops of this size the contact line assumes an almost square shape which reflects the surface geometry. Accordingly, as the drop size increases (Fig. 10(b)) the contact line tends to become more circular. Theoretical treatment of drop perturbation behaviour for similar model roughnesses[33] is consistent with these qualitative findings.

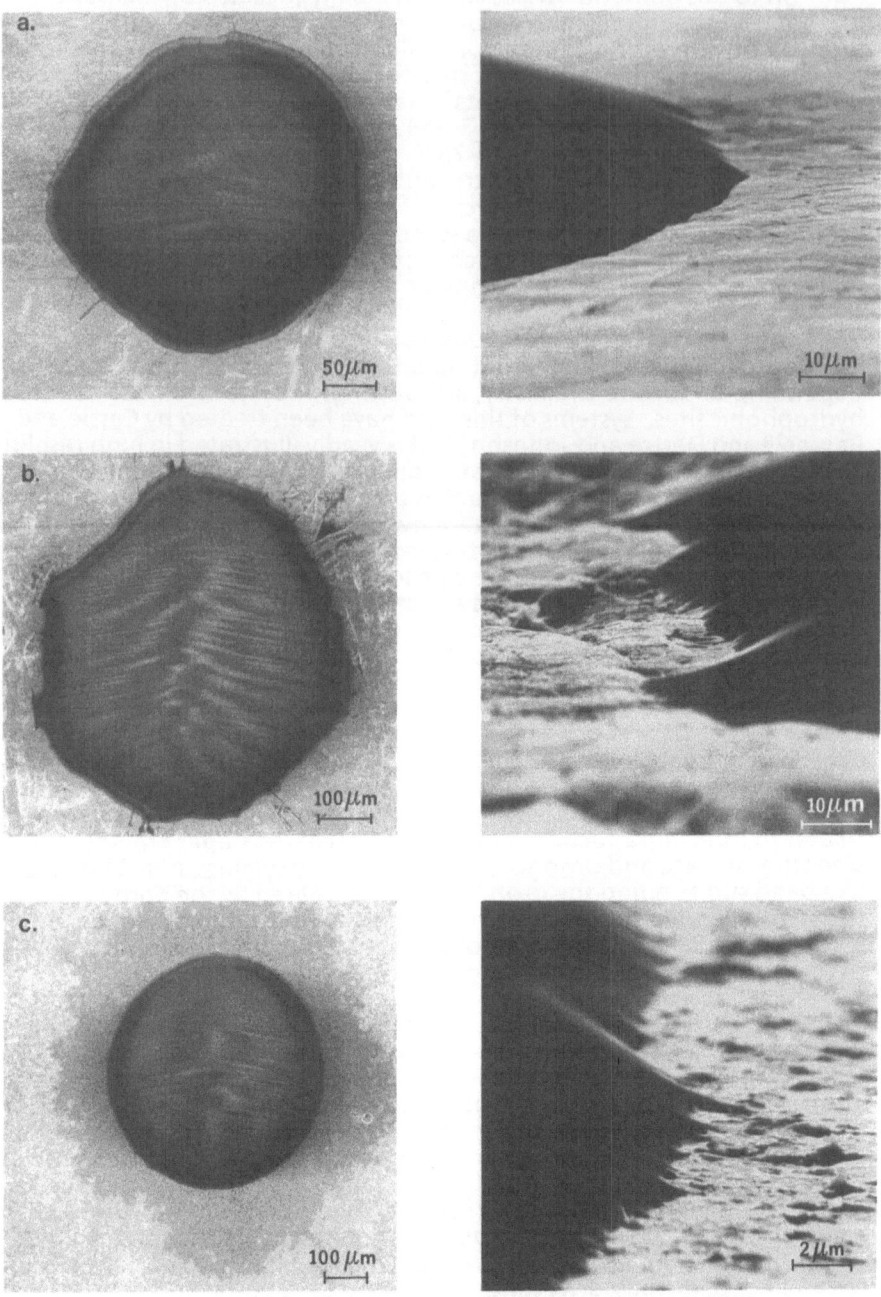

Fig. 8.   SEM graphs of liquid polyphenylether drops deposited on: (a) transparent glassine; (b) opaque glassine; and (c) cast-coated papers, showing top (left) and side (right) views respectively (Reproduced with permission from Ref. 24. Copyright 1983, The British Paper and Board Industry Federation).

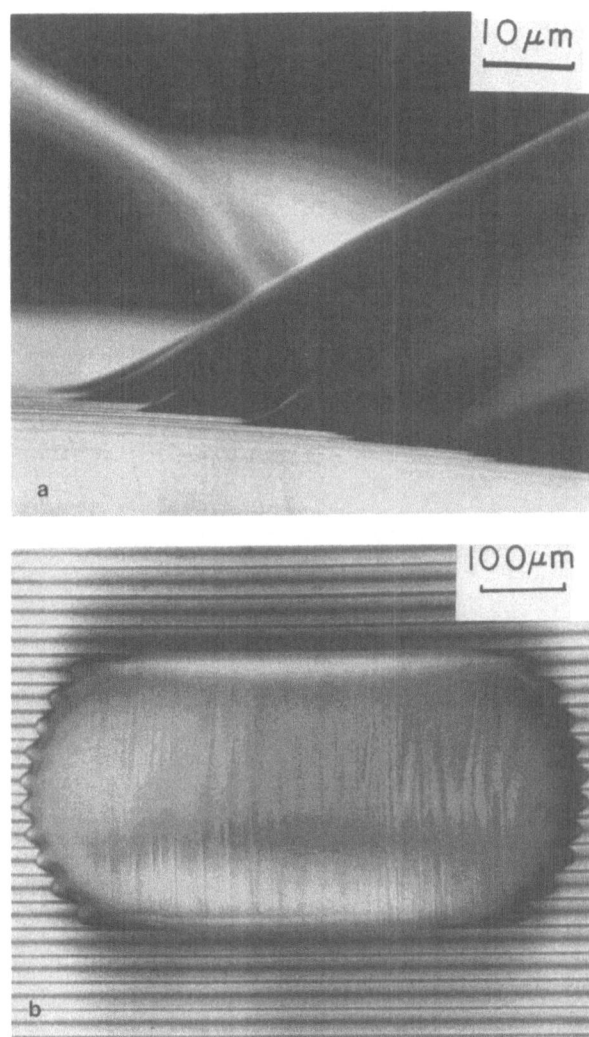

Fig. 9.   SEM graphs of a cylindrical polyphenylether drop formed on a parallel vee-grooved nitrocellulose surface, viewed (a) from the top; and (b) across the grooves (Reproduced with permission from Ref. 31. Copyright 1977, Gordon and Breach Science Publishers Ltd., England).

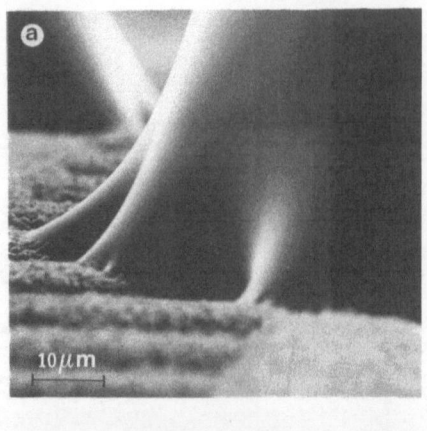

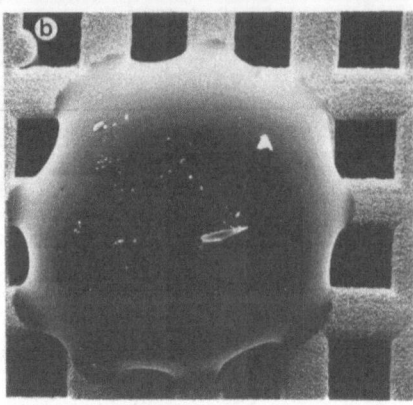

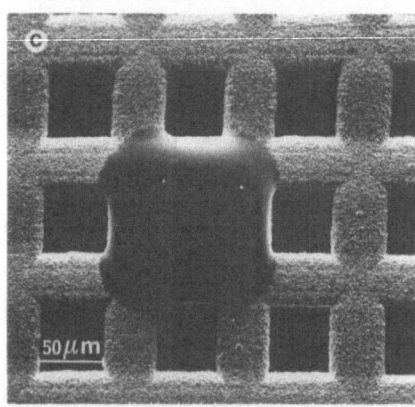

Fig. 10. Solidified PMMA drops after spreading on electroformed nickel mesh surface which serves as an idealized model of a rough and porous surface. (a) Side view which reveals a large difference between the macroscopic θ and the microscopic value (measured within 1 μm of the contact line), which varies according to the contact line location. (b, c) Top views showing that holes prevent spreading and that as the drop size increases the contact line contour becomes more circular (Reproduced with permission from Ref. 32. Copyright 1977, Academic Press Inc.).

Apart from physical ink/paper interactions the role of chemical and optical interactions are also very important particularly in color IJP. For instance, the light fastness of the dye affects print archival properties[34] and may vary according to the type of paper[35]. Certain papermaking additives have a dramatic effect on the chemical affinity of some IJ dyes[36]. In order to be water-fast the dye should have a high specificity for the paper surface chemistry. Also, upon printing, to achieve high optical density and color saturation, ideally the ink vehicle should penetrate paper rapidly leaving the dye immobilized on the surface.

Apart from the color quality of IJ dyes, their color reproduction when imaged on paper may diminish considerably depending upon the paper surface composition and the manner in which it optically degrades a colored image[37]. Of major importance in this respect are the substrate smoothness and scattering coefficient. Furthermore, depending upon the colorimetric properties of the dye, the surface pH of paper may also have an overriding influence on the eventual image color.

ACKNOWLEDGEMENTS

The complex structure of paper creates many subtle effects on drying and image quality in ink jet printing. Use of dynamic high resolution techniques enable a more meaningful understanding of some of the important ink/paper interaction phenomena. Recognition of the importance of these phenomena will lead to a more realistic model of the inter-relationship between paper structure, drying and image quality.

CONCLUSION

The author is grateful to Xerox Corporation for granting permission to publish this article. Special thanks are also due to the Microscopy Department of the Pulp and Paper Research Institute of Canada for performing the SEM studies.

REFERENCES

1.   M.B. Lyne and J.S. Aspler, in "Colloids and Surfaces in Reprographic Technology", M.L. Hair, editor, ACS Symposium Series, No. 200, 385 (1982).
2.   D.G. Gray, ibid, p.421.
3.   H. Corte, in Handbook of Paper Science, Vol. 2 "The Structure and Physical Properties of Paper", p.175, editor, H.F. Rance, Elsevier Scientific Publishing Co., 1982,.
4.   R.H. Darling, C.H. Lee and L. Kuhn, IBM J. Res. Develop. $\underline{28}$, 300 (1984).
5.   A.B. Jaffe and R.N. Mills, Proc. SID $\underline{24}$ (3), 219 (1983).
6.   C.T. Ashley, K. Edds and D. Elbert, IBM J. Res. Develop. $\underline{21}$, 69 (1977).
7.   Proc. SPSE Second International Conference on Advances in Non-Impact Printing Technologies, Arlington, VA, November 4-8, 1984.
8.   J.J. Stone, TAPPI J. $\underline{61}$ (10), 112 (1978).
9.   G. Rosenstock, TAPPI J. $\underline{62}$ (9), 87 (1979).
10. T. Gillespie, J. Colloid Interface Sci. $\underline{14}$, 123 (1959).
11. R.W. Hoyland in "Fibre-Water Interactions in Papermaking", p.557, Technical Division, British Paper and Board Industry Frederation, London, 1978,

12. J.A. Bristow. Svensk Paperstid. 70, 623 (1967).
13. R.W. Hoyland and R. Field, Paper Techn. Ind. 18, 7 (1977).
14. M.B. Lyne and J.S. Aspler, TAPPI J. 65 (12), 98 (1982).
15. J.F. Oliver, TAPPI J. 67 (10), 90 (1984).
16. G.E. Charles and S.G. Mason, J. Colloid Interface Sci. 15, 236 (1960).
17. J.F. Oliver in "The Role of Research in Papermaking", p.851, Technical Division, British Paper and Board Industry Frederation (London), 1983.
18. M. Carrol and S.G. Mason, Can. J. Tech. 30, 321 (1952).
19. G.D. Cheever, in "Interface Conversion for Polymer Coatings", p.150, G. Weiss, editor, Elsevier 1969
20. W.J. Frederick and E.G. Bobalek, Ind. Eng. Chem. Fundam. 14 (1) 40 (1975).
21. B. Miller and D.B. Clark, Textile Res. J. 48, 150 (1978).
22. J.F. Oliver, ACS Symposium Series 200, 435 (1982).
23. M.B. Lyne and J.S. Aspler, TAPPI J. 68 (5), 106 (1985).
24. J.F. Oliver and S.G. Mason, in "Fundamental Properties of Paper Related to its Uses", p.428, Technical Division, British Paper and Board Industry Frederation (London) 1976.
25. D. Tollenaar, "Surfaces and Coatings Related to Paper and Wood", p.195. R.H. Marchessault and C. Skaar editors, Syracuse Univ. Press 1967.
26. E. Back, Svensk Papperstidn. 69, 219 (1966).
27. G.R. Sears, J.A. Van den Akker, M.H. Aprison, N.J. Beckman and C.W. Denzer, Pulp Paper Mag. Canada 59, 94 (1954).
28. A.B.D. Cassie and S. Baxter, Trans. Farad. Soc. 40, 546 (1944).
29   R.H. Dettre and R.E. Johnson, Society of Chemical Industry Monograph, "Wetting" 25, 144 (1967).
30. J.F. Oliver, C. Huh and S.G. Mason, J. Colloid Interface Sci. 59, 568 (1977).
31. J.F. Oliver, C. Huh and S.G. Mason, J. Adhesion 8, 223 (1977).
32. J.F. Oliver and S.G. Mason, J. Colloid Interface Sci. 60, 480 (1977).
33. C. Huh and S.G. Mason, J. Colloid Interface Sci. 60, 11 (1977).
34. A.B. Jaffe, E.W. Luttmen and W. Crooks, ACS Symposium Series 200, 531 (1982).
35. P.A. McManus, C.W. Jaeger, H.P. Le and D.R. Titterington, TAPPI J. 66 (7), 81 (1983).
36. J.F. Oliver and L. Kale, in "Proc. TAPPI Papermakers Conference Proceedings," p. 217, Denver, CO, April 1985.
37. R.M. Leekley, R.F. Tyler and J.D. Hultman, TAPPI J. 61 (1), 108 (1978).

ABOUT THE CONTRIBUTORS

Ilhan A. Aksay is Professor at the University of Washington, Seattle. He received his Ph.D. degree in Materials Science in 1973 from the University of California, Berkeley. In 1975, he returned to his native country, Turkey, to accept a teaching position at the Middle East Technical University in Ankara where he stayed until 1981. In 1981 he joined UCLA's Materials Science and Engineering Department as a Visiting Associate Professor, and 1983 joined the faculty of the University of Washington. Hes research interests and teaching activities include the processing science of ceramics; ionic systems; and interfacial reactions and capillarity phenomena. Most recently he has been active in the utilization of colloidal techniques to ceramic processing.

S.V. Babu is currently an Associate Professor of Chemical Engineering at Clarkson University which he joined in 1981. He holds a Ph.D. in Physics from the State University of New York at Stony Brook, and was a visiting scholar at the Niels Bohr Institute in Copenhagen, Denmark and The International Center for Theoretical Physics at Trieste, Italy from 1969 to 1970. Before coming to Clarkson, he was on the Chemical Engineering Faculty at the Indian Institute of Technology, Kanpur, India. He has spent the last four summers as a summer faculty member/visiting professor with IBM in Endicott.

John Baglin is a research physicist in Materials Science at the IBM Thomas J. Watson Research Center, Yorktown Hts., NY. He received his Ph.D. degree in 1963 from the University of Melbourne, Australia, and he taught at Iowa State University and Yale University before joining IBM in 1972. His current interests lie in the field of thin film interactions and interface behavior, and in the interaction of radiation with solids. He is currently a Vice President of the Materials Research Society.

Andre Barraud is presently Head of the Laboratory on Physico-Chemistry of the Oraganic Solid State, and Co-manager of the Service de Chimie Moleculaire at C.E.N. Saclay, France. He carried out his Thesis of Doctorat d'Etat on instabilities and oscillations in Gallium Arsenide in Paris in 1967. He has been doing research on various aspects of Langmuir-Blodgett films since 1967 and some of the salient results have been: L-B film capacitors, 1970; M.I.S. structures, 1971; L-B photoresists, 1976; and L-B conducting films, 1984.

F. James Boerio has been Professor of Materials Science at the University of Cincinnati since 1977, which he had joined as Assistant Professor in 1970. He received his PhD. degree in Macromolecular Science from Case Western Reserve University in 1971. His current research interests include adhesive bonding of metals and polymers and surface analysis, especially using infrared, Raman and X-ray photoelectron spectroscopy. He has published approximately 60 papers describing his research results.

Francis J. Bonner is Professor of Chemical Engineering at Vanderbilt University, Nashville, TN. He earned his PhD. in Chemical Engineering from the University of Delaware, and Fil. lic. and Fil. dr. in Physical Chemistry from Uppsala University, Sweden. He has had a number of industrial and accedemic appointments, and before his present position he was Oberassistant (Associate Professor) at Swiss Federal Institute of Technology, Zurich, Switzerland. He has numerous patents and publications to his credit, and his research interests relate to transport, macromolecular and colloidal phenomena, partly with a bio-orientation.

Jens Borch is an Advisory Engineer with the polymer and paper engineering function of the supplies products group, Information Products Division, IBM Corp., Boulder, CO. He joined IBM in San Jose in 1979, and prior to joining IBM, he had worked in several research and development areas of polymer chemistry, cellulose and paper science. He received his MSChE from the Technical University of Denmark, Copenhagen, and completed his Ph.D. degree in Physical Chemistry at the State University of New York, Syracuse, in 1970.

Charles A. Brown is a Research Staff Member at the IBM Almaden Research Center in San Jose, CA, doing organic chemistry work.

Joseph Cesarano III is currently a graduate student at the University of Washington, Seattle, pursuing a Ph.D. in Ceramic Engineering with interest in ceramic processing. In 1985 he received an M.S. in Ceramic Engineering at the University of Washington. Research completed was concerned with polyelectrolyte stabilization of aqueous oxide suspensions and it was partly completed at Oak Ridge National Laboratory.

C. Chauvin is a graduate student in the Department de Genie Physique, Ecole Polytechnique, Montreal, Canada.

Dong Lyun Cho is a graduate student in chemical engineering at the University of Missouri-Rolla. He received his B.S. in Chemical Engineering at Seoul National University, Seoul, Korea in 1976, and M.S. in Chemical Engineering at the University of Missouri-Rolla in 1984. During 1976-81 he was employed by the Agency for Defence Development, Daejeon, Korea. His research interests are plasma polymerization, adhesion of polymer Coatings, modification of surfaces, and thin film technology.

Ned J. Chou is a Member of the Research Staff at the IBM T.J. Watson Research Center, Yorktown Hts., N.Y. He received his Ph.D. in Materials Science from New York University in 1967. He is currently engaged in various research projects involving advanced packaging materials in VLSI technology.

Lawrence B. Cohen has been Vice President - Technical at Cavedon Chemical Co., Woonsocket, R.I., since 1983 and has continued to direct zircoaluminate R&D activities while defining and guiding the marketing program for these coupling agents. Before coming to Cavedon Chemical Co. in 1982, he had had a number of industrial positions. He received M.S. degree in organometallic chemistry from Boston University in 1973 where he was a U.S. Department of Education Fellow and remained as a member of the chemistry department faculty (1973-1976). He has authored several articles on zircoaluminate application and chemistry, and has a number of patents to his credit.

Melvin D. Croucher is an Area Manager with Xerox Research Centre of Canada, Mississauga, Ontario, Canada, where he has been since 1975 after graduating with a Ph.D. from McGill University. His research has emphasized the areas of the rheology, stability and polymerization of sterically stabilized polymer colloids, in which he has published extensively. He received the Charles Ives award from the Society of Photographic Science and Engineering in 1982. He has been invited speaker at several Gordon Conferences and other notable scientific conferences, and has over 50 publications and numerous patents.

George Czornyj is currently with IBM Corporation, Hopewell Junction, N.Y, which he joined in 1977. He did his graduate studies in Chemistry at Rensselaer Polytechnic Institute in Troy. He was a postdoctoral fellow at IBM San Jose during 1976-1977 developing fluorocarbon films by plasma polymerization. He has been working in the area of lithographic materials, and his current interest is in high temperature polymers for packaging application for electronics.

S.C. Danforth is in the faculty of Rutgers University where he joined in 1982 as an Assistant Professor in the Department of Ceramics. Prior to coming to Rutgers, he was a staff scientist in the Ceramic Processing Laboratory at MIT where he worked in the areas of laser synthesis and processing of ultrafine, ideal ceramic powders. He received his Ph.D. in Materials Science from Brown University in 1978. His current research interests center on the synthesis and consolidation behavior of uniform fine powders such as silicon, silica, silicon nitride etc. He has authored over 20 publications and many reports and is joint inventor on one patent and two disclosures.

James M. Duff is a Principal scientist at the Xerox Research Centre of Canada, Mississauga, Ontario, Canada, where he has been employed since graduating from the University of Toronto in 1975. His research has emphasized the areas of photoreceptor chemicals and liquid marking materials. He is the author of more than 40 papers and has numerous patents.

John F. Evans is presently Associate Professor of Chemistry at the University of Minnesota, Minneapolis. He received his Ph.D. in Chemistry from the University of Delaware in 1977. His research interests are in the areas of surface modification and analysis, including nonequilibrium plasma-surface interections and the use of electroactive thin polymer films in catalytic and photocatalytic applications. He has authored over 40 publications in these fields and has been the recipient of a Sloan Fellowship for the period 1983-1985.

F.M. Fowkes is Professor of Chemistry, Lehigh University, Bethlehem, PA, where he has been since 1968. Prior to coming to L.U. he was Director of Research, Sprague Electric Company, 1962-1968. Before that, Dr. Fowkes headed the Surface and Chemistry Research for Shell Development Company and Shell Oil Company. He has been very active in the American Chemical Society, Electrochemical Society, and was chairman of the 1971 Gordon Research Conference on "Chemistry at Interfaces", and chairman of the 1973 Gordon Research Conference on "Science of Adhesion". Dr. Fowkes is an internationally recognized scientist in the area of Surface Chemistry and Adhesion. His publications include over 60 technical articles, chapters in books, and he is the editor of 2 books.

R.L. Geary is currently employed by DuPont Company and works at Towanda, PA. He completed his MS in Chemical Engineering at Clarkson University in 1985.

James H. Gibson is presently employed at the Standard Oil Co., Research Center, Cleveland, OH where he is a staff member responsible for mass spectrometric analysis and methods development. He joined SOHIO after completing an MS degree in Chemistry at the University of Minnesota in 1985.

Rejean Groleau is a partner in a consulting firm since 1985. He obtained a Ph.D. in Nuclear Physics from Yale University in 1980. From 1980 to 1985, he held a research associate position at the Laboratoire de Physique Nucleaire of the University of Montreal, where he used the elastic recoil detection (E.R.D.) method to study light elements in thin films. He also improved the technique by applying the time-of-flight method to E.R.D.

Subhash Gujrathi is since 1984 on the permanent staff at Vaniar College and also works as an invited scientist in the Department of Physics, University of Montreal. He received his Ph.D. degree from Calcutta University, India, in 1968. In 1968 he came to Simon Fraser University, Canada, on a postdoctoral fellowship and, after 3 years, joined the research group of the Foster Radiation Laboratory, McGill University as a research associate. From 1974-1984 he taught at Dawson College and actively participated in various research projects of the universities in Montreal. He has published over 40 articles in the field of nuclear spectroscopy and scattering. His current research interests are in the field of thin film surface analysis, using nuclear scattering techniques.

R.L. Headrick is graduate student, Department of Materials Science and Engineering, University of Pennsylvania, Philadelphia.

John N. Helbert is currently a member of the technical staff and team leader in resist processes of the Motorola SRDL submicron lithography laboratories. He received his Ph.D. degree in Physical Chemistry from Wayne State University. He has worked extensively in the area of radiation chemistry of organic and polymeric systems. His current research studies involve E-beam and photoresist development and resist dry-process compatibility studies.

Guy F. Hudson is pursuing a master's degree in the Department of Materials Science and Engineering at the University of Arizona in Tucson. He obtained his bachelor's degree from the same university in May of 1984.

William B. Jensen is currently at the University of Cincinnati where he holds the Oesper position in Chemical Education and the History of Chemistry. In addition to his activities in chemical education and history, his research interests include the theory and application of the Lewis acid-base concepts, the development of empirical structure-reactivity sorting maps in organic chemistry, inorganic crystal chemistry, and qualitative bonding theory. He received his Ph.D. in Inorganic Chemistry in 1982 from the University of Wisconsin. In the area of Lewis acid-base concepts, he has authored numerous review articles and a book entitled The Lewis Acid-Base Concepts: An Overview published by Wiley Interscience in 1980. Prior to his current position, he wäs at the Rochester Institute of Technology.

B.M.J. Kellner has been with IBM Corp. in Hopewell Junction, NY since 1978, where he is presently an Advisory Engineer in Via Technology Department. He received his Ph.D. in Physical Chemistry from SUNY-Buffalo in 1977.

Gregory J. Kovacs is a research scientist at Xerox Research Centre of Canada, Mississauga, Ontario, Canada, which he joined in 1979. He received his Ph.D. degree in Physics from the University of Toronto, where his doctoral work concerned the optical excitation of surface plasmons in thin films. As a NATO Postdoctoral Fellow at the University of Dusseldorf he studied the Surface Enhanced Raman Effect. At Xerox, his research activities have principally involved the structure and properties of thin films, which include the formation mechanism of subsurface particulate monolayers, and more recently, the formation and applications of Langmuir-Blodgett monomolecular films.

Robert H. Lacombe is currently employed by the IBM Corp. in Hopewell Jct., N.Y. doing experimental characterization of the thermal-mechanical properties of high temperature polymers, laboratory automation, and theoretical modelling of the stress distribution in multilevel microelectronic structures. He received his Ph.D. degree in the Deaprtment of Macromolecular Science at Case Western Reserve University with a theoretical thesis on polymer chain dynamics. His theoretical studies continued in the Department of Polymer Science and Engineering at the University of Massachusetts, where he was mainly interested in the problem of polymer solution thermodynamics and polymer-polymer compatibility.

Kashmiri Lal Mittal* is presently a senior Institute staff member at the IBM Corporate Technical Institutes, Thornwood, N.Y. He received his M.Sc. (First Class First) in 1966 from Indian Institute of Technology, New Delhi, and Ph.D. in Colloid Chemistry in 1970 from the University of Southern California. In the last thirteen years, he has organized and chaired a number of very successful international symposia and in addition to this volume, he has edited 24 more volumes as follows: Adsorption at Interfaces, and Colloidal Dispersions and Micellar Behavior (1975); Micellization, Solubilization, and Microemulsions, Volumes 1 & 2 (1977); Adhesion Measurement of Thin Films, Thick Films and Bulk Coatings (1978); Surface Contamination: Genesis, Detection, and Control, Volumes 1 & 2 (1979); Solution Chemistry of Surfactants, Volumes 1 & 2 (1979); Solution Behavior of Surfactants - Theoretical and Applied Aspects, Volumes 1 & 2 (1982); Physicochemical Aspects of Polymer Surfaces, Volumes 1 & 2 (1983); Adhesion Aspects of Polymeric Coatings, (1983); Surfactants in Solution, Volumes 1, 2 & 3 (1984), Volumes 4, 5 & 6 (1986); and Polyimides: Synthesis, Characterization and Applications, Volumes 1 & 2 (1984). Also he is Editor of the series, Treatise on Clean Surface Technology, a multi-volume work, the premier volume published in 1987. In addition to these books, he has published more than 50 papers in the areas of surface and colloid chemistry, adhesion, polymers, etc. He has given many invited talks on the multifarious facets of surface science, particularly adhesion, on the invitation of various societies and organizations in many countries all over the world, and is always a sought-after speaker. He is Fellow of the American Institute of Chemists and Indian Chemical Society, and is listed in many biographical reference works. He is or has been a member of the Editorial Boards of a number of scientific and technical journals and is the Editor of the new Journal of Adhesion Science and Technology. In August 1986, at the time of the Sixth International Symposium on Surfactants in Solution held in New Delhi, he was honored with a Recognition Plaque by the International Surface and Colloid Science Community (prominent scientists from 53 countries) for his Continued Leadership and Distinguished Professional Service.

---

*As the Editor of this Volume.

John G. Newman has since 1984 been employed by the Perkin-Elmer Corp., Physical Electronics Division, Eden Prairie, MN, where he is a specialist in secondary ion mass spectrometry in the analytical laboratory at that facility. He completed his MS degree at the University of Minnesota in 1984.

John F. Oliver is a project leader in the Paper Science Laboratory of Xerox Research Centre of Canada, where he is responsible for research in non-impact printing, particularly ink jet printing. He received his Ph.D. in Physical Chemistry at McGill University, Montreal. Prior to joining Xerox, he was a scientist at U.K. A.E.A., Harwell (England) studying the surface chemistry of ceramic materials, and Pulp and Paper Research Institute of Canada when he conducted research on paper adhesion during paper making. He has published widely and presented much of his work internationally in the areas of wetting and adhesion phenomena on rough surfaces.

D.J. Ondrus is presently working towards his Ph.D. degree at the University of Cincinnati, where he received his M.S. degree in Materials Science in 1985. His research interests include adhesive bonding and surface analysis using X-ray photoelectron spectroscopy and infrared spectroscopy.

Stephen V. Pepper is a senior research scientist and Deputy Chief in the Surface Science Branch at NASA Lewis Research Center, Cleveland, OH. He holds a Ph.D. in Physics from the University of Rochester. He has published in the areas of tribology, electron spectroscopy and surface analysis.

Edwin P. Plueddemann is a Scientist in the Corporate Research of Dow Corning Corp., Midland, MI, which he joined in 1955. He received his Ph.D. degree in Chemistry from Ohio State University in 1942 with work on organic compounds of fluorine. Subsequently, he was employed at Westvaco Chlorine Products Corp. (1942-1947) and Plaskon Division of Libby-Owens-Ford Glass Co. (1947-1955). His publications include over 70 U.S. patents, about 65 technical articles, two encyclopedia articles, chapters in technical books and is the editor of the book, Interfaces in Polymer Matrix Composites and author of the book Silane Coupling Agents. He has been the recipient of five SPI Best Paper Awards, chosen SPI "Man of the Year" (1971), and American Chemical Society "Award for Creative Invention" (1984). His current research deals with organosilicone compounds and their applications in adhesion and surface modification of minerals, and theory of adhesion.

Melvin Pomerantz has been a Staff Member at the IBM Research Center, Yorktown Hts. since 1960 with the exception of 1969 when he was a Visiting Lecturer at the University of California, Berkeley. He received his Ph.D. degree in 1959 at the University of California, Berkeley. With his thesis advisor, Prof. W.D. Knight, and R. Hewitt he was codiscoverer of pure nuclear quadrupole resonance in metals. He spent a postdoctoral year (1958-1959) on a Fulbright Fellowship at CEN Saclay, France. His early interests were in nuclear quadrupole interactions in metals, microwave ultrasonics, and heat pulses. More recently, he has studied organic thin films, which he used to construct the first large scale literally two-dimensional magnet. He has also been working on granular semiconductors and coupled ferromagnetic films. He is a Fellow of the American Physical Society.

J.P. Rabe is currently at the Max-Planck-Institute of Polymer Research in Mainz, W. Germany. He was an IBM Postdoctoral Fellow at the time this work was done.

John F. Rabolt is a Research Staff Member at the IBM Almaden Research Center in San Jose, CA. He received his Ph.D. in Polymer Physics from the University of Michigan. He does research on spectroscopy of thin polymeric and L-B films.

Srini Raghavan is an associate professor of materials science and engineering at the University of Arizona. His current research interests are in the areas of non-aqueous dispersions, applied surface chemistry and coatings.

Francine Y. Robb is currently a Principal Staff Scientist and a member of the Technical Ladder at Motorola in Phoenix, AZ where she has been employed since 1977. She received her M.S. degree in Solid State Chemistry from Arizona State Universityin 1977. Her activities include the development of dry etching and fine line measurement techniques for submicron processing.

Donald A. Roylance is a development engineer in the Advanced Media Technology Group of IBM (GPD) in Tucson. He has a master's degree in fuels engineering from the University of Utah.

Edward Sacher is a member of the Department of Engineering Physics of the Ecole Polytechnique, Montreal, Quebec, Canada. He is also director of the Surface Laboratory and a senior member of the Groupe des couches minces (Thin Film Group). A chemist, he received his doctorate from Penn State. After several postdoctoral fellowships in the United States and Canada, he spent nineteen years in industrial research before coming to his present post. His research interests are in the area of structure and properties of surfaces and thin films.

Naresh C. Saha is a Senior Staff Scientist in the Semiconductor Research and Development Laboratories at Motorola in Phoenix, AZ. He received a Ph.D. degree in Physical Chemistry in 1979 from the University of Calcutta, India. Before joining Motorola, he was a Postdoctoral Research Associate for five years at the University of Notre Dame. His research interests include radiation chemistry, and the application of ESCA, Auger, and electron spectroscopies to different areas of surface and interface chemistry.

L.E. Seiberling is an Assistant Professor, Department of Physics, University of Pennsylvania, Philadelphia. She received her Ph.D. degree in 1980 from Caltech.

J.G. Stephanie has since 1982 been with IBM Corp. in Endicott, NY and is working primarily in the area of photoresists and process and product design of new chip carriers. Before coming to IBM, he was with the Photographic Division of GAF Corp., where he was instrumental in developing new medical x-ray films and also in developing new photographic emulsions for the graphics arts industry. He received his Ph.D. in organic chemistry from the University of Iowa.

Jerome D. Swalen, manager of the Molecular Films Project, has been with the IBM Research Laboratory in San Jose since 1962, where he has held several managerial positions. During his 22 years with IBM, he has made contributions in a variety of research areas, some of which are EPR of transition metal complexes, computer analysis of NMR spectra of organic molecules, and laser spectroscopy of gases and of thin films by guided and surface waves. During a sabbatical year (1972-1973) he was a visiting professor at the Physical Chemistry Institute at the University of Zurich, Switzerland. His current interests center around the use of optical techniques to study molecular interactions at surfaces and their relation to structure, orientation and reactivity. He received his Ph.D. in Chemical Physics in 1956 from Harvard University. He has been vary active in the American Physical Society, and is a Fellow of the American Physical Society, and American Association for the Advancement of Science.

Bridget R. Svechovsky is currently a process engineer in the Semiconductor Research and Development Laboratory of Motorola SPS, in Phoenix, AZ. She received a B.S. degree in Chemical Engineering in May 1984 from the University of Colorado in Boulder.

Chwan-Hsin Tang is Manager of Materials Characterization at the Gould Research Center, Rolling Meadows, IL. He received his Ph.D. in Materials Science and Engineering from Northwestern University in 1982. From 1983 to 1984, he was involved with interfacial reaction studies of packaging-materials at the IBM T.J. Watson Research Center.

M. Velazquez is currently with Corning Electronics in Raleigh, NC, which he joined in Nov. 1984. He received his M.S. in 1984 in ceramics in the Department of Ceramics, Rutgers University. He is coauther of several papers, presentations and progress reports.

Paul S. Vincett manages a thin film group whose efforts extend from basic research right through advanced business development at Xerox Research Center of Canada which he joined in 1974. He received his degrees in Physics from Cambridge University, England, and was a Foundation Scholar at Corpus Christi College. His research interests especially involve the structure and properties of thin films and he is a member of the Editorial Board of Thin Solid Films. In addition to his L-B work, he was a codiscoverer of the "Critical Optimization" effect in vacuum-deposited films, and for several years has been responsible for the business and technical development of a major thin-film imaging technology. He has published about 50 scientific papers and holds a number of patents.

Donald R. Wheeler is a senior research scientist in the Surface Science Branch at NASA Lewis Research Center, Cleveland, OH. He holds a Ph.D. in Physics from the University of New Mexico. For the past ten years he has published in the areas of tribology, surface science and solid interfaces.

Raymond W. Wong is a Member of Research Staff at the Xerox Research Centre of Canada, Mississauga, Ontario, Canada, where he has been employed since graduating from the University of Waterloo in 1979. His interests lie in materials for liquid marking. He is the author of 10 publications and has numerous patents.

Hirotsugu K. Yasuda is Professor of Chemical Engineering and Director, Institute for Thin Film Processing Science at the University of Missouri-Rolla. He received his Ph.D. degree in Physical & Polymer Chemistry in 1961 at the State University of New York, Syracuse. Before his current position, he had had a number of academic and nonacademic appointments including Guest Scientist, Department of Polymer Technology, Royal Institute of Technology, Stockholm, Sweden (1965-1966), and various positions at Research Triangle Institute, Research Triangle Park, North Carolina. He has approximately 170 publications in scientific journals, government publications, books and encyclopedias on a number of topics. His research interests include polymer membrane technology, biomedical materials, thin film technology, plasma polymerization, adhesion of coatings, and modification of surfaces.

<u>Arthur Yelon</u> is Professor of Engineering Physics at the Ecole Polytechnique, Montreal, Quebec, Canada. He is also currently the director of the Groupe des couches minces (Thin Film Group) of the Ecole and the University of Montreal. After receiving his Ph.D. in Physics from Case Institute of Technology he worked at IBM, Yorktown Heights, CNRS, Grenoble, and Yale University before coming to Montreal in 1972. His research has been in various aspects of thin film studies: growth and structure, magnetic properties, electron tunneling, insulating materials, especially polymers, and amorphous semoconductors. He is particularly interested in structure-property relations, and in applications.

INDEX